# Think Like a Chemist: Compute Like a Chemist

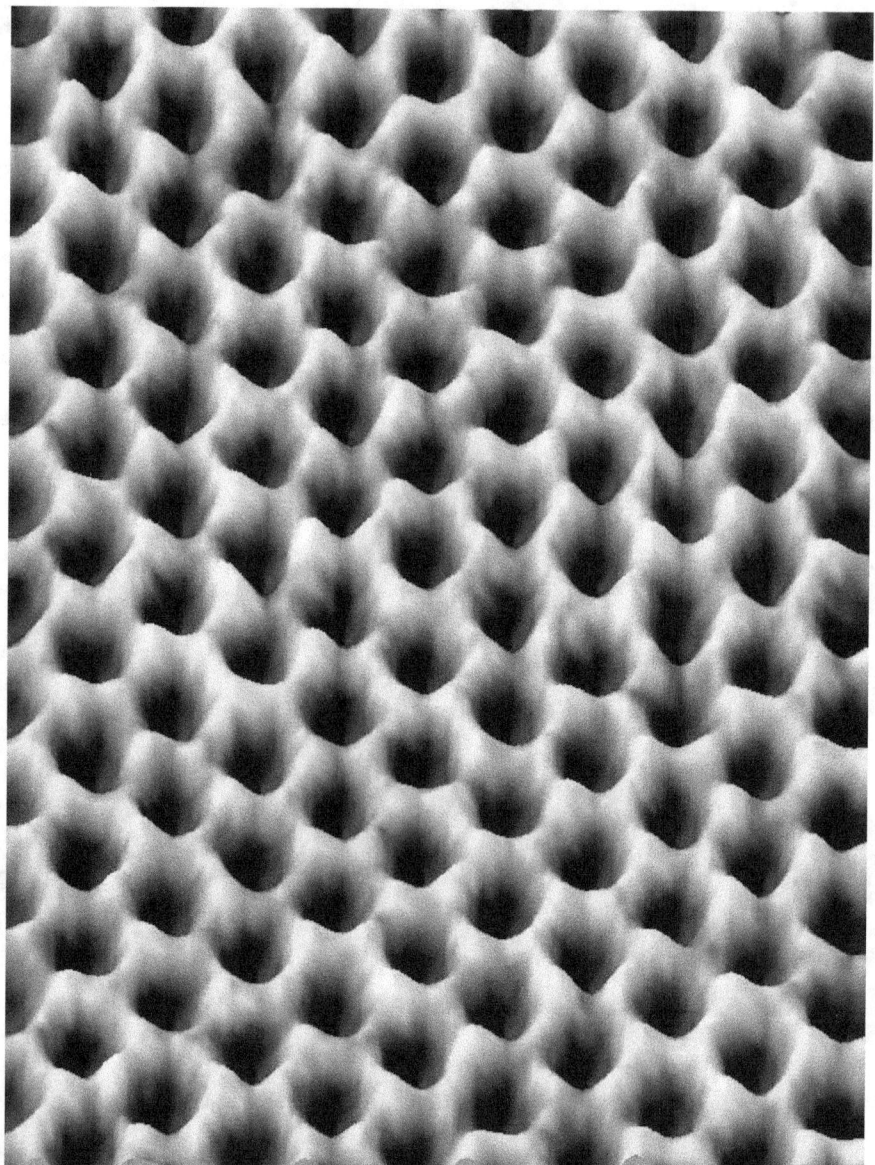

**Scanning Electron Microscope (SEM) Photo of Graphene**: Graphene is a form of carbon consisting of a single layer of carbon atoms arranged in an hexagonal (six sided) pattern. It's hard to identify them, but the little bumps are where the carbon atoms sit on the surface. Graphene is a giant array (see p. 56) that is flexible, transparent, extremely strong and an exceptional electrical and heat conductor. Much research is being done with it for semiconductors, electronics, computers (https://bit.ly/3ebr4YF), battery components, and composites.

*Photo courtesy of U.S. Army Materiel Command - https://www.flickr.com/photos/armymaterielcommand/6795812766, CC BY 2.0, https://commons.wikimedia.org/w/index.php?curid=37863884*

# Think Like a Chemist: Compute Like a Chemist

## Learn the Good Stuff and Get Ready for General Chemistry

### Harvey F. Carroll
Professor Emeritus of Physical Sciences
Kingsborough Community College
The City University of New York

Cover photo: **Cassiopeia A supernova remnant**. For more information see page 400. *Photo courtesy of NASA/CXC/SAO*

Copyright © 2021, 2023 by Harvey F. Carroll
All rights reserved. Exceptions are material in the public domain and Creative Commons ShareAlike for commercial purposes.

**ISBN:** 9798678981899
**Imprint:** Independently published.

*To Linda and Adam who have brought love, support and joy to my life, and in the memory of my parents with love and gratitude who supported my interest in science and, with some misgivings, the experiments I did.*

## Quotes by Famous People about Science and Life
See pp. viii, xv, 371 for more quotes. The lives of these people are *really* interesting. Check out the Wikipedia articles about them.

Chemistry is necessarily an experimental science: its conclusions are drawn from data, and its principles supported by evidence from facts.
-**Michael Faraday** (1791-1867)

We must trust to nothing but facts: these are presented to us by nature and cannot deceive. We ought, in every instance, to submit our reasoning to the test of experiment, and never to search for truth but by the natural road of experiment and observation.      -**Antoine Lavoisier** (1743-1794)

It doesn't matter how beautiful your theory is, it doesn't matter how smart you are. If it doesn't agree with experiment, it's wrong.
-**Richard P. Feynman** (see pages 415-416 in this book)

We need to teach how doubt is not to be feared but welcomed. It's OK to say, "I don't know."      -**Richard P. Feynman**

We never are definitely right; we can only be sure we are wrong.
-**Richard P. Feynman**

Scientific knowledge is a body of statements of varying degrees of certainty — some most unsure, some nearly sure, none absolutely certain.
-**Richard P. Feynman**

Progress in science comes when experiments contradict theory.
-**Richard P. Feynman**

Knowledge is the intellectual manipulation of carefully verified observations.      -**Sigmund Freud** (1856-1939)

There must be no barriers to freedom of inquiry. There is no place for dogma in science. The scientist is free, and must be free to ask any question, to doubt any assertion, to seek for any evidence, to correct any errors.
-**J. Robert Oppenheimer** (1904-1967)

We won't have a society if we destroy the environment.
-**Margaret Mead** (1901-1978)

Prediction is very difficult, especially if it's about the future.
-**Niels Bohr** (see pages 16, 223 and 245 in this book)

# BRIEF TABLE OF CONTENTS

Detailed Table of Contents   ix
About this Book   xv
Some Hints on Studying Chemistry   xviii
Photos and Essays   xxi
Acknowledgements   xxiii
Photo and Image Credits   xxiv
Roman Numerals   xxv
Quotes   vi, viii, xv, 371, 401, 414-416

| | | |
|---|---|---|
| Chapter 1 | ATOMS, ISOTOPES AND ELEMENTS   1 | |
| Chapter 2 | RELATIVE ATOMIC WEIGHT   21 | |
| Chapter 3 | MOLECULES AND THE BALANCED CHEMICAL EQUATION   41 | |
| Chapter 4 | ELEMENTS, COMPOUNDS, AND MIXTURES   55 | |
| Chapter 5 | SCIENTIFIC NOTATION   63 | |
| Chapter 6 | SIGNIFICANT FIGURES   93 | |
| Chapter 7 | UNITS AND UNIT CONVERSIONS   109 | |
| Chapter 8 | AVOGADRO'S NUMBER, THE MOLE, AND MOLECULAR WEIGHT   123 | |
| Chapter 9 | STOICHIOMETRY   143 | |
| Chapter 10 | PERCENT COMPOSITION, EMPIRICAL FORMULAS, AND MOLECULAR FORMULAS   159 | |
| Chapter 11 | MOLARITY AND SOLUTION STOICHIOMETRY   171 | |
| Chapter 12 | GASES AND THE IDEAL GAS LAW   189 | |
| Chapter 13 | ATOMIC ORBITALS   219 | |
| Chapter 14 | ORBITALS IN ATOMS AND THE PERIODIC TABLE   233 | |
| Chapter 15 | THE CHEMICAL BOND   259 | |
| Chapter 16 | NOMENCLATURE OF SIMPLE INORGANIC COMPOUNDS   287 | |
| Chapter 17 | OXIDATION NUMBERS   305 | |
| Chapter 18 | BALANCING OXIDATION-REDUCTION EQUATIONS   325 | |
| Chapter 19 | LOGARITHMS AND pH   347 | |
| | GLOSSARY   359 | |
| | SOLUTIONS TO PROBLEMS   372 | |
| | APPENDIX   399 | INDEX   417 |

## Quotes by Famous People about Science and Life
See pp. vi, xv, 371 for more quotes. The lives of these people are *really* interesting. Check out the Wikipedia articles about them.

The country which is in advance of the rest of the world in chemistry will also be foremost in wealth and in general prosperity.
 -**William Ramsay** (1852-1916)

Education is the most powerful weapon which you can use to change the world.  -**Nelson Mandela** (1918-2013)

The reward of the young scientist is the emotional thrill of being the first person in the history of the world to see something or understand something.  -**Cecilia Payne-Gaposchkin** (1900-1979)

Body concentrates order. It continuously self-repairs. Every five days you get a new stomach lining. You get a new liver every two months. Your skin replaces itself every six weeks. Every year, 98 percent of the atoms of your body are replaced. This non-stop chemical replacement, metabolism, is a sure sign of life.  -**Lynn Margulis** (1938-2011)

Chlorine is a deadly poison gas employed on European battlefields in World War I. Sodium is a corrosive metal which burns upon contact with water. Together they make a placid and unpoisonous material, table salt. Why each of these substances has the properties it does is a subject called chemistry.  -**Carl Sagan** (see page 401 in this book.)

One of the saddest lessons of history is this: If we've been bamboozled long enough, we tend to reject any evidence of the bamboozle. We're no longer interested in finding out the truth. The bamboozle has captured us. It's simply too painful to acknowledge, even to ourselves, that we've been taken. Once you give a charlatan power over you, you almost never get it back.  -**Carl Sagan**

We live in a society exquisitely dependent on science and technology, in which hardly anyone knows anything about science and technology.
 -**Carl Sagan**

If we continue to accumulate only power and not wisdom, we will surely destroy ourselves.  -**Carl Sagan**

# DETAILED TABLE OF CONTENTS

About this Book  xv
Some Hints on Studying Chemistry  xviii
Photos and Essays  xxi
Acknowledgements  xxiii
Photo and Image Credits  xxiv
Roman Numerals  xxv
Quotes  vi, viii, xv, 371, 401, 414-416

## Chapter 1 ATOMS, ISOTOPES AND ELEMENTS  1
1-1. Atoms Consist of a Positive Nucleus and Negative Electrons  1
1-2. The Atom Stays Together Because of Electrical Attraction  2
1-3. The Properties of Electrons, Protons, and Neutrons are Related to Their Mass and Charge  3
1-4. There Are Three Kinds of Hydrogen Atoms: Protium, Deuterium, and Tritium  5
1-5. The Three Isotopes of Hydrogen Make Up the Lightest Element  6
1-6. The Notation for Writing Isotopes Tells us the Number of Protons and Neutrons in the Nucleus  8
1-7. The Atomic Number and the Mass Number Are Used in Writing Isotopic Symbols  10
1-8. Atoms of Elements Heavier Than Hydrogen Have More Protons, Neutrons, and Electrons  10
  *Appendix 1-1. A brief history of the structure of the atom*  16
  *Problems 17 Keyed Problems 17 Supplemental Problems*  17

## Chapter 2 RELATIVE ATOMIC WEIGHT  21
2-1. The Atomic Mass Unit is a Unit Used to Measure the Mass of an Atom  21
2-2. The Relative Atomic Weight Takes into Account All the Naturally Occurring Isotopes of an Element  22
2-3. The Weighted Average Gives More Weight to Numbers That Appear More Often  22
2-4. Calculating Relative Atomic Weights Uses the Weighted Average  25
2-5. Binding Energy Holds the Nucleus Together  32
  *Appendix 2-1. The distributive law*  34
  *Appendix 2-2. Using a calculator*  36
  *Problems  37          Keyed Problems  37*
  *Supplemental Problems  38     Special Percentage Problems  39*

**Chapter 3  MOLECULES AND THE BALANCED CHEMICAL EQUATION  41**
3-1. Molecules Consist of Two or More Atoms  41
3-2. Balancing Chemical Equations Conserves Atoms  42
3-3. Balance Groups as a Whole in Equations With More Complex Molecular Formulas  50
   *Problems 51    Keyed Problems 51    Supplemental Problems 52*

**Chapter 4  ELEMENTS, COMPOUNDS, AND MIXTURES  55**
4-1. Each Element Consists of a Single Kind of Atom  55
4-2. Compounds Consist of Different Kinds of Atoms  55
4-3. Mixtures Consist of Two or More Substances  58
   *Problems 59    Keyed Problems 59    Supplemental Problems 60*

**Chapter 5  SCIENTIFIC NOTATION  63**
5-1. An Exponent Tells How Many Times to Multiply a Number by Itself  63
5-2. A Ten Written With Exponents Allows us to Write Numbers in Scientific Notation  64
5-3. Some Detials of Multiplication and Fractions  66
5-4. Scientific Notation Using Positive Exponents  69
5-5. Scientific Notation Using Negative Exponents  75
5-6. Using a Calculator in Scientific Notation  80
5-7. Powers and Roots of Numbers  82
   *Problems 88    Keyed Problems 88    Supplemental Problems 90*

**Chapter 6  SIGNIFICANT FIGURES  93**
6-1. The Percent Uncertainty of a Measurement is the Key to Understanding Significant Figures  93
6-2. Recognizing Significant Figures is Easiest When Numbers are Expressed in Scientific Notation  95
6-3. Accuracy and Precision Tell Us How Good Our Measurements Are  98
6-4. Determining Significant Figures in Addition and Subtraction is Different From That in Multiplication and Division  101
6-5. Rounding a Number Preserves the Proper Number of Significant Figures  104
6-6. Exact Numbers Have an Infinite Number of Significant Figures  105
6-7. The Moral of This Chapter is to Always Present Your Results in the Proper Number of Significant Figures  106
   *Problems 106    Keyed Problems 106    Supplemental Problems 107*

## Chapter 7  UNITS AND UNIT CONVERSIONS  109

7-1. Units and Their Abbreviations Are Used to Describe Physical Quantities  109
7-2. "One" is the Identity Element in Multipilcation  110
7-3. Units Divide in a Way Similar to Any Algebraic Quantity  111
7-4. Identities and Their Ratios Allow Us to Convert From One Unit to Another  111
7-5. Converting From One Unit to Another Uses Ratio Identities  113
7-6. The Metric System Uses Prefixes in Front of a Unit to Indicate a Power of 10  114
7-7. When Converting From One Unit to Another, Be Sure That the Units Divide in the Correct Way  116
*Problems 120    Keyed Problems 120    Supplemental Problems 120*

## Chapter 8  AVOGADRO'S NUMBER, THE MOLE, AND MOLECULAR WEIGHT  123

8-1. Atomic Weights are Related to Numbers of Atoms  123
8-2. The Gram is Used as the Unit of Atomic Weight  124
8-3. One Atomic Weight of Atoms Contains One Avogadro's Number of Atoms  125
8-4. A Mole is One Atomic Weight or one Avogadro's Number of Atoms  126
8-5. Calculating Moles and Numbers of Atoms Uses Atomic Weights and Avogadro's Number  128
8-6. One Mole of Molecules Contains an Avogadro's Number of Molecules Whose Mass is One Molecular Weight  130
8-7. Calculating Moles and Numbers of Molecules Uses Molecular Weights and Avogadro's Number  131
8-8. Calculating Moles of Atoms in Molecules Uses the Molecular Formula  132
8-9. The Mole is Defined in Terms of a Certain Quantity of a Substance  136
8-10. The Mole Interpretation of Chemical Equations Uses the Coefficients of the Substances in the Balanced Equation  137
*Problems 137    Keyed Problems 137    Supplemental Problems 139*

## Chapter 9  STOICHIOMETRY  143

9-1. Identities From the Balanced Chemical Equation Use the Coefficients of the Substances in the Equation  143
9-2. Mole Relationships From the Balanced Chemical Equation Use the Coefficients of the Substances in the Equation  145
9-3. Mass Relationships From the Balanced Chemical Equation Use Both the Coefficients and the Atomic or Molecular Weights  145
9-4. The Limiting Reagent is the Reactant That is Used Up First in a Chemical Reaction  149
*Problems 154    Keyed Problems 154    Supplemental Problems 155*

**Chapter 10 PERCENT COMPOSITION, EMPIRICAL FORMULAS, AND MOLECULAR FORMULAS   159**
10-1. Percent is Based on 100   159
10-2. Percent Composition is the Mass Percent of the Elements in a Compound   161
10-3. An Empirical Formula and a Molecular Formula Can Be Calculated From the Percent Composition and Molecular Weight of a Compound   163
10-4. The Formula Weight of a Giant Array is the Mass of the Simplest Ratio of Atoms in the Array   166
   *Problems 167   Keyed Problems 167   Supplemental Problems 167*

**Chapter 11 MOLARITY AND SOLUTION STOICHIOMETRY   171**
11-1. Molarity is a Unit of Concentration Involving Moles and Liters   171
11-2. Solution Stoichiometry Uses Molarity   176
11-3. When a Solution is Diluted, the Number of Moles of Solute Doesn't Change   182
   *Problems 184   Keyed problems 184   Supplemental problems 185*

**Chapter 12 GASES AND THE IDEAL GAS LAW   189**
12-1. Molecules of a Gas Collide and Change Their Speed and Direction   189
12-2. Pressure is the Force Per Unit Area   190
12-3. Degrees Celsius and Degrees Fahrenheit Are Two Units of Temperature   193
12-4. As the Temperature of a Gas Decreases, the Volume of the Gas Decreases   196
12-5. The Lowest Possible Temperature is Zero Kelvins   198
12-6. A Note About Proportionality and Proportionality Constants   199
12-7. The Volume of a Gas is Directly Proportional to the Temperature in Kelvins   200
12-8. The Pressure of a Gas is Inversely Proportional to the Volume of the Gas   200
12-9. The Pressure of a Gas is Directly Proportional to the Temperature in Kelvins   201
12-10. The Pressure of a Gas is Directly Proportional to the Amount of Gas   203
12-11. The Volume of a Gas is Directly Proportional to the Amount of Gas   204
12-12. The Ideal Gas Law Relates Pressure, Volume, Moles, and Temperature   204
12-13. Changing $P$, $V$, and $T$ While Keeping $n$ Constant: The Combined Gas Law   209
12-14. Dalton's Law of Partial Pressures Says That the Sum of the Pressures of the Individual Gases Equals the Total Gas Pressure   211
   *Problems 214   Keyed Problems 214   Supplemental Problems 215*

**Chapter 13 ATOMIC ORBITALS   219**
13-1. The Problem of Seeing Things is Related to Their Size and How You Look at Them   219
13-2. The Uncertainty Principle Limits Our Ability to "See" an Electron   220
13-3. The Properties of Light Depend on its Wavelength   220
13-4. We Cannot See an Electron Because of the Uncertainty Principle   222

13-5. The Electron is Described with Fuzzy Pictures Called Orbitals  223
13-6. The Rules for Placing Electrons in Orbitals Depend on the Properties of the Electron  224
13-7. Orbitals are Arranged in Energy Levels  225
13-8. Spin is a Property of Electrons  226
13-9. The Four Basic Types of Orbitals are the s, p, d, and f Orbitals  227
   Problems 230

## Chapter 14   ORBITALS IN ATOMS AND THE PERIODIC TABLE   233

14-1. Orbitals are Arranged in Energy Levels in the Order s, p, d, and f  233
14-2. Atoms of Different Elements Have Their Orbitals Filled in a Specific Order  234
14-3. Orbital Diagrams and Electron Configurations Are Ways of Representing the Arrangement of Electrons in Atoms  238
14-4. The Periodic Table Arranges the Elements According to Their Orbital Configuration and Chemical Properties  245
14-5. Chemical Families Have Similar Chemical Properties  248
   Appendix 14-1. Orbital spacing in multielectron atoms  251
   Problems 254   Keyed Problems 254   Supplemental Problems 255

## Chapter 15   THE CHEMICAL BOND   259

15-1. The Covalent Bond of the Hydrogen Molecule is the Simplest Covalent Bond  259
15-2. Lewis Formulas of Atoms in the First and Second Periods Show the Valence Electrons  261
15-3. Covalent Bonding for Some Molecules in the First and Second Periods: The Elements C, N, O, and F Obey the Octet Rule  264
15-4. Covalent Bonding in Other Molecules of the Second-Period Elements: The Elements C, N, O, and F Obey the Octet Rule  272
15-5. Lewis Formulas Using Lines Instead of Dots is Common  274
15-6. Covalent Bonding in the Third-Period Elements: The Elements P, S, and Cl May or May Not Obey the Octet Rule  275
15-7. The Strong Electrical Attraction Between Oppositely Charged Ions Forms the Ionic Bond  278
   Problems 283   Keyed Problems 283   Supplemental Problems 283

## Chapter 16   NOMENCLATURE OF SIMPLE INORGANIC COMPOUNDS   287

16-1. Naming Ionic Compounds Formed Between a Metal and a Nonmetal  287
16-2. Naming Covalent Compounds Formed Between Nonmetals  297
16-3. Naming Acids and Oxyacids  299
16-4. Chemical Names Should be as Simple as Possible But Completely Clear  300
   Problems 301   Keyed Problems 301   Supplemental Problems 301

## Chapter 17 OXIDATION NUMBERS 305

17-1. Good Battery, Bad Battery: Oxidation and an Environmental Lesson 305
17-2. Electronegativity is a Measure of the Ability of an Atom to Attract Electrons Toward Itself in a Covalent Bond 306
17-3. Calculating Oxidation Numbers is Based on the Electronegativities of the Elements 309
17-4. Calculating Oxidation Numbers of Atoms in Carbon Compounds Uses the Electronegativities of the Atoms 316
*Problems 319  Keyed Problems 319  Supplemental Problems 320*

## Chapter 18 BALANCING OXIDATION-REDUCTION EQUATIONS 325

18-1. Oxidation and Reduction in Chemical Reactions Involve the Loss and Gain of Electrons 326
18-2. When Balancing Redox Equations Using Half-Reactions, Leave Out the Spectator Ions 327
18-3. The Rules for Balancing Redox Equations Are Based on Mass and Electrical Balance 330
18-4. Balance Redox Equations in Acidic Solution by Using $H^+$ and $H_2O$ to Balance Hydrogen and Oxygen Atoms 331
18-5. Balance Redox Equations in Basic Solution by Using $OH^-$ and $H_2O$ to Balance Hydrogen and Oxygen Atoms 333
18-6. Check the Balanced Redox Equation for Mass and Electrical Balance in the Following Four Examples 335
18-7. Oxidizing and Reducing Agents Cause Oxidation and Reduction 340
*Problems 342  Keyed Problems 342  Supplemental Problems 343*

## Chapter 19 LOGARITHMS AND pH 347

19-1. The Logarithm of a Number is the Power to Which 10 Must be Raised to Give That Number 348
19-2. An Antilogarithm is the Number Corresponding to a Given Logarithm 351
19-3. The Natural Logarithm is Useful in Advanced Scientific Work 353
19-4. pH is an Application of Logarithms to Acidic and Basic Solutions 353
*Problems 357  Keyed Problems 357  Supplemental Problems 357*

**GLOSSARY 359**
**SOLUTIONS TO PROBLEMS 372**
**INDEX 417**

**APPENDIX 399**
About the Author 399
The Front Cover Explained 400
Scientific Calculators 403
Constants, Units and Prefixes 403
Rechargeable Battery 405
Double Slit Experiment 412

# ABOUT THIS BOOK

Every aspect of the world today, even politics and international relations, is affected by chemistry.    —**Linus C. Pauling** Priestley Medal Address, *Chemistry and the World of Tomorrow* (1984) https://bit.ly/2IrthAm

Future generations are unlikely to condone our lack of prudent concern for the integrity of the natural world that supports all life. —**Rachel Carson**, p. 13 in her 1962 book, *Silent Spring*, that began the modern environmental movement!

Scientists proceed by common sense and ingenuity. There are no rules, only the principles of integrity and objectivity, with a complete rejection of all authority except that of fact.    — **Joel H. Hildebrand** *Science in the Making* (1957), p 8.

The first principle is that you must not fool yourself — and you are the easiest person to fool. (In *Cargo Cult Science: Some remarks on science, pseudoscience, and learning how to not fool yourself* [1947]. https://bit.ly/3a4fg87).

For a successful technology, reality must take precedence over public relations, for nature cannot be fooled. (In *Report of the Presidential Commission on the Space Shuttle Challenger Accident*. NASA. 6 June 1986. Volume 2, Appendix F: "Personal observations on the reliability of the Shuttle, by R. P. Feynman." https://go.nasa.gov/3HOVYEW)    —**Richard P. Feynman**

**NOTE:** See pp. vi, viii, and 371 for more quotes. Many of the quotes in this book are about honesty in science. You should take them very seriously as without honesty in science, scientific discoveries would be meaningless. An example of scientific dishonesty was published in the journal *Science* in 2016. The authors claimed that the larvae of European perch (*Perca fluviatilis*) get addicted to eating microplastics, which stunts their growth, makes them more vulnerable to predators, and increases their mortality. It appears that they never did the study as described in the *Science* article and the article was retracted. Details can be found at https://bit.ly/3xp33Ja. Notice the problems that both the accusers and the accused had after this dishonesty was reported. Fortunately, scientific dishonesty is exceedingly rare. Between 1950 and 2004, only about 6 out of every 100,000 papers published in over 4000 different journals were retracted (https://bit.ly/3zwLx7j). **As a rough estimate, over 99.99% of scientists are honest, very careful, and of the ones I have known, very hard working.**

**NOTE:** The shortened **bit.ly** web addresses are **case sentitive**, unlike full web addresses (URLs) which are not case sensitive. **WARNING:** Because of *link rot*, URLs sometimes change or are taken down, so not all of the web addresses in this book may work.

*Think Like a Chemist: Compute Like a Chemist* is designed to help you acquire the conceptual understanding of atoms and molecules, and the problem-solving skills, needed to prepare you to take a two-semester or a three-quarter general chemistry course. These courses are generally intended for chemistry majors, pre-professional students in medical fields, and students in engineering, biology, physics, geology, and other science majors. **You can use this book for self-study (complete solutions to all end-of-chapter problems are provided) or as the textbook in a course.**

General chemistry can be challenging because of the abstraction and problem solving involved. If you come into a general chemistry course with good problem-solving techniques and some feeling for the abstract nature of chemistry, you can then concentrate on getting a deeper understanding of the subject. This book will get you on your way to thinking like a chemist and computing like a chemist.

As a result, you won't get bogged down puzzling over the following topics: *atoms and isotopes, atomic weight, molecules and balancing chemical equations, elements, compounds, and mixtures, scientific notation, significant figures, units and unit conversions, Avogadro's number and constant, moles and stoichiometry, percent composition, empirical formulas and molecular formulas, molarity and solution stoichiometry, gases and the ideal gas law, the simpler aspects of atomic structure, orbitals and the periodic table, chemical bonding and molecular structure, chemical nomenclature, oxidation numbers, balancing oxidation-reduction equations, and pH and logarithms.*

This book purposely covers these topics in great detail, and the order is intentionally somewhat different from that in most books. For instance, I believe that since you are in a chemistry course, you should start learning about chemistry. Therefore, the first chapter is called "Atoms, Isotopes and Elements." Atoms are the building blocks of chemistry and most naturally-occurring elements have more than one isotope.

Computation starts in the second chapter called "Relative Atomic Weight." Computation is a skill that is necessary to learning and "doing" chemistry. Then Chapters 3 and 4 discuss molecules and balancing equations, and elements, compounds and mixtures, respectively. Computation returns in Chapter 5 with "Scientific Notation."

The math skills that this book uses are: fractions, decimals, percentages, exponents, logarithms, and elementary algebra. In case you are a bit rusty about some of these topics, this book reviews all the necessary material in great detail at the place where it is used in the text. **No previous background in chemistry is assumed.** On page 402 there is a discussion of scientific calculators. You will need one to do the calculations in this book. Detailed calculator keystrokes are given in the text where appropriate.

In addition to the abstraction and problem solving involved in studying chemistry, authors sometimes use a different notation than other authors for the same thing. Each author is convinced that his or her notation is the best, or they use the notation they are familiar with. As far as I could, I have tried to point out where this can occur, and what these important organizations recommend: The International Union of Pure and Applied Chemistry (IUPAC) and the National Institute of Standards and Technology (NIST). See https://bit.ly/2OBuM0r for some of the details. I have also explained the few cases where I use a notation that is not the latest recommendation of these organizations.

*If you are taking a chemistry class, use the notation your instructor uses.*

In this book I have presented many applications of chemistry in the examples and problems. To give you a feeling for some famous scientists, a bit of chemical history, and some modern topics in chemistry, I have included photos and essays. The photos have extensive discussions as captions. They are listed on pages **xxi** and **xxii** under the title "Photos and Essays."

Although there is some history included in this book, the approach is not historical. In my many years of teaching, I concluded that the historical approach is incomprehensible to beginning students. To give one example, read the article about the **phlogiston theory** at https://bit.ly/3NGV92y. This theory was very popular in the 1700s but was finally discarded around 1800 as new experiments were done which the theory could not explain. (Or maybe because as **Max Planck** (see page 20) said, "Science advances one funeral at a time.")

Some topics I have included are not usually found in a book at this level, but I feel that they can give you greater insight into the strange world of atoms and molecules. The discussion is descriptive for the most part, and the little math used is arithmetic and simple algebra. They are:

—The discussion of binding energy in Chapter 2.

—The discussion in Chapter 14 of the reason why orbital spacing changes in multielectron atoms as compared to the orbital spacing in the hydrogen atom.

—The discussion in Chapter 17 of the oxidation numbers of atoms in organic molecules using structural formulas.

—The discussion on page 231 of momentum and angular momentum.

—The discussion on page 257 of waves and particles as applied to light, electrons and baseballs, and the de Broglie concept that all moving particles have an associated wave.

—The discussion on pages 285 and 322 of wave diffraction and interference, and the application of all this to cryo-Electron Microscopy (cryo-EM) and x-ray crystallography.

—A discussion of where elements are formed in the universe on pages 400-401.

—A discussion of how a simple rechargeable battery works on pages 405-411

—A discussion of the double slit experiment using electrons on pages 412-416.

The Appendix starting on page 399 includes the last 3 essays described above and other material that you might find interesting and useful. It also includes a short biography of the author. Enjoy the essays and get inspired to learn more about chemistry and science in general.

There are many worked examples in the text. In Chapters 7 through 9, I use the *given* (G) and *asked for* (AF) notation to help you understand the problem. After that, you should have enough problem-solving ability to determine what is given and what is asked for in a problem.

Many famous scientists are mentioned in this book, along with their discoveries. I could not possibly discuss the detailed story of how they made their discoveries. They were (are) very smart people and it took years of hard work. We are the beneficiaries of their genius and their discoveries have changed the world!

Harvey F. Carroll<br>
Seattle, WA

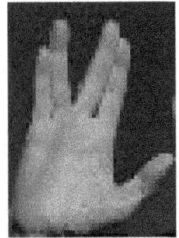

*Dif-tor heh smusma.*

# SOME HINTS ON STUDYING CHEMISTRY

You cannot learn problem solving by reading a book or watching an instructor lecturing. Just as in learning how to drive, ski, ice skate, play chess, play a musical instrument, play basketball, be a gymnast, or to become proficient in any other complex human activity, you acquire skill only through repeated practice. **The only way to learn how to solve problems in chemistry (or in any other subject) is to solve a lot of them.**

Therefore, to get the most benefit from this book, you should solve as many of the problems as possible. To help you in this endeavor, the problems at the end of each chapter (except Chapter 13) are divided into two parts. The Keyed Problems are "keyed" to the almost identical Example of the same number in the body of the chapter. The Supplemental Problems cover all the topics in the chapter in a random order. You should try to solve all the Keyed Problems and as many of the Supplemental Problems as possible.

**The best way I have found to solve the problems in this book is to follow exactly the way I present the solutions in the discussion and the examples. At least when you are first learning to solve problems, don't try to cut corners. Correct units and chemical notation are your friends.** *Use them wisely and save yourself a lot of grief.*

Of course, after you have mastered the solutions in this book, you can strike out on your own and decide what is the best way for you to solve problems, especially ones that you have never seen before.

To show you how **seriously** the international scientific community, industry, and governments take a consistent set of units (Chapter 7), chemical symbols (Chapter 1) and chemical names (nomenclature) (Chapter 16), here is a bit of history on how all this was set up.

---

The **metric system** was adopted in France in the 1790s in an attempt to standardize units of measurement. At the time there were seven or eight hundred different names for the various units of measure. On the eve of the French Revolution (1789) a quarter of a million different units of measure were in use in different towns and trades in France. It must have been total chaos when traveling around and/or trying to buy things!

In Paris in 1875, a set of international committees was established to set new standards and regulations as the ones set up in the 1790s became inadequate because of the advances in science. This is the **International System of Units** (called the **SI system** from the French, Le Système International d'Unités – see Section 7-6). Now all scientists in the world use the recommendations of these international committees. The SI system is informally still called the metric system.

"The SI is regulated and continually developed by three international organizations that were established in 1875 under the terms of the Metre Convention (which was ratified by 17 countries). They are the General Conference on Weights and Measures (Conférence générale des poids et mesures – CGPM), the International Committee for Weights and Measures (Comité international des poids et mesures – CIPM), and the International Bureau of Weights and Measures (Bureau international des poids et mesures – BIPM). The ultimate authority rests with the CGPM, which is a plenary body through which its Member States act together on matters related to measurement science and measurement standards; it usually convenes every four years." Quote from https://bit.ly/2F12j3Y.* See page 404 for all the SI prefixes, including 4 new ones approved at the November 2022 meeting of the CGPM.

Two other important organizations are mentioned in this book.

In 1901, the National Bureau of Standards (NBS) was founded with the mandate to provide standard weights and measures, and to serve as the national physical laboratory for the United States. Congress established the agency to remove a major challenge to U.S. industrial competitiveness at the time — a second-rate measurement infrastructure that lagged behind the capabilities of the United Kingdom, Germany, and other economic rivals. Due to its changing mission, especially after World War II, the National Bureau of Standards became the National Institute of Standards and Technology (**NIST**) in 1988 (https://bit.ly/2sl22CI).*

"The **IUPAC** [International Union of Pure and Applied Chemistry] was formed in 1919 by chemists from industry and academia, who recognized the need for international standardization in chemistry. The standardization of weights, measures, names and symbols is essential to the well-being and continued success of the scientific enterprise and to the smooth development and growth of international trade and commerce." Quote from https://iupac.org/who-we-are/our-history/.

---

**Now that you are convinced on how important it is to correctly use units, chemical symbols and chemical names (nomenclature)**, here are some more hints on studying chemistry.

**One of the best ways to learn is to teach.** Try to explain concepts and solutions to the problems to people who are also studying chemistry (or annoy your friends and/or relatives who aren't taking chemistry). *If you are taking a chemistry class, don't be shy about going to your instructor (or TA) for help.*

But whatever your study techniques, solve a lot of problems. As a study aid, solutions to all the end-of-chapter problems can be found at the back of the book. These are preceded by a glossary in which you can look up the definitions of terms.

There are many links to the web in this book that may help you to understand the material. If you have time, check them out. Also, just for fun, there are seven Easter eggs in the guise of bit.ly links scattered around this book.

Of course, an online search will also give you definitions and much additional reliable information. Wikipedia articles are very good, but some are more advanced than the material in this book. Sites from the National Institute of Standards and Technology (NIST), the National Aeronautics and Space Administration (NASA), the International Union of Pure and Applied Chemistry (IUPAC), and other governmental agencies are very reliable. Many universities also have good web sites about chemistry. The web site, https://sciencenotes.org/, has a wealth of good information, including printable periodic tables in color (pick one and print it as large as possible) and a collection of chemistry jokes. A TED talk about the history of the periodic table, which is discussed on pages 232 and 245-251, is at https://bit.ly/30wq2jp.* One list of chemistry apps for smartphones is at https://bit.ly/36UYUhA*.

There are many YouTube videos online that teach various aspects of chemistry. Sites that seem appropriate are:

- https://www.youtube.com/user/tdewitt451
- https://www.khanacademy.org/science/chemistry
- https://socratic.org/chemistry
- https://bit.ly/3alLPi8.*

A very nice site that has short videos (a few minutes) on many topics in physics, uses very little math, and is designed for non-science majors, is at www.hewittdrewit.com/title_list.htm. Some of these videos will be very useful when you read pages 188 (#59-62, 67), 231 (#24-27, 43), 257 (#82-86, 111, 112), and 412 (#123-126). There are many other videos that cover material mentioned in this book.

**DISCLAIMER:** I have only watched a few samples of the videos at each web site. **NOTE:** The chemistry sites use the term *atomic mass* or *relative atomic mass* for *atomic weight*. They also use the term *molar mass* for *molecular weight* and *formula weight*.

There is a free complete general chemistry text at https://openstax.org/details/books/chemistry-2e.

**DISCLAIMER:** I have not read the book, only looked at sample pages.

*The **bit.ly** shortened URLs (web addresses) are **case sensitive**, unlike full URLs, which are not case sensitive.

---

There is an old saying by Voltaire (1694-1778): *Le mieux est l'ennemi du bien,* translated from the French as: "The best is the enemy of the good." In the news recently, Voltaire's saying is being quoted more and more frequently. It is usually translated (somewhat incorrectly) this way: "The perfect is the enemy of the good." The idea behind this saying is that if you try for the best or the perfect, you may get so bogged down that you won't accomplish much. After all, the best is *really* hard and perfection is essentially impossible to attain. In other words, if you try for the best or the perfect and this causes you to get bogged down while studying, it might be a good idea to move on or take a break and return to the bottleneck later.

# PHOTOS AND ESSAYS

| | |
|---|---|
| Graphene | Opposite Title Page |
| About this Book | xv |
| Some Hints on Studying Chemistry | xviii |
| A Brief History of the Metric System, the Modern SI, the IUPAC and NIST | xviii |
| Roman Numerals | xxv |
| Lightning Striking Skyscrapers | Opposite 1st Page of Chapter 1 |
| Radioactive Isotopes | 7 |
| Neutrons, Neutrinos, Antimatter, and Enrico Fermi | 8 |
| A Brief History of the Structure of the Atom | 16 |
| The Unified Atomic Mass Unit (u) | 19 |
| The 1927 Solvay Conference and Marie Curie | 20 |
| Discovery of Nuclear Fission | 34 |
| Update: National Average Salary for a Chemist in the US | 39 |
| A Forest Fire | 40 |
| Rubies, Sapphires, Watch Crystals and the First Laser | 53 |
| An Ice Crystal (a snowflake) | 54 |
| Salt and Salt Crystals | 57 |
| Element Update – 104 to 118 | 61 |
| The Andromeda Galaxy | 62 |
| Fun With Large Numbers | 91 |
| Significant Figures: Accuracy and Precision | 92 |
| Digital Analytical Balances | 101 |
| Astronaut Stephanie Diana Wilson and the Harmony Utility Hub for the Space Shuttle | 108 |
| The Avogadro Project (Part 1) | 122 |
| Amedeo Avogadro: An Image and His Character | 140 |
| Remarks about Some Names Used in This Book | 141 |
| A Space Shuttle Liftoff and the Solid Composite Propellant used in its Booster Rockets | 142 |
| Mole Day and Pi Day | 156 |
| The Avogadro Project (Part 2) | 157 |
| Superconductors and MRI Machines | 158 |
| Proof That the Element Lead Has at Least Two Isotopes: A Real-Life Stoichiometry Problem | 169 |
| Performing a Titration with a Buret | 170, 177 |
| Acids, Bases and Salts | 187 |
| Hot Air Balloons | 188 |

| | |
|---|---:|
| How Normal Body Temperature Changed Since 1851 | 193 |
| A SCUBA Diver | 202 |
| Sealab II and Gas Mixtures for Deep Diving | 213 |
| Avogadro's Law and The Ideal Gas Law | 217 |
| Lasers and Some Recent Nobel Prize Winners | 218 |
| Schrödinger and His Equation (footnote) | 223 |
| Momentum and Angular Momentum | 231 |
| Dmitri Mendeleev's 1871 Periodic Table | 232 |
| How the Atomic Number Was Used in Arranging the Periodic Table and How Atomic Orbitals Became Assigned to Various Parts of the Periodic Table (footnote) | 245 |
| A Modern Periodic Table | 246 |
| Periodic Table Day and Two Interactive Periodic Tables | 250 |
| Henry Gwyn Jeffreys Moseley | 256 |
| Waves and Particles: Light, Electrons, and Baseballs | 257 |
| Very Large-Scale Integrated Circuits | 258 |
| Gilbert Newton Lewis | 262 |
| Linus Carl Pauling | 265 |
| Phosphine (pronounced "faas-feen") on Venus? | 284 |
| Wave Diffraction and Interference, and the Structure of Small Complex Molecules (Part 1) | 285 |
| OTC Items from the Shelves of a Pharmacy | 286 |
| Update on Naming Nonmetal – Nonmetal Compounds | 303 |
| Rechargeable Zinc-Air Batteries | 304 |
| The Structure of Small Complex Molecules (Part 2) | 322 |
| A Lithium-Ion Battery in a Smartphone | 324 |
| How Two 26-Year Old Men Changed the World (continued on page 358) | 345 |
| pH and a Scanning Electron Microscope Photo of Blood Cells | 346 |
| Magnesium compounds as Antacids, Laxatives and Supplements (footnote) | 354 |
| Digital pH Meters | 355 |
| How Two 26-Year-Old Men Changed the World | 358 |
| Two Riddles that Require Logical Thinking | 398 |
| Additional SI Prefixes proposed to the BIPM | 404 |

**APPENDIX** 399
    About the Author 399
    The Front Cover Photograph Explained 400
    Scientific Calculators 402
    Fundamental Physical Constants, SI Base Units, and SI Prefixes 403
    How a Rechargeable Battery Works 405
    Resistance, Conductors, Semiconductors, and Insulators 411
    The Double Slit Experiment with Electrons 412

# ACKNOWLEDGEMENTS

I wish to thank the hundreds of students who used the earlier version of this book and made many useful suggestions. Also, I wish to thank the thousands of students in my general chemistry classes who insisted that I explain difficult concepts clearly.

Sir Rudolf E. Peierls, when he was at the University of Washington, read Chapter 13 and agreed that my treatment of the uncertainly principle is reasonable at this level. He also suggested using a virus image as shown in Figure 13-2.

Professor Roald Hoffmann of Cornell University was kind enough to read Appendix 14-1 on orbital spacing. His acknowledgment that my treatment is reasonable at this level was very encouraging.

Professor Michael Schick of the University of Washington read the material on momentum and angular momentum (page 231), on the wave-particle nature of matter (page 257), and on the double slit experiment (page 412). His comments were most helpful in clearing up some oversimplifications I had made that were misleading or confusing. Professor Ron Lifshitz of Tel Aviv University pointed out some misleading statements I had made about the double slit experiment. Professor Tamir Gonen of UCLA read my discussion of his research on cryo-EM. He cleared up some details of his research that were not clear to me and gave me permission to use the graphic on page 322. Professor Sharon Cantor of UMass Medical School clarified my language about viral proteases on page 323.

Professor Sidney Emerman of Kingsborough Community College, CUNY used the earlier version of this book, and his advice and counsel were most valuable. Professor Patrick Lloyd, also of Kingsborough, was most helpful in discussions about bonding in third period elements.

Professor Gary D. Christian of the University of Washington was very helpful in our discussions about error analysis in analytical chemistry.

I offer my sincere thanks to these distinguished scientists, but any errors are mine alone.

The earlier version of this book, now out of print, was published by John Wiley & Sons. They were a group of highly competent professionals and it was a pleasure working with them. Wiley has turned over the copyright to the author and I thank them for their generosity. Much of the material, layout and style of that version is reflected in the current book. The comments of the outside reviewers of that version are much appreciated.

Finally, I thank my wife Linda and my sister Florence Lebenberg. Linda proofread all the new material of the current book. Her expertise in grammar, logical development and precise language made this a much better book. Flo advised me on the nuances of translating French into English. But again, any errors are mine alone.

# PHOTO AND IMAGE CREDITS

All photo and image credits and attributions in this book, with a few exceptions noted below, are listed on the page with the photo or image.

For permission to use photos, diagrams, or images:
- In the public domain.
- Under a Creative Commons CC BY-SA license. If a photo or image is under a Creative Commons CC BY-SA license, the type of ShareAlike has been listed and the attribution or web link to the attribution, if needed, has been stated.
- Permission has been granted to the author by the copywrite holder and attribution is stated.

The image of Leonard Nimoy's hand on page xvii is at https://bit.ly/3gnxvdb. It is under a CC by 2.0 license. The image of his hand has been cropped from the full photo and converted to grey scale.

The image of Avogadro on page 140 is in the public domain. The photo of a living mole on page 156 has been slightly cropped and converted to grey scale.

The term "Atomic Mass" has been replaced by the term used in this book, "Atomic Weight," in the large **Au** box of the periodic table in Figure 14-10. Also, orbital names (*s, p, d, f*) have been added and the electronic shells table has been removed.

For the images in Figure 15-2, the atomic orbitals are from File:Sr core-electron orbitals for Wiki.jpg - Wikimedia Commons (CC by SA-4.0). The author cropped the image of "Sr 1s," and converted it to greyscale. The molecular orbital is from File:Sigma bond.svg - Wikimedia Commons (CC by SA-3.0 Unported). The author removed the dotted elliptical line, converted the image to greyscale, cropped the image, and changed the size and position of the black dots that represent the nuclei.

For the images in Figures 15-3 thru 15-8: The author owns the copyright for the geometric figures. The wonderful ball and stick models are provided free of charge by Millipore Sigma. The models and more about the theory used in explaining these molecular structures can be found at https://bit.ly/2L939vy. The author added element symbols to the balls and dots to the electron clouds. Images were converted to grey scale.

For the diagrams of the Daniell cell on pages 406 and 407, the original diagram can be found at https://bit.ly/2SEwZvW (CC by SA-3.0 Unported). The author wishes to thank the artist for this exceptionally clear and informative diagram. The author modified both the original diagram (Figure 1) and a copy to show the cell in charging mode (Figure 2). Diagrams have been converted to grey scale, and sulfate ions have been added to the solutions. For the Discharge Mode (as a battery): Captions for the resistance and salt bridge have been added or modified. For the Charging Mode: A "**G**" has replaced the resistance symbol, captions have been added or modified for the generator and the salt bridge, the ion flow and direction in the salt bridge has been changed to reflect the charging state, and the captions for the anode/oxidation and cathode/reduction designations have been transposed, along with the (+) and (−) signs.

# Roman Numerals

A Clock with Roman Numerals on the Western Façade of the *Collegiate Church of Saint Peter at Westminster* (commonly known as *Westminster Abbey*) in London. The number 4 in this clock face is written as IIII. In the table below it is written as IV. The IIII notation seems to be used mostly on clocks and watches, even today. An exception for an old clock is Big Ben in London which uses IV.

*Photo by Author.*

Roman numerals are used in this book to number the "front matter," the pages before "page zero," the left page facing the first page of Chapter 1. The numbering of the "front matter" begins on the first right-facing page after the cover. This page and the next 5 have no page numbers written on them but they are counted. So the Roman numerals that paginate the "front matter" begin with VI. Roman numerals are also used in Chapter 14 in the discussion of the periodic table and in Chapter 16 in the discussion of naming compounds.

For all about Roman numerals and how they are used today, look at the Wikipedia article "Roman numerals." Below is a chart listing the first 30 Roman numerals.

## The First 30 Roman Numerals

| 1 | I | 16 | XVI |
|---|---|---|---|
| 2 | II | 17 | XVII |
| 3 | III | 18 | XVIII |
| 4 | IV | 19 | XIX |
| 5 | V | 20 | XX |
| 6 | VI | 21 | XXI |
| 7 | VII | 22 | XXII |
| 8 | VIII | 23 | XXIII |
| 9 | IX | 24 | XIV |
| 10 | X | 25 | XV |
| 11 | XI | 26 | XVI |
| 12 | XII | 27 | XVII |
| 13 | XIII | 28 | XVIII |
| 14 | XIV | 29 | XXIX |
| 15 | XV | 30 | XXX |

**Lightning Striking Skyscrapers:** Lightning occurs when electrons are separated from atoms. Eventually, because the air cannot support this charge separation, there is a giant spark to unite the negative electrons and the positive atoms that have had some of their electrons removed. Note that positive and negative charges attract each other. A lightning bolt can generate over 1 billion volts. Contrast this with normal house voltage in the United States which is around 120 volts. **NOTE:** Many of the essays that accompany photos at the beginning of chapters discuss material that is covered in that chapter. To get full value from these essays, it would be best if you returned to them after reading the chapter.     *Photo is in the Public Domain.*

# ATOMS, ISOTOPES AND ELEMENTS

Suppose that you took a small piece of iron and chopped it up as finely as you could. Then, somehow, you found someone who could chop it even more finely. The pieces would become finer and finer, finer than it is really possible to imagine. What would eventually happen?

At some point in the chopping, you would find that you couldn't chop anymore and still have iron. This is because you already had the smallest particles of iron possible. These particles are called **atoms**.[1] If you chopped up the iron atoms, they would no longer be iron atoms.

In this book we will discuss some of the properties and reactions of atoms. We will also look at the structure of atoms in some detail. Since atoms are very important in the study of chemistry, let's start our discussion with the structure of atoms.

## 1-1
### ATOMS CONSIST OF A POSITIVE NUCLEUS AND NEGATIVE ELECTRONS

One of the basic particles of matter is called the **atom**. The atom consists of two parts. In the center is a **nucleus**, which is small and has a positive electrical charge. The nucleus also contains almost all of the mass of the atom.[2]

**Mass** is the inherent amount of material of an object. When the force of gravity pulls an object down, we say that the object has weight. An object always has the same mass, but its weight can vary. The weight depends on the gravity force. Thus the astronauts who went to the moon had the same mass as they had on earth, but they weighed only one-sixth of their weight

---

[1] "The idea that matter is made up of discrete units is a very old one, appearing in many ancient cultures such as Greece and India. The word *atomos*, meaning 'uncuttable,' was coined by the ancient Greek philosophers **Leucippus** and his pupil **Democritus** (c. 460 – c. 370 BC)." (https://bit.ly/2TwXXGH)

[2] As discussed on the next page, the other part of the atom consists of negative electrons that are outside the nucleus. (**NOTE:** There are only two kinds of electrical charge, positive and negative.) **Benjamin Franklin** (1706-1790) was the scientist who decided around 1750 what was positive and what was negative electricity. His choice was arbitrary, but because of it, electrons (from the Greek word for amber) became negative, and thus the nucleus had to be positive. See https://go.nasa.gov/2R8WWCI and https://bit.ly/3bN0AMe for more information. In the second link, it is likely that Franklin used amber instead of rubber.

# 2    ATOMS

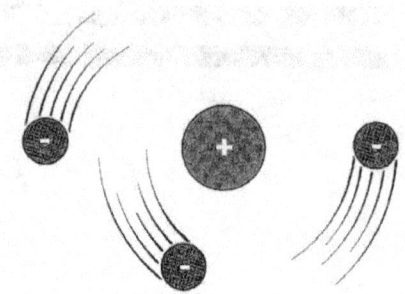

**FIGURE 1-1**  A *very* rough picture of the atom. The large ball with the plus sign represents the nucleus. The smaller balls with the minus sign represent electrons.

on earth. The gravity force on the moon is only one-sixth the gravity force on the earth. That's why they could jump around so easily.

The second part of the atom surrounds the nucleus and consists of one or more electrons. The way the electrons surround the nucleus will be discussed in detail in Chapter 13.

The electrons have a negative electrical charge and have very little mass. Compared to the mass of the nucleus, the electron is a real lightweight. Figure 1-1 shows a very rough picture of an atom. It will be modified later as you learn more.

In Figure 1-1 the large ball with the plus sign stands for the nucleus with its positive charge. Each small ball with a minus sign stands for an electron. The minus sign shows that the electron has a negative charge. This negative charge turns out to be the smallest unit of negative charge that can exist.

## 1-2
### THE ATOM STAYS TOGETHER BECAUSE OF ELECTRICAL ATTRACTION

A very important rule of nature is that **opposite electrical charges attract each other** and **like electrical charges repel each other**. Therefore, the positive charge of the nucleus attracts the negative charge of the electron. It is this electrical attraction that holds the atom together.

You might ask why the nucleus doesn't attract the electron with so much force that the electron falls into the nucleus. One reason is that the electron is in constant motion "around" the nucleus. This constant motion "around" the nucleus creates a force called centrifugal force. The electrical attraction and the centrifugal force exactly balance each other, as is shown in Figure 1-2. Thus the electron doesn't fall into the nucleus. There is another reason why the electron doesn't fall into the nucleus that is discussed in Appendix 1-1 on page 16.

You are probably familiar with centrifugal force. As pictured in Figure 1-3, if you tie a rock to a string and whirl it around, the centrifugal force on the rock keeps the string pulled tight.

If you drive around a corner too fast, the centrifugal force can become greater than the force holding the tires on the road, and the car skids or even

## 1-3  The Properties of Electrons, Protons, and Neutrons

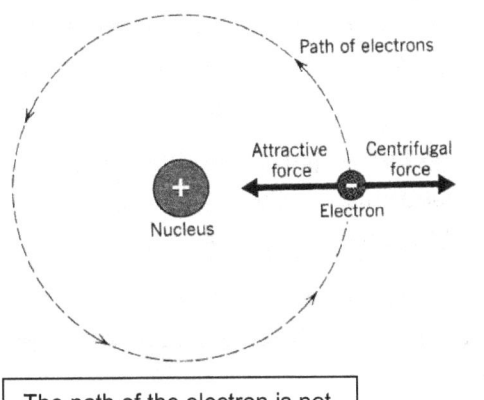

The path of the electron is not really circular. The reason is explained in Chapter 13.

**FIGURE 1-2** The forces on the electron is one reason why the electron doesn't fall into the nucleus. The inward-pulling force is the electrical attraction between the plus charge of the nucleus and the minus charge of the electron. It is exactly balanced by the outward-pulling centrifugal force.

In Fig. 1-2, if we replace the nucleus with our sun, and the electron with the earth, the electrical attractive force would be replaced with a gravitational attractive force. **NOTE:** Gravitational force is always attractive. There is no repulsive gravitational force.

In Figure 1-3 below, the inward-pulling force on the rock is provided by the string. The outward-pulling force is centrifugal force.

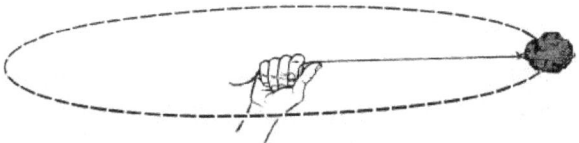

**FIGURE 1-3** The rock moves in a circular path because of the forces on it.

turns over. A centrifugal force is generated *only* when you change the direction of the motion of an object.

## 1-3
### THE PROPERTIES OF ELECTRONS, PROTONS, AND NEUTRONS ARE RELATED TO THEIR MASS AND CHARGE

Although each isolated electron is small, when electrons surround the nucleus they take up a lot of space. For instance, if the nucleus were the size of a basketball, the electrons would surround the nucleus up to 10 miles from it. This is just an example to give you a feeling for the relative size; but remember, the atom is *very, very* small.

Now we will go back and look at what makes up the nucleus. The nucleus of an atom contains **protons** and **neutrons.** The exception is the simplest nucleus, which is just a proton. The proton has one unit of positive electrical charge. This positive electrical charge is equal to the negative charge on the electron outside the nucleus but has the opposite sign. The proton and the neutron are very heavy compared to the electron. However, the neutron has *no* electrical charge. It is neutral, which is why it is called a neutron.

The proton and the neutron have very nearly the same mass. Since these particles are so basic to the structure of the atom, they each are

# 4  ATOMS

casually referred as having one **unified atomic mass unit**. But their actual mass has been measured very carefully as shown in Table 1-1. (The reason that their mass is not exactly one unified atomic mass unit will have to wait until we define the unified atomic mass unit in Chapter 2.) The abbreviation for unified atomic mass unit is u. The electron, on the other hand, is very light and has a mass of only about 1/1823 u, but again, its actual mass has been measured very carefully as shown in Table 1-1. An electron is almost 2000 times lighter than a proton or a neutron. See page 19 for a history of the unified atomic mass unit.

Since the space that the atom takes up is mostly filled with very light electrons, it is interesting to think what matter would be like if we could collect a sample of just the nuclei alone. If we could do this, and the nuclei were touching each other like marbles in a jar, we would have a material that is very dense. In fact, a piece of it the size of a marble (about 1 cubic centimeter, or 1/5 of a US teaspoon) would weigh around as much as Mt. Everest.

There are stars where this compression of matter has taken place. Normally, the burning fuel in a star creates enough outward pressure to keep it from collapsing by gravity. (**NOTE:** Astronomers use the term "burning" when a star is undergoing nuclear fusion. See page 34.) When the star begins die because it is running out of fuel, there is not enough outward "burning pressure," and the star begins to collapse from gravity. If the star is big enough, the tremendous gravitational force that makes it smaller is so great that the electrons of the atoms are forced into the protons of the nuclei to make neutrons. Since neutrons are neutral, they can get very close to each other, like the marbles in the jar discussed above. The resulting star is, naturally, called a **neutron star**. Our star, the sun, will not become a neutron star when it dies in five or six billion years; it is too small. See page 400 for more on dying stars.

Most of the time matter is electrically neutral. The reason for this is that atoms have the same number of protons in the nucleus as they have electrons surrounding the nucleus. Since the positive charge on the proton is the same size as the negative charge on the electron, the two charges cancel each other and the atom is neutral. When an atom isn't neutral, it is because one or more of its electrons have been removed, leaving more protons than electrons. Then there are one or more positive charges than negative ones, and the atom as a whole is positive.

TABLE 1-1
THE PROPERTIES OF THE PROTON, NEUTRON, AND ELECTRON

| Particle | Approximate and Measured Mass* in Unified Atomic Mass Units | Charge | Location in Atom | Amount of Space Occupied in the Atom |
|---|---|---|---|---|
| Proton | 1 u<br>1.007276466621 u | +1 | Nucleus | A little |
| Neutron | 1 u<br>1.00866491588 u | 0 | Nucleus | A little |
| Electron | 1/1823 u<br>0.000548579909070 u | −1 | Space around the nucleus | A lot |

*Consider the time and effort put into measuring the mass of these particles. That's because these numbers are very important to scientists who need them for their research.

The torn-off electrons can, among other things, make a spark of electricity or an electric current. This separation of charge occurs only under special conditions. It is obvious when you rub your feet along a carpet and then touch a metal object or see a flash of lightning.

To complete the story, we should mention that one or more electrons can also add to some atoms, making them negative. We will discuss positively and negatively charged atoms in Chapter 15.

The appendix at the end of this chapter gives a very brief history of some of the problems scientists have had in figuring out the structure of the atom. Chapters 13 and 14 discuss atomic structure in more detail.

## 1-4
### THERE ARE THREE KINDS OF HYDROGEN ATOMS: PROTIUM, DEUTERIUM, AND TRITIUM

We can now start to describe some real atoms. The simplest possible arrangement is an atom made of one proton in the nucleus with one electron around the nucleus. A rough diagram of this atom is shown in Figure 1-4. The "1p$^+$" represents the one proton (which has a single positive charge) in the nucleus. The "1e$^-$" shows that the atom has one electron. The half-circle in the diagram separating the "1" and the "e$^-$" represents the electron surrounding the nucleus. The reason that the electron is not shown going around the nucleus in a full circle is that scientists do not believe that it goes around in a circle like the rock in Figure 1-3. Just think of it as surrounding the nucleus in some manner. Chapter 13 discusses this in more detail.

This simple atom, consisting of one proton and one electron, is called a **hydrogen atom**. It has a mass of about 1 u, which comes almost entirely from the proton. The mass of the electron is so small that we can ignore it for now.

**All the atoms with the same number of protons in their nucleus are given a name, a symbol that stands for the name, and a number.** This number is called the **atomic number** and is the number of protons in the nucleus. **All the atoms with the same atomic number make up an element.** As of 2021, there are 118 known elements. For example, the name of the element described above is hydrogen; the symbol for the element hydrogen is H. Since hydrogen has one proton in the nucleus, its atomic number is 1. Atomic numbers for most of the elements are listed in Tables 1-2 and 2-2.

Please note that the atom with one proton and one electron is sometimes called **protium**, to distinguish it from the atoms of deuterium and tritium, which are discussed below.

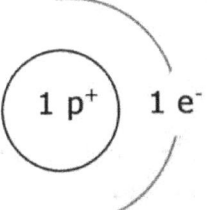

**Figure 1-4** A diagram of protium, the simplest atom.

6 ATOMS

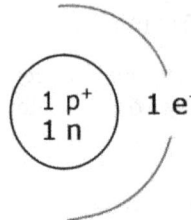 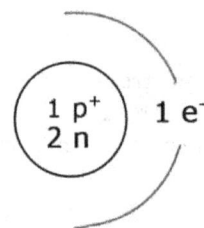

**FIGURE 1-5** A diagram of a deuterium atom.

**FIGURE 1-6** A diagram of a tritium atom.

The next simplest atom has one proton and one neutron in the nucleus and one electron surrounding the nucleus. (Remember that for neutral atoms, the number of protons equals the number of electrons.) It has a mass of about 2 u, one each from the proton and the neutron. It is called **deuterium** and is usually given the symbol D. The atomic number of deuterium is also 1, since there is one proton in the nucleus. To sketch a picture of the deuterium nucleus (Figure 1-5), put a 1n under $1p^+$. This 1n stands for the one neutron.

With one proton and two neutrons in the nucleus and one electron outside, we have **tritium**. It is usually given the symbol T. The tritium atom has a mass of about 3 u (1 proton + 2 neutrons). The atomic number of tritium is also 1. A diagram of tritium is shown in Figure 1-6.

**Since protium, deuterium and tritium all have same number of protons in the nucleus, they are all atoms of the same element – hydrogen.**

The process of adding another neutron to the nucleus has not been successful: If we try to make a nucleus with one proton and three neutrons, we cannot; they do not stick together. So for elements heavier than hydrogen, there must be at least two protons in the nucleus. But before we continue, let's go back and look at protium, deuterium, and tritium a little more closely.

## 1-5
### THE THREE ISOTOPES OF HYDROGEN MAKE UP THE LIGHTEST ELEMENT

Protium, deuterium, and tritium atoms all have the same number of protons and electrons but a different number of neutrons in the nucleus. The atomic number of all three atoms is the same, namely, 1. Atoms like this are called isotopes. **Isotopes of an element are atoms with the same number of protons (and, of course, electrons) but different numbers of neutrons.**

Chemical properties are determined primarily by the electrons that surround the nucleus and depend only a little on the mass of the nucleus. Since the number of electrons is the same in each atom of the different isotopes of hydrogen – protium, deuterium, and tritium – they have similar, but not exactly, the same chemical properties. This is called the **isotope effect**. The atoms protium, deuterium, and tritium are *isotopes of the element that has one proton in its nucleus.*

**To repeat, an *element* is all atoms with a given atomic number, or number of protons in the nucleus.** If there is only one proton in the nucleus of an atom, the element is called **hydrogen**. The atoms of an element are the

## 1-5 The Three Isotopes of Hydrogen Make Up the Lightest Element

> **RADIOACTIVE ISOTOPES:** Their existence was first suggested in 1913 by the British radiochemist **Frederick Soddy** (1877-1956). The term "isotope," Greek for "at the same place" (meaning at the same place *in the periodic table—in other words, the same element*) was suggested to Soddy by **Margaret Todd** (1859-1918), a Scottish physician and family friend, during a conversation in which he explained his ideas to her. Soddy won the 1921 Nobel Prize in chemistry in part for his work on isotopes. The first evidence for multiple isotopes of the stable (non-radioactive) element neon was found by the British physicist **J.J. Thomson** (1856-1940) in 1913. (He won the Nobel Prize in physics in 1906 for discovering the electron.) The British chemist and physicist **F. W. Aston** (1877-1945) in 1921 discovered multiple stable isotopes for many elements. He won the Nobel Prize in 1922 for this work. See p. 169 for the first proof of lead isotopes in 1914. Only thirty years later, scientists would be well on their way to building the first atomic bombs, which used an isotope of either uranium or plutonium (see https://bit.ly/3QtJ2bL for information on nuclear weapons).

smallest particles in nature that have a "'chemical" identity. If you chop up atoms, you have protons, neutrons, and electrons. These particles do not have a chemical identity. Of the 118 known elements (as of 2021), only hydrogen (with one proton in its nucleus) has isotopes that are given special names. The special names are used because these isotopes are so useful in many fields of research. When a chemist says "'hydrogen," the reference is to the naturally-occurring mixture of isotopes of the element with atomic number 1 (or sometimes to the lightest isotope of that element [namely, protium], since it makes up 99.985% of all the hydrogen atoms in a sample). Historically, the element hydrogen was known long before anybody knew about isotopes or, for that matter, protons, neutrons, and electrons. Hydrogen was discovered in 1766 by the British scientist **Henry Cavendish** (1731-1810) and named in 1783 by the French chemist **Antoine Lavoisier** (1743-1794). You might enjoy reading about him at https://bit.ly/2G2Mhne. Be sure to note how he died and ignore the discussion of phlogiston. His highly intelligent wife **Marie Anne** helped him with his experiments.

**REMEMBER: When scientists are talking about an element by name, they almost always mean the mixture of all the naturally-occurring isotopes of that element. If they are talking about a specific isotope of an element, they will name that isotope.** See Table 2-1 for some elements and percentages of each isotope present in the naturally-occurring element.

The word "isotope" is used in two different ways. As we have been discussing with the three isotopes of the element hydrogen, the word isotope can refer to different nuclei of the same element. You may also have heard the expression "radioactive isotopes can be dangerous." In this use, no particular element is referred to, just various nuclei that happen to be **radioactive**. Nuclei that are unstable and fly apart are said to be radioactive. A shorthand term for these radioactive nuclei is **radionuclide** (singular) or **radionuclides** (plural).

Naturally-occurring hydrogen consists mostly of the light isotope protium, the one we usually call "hydrogen." On earth, for about every 5000 protium atoms, there is one deuterium atom. There is also an *extremely* small amount of tritium in nature, although large quantities of tritium have been made in nuclear reactors because it is very useful in research.

# 8 ATOMS

> **NEUTRONS, NEUTRINOS, ANTIMATTER, AND ENRICO FERMI:** Protons and electrons are stable. Neutrons are a different story. When a neutron is bound in the nucleus of an atom, it is stable. Free neutrons are not stable and after about 10.2 minutes, half of them decay into a proton, an electron, and an **antineutrino**.[3] **Neutrino** means "neutral little one" in Italian. The Nobel Prize winning Italian American physicist **Enrico Fermi**[4] (1901-1954) named it. Element 100, fermium (Fm), is named in his honor. See page 34 for more on Fermi. See the web site from Fermilab—https://bit.ly/3NMpgXg—for more on neutrinos. See https://bit.ly/3tvp0DO for more about neutrons. You may have seen a discussion of antimatter that provides the power source for the starship Enterprise's engines on *Star Trek*. When matter and antimatter collide, they convert almost completely into pure energy. This is the greatest source of energy and the source of energy for photon torpedoes and warp drive on the show. See https://bit.ly/3xt8tS1 and https://binged.it/3QruzgA.
>
> [3]Every particle has an antiparticle. For example, the antiparticle of an electron is an antielectron. It has a positive charge and is called a **positron**. When an electron and a positron collide, they turn into two gamma rays ( see page 221). This reaction is used in medicine for research and diagnosis. It is called positron emission tomography (PET scan — see https://mayocl.in/3zJyUpl).
>
> [4]A fascinating story of Fermi's life can be found at https://bit.ly/3zAVtwk. After Mussolini's Fascist Italy passed the anti-Jewish Racial Laws in 1938, Fermi, and his wife, Laura (1907-1977), who was Jewish, began thinking about leaving Italy. Fermi's winning of the 1938 Nobel Prize in Physics, which came with over $300,000 in today's dollars, gave the Fermis their chance to escape. Without telling anyone, not selling their house, only taking money and luggage appropriate for the trip to Stockholm, they and their two children left Italy, eventually ending up at Columbia University in New York City. (I heard these details at a Cornell University symposium commemorating Fermi's life and research. Carl Sagan [pp viii, 371, 401] was master of ceremonies.)

The reason that hardly any tritium exists in nature is that its nucleus is unstable and flies apart. Tritium is radioactive and is the lightest radioactive isotope that exists. Since half the tritium atoms in any sample would decay in 12.3 years[5] (**decay** is the technical term for the decomposition of the nucleus of an atom) and since the earth is about 4.5 billion years old, you might wonder why there is any tritium at all in nature. The reason is that a very little bit is always being made in the upper atmosphere by **cosmic rays** when they collide with the nuclei of air molecules. Scientists can make tritium for experimental use by putting lithium metal in a nuclear reactor. The neutrons that are produced by the reactor react with the lithium to produce tritium. **NOTE:** Cosmic rays are very high energy nuclei that come from outer space. About 90% are protons, the rest being heavier nuclei.

## 1-6
### THE NOTATION FOR WRITING ISOTOPES TELLS US THE NUMBER OF PROTONS AND NEUTRONS IN THE NUCLEUS

There is a notation for atoms that makes it very easy to tell what element and what isotope we are dealing with. Figure 1-7 shows diagrams of the three isotopes of hydrogen. Consider protium.

[5]If half of the tritium atoms in a sample decay in 12.3 years, and we start with 100 tritium atoms, in 12.3 years there would be 50 tritium atoms left. After another 12.3 years, there would be 25 tritium atoms left. And so on until all the tritium is gone.

## 1-7 The Atomic Number, the Mass Number and Writing Isotopic Symbols

If we write the **number of protons in the nucleus in the lower left corner** of the H symbol (i.e., $_1$H) and write the **total number of protons plus neutrons** (for this one isotope there are no neutrons) in the top left corner (i.e., $^1$H) and combine both, we have

$$^1_1H$$

This is the **isotopic symbol** for the lightest isotope of hydrogen.

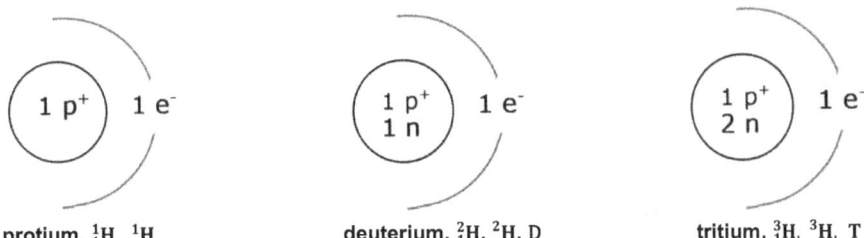

protium, $^1_1$H, $^1$H      deuterium, $^2_1$H, $^2$H, D      tritium, $^3_1$H, $^3$H, T

**Figure 1-7** The diagrams, names and isotopic symbols for the isotopes of hydrogen.

Since the number of electrons in the neutral atom equals the number of protons, the bottom number also tells us the number of electrons. The bottom number in an isotopic symbol is called the **atomic number** and the top number is called the **mass number**.

Now let's look at deuterium in Figure 1-7. The isotopic symbol would be $^2_1$H, where the mass number equals 2 and represents the sum of 1 proton + 1 neutron. A diagram of tritium is also shown in Figure 1-7. Tritium has the isotopic symbol $^3_1$H, where the mass number equals 3 and represents the sum of 1 proton + 2 neutrons.

Above we wrote the isotopic symbols for deuterium as $^2_1$H and tritium as $^3_1$H. However, it is very common to use the symbols D and T for deuterium and tritium. When D and T are used, they are usually written without numbers. These are the only isotopes where this is reasonable as everybody knows the atomic numbers and mass numbers for D and T. **NOTE:** P is *never* written for protium (hydrogen-1) as P is the chemical symbol for the element phosphorus. However, p, p$^+$, or $^1_1$p are used for a proton. Also, n or $^1_0$n are used for a neutron, and e, e$^-$ or $^{\ 0}_{-1}$e are used for an electron. It is the writer's or editor's choice.

It is common to write isotopic symbols without the atomic number, as the element determines the atomic number. You may see $^1$H, $^2$H or $^3$H written instead of $^1_1$H, $^2_1$H or $^3_1$H. In this book, isotopic symbols will be written with both notations. You should be comfortable using all the names and isotopic symbols mentioned on this page. Figure 1-7 summarizes all the possibilities.

Let's summarize the notation for isotopic symbols. If **E** stands for the symbol of any element, then the isotopic symbol for this imaginary element **E** would be

$$^{\text{number of protons + number of neutrons}}_{\text{number of protons}}E$$

**EXAMPLE 1** The most common isotope of the element oxygen has 8 protons and 8 neutrons in the nucleus. The symbol for oxygen is O. Write the isotopic symbol for this isotope.

*Solution:* The isotopic symbol is $^{16}_{8}$O or $^{16}$O, where 8 protons + 8 neutrons = 16. ■

## 1-7
### THE ATOMIC NUMBER AND THE MASS NUMBER ARE USED IN WRITING ISOTOPIC SYMBOLS

There are special names given to the numbers that in front of the letter symbols for the isotopes. As you already know, the bottom number is called the **atomic number** and is the number of protons in the nucleus. The top number is called the **mass number** and is the sum of the protons and neutrons in the nucleus.

Since the mass of each proton and neutron is very close to 1 u, the mass number gives us an approximate total mass of the atom (remember that the mass of the electrons is very small compared to the mass of the protons and neutrons). Referring to our symbol representing any element, E, we can write

$$^{\text{mass number}}_{\text{atomic number}} E$$

**A REMINDER:** *The actual mass of an atom in unified atomic mass units (u) is not the same as the mass number.*

**EXAMPLE 2** What is the atomic number and the mass number of the oxygen isotope $^{16}_{8}O$?

*Solution:* The atomic number is 8, and the mass number is 16. ∎

**EXAMPLE 3** If an atom has a mass number of 25 and an atomic number of 12, how many neutrons are in the nucleus? Also, how many electrons surround the nucleus?

*Solution:* Since the mass number is equal to the protons + neutrons, we can write the simple equation

$$\text{mass number} = \text{number of protons} + \text{number of neutrons}$$

or

$$\text{number of neutrons} = \text{mass number} - \text{number of protons}$$

Substituting numbers from this example, we can write

$$\text{number of neutrons} = 25 - 12 = 13$$

Since the number of electrons equals the number of protons, the atom has 12 electrons. ∎

## 1-8
### ATOMS OF ELEMENTS HEAVIER THAN HYDROGEN HAVE MORE PROTONS, NEUTRONS, AND ELECTRONS

Now we can examine atoms with more than one proton in the nucleus. A nucleus with just two protons cannot exist. Both protons have a positive

## 1-8 Atoms of Elements Heavier Than Hydrogen

charge, and positive charges repel each other. A neutron is needed to stabilize the two protons so that they stick together. Neutrons act as a sort of nuclear "glue" to hold the nucleus together. However, the story is much more complicated than just calling neutrons nuclear "glue" — too many neutrons can make a nucleus unstable.

The element that contains two protons in the nucleus is helium (symbol He). Two isotopes of helium exist in nature. Both are stable. The first has two protons and one neutron; it has a mass number of 3. This isotope very rare in nature (0.0002% of all the helium).

**EXAMPLE 4** Sketch a diagram and write down the isotopic symbol for the helium isotope with a mass number of 3.

*Solution:* The diagram and symbol are shown in Figure 1-8. ∎

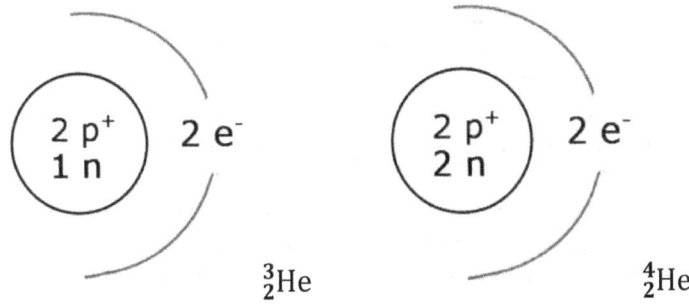

**FIGURE 1-8** The diagrams and symbols of the two isotopes of helium.

The second isotope of helium has two protons and two neutrons in the nucleus; its mass number is 4. This is by far the most abundant isotope of He in nature (99.9998% of all helium). The diagram and isotopic symbol of the atom is also shown in Figure 1-8.

If more neutrons are added to a $^4_2$He nucleus, it becomes unstable and flies apart.

For light nuclei, a combination of about half neutrons and half protons results in a stable nucleus. For heavier nuclei, the number of neutrons is larger than the number of protons.

After helium, the next stable nucleus contains three protons. The element with three protons in the nucleus is lithium (symbol Li). In nature, lithium exists in two isotopic forms. One form has 3 neutrons in the nucleus, and the other has 4 neutrons in the nucleus.

**EXAMPLE 5** Sketch both isotopes of lithium and write down their isotopic symbols.

*Solution:* The solution is shown in Figure 1-9. ∎

Isotopes are sometimes referred to with the mass number following the name, such as lithium-6 and lithium-7. When naming isotopes by this method, the element name is not abbreviated. Writing Li-6 for lithium-6 is not correct. $^3_2$He would be called helium-3, $^4_2$He would be called helium-4, and $^2_1$H would be called hydrogen-2 (most people would call it deuterium).

## 12 ATOMS

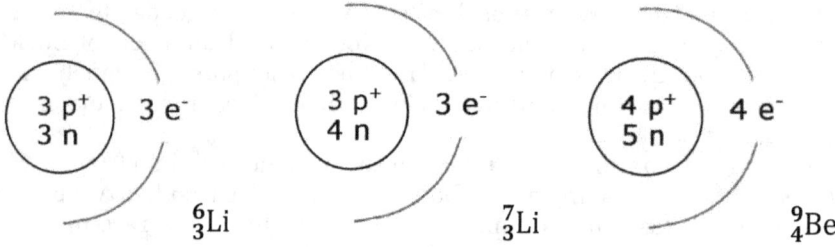

**FIGURE 1-9** The diagrams and symbols of the naturally occurring isotopes of lithium and beryllium.

After lithium comes beryllium, which has the symbol Be. Beryllium has only one naturally occurring isotope, which has a mass number of 9. The symbol and diagram are shown in Figure 1-9.

More and more protons and neutrons can be added (along with the appropriate number of electrons), giving many more elements and their isotopes. As of 2021, 118 different elements are known. From these elements, 339 isotopes occur in nature. Thirty-four of these isotopes are radioactive. In addition to the naturally-occurring isotopes, nuclear scientists have made about 950 radioactive isotopes.

Table 1-2 lists the elements, along with their naturally-occurring isotopes. A few of the elements have no stable isotopes and some of them do not exist in nature (at least not on the earth because they decay so fast) but have been made by in the laboratory by scientists. Elements that don't exist naturally on earth are not listed in the Table 1-2 but are discussed on page 61.

Usually, but not all the time as discussed in the following paragraph, isotopes that occur in nature are not radioactive, or if they are radioactive they decay very slowly. The reason is that most of the elements and their isotopes on earth were all formed in some manner long before our star, the sun, and its planets were made. See page 400 for more details. These elements then became part of our sun and its planets. Since our earth is about 4.5 billion years old, any radioactive elements that existed at the beginning of the earth must not have a **half-life**[3] too much shorter than 4.5 billion years. Any isotope that decayed in a much shorter time than 4.5 billion years would simply have disappeared by now.

A few isotopes are younger than the earth. This is because they have been formed since the earth was made. There are three ways that this can happen. The first way is that these young isotopes are formed from the radioactive decay of other isotopes. The second way is that isotopes are made by cosmic rays hitting the earth's atmosphere. Examples of isotopes that formed by means of the first way are radioactive radium-226 and a stable isotope of lead, namely, lead-206.

[3]**Half-life** is defined as the time it takes for ½ of a radioactive sample to disappear. Thus if the half-life of uranium-238 is 4.5 billion years, ½ of it will have disappeared in 4.5 billion years.

TABLE 1-2
**THE NATURAL ISOTOPES OF THE ELEMENTS** (continued on pages 14 and 15)

This table lists the mass numbers for the stable isotopes, and the mass number of the longest-lived isotope for elements without a stable nucleus (they are radioactive). Isotopes that are radioactive but present on earth in small to fairly large amounts are marked with a hash mark (#) (as in hashtag), or as known in North America, a pound sign. Isotopes marked with an asterisk (*) are present on earth in *extremely* small amounts and are radioactive. The data below were taken from https://bit.ly/31WUdjX which also includes the exact mass of each isotope and its percent abundance.

| Atomic Number | Symbol | and Mass Numbers |
|---|---|---|
| 1 | H | 1, 2, 3* |
| 2 | He | 3, 4 |
| 3 | Li | 6, 7 |
| 4 | Be | 9 |
| 5 | B | 10, 11 |
| 6 | C | 12, 13, 14* |
| 7 | N | 14, 15 |
| 8 | O | 16, 17, 18 |
| 9 | F | 19 |
| 10 | Ne | 20, 21, 22, |
| 11 | Na | 23 |
| 12 | Mg | 24, 25, 26 |
| 13 | Al | 27 |
| 14 | Si | 28, 29, 30 |
| 15 | P | 31 |
| 16 | S | 32, 33, 34, 36 |
| 17 | Cl | 35, 37 |
| 18 | Ar | 36, 38, 40 |
| 19 | K | 39, 40, 41 |
| 20 | Ca | 40, 42, 43, 44, 46, 48 |
| 21 | Sc | 45 |
| 22 | Ti | 46, 47, 48, 49, 50 |
| 23 | V | 50, 51 |
| 24 | Cr | 50, 52, 53, 54 |
| 25 | Mn | 55 |
| 26 | Fe | 54, 56, 57, 58 |
| 27 | Co | 59 |

# 14 ATOMS

| Atomic Number | Symbol | and Mass Numbers |
|---|---|---|
| 28 | Ni | 58, 60, 61, 62, 64 |
| 29 | Cu | 63, 65 |
| 30 | Zn | 64, 66, 67, 68, 70 |
| 31 | Ga | 69, 71 |
| 32 | Ge | 70, 72, 73, 74, 76 |
| 33 | As | 75 |
| 34 | Se | 74, 76, 77, 78, 80, 82 |
| 35 | Br | 79, 81 |
| 36 | Kr | 78, 80, 82, 83, 84, 86 |
| 37 | Rb | 85, 87 |
| 38 | Sr | 84, 86, 87, 88 |
| 39 | Y | 89 |
| 40 | Zr | 90, 91, 92, 94, 96 |
| 41 | Nb | 93 |
| 42 | Mo | 92, 94, 95, 96, 97, 98, 100 |
| 43 | Tc | 99* |
| 44 | Ru | 96, 98, 99, 100, 101, 102, 104 |
| 45 | Rh | 103 |
| 46 | Pd | 102, 104, 105, 106, 108, 110 |
| 47 | Ag | 107, 109 |
| 48 | Cd | 106, 108, 110, 111, 112, 113, 114, 116 |
| 49 | In | 113, 115 |
| 50 | Sn | 112, 114, 115, 116, 117, 118, 119, 120, 122, 124 |
| 51 | Sb | 121, 123 |
| 52 | Te | 120, 122, 123, 124, 125, 126, 128, 130 |
| 53 | I | 127 |
| 54 | Xe | 124, 126, 128, 129, 130, 131, 132, 134, 136 |
| 55 | Cs | 133 |
| 56 | Ba | 130, 132, 134, 135, 136, 137, 138 |
| 57 | La | 138, 139 |
| 58 | Ce | 136, 138, 140, 142 |
| 59 | Pr | 141 |
| 60 | Nd | 142, 143, 144, 145, 146, 148, 150 |
| 61 | Pm | 145* |
| 62 | Sm | 144, 147, 148, 149, 150, 152, 154 |
| 63 | Eu | 151, 153 |
| 64 | Gd | 152, 154, 155, 156, 157, 158, 160 |
| 65 | Tb | 159 |

The link to this table on page 13 lists $^{98}$Tc. It is only made in the laboratory. $^{99}$Tc is naturally-occurring in extremely small amounts in the ore of uranium called pitchblende. It comes from the spontaneous fission of uranium-238. It is also made in red giant stars near the end of their life. See page 401. A form of $^{99}$Tc, $^{99m}$Tc, has a half-life of 6 hours and is the most widely used radionuclide for diagnosis in nuclear medicine.

Even though xenon-124 is not listed in Table 1-2 as being radioactive, it actually is. Its half-life is very long at $1.8 \times 10^{22}$ years. This is about a trillion times longer than the age of the universe. Radioactive isotopes with a half-life longer than the age of the universe are generally *not* considered radioactive for most purposes.

## 1-8 Atoms of Elements Heavier Than Hydrogen

| Atomic Number | Symbol and Mass Numbers | |
|---|---|---|
| 66 | Dy | 156, 158, 160, 161, 162, 163, 164 |
| 67 | Ho | 165 |
| 68 | Er | 162, 164, 166, 167, 168, 170 |
| 69 | Tm | 169 |
| 70 | Yb | 168, 170, 171, 172, 173, 174, 176 |
| 71 | Lu | 175, 176 |
| 72 | Hf | 174, 176, 177, 178, 179, 180 |
| 73 | Ta | 180, 181 |
| 74 | W | 180, 182, 183, 184, 186 |
| 75 | Re | 185, 187 |
| 76 | Os | 184, 186, 187, 188, 189, 190, 192 |
| 77 | Ir | 191, 193 |
| 78 | Pt | 190, 192, 194, 195, 196, 198 |
| 79 | Au | 197 |
| 80 | Hg | 196, 198, 199, 200, 201, 202, 204 |
| 81 | Tl | 203, 205 |
| 82 | Pb | 204, 206, 207, 208 |
| 83 | Bi | 209 |
| 84 | Po | 209* |
| 85 | At | 210* |
| 86 | Rn | 222* |
| 87 | Fr | 223* |
| 88 | Ra | 226* |
| 89 | Ac | 227* |
| 90 | Th | 232# |
| 91 | Pa | 231# |
| 92 | U | 234#, 235#, 238# |
| 93 | Np | 237*, 239* |
| 94 | Pu | 244* |

Elements with atomic numbers 95-118 are not found naturally on earth but are made in the laboratory. As of 2021, elements with atomic numbers higher than 118 have not been made. See Table 2-2 and page 61 for details of elements 95-118.

Both of these elements come from the radioactive decay of uranium-238. Examples of isotopes that formed by means of the second way are radioactive tritium and radioactive carbon-14. The third way is by interstellar dust particles falling on earth. An example is radioactive plutonium-244, *extremely* small amounts of which have been identified in undersea deposits, https://bit.ly/2Xj7LI4. It is made during the collision of two neutron stars. **NOTE:** *Extremely* small amounts of neptunium-237 are found from the transmutation (see "Glossary" p. 370 for definition) of uranium-238 (it captures a neutron) in pitchblende, an ore of uranium.

16　ATOMS

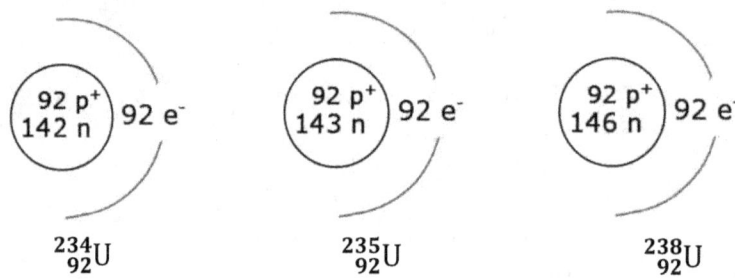

**Figure 1-10**  The naturally-occurring isotopes of uranium.

**EXAMPLE 6**  Referring to Table 1-2, write the diagrams and the symbols for the three isotopes of uranium, atomic number 92.

*Solution:* The solution is shown in figure 1-10.  ■

## APPENDIX 1-1
## A BRIEF HISTORY OF THE STRUCTURE OF THE ATOM

The electron was discovered in 1897 by the British physicist **J. J. Thomson** (1856-1940). Some positively charged particles associated with atoms were also discovered at about the same time. These discoveries led to the following question: How are the negative electrons and the positive particles arranged to form an atom?

We now believe that the electrons surround a small, heavy, positively charged nucleus. The positive charge of the nucleus attracts the negative charge of the electrons; this attraction is what holds the atom together. But around 1900 the situation wasn't that clear. Scientists of that era knew that when electrons traveled in a curved path, such as circles, they would radiate (give off) energy. This was a basic rule of the physics at that time. This energy loss would quickly cause the electrons to "spiral" into the nucleus, as an earth satellite would do when it loses energy in the earth's atmosphere and spirals into the ground. Scientists estimated that if electrons were really outside the nucleus, all the atoms in the universe would collapse in about one second. The scientists of 1900 felt that it was impossible for electrons to be outside the nucleus.

To try to answer these problems, J. J. Thomson proposed a model of the atom that had the electrons embedded in a large blob of nuclear "goo." It was called the "plum-pudding model." At the time, no one really thought that this was a very good model, but nobody could think of anything better.*

Then, in 1911, after a long series of experiments, **Ernest Rutherford** (1871-1937), a New Zealander working in England, found that the atom really did have a small, heavy positive nucleus with the electrons outside the nucleus. But as described above, this model, called the planetary model, couldn't be explained by the laws of physics of 1911.

In 1913, the Danish physicist, **Niels Bohr** (1885-1962), proposed a way out of the dilemma.  He said that the electrons didn't spiral into the nucleus because they couldn't. He assumed that new laws of physics had to be used to explain the structure of the atom.  His revolutionary ideas agreed with certain

---

*Personal communication from Prof. **Peter Debye** (1884-1966) of the Department of Chemistry & Chemical Biology, Cornell University. He received the Nobel Prize in Chemistry in 1936. If you open the link on page 20 to the full picture with names of the Solvay Conference of 1927, you can find Bohr and Debye at opposite ends of the second row.

experiments and were the start of a new theory of the atom, called quantum theory. This theory was developed in its present form in the 1920s by many people. Our present description of the atom is based on the results of quantum theory. We will briefly discuss some of its ideas and applications to chemistry in Chapters 13 through 15.

## PROBLEMS[1]

### KEYED PROBLEMS

1. An isotope of rubidium has 37 protons and 48 neutrons in the nucleus. The symbol for rubidium is Rb. Write the isotopic symbol for rubidium.

2. What is the atomic number and the mass number of the gold isotope $^{197}_{79}Au$?

3. If an atom has a mass number of 41 and an atomic number of 19, how many neutrons are in the nucleus. Also, how many electrons surround the nucleus?

4. Sketch a diagram and write down the isotopic symbol for the fluorine isotope with a mass number of 19. The atomic number of fluorine is 9.

5. The atomic number of nitrogen is 7. Nitrogen has two naturally-occurring isotopes with mass numbers 14 and 15. Sketch both isotopes of nitrogen and write down their isotopic symbols.

6. Referring to Table 1-2, sketch the diagram and write the symbol for the one naturally-occurring isotope of radon, element number 86. This isotope has a half-life of only 3.8 days, but since it is the decay product of radium-226, some of it is continuously being made. Radon is a noble gas and is fairly unreactive but it is radioactive and can be very dangerous. See https://bit.ly/32McBRD for details. Home radon test kits are sold online.

### SUPPLEMENTAL PROBLEMS

7. Define the following terms:
   a. Atomic number    b. mass number    c. element    d. isotope
   e. radioactive       f. deuterium      g. tritium

8. Sketch the diagram and write the isotopic symbol for all the naturally-occurring isotopes of the following elements (refer to Table 1-2).
   a. Ne   b. S   c. Ti   d. Cu   e. Se   f. Y   g. I   h. Eu   i. Ir   j. Bi

   **NOTE:** For Problems 9-11 refer to Table 1-2, Table 2-2 and/or Figure 14-10.

9. Using very fast chemical separation techniques, a team of chemists at the Lawrence Berkeley National Laboratory has observed the chemical

[1]Solutions to all problems in this book can be found starting on page 372.

# 18 ATOMS

properties of the isotope seaborgium-263, element 106 (Sg). American nuclear chemist **Darleane C. Hoffman** (b. 1926), a chemistry professor at UC Berkeley, bombarded californium-249 (Cf, element 98) with oxygen-18 atoms to make seaborgium-263 (four neutrons are also made), which decays in a few seconds. Sketch the diagram and write the isotopic symbol for each of the three isotopes used in this experiment.

**NOTE:** Seaborgium (Sg) is named after **Glenn T. Seaborg** (1912-1999), an American nuclear chemist, winner of the 1951 Nobel Prize in Chemistry, and professor at UC Berkeley. Prof. Hoffman won the National Medal of Science (considered the American "Nobel") in 1997 and the Priestly Medal in 2000, the highest honor of the American Chemical Society, for this and other research.

10. Scientists at the Lawrence Berkeley National Laboratory have made some long-lived isotopes of lawrencium. The isotopes, namely, lawrencium-261 and lawrencium-262 (element 103), were made by bombarding einsteinium-254 (Es, element 99) with neon-22 in a cyclotron. Sketch the diagram and write the isotopic symbol for these isotopes. The chemical symbol for lawrencium is Lr. It is named after the American nuclear physicist **Ernest Orlando Lawrence** (1901-1958) who invented the cyclotron. He won the 1939 Nobel Prize in Physics for this invention.

11. Fill in the blank spaces in the following chart.

| ISOTOPIC SYMBOL | NUMBER OF | | | ATOMIC NUMBER | MASS NUMBER |
|---|---|---|---|---|---|
| | PROTONS | NEUTRONS | ELECTRONS | | |
| $^{24}_{12}Mg$ | | | | | |
| | 15 | 16 | | | |
| | | | | 20 | 40 |
| | | | 17 | | 37 |
| $_{18}Ar$ | | 22 | | | |
| | | 40 | | 32 | |

## THE UNIFIED ATOMIC MASS UNIT

(This essay may be more meaningful after you have read Chaps. 2, 7, and Sec. 8-6.)

The term "unified atomic mass unit" introduced on page 4 was called the "atomic mass unit" before 1961. Here is the reason why the name was changed in 1961.

If you look at Table 1-2, you can see that oxygen has three naturally-occurring isotopes whose mass numbers are 16, 17, and 18. (Remember that the actual mass of an isotope is not the same as the mass number.) The discovery of these isotopes in 1929 led to a serious problem. The chemists based the value of one atomic mass unit ("amu" was the abbreviation that was then used) on the average mass of the atoms of all three isotopes. (How this average mass is calculated will be discussed in Chapter 2.) The physicists based their value on just the mass of an oxygen-16 atom. Clearly, the values they got for the atomic mass unit were different. The value of the atomic mass unit determines the mass of the electron, the proton, the neutron, and the mass of the isotopes of all the elements.

Chemists and physicists were using different numbers in their tables of isotopic masses (like Table 2-1 but including all the elements). This also meant that they got different atomic weights for the elements, so there were 2 different tables of atomic weights. See Chapter 2 for a discussion of atomic weight. *One set of tables was used by chemists, and the other set was used by physicists.* This created great confusion and led to errors in communication unless scientists were really careful about stating which tables they used. (The difference in the two sets of tables was small [less than 0.1 %] and didn't affect most calculations done in undergraduate science courses. But it was significant enough for researchers so that they agreed on the change discussed below.)

As chemists and physicists started working together, especially during and after World War II, they saw the need to come up with one version of all the values and tables mentioned above. In 1961 they got together and decided to **unify** the two atomic mass units so that there was only one value. The standard they chose was the isotope of carbon, carbon-12. The new name of the atomic mass unit based on carbon-12 was called, of course, the **unified atomic mass unit** and given the abbreviation "u." With this new definition of "u" they calculated: (1) one set of values for the mass the proton, neutron and electron, (2) one table of isotopic masses, and (3) one table of atomic weights (page 27).

The **dalton (Da)** is another name for the unified atomic mass unit.[1] The dalton is used especially for very large molecules such as proteins and other polymers, such as plastics and rubbers. That's because it can easily take metric prefixes (see page 404). For example, the largest known protein, **titin**, is found in human muscle tissue and has a molecular weight (discussed in Section 8.6) of about 3 million daltons, or 3 MDa. (Scientists don't seem to use metric prefixes with the unified atomic mass unit, "u.") **Polymers** are long chains of atoms with a very high molecular weight. The dalton is used only briefly in this book.

You should now only use the abbreviation "amu" if you are referring to research done before 1961.

---

[1]**John Dalton** (1766–1844) was an English chemist, physicist, and meteorologist. He is best known for introducing the atomic theory into chemistry, and for his research into color blindness. He also contributed to our understanding of gases, and discovered Dalton's Law of Partial Pressures, which is discussed in Section 12-14.

**The 1927 Solvay Conference** on *Electrons and Photons* in **Brussels:** This is a cropped portion of the full photo. To see the full photo in color, look at https://bit.ly/3lbhdE5. This link also gives the names of all the attendees, who except for Madam Curie, were men. These were the greatest physicists at that time and met to discuss the new theory of quantum mechanics (see Chap 13 and p 412). Some are mentioned in this book.

Sitting in the first row with the white beard is **Hendrik Antoon Lorentz** (1853–1928), one of the great physicists of his time. To his left is **Albert Einstein** (1879–1955). To Lorentz's right is **Marie Sklodowska Curie** (1867–1934). To Madam Curie's right is **Max Planck** (1858-1947) who in 1900 introduced the idea of the quantum of light, what we now call a photon (see Chapter 12 and page 257).

Madam Curie won the Nobel Prize for Physics in 1903 and the Nobel Prize for Chemistry in 1911. They were for the development of the theory of *radioactivity* (a term that she coined), techniques for isolating radioactive isotopes, and the discovery of two elements, polonium and radium. She is the only person to ever win Nobel prizes in both chemistry and physics. She and her husband **Pierre Curie** (1859-1906)* are the first persons to have an element named after them, element 96, curium (Cm). Under her direction, the world's first studies using radioactive isotopes for the treatment of cancer were conducted. She founded the Curie Institutes in Paris and in Warsaw, which remain major centers of medical research today. During World War I, she developed mobile x-ray units for use in field hospitals (like today's Mobile Army Surgical Hospitals [MASH]).

*Photo is in the Public Domain*

*Pierre Curie shared the 1903 Nobel Prize in Physics with his wife Marie and **Henri Becquerel** (1852-1908, their mentor) "in recognition of the extraordinary services they have rendered by their joint research on the radiation phenomena [radioactivity] discovered by Professor Henri Becquerel."

# RELATIVE ATOMIC WEIGHT

In our discussion so far, we have referred to the mass number of an atom. As you know, the mass number is the sum of the number of protons and neutrons in the nucleus. But the mass number is not the exact mass of an atom. The following discussion will explain why this is true and then introduce the concept of the relative atomic weight.

## 2-1
### THE UNIFIED ATOMIC MASS UNIT IS A UNIT USED TO MEASURE THE MASS OF AN ATOM

Scientists find it necessary to know the exact mass of an atom. But because they cannot directly weigh an individual atom, they have had to come up with the concept of relative masses of atoms. Of course, to talk about relative masses we must have a standard with which to compare one mass with another. As discussed on page 19, since 1961 the standard of comparison that scientists use for atoms is one of the isotopes of carbon, namely carbon-12. An atom of carbon-12 is assigned a mass of *exactly* 12 unified atomic mass units (u). This unit was first introduced on page 4. When we say that an atom of carbon-12 is assigned a mass of 12 u, we are including both the nucleus and the electrons.

The definition of the unified atomic mass unit is as follows:

**One unified atomic mass unit is exactly one-twelfth the mass of an atom of one of the isotopes of carbon, namely, carbon-12.**

Based on this definition, the mass of any proton is found by experiment to be 1.007276466621 u. The mass of any neutron is found by experiment to be 1.00866491588 u. The mass of any electron is found by experiment to be 0.000548579909070 u.

The reason for the masses of the proton and neutron not equaling exactly 1u will be discussed in Section 2-5, when we discuss binding energy.

Now let's discuss the experimentally determined masses of two atoms.

(Opposite) The Solvay Conferences were founded in 1911 by the Belgian industrialist **Ernest Solvay** (1838-1922) and are located in Brussels. Recent Solvay Conferences usually go through a three-year cycle. There are both physics and chemistry conferences. Ernest Solvay was a chemist who discovered a way to commercially make inexpensive sodium carbonate (washing soda, soda ash) that has many industrial uses, https://bit.ly/35uhGQ7. Solvay became very wealthy from his invention and funded the conferences named after him. I chose to show the 1927 Physics Conference because two famous scientists that you may have heard of, Marie Curie and Albert Einstein, are sitting front row center.

## 22  RELATIVE ATOMIC WEIGHT

Consider oxygen-16, the most common isotope of oxygen. From Table 2-1 the mass of an oxygen-16 atom is, from experiment, 15.99491 u. Also consider boron-10, which is one of two stable isotopes of boron. The mass of an atom of boron-10 is, from experiment, 10.01294 u. Notice that oxygen-16's mass is *less* than its mass number, and boron-10's mass is *greater* than its mass number. The reason that you cannot determine an atom's mass by adding up the masses of its protons, neutrons, and electrons will be discussed in Section 2-5, when we talk about binding energy. **REMEMBER:** *Because it is defined that way, an atom of carbon-12 is the only isotope whose mass number is equal to its actual mass. The mass number and actual mass of all the other atoms are not equal.*

## 2-2
### THE RELATIVE ATOMIC WEIGHT TAKES INTO ACCOUNT ALL THE NATURALLY OCCURRING ISOTOPES OF AN ELEMENT

From our discussion in Chapter 1, we have seen that most elements have more than one naturally occurring isotope. If you look at Table 1-2, you will see that the element with the most naturally occurring isotopes is tin (symbol Sn, atomic number 50); tin has ten naturally occurring isotopes.

The naturally occurring isotopes of an element do not usually occur in equal amounts. The average mass of all the naturally occurring isotopes that make up an element is called the relative atomic weight of that element. The word "relative" is used because all the masses are compared to (or relative to) carbon-12. The unit of relative atomic weight we will use is the unified atomic mass unit (u).[1]

The next two sections will show you how to compute the relative atomic weight of an element.

## 2-3
### THE WEIGHTED AVERAGE GIVES MORE WEIGHT TO NUMBERS THAT APPEAR MORE OFTEN

In the previous section, the relative atomic weight was defined as the average mass of all the naturally occurring isotopes that make up an element. The way that this average mass is calculated is the subject of this and the following section.

First, let's review how to calculate the usual average with which you are familiar. As an example, consider how you would average the weight of two people who weigh 120 pounds and 130 pounds. (The abbreviation for pound is lb.) The average weight is the sum of the two weights divided by 2:

$$\frac{120 \text{ lb} + 130 \text{ lb}}{2} = \frac{250 \text{ lb}}{2} = 125 \text{ lb}$$

[1]Actually, relative atomic weight is dimensionless, which means it does not have a unit assigned to it. This is because it is a ratio of two numbers whose units divide out. But it is easier to understand the material in this chapter if it is given a unit. This is a case where I have taken liberty with the official definition. Also, the term "relative atomic weight" is not an official name but is used here to make things easier to understand.

Another example would be to average the weights of six people. Assume the weights are 120, 120, 130, 130, 130, and 150 lb. The average is

$$\begin{array}{r} 120 \text{ lb} \\ 120 \text{ lb} \\ 130 \text{ lb} \\ 130 \text{ lb} \\ 130 \text{ lb} \\ +150 \text{ lb} \\ \hline 780 \text{ lb} \end{array} \qquad \frac{780 \text{ lb}}{6} = 130 \text{ lb}$$

Or, in one equation, we have

$$\frac{120 \text{ lb} + 120 \text{ lb} + 130 \text{ lb} + 130 \text{ lb} + 130 \text{ lb} + 150 \text{ lb}}{6} = \frac{780 \text{ lb}}{6} = 130 \text{ lb}$$

There is another way of writing

$$120 + 120 + 130 + 130 + 130 + 150.$$

That way is

$$(2)(120) + (3)(130) + (1)(150)$$

where (2)(120) is the same as 2 × 120, (3)(130) is the same as 3 × 130, and (1)(150) is the same as 1 × 150. Thus our expression for the average weight can be written as

$$\frac{(2)(120) + (3)(130) + (1)(150)}{6}$$

Applying the distributive law of algebra (see Appendix 2-1 at the end of this chapter) to this expression, we get

$$\frac{(2)(120) + (3)(130) + (1)(150)}{6} = \frac{(2)(120)}{6} + \frac{(3)(130)}{6} + \frac{(1)(150)}{6}$$

This can be rearranged slightly to give

$$(\tfrac{2}{6})(120) + (\tfrac{3}{6})(130) + (\tfrac{1}{6})(150)$$

Let's now do the division of each fraction to get a decimal.

$$\tfrac{2}{6} = \tfrac{1}{3} = 0.333 \qquad \tfrac{3}{6} = \tfrac{1}{2} = 0.500 \qquad \tfrac{1}{6} = 0.167$$

Our expression for the average weight thus becomes

$$(0.333)(120) + (0.500)(130) + (0.167)(150)$$

Thus we have shown that

$$\frac{120 + 120 + 130 + 130 + 130 + 150}{6}$$
$$= (0.333)(120) + (0.500)(130) + (0.167)(150)$$

Since the left side of this equation gives the average weight of the six people, the right side must also give the average weight of the six people, which is 130 lb.

Let's go back to the original group of six people. Their weights were 120 lb, 120 lb, 130 lb, 130 lb, 130 lb, and 150 lb. Note that 2 of the 6 people weighed 120 lb, 3 of the 6 weighed 130 lb, and 1 of the 6 weighed 150 lb. The *percentage* of the people who had each weight is given in the following table.

| WEIGHT | FRACTION | DECIMAL | PERCENTAGE |
|---|---|---|---|
| 120 lb | $\frac{2}{6}$ | 0.333 | 33.3% |
| 130 lb | $\frac{3}{6}$ | 0.500 | 50.0% |
| 150 lb | $\frac{1}{6}$ | 0.167 | 16.7% |

The percentage is formed from the decimal by moving the decimal point two places to the right.

We have discovered an interesting thing. All you have to do to take an average weight is multiply each weight by the decimal equivalent of the people who had that weight and add all the terms. This way of taking an average is called a **"weighted" average** because it gives more "weight" to the numbers that appear more often. (Don't confuse the word "weight" in *weighted average* with weight in pounds. The term *weighted average* is used for *all* kinds of averaging, not just those involving pounds.)

Example 1 gives an example of the use of the weighted average for averaging money. Try to work it out yourself before looking at the solution. Appendix 2-2 at the end of this chapter will show you how to work out the examples in this chapter by using a calculator.

**EXAMPLE 1** Using the method of the weighted average, find the average annual salary of Ph.D. chemistry professors in the United States in 1986 from the following data.

| SALARY | THOSE MAKING THIS SALARY, percent |
|---|---|
| $50,000 | 50% |
| $38,000 | 30% |
| $30,000 | 10% |
| $25,000 | 10% |

*Solution:* First we must calculate the decimal equivalent of the four percentages. This is done by moving the decimal point two places to the left. If you do not see a decimal point in a number, assume it is after the last digit. For example, 10% could be written as 10.%, 40% as 40.%, and so on.

| PERCENT | DECIMAL EQUIVALENT |
|---------|--------------------|
| 50%     | 0.50               |
| 30%     | 0.30               |
| 10%     | 0.10               |
| 10%     | 0.10               |

The average salary is calculated as follows:

$$(0.50)(50{,}000) + (0.30)(38{,}000) + (0.10)(30{,}000) + (0.10)(25{,}000)$$
$$= 25{,}000 + 11{,}400 + 3{,}000 + 2{,}500 = \$41{,}900$$

The average annual salary of the chemistry professors is $41,900. ∎

The nice thing about taking the weighted average in Example 1 is that we didn't have to know how many professors we were talking about. All we had to know were the salaries and the percent of the professors making that salary. As you shall soon see, this is the kind of information we usually have available about atoms. We know the isotopic masses and the percentage of the atoms with a particular mass. We do not know how many atoms we have.[2] The weighted average is thus an ideal method for calculating the relative atomic weights of the elements.

## 2-4
### CALCULATING RELATIVE ATOMIC WEIGHTS USES THE WEIGHTED AVERAGE

We can now define relative atomic weights in terms of weighted averages. The relative atomic weight is the weighted average mass of all the naturally-occurring isotopes of an element.

Table 2-1 lists the percent of each isotope of some elements and the corresponding isotopic masses. We can use the information in Table 2-1 to calculate some relative atomic weights. Table 2-2 lists the relative atomic weights for 118 elements. The relative atomic weights calculated from Table 2-1 should agree closely with those listed in Table 2-2. **NOTE:** The official title of Table 2-2 is "Table of Atomic Weights." The reason it is not called a table of relative atomic weights is explained in the footnote on page 22 and in Section 8-2.

[2] As shown in section 8-5, we can *calculate* the number of atoms of an element in a sample, but only *after* we know the atomic weight of that element.

## TABLE 2-1
PERCENT ABUNDANCES AND MASSES OF SOME ISOTOPES

| ISOTOPE | ABUNDANCE IN NATURE, % | MASS, u |
|---|---|---|
| Hydrogen-1 | 99.985 | 1.007825 |
| Hydrogen-2 | 0.015 | 2.01410 |
| Boron-10 | 20.0 | 10.01294 |
| Boron-11 | 80.0 | 11.00931 |
| Carbon-12 | 98.89 | 12.00000 |
| Carbon-13 | 1.11 | 13.00335 |
| Nitrogen-14 | 99.64 | 14.00307 |
| Nitrogen-15 | 0.36 | 15.00011 |
| Oxygen-16 | 99.76 | 15.99491 |
| Oxygen-17 | 0.04 | 16.99913 |
| Oxygen-18 | 0.2 | 17.99916 |
| Fluorine-19 | 100.000 | 18.9984 |
| Neon-20 | 90.51 | 19.99244 |
| Neon-21 | 0.27 | 20.99385 |
| Neon-22 | 9.22 | 21.99138 |
| Chlorine-35 | 75.77 | 34.96885 |
| Chlorine-37 | 24.23 | 36.96590 |
| Iron-54 | 5.8 | 53.9396 |
| Iron-56 | 91.8 | 55.9349 |
| Iron-57 | 2.1 | 56.9394 |
| Iron-58 | 0.3 | 57.9333 |
| Tin-112 | 1.0 | 111.9040 |
| Tin-114 | 0.7 | 113.9030 |
| Tin-115 | 0.4 | 114.9035 |
| Tin-116 | 14.7 | 115.9021 |
| Tin-117 | 7.7 | 116.9031 |
| Tin-118 | 24.3 | 117.9018 |
| Tin-119 | 8.6 | 118.9034 |
| Tin-120 | 32.4 | 119.9021 |
| Tin-122 | 4.6 | 121.9034 |
| Tin-124 | 5.6 | 123.9052 |
| Uranium-234 | 0.005 | 234.0409 |
| Uranium-235 | 0.720 | 235.0439 |
| Uranium-238 | 99.275 | 238.0508 |

## Table 2-2

**Table of Atomic Weights**

Values are compared to the isotopic mass of one atom of carbon-12 which equals exactly 12 u. The data in this table was taken from *Pure and Applied Chemistry*, 2021. The authors at https://bit.ly/3e31VUl modified it and also presented a table in numerical order by element. I further modified and clarified the table and the footnotes below.

The atomic weight of many elements can change slightly, depending on the origin and treatment of the material. The footnotes to this table explain the types of variations to be expected for individual elements. The values of the atomic weight given here apply to elements as they exist naturally on earth. Values in brackets are used for certain radioactive elements; the value given is the mass number of the isotope of that element with the longest known half-life. The spaces after the decimal point in some numbers are for clarity.

| Atomic Number | Symbol | Name | Atomic Weight | Notes |
|---|---|---|---|---|
| 89 | Ac | Actinium | [227] | 4 |
| 13 | Al | Aluminum | 26.981 5384 | |
| 95 | Am | Americium | [243] | 4 |
| 51 | Sb | Antimony | 121.760 | 1 |
| 18 | Ar | Argon | 39.948 | 1, 2 |
| 33 | As | Arsenic | 74.921 595 | |
| 85 | At | Astatine | [210] | 4 |
| 56 | Ba | Barium | 137.327 | |
| 97 | Bk | Berkelium | [247] | 4 |
| 4 | Be | Beryllium | 9.012 1831 | |
| 83 | Bi | Bismuth | 208.980 40 | |
| 107 | Bh | Bohrium | [270] | 4 |
| 5 | B | Boron | 10.81 | 3 |
| 35 | Br | Bromine | 79.904 | |
| 48 | Cd | Cadmium | 112.414 | 1 |
| 55 | Cs | Cesium | 132.905 451 96 | |
| 20 | Ca | Calcium | 40.078 | 1 |
| 98 | Cf | Californium | [251] | 4 |
| 6 | C | Carbon | 12.011 | |
| 58 | Ce | Cerium | 140.116 | 1 |
| 17 | Cl | Chlorine | 35.45 | 3 |
| 24 | Cr | Chromium | 51.996 1 | |
| 27 | Co | Cobalt | 58.933 194 | |
| 112 | Cn | Copernicium | [285] | 4 |
| 29 | Cu | Copper | 63.546 | 2 |
| 96 | Cm | Curium | [247] | 4 |

## RELATIVE ATOMIC WEIGHT (Table 2-2 continued)

| 105 | Db | Dubnium | [270] | 4 |
|---|---|---|---|---|
| 66 | Dy | Dysprosium | 162.500 | 1 |
| 99 | Es | Einsteinium | [252] | 4 |
| 68 | Er | Erbium | 167.259 | 1 |
| 63 | Eu | Europium | 151.964 | 1 |
| 100 | Fm | Fermium | [257] | 4 |
| 114 | Fl | Flerovium | [289] | 4 |
| 9 | F | Fluorine | 18.998 403 162 | |
| 87 | Fr | Francium | [223] | 4 |
| 64 | Gd | Gadolinium | 157.25 | 1 |
| 31 | Ga | Gallium | 69.723 | |
| 32 | Ge | Germanium | 72.630 | |
| 79 | Au | Gold | 196.966 570 | |
| 72 | Hf | Hafnium | 178.486 | |
| 108 | Hs | Hassium | [270] | 4 |
| 2 | He | Helium | 4.002 602 | 1, 2 |
| 67 | Ho | Holmium | 164.930 329 | |
| 1 | H | Hydrogen | 1.008 | 3 |
| 49 | In | Indium | 114.81 | |
| 53 | I | Iodine | 126.904 4 | |
| 77 | Ir | Iridium | 192.217 | |
| 26 | Fe | Iron | 55.845 | |
| 36 | Kr | Krypton | 83.79 | 1, 3 |
| 57 | La | Lanthanum | 138.905 47 | 1 |
| 103 | Lr | Lawrencium | [262] | 4 |
| 82 | Pb | Lead | 207.2 | 1, 2 |
| 3 | Li | Lithium | 6.94 | 3 |
| 116 | Lv | Livermorium | [293] | 4 |
| 71 | Lu | Lutetium | 174.9668 | 1 |
| 12 | Mg | Magnesium | 24.305 | |
| 25 | Mn | Manganese | 54.938 043 | |
| 109 | Mt | Meitnerium | [278] | 4 |
| 101 | Md | Mendelevium | [258] | 4 |
| 80 | Hg | Mercury | 200.592 | |
| 42 | Mo | Molybdenum | 95.95 | 1 |
| 115 | Mc | Moscovium | [289] | 4 |
| 60 | Nd | Neodymium | 144.242 | 1 |
| 10 | Ne | Neon | 20.1797 | 1, 3 |
| 93 | Np | Neptunium | [237] | 4 |
| 28 | Ni | Nickel | 58.6934 | |
| 113 | Nh | Nihonium | [286] | 4 |
| 41 | Nb | Niobium | 92.906 37 | |
| 7 | N | Nitrogen | 14.007 | |
| 102 | No | Nobelium | [259] | 4 |
| 118 | Og | Oganesson | [294] | 4 |
| 76 | Os | Osmium | 190.23 | 1 |

## 2-4 Calculating Relative Atomic Weights uses the Weighted Average

| | | | | |
|---|---|---|---|---|
| 8 | O | Oxygen | 15.999 | |
| 46 | Pd | Palladium | 106.42 | 1 |
| 15 | P | Phosphorus | 30.973 761 998 | |
| 78 | Pt | Platinum | 195.084 | |
| 94 | Pu | Plutonium | [244] | 4 |
| 84 | Po | Polonium | [209] | 4 |
| 19 | K | Potassium | 39.0983 | |
| 59 | Pr | Praseodymium | 140.907 66 | |
| 61 | Pm | Promethium | [145] | 4 |
| 91 | Pa | Protactinium | 231.035 88 | 4 |
| 88 | Ra | Radium | [226] | 4 |
| 86 | Rn | Radon | [222] | 4 |
| 75 | Re | Rhenium | 186.207 | |
| 45 | Rh | Rhodium | 102.905 49 | |
| 111 | Rg | Roentgenium | [281] | 4 |
| 37 | Rb | Rubidium | 85.4678 | 1 |
| 44 | Ru | Ruthenium | 101.07 | 1 |
| 104 | Rf | Rutherfordium | [267] | 4 |
| 62 | Sm | Samarium | 150.36 | 1 |
| 21 | Sc | Scandium | 44.955 907 | |
| 106 | Sg | Seaborgium | [269] | 4 |
| 34 | Se | Selenium | 78.971 | |
| 14 | Si | Silicon | 28.085 | |
| 47 | Ag | Silver | 107.8682 | 1 |
| 11 | Na | Sodium | 22.989 769 28 | |
| 38 | Sr | Strontium | 87.62 | 1, 2 |
| 16 | S | Sulfur | 32.06 | |
| 73 | Ta | Tantalum | 180.947 88 | |
| 43 | Tc | Technetium | [97] | 4 |
| 52 | Te | Tellurium | 127.60 | 1 |
| 117 | Ts | Tennessine | [293] | 4 |
| 65 | Tb | Terbium | 158.925 354 | |
| 81 | Tl | Thallium | 204.38 | |
| 90 | Th | Thorium | 232.0377 | 1, 4 |
| 69 | Tm | Thulium | 168.934 219 | |
| 50 | Sn | Tin | 118.710 | 1 |
| 22 | Ti | Titanium | 47.867 | |
| 74 | W | Tungsten | 183.84 | |
| 92 | U | Uranium | 238.028 91 | 1, 3, 4 |
| 23 | V | Vanadium | 50.941 | |
| 54 | Xe | Xenon | 131.29 | 1, 3 |
| 70 | Yb | Ytterbium | 173.045 | 1 |
| 39 | Y | Yttrium | 88.905 838 | |
| 30 | Zn | Zinc | 65.38 | 2 |
| 40 | Zr | Zirconium | 91.224 | 1 |

(See next page for Notes related to this table.)

# 30 RELATIVE ATOMIC WEIGHT (Table 2-2 continued)

## Notes related to Table 2-2.

1. Geological specimens are known in which the element has an isotopic composition that is significantly different than for normal material. Thus the atomic weight of the element in those geological specimens would be somewhat different than that listed in the Table.

2. The range in isotopic composition of normal terrestrial (the earth's crust and the atmosphere) material prevents a more precise value being given (see Sec 6-3 for a discussion of precision); the values in the Table should apply to any normal material.

3. Modified isotopic compositions may be found in commercially available material because it has been subject to an undisclosed or inadvertent isotopic separation. Substantial deviations in atomic weight of the element from that given in the Table can occur.

4. This element has no stable isotopes. The value enclosed in brackets, such as [209], indicates the mass number of the longest-lived isotope of the element. There are three elements which have no stable isotopes (Th, Pa, and U) but do have a characteristic terrestrial isotopic composition, and for these an atomic weight is listed in the Table. The reason is that the half-life of the isotopes of Th and U are very long. The half-life of the isotopes of Pa are much shorter and they all come from the radioactive decay of very long-lived U isotopes. Also, there are very large amounts of U isotopes on earth and very little of the isotopes of Pa. Both these facts combine to give a constant percent abundance of the Pa isotopes, thus allowing an atomic weight of Pa to be listed in the Table.

---

A word about nomenclature. The percent, by atoms, of each isotope of an element is called the **percent abundance.** The decimal equivalent of the percent abundance is called the **fractional abundance.**

**EXAMPLE 2** From Table 2-1 we see that nitrogen has two naturally occurring isotopes, namely, nitrogen-14 and nitrogen-15. Calculate the relative atomic weight of nitrogen.

*Solution:* From Table 2-1 we have the following information.

| ISOTOPE | MASS, u | ABUNDANCE, % | FRACTIONAL ABUNDANCE |
|---|---|---|---|
| nitrogen-14 | 14.00307 | 99.64 | 0.9964 |
| nitrogen-15 | 15.00011 | 0.36 | 0.0036 |

## 2-4 Calculating Relative Atomic Weights Uses the Weighted Average

The weighted average is thus

$$(0.9964)(14.00307) + (0.0036)(15.00011)$$
$$= 13.952659 + 0.0540005 = 14.0067 \text{ u}$$

The value for the relative atomic weight listed in Table 2-2 is 14.007 u. This is the same as our calculated value rounded to five digits. Rules for rounding numbers is discussed in Section 6-5. ■

**EXAMPLE 3** Oxygen has three naturally occurring isotopes: oxygen-16, oxygen-17, and oxygen-18. Calculate the relative atomic weight of oxygen using the values listed in Table 2-1.

*Solution:* From Table 2-1 we have the following information.

| ISOTOPE | MASS, u | ABUNDANCE, % | FRACTIONAL ABUNDANCE |
|---|---|---|---|
| oxygen-16 | 15.99491 | 99.76 | 0.9976 |
| oxygen-17 | 16.99913 | 0.04 | 0.0004 |
| oxygen-18 | 17.99916 | 0.2 | 0.002 |

The weighted average is

$$(0.9976)(15.99491) + (0.0004)(16.99913) + (0.002)(17.99916)$$
$$= 15.956522 + 0.0067997 + 0.0359983 = 15.99932 \text{ u}$$

This is within the error limits of the value found in Table 2-2. ■

You might wonder why some of the relative atomic weights in Table 2-2 are listed to four or five decimal places, whereas some are listed to only two or three decimal places. The reason is that the isotopic abundances of some elements are not known to the same accuracy as are others. The footnotes to Table 2-2 describe the problems encountered with these elements.

For those of you without a calculator, you might want to try calculating an approximate relative atomic weight. You can get a reasonably close answer by using the mass number instead of the isotopic mass.

**EXAMPLE 4** Calculate the approximate relative atomic weight of nitrogen by using the mass number instead of the isotopic mass.

**Solution:** Using the information from Example 2, we have the following.

| ISOTOPE | MASS NUMBER | FRACTIONAL ABUNDANCE |
|---|---|---|
| Nitrogen-14 | 14 | 0.9964 |
| Nitrogen-15 | 15 | 0.0036 |

The approximate relative atomic weight is

$$(14)(0.9964) + (15)(0.0036) = 14.0036 \text{ u}$$

Compare this to the actual relative atomic weight of 14.007 u. ∎

The next example is just a bit of practice in reading Table 2-2.

**EXAMPLE 5** From Table 2-2, what are the relative atomic weights of the following elements: hydrogen, copper, gold, tantalum, and zirconium? Digits after the decimal point are spaced in groups of three for clarity.

*Solution:*

    hydrogen    1.008 u
    copper    63.546 u
    gold    196.966 570 u
    tantalum    180.947 88 u
    zirconium    91.224 u ∎

## 2-5
### BINDING ENERGY HOLDS THE NUCLEUS TOGETHER

This section will help explain why the mass of a proton and a neutron is not exactly 1 u. It will also explain why the mass of an atom is not exactly the mass number of that atom.

To begin, there is a natural tendency for things to become more stable by releasing energy if they can. One example is the explosion of TNT. The energy released makes the products of the explosion more stable than the TNT was. This tendency to become more stable by releasing energy is central to the discussion of binding energy.

Binding energy is responsible for the fact that a nucleus made up of more than one nucleon (a nucleon is a nuclear particle, i.e., a proton or a neutron) weighs less than the sum of the weights of its uncombined protons and neutrons. This is because some energy is released to stabilize the nucleus when it is formed from its nucleons. This energy comes from the mass of the nucleons. A little bit of the mass has been converted into energy. This energy is the binding energy of the nucleus.

## 2-5 Binding Energy Holds the Nucleus Together

The amount of energy can be calculated by using the Einstein mass—energy equation, which is $E = mc^2$. In this equation, $E$ is the energy, $m$ is the mass, and $c$ is the speed of light. Since $c$ = 299,792,458 meters/second or 186,282 miles/second, the speed of light is a large number. The formula shows that a small amount of mass is equivalent to a very large amount of energy.

Remember that an atom of carbon-12 is assigned a mass of exactly 12 u. It is instructive to compare some other atoms with carbon-12 as far as binding energy is concerned. Atoms that weigh less than carbon-12 should have less binding energy than carbon-12. Atoms that weigh more than carbon-12 should have more binding energy than carbon-12.

If you look up fluorine in Table 1-2, you see that it only has one naturally occurring isotope, fluorine-19. The mass number of this isotope is 19. If you look up fluorine's atomic weight (which is the same as the isotopic mass of fluorine-19 because fluorine has only one naturally occurring isotope) in Table 2-2, you find that it is 18.998403 u. Thus, even with the mass of the nine electrons counted in, the atomic weight is a bit less than 19. Fluorine-19 has more nucleons and more binding energy than carbon-12.

Now look at Table 1-1. Notice that because carbon-12 is used as a mass reference, each separate proton and neutron has a mass slightly greater than 1 u. The experimentally determined mass of a proton is 1.007276466621 u, and the experimentally determined mass of a neutron is 1.00866491588 u. This is to be expected since protons and neutrons do not have a nuclear binding energy when they are not in a nucleus. There is no need for you to memorize the exact masses of the proton and neutron. Curiously, in the mid-1950s a 16-year-old young woman who was an expert on atomic energy didn't win the bonus $64,000 on a TV quiz show because she could not remember the mass of a neutron. I think they only asked her to give the mass to five digits, in other words, 1.0087 u. But she had already won $64,000 so her appearance on the show wasn't a total loss. Due to inflation, $64,000 in 1955 was equivalent to about $622,000 in 2020 purchasing power.

Looking again at Table 1-2, you can see that beryllium has one naturally occurring isotope, $^9$Be. Beryllium's relative atomic weight is 9.01218 u. This is larger than 9 and is expected, since beryllium-9 has three fewer nucleons than carbon-12. Thus beryllium-9 has less binding energy than carbon-12.

In Table 2-1, you will find that, in general, the isotopic masses for elements heavier than carbon-12 are less than the atomic mass numbers. Elements lighter than carbon-12 have isotopic masses greater than the mass number. This is consistent with what we have said about binding energy.

You might notice that the isotopes carbon-13, nitrogen-14, and nitrogen-15 seem not to be consistent with what we have said in the previous paragraph. In fact, these isotopes do have more binding energy than carbon-12. However, it is not enough additional binding energy to reduce the isotopic mass below the mass number.

Although binding energy involves only very small mass changes, the energy that this mass becomes is very large. You can see and feel this energy every day. In fact, without this energy, you would not be alive. This is because the binding energy that is released when four $^1$H nuclei (protons) form a helium-4 nucleus

> Based on experiments by the Italian physicist **Enrico Fermi** (1901-1954, see page 8 for more about Fermi), who bombarded uranium with neutrons, in 1934 the German chemist **Ida Noddack** (1896-1978) suggested that "it is conceivable that the nucleus breaks up into several large fragments." (She discovered the element rhenium.) But there was no experimental proof until barium was found in experiments bombarding uranium with neutrons performed by chemists **Otto Hahn** (1879-1968) and **Fritz Strassmann** (1902-1980) in Berlin. Physicist **Lise Meitner** (1878-1968), an Austrian Jew, had to flee Hahn's lab in July 1938, going to Sweden. Hahn and Strassmann couldn't explain why barium was found, but when Hahn notified Meitner about it, she, in collaboration with her physicist cousin **Otto Frisch** (1904-1979), realized that nuclear fission was taking place and she explained it theoretically. Element 109, meitnerium, is named in her honor.

is the energy that powers our sun and many other stars. This combining of light nuclei to make heavier nuclei is called **nuclear fusion**. On earth, scientists have used a similar reaction (deuterium plus tritium to give helium plus energy) in the hydrogen bomb. And this same fusion reaction, when controlled, may lead to almost unlimited power generation on earth without generating any greenhouse gases. Scientists have been trying to build a successful fusion machine since the 1950s but have succeeded only recently as it is *very* difficult. The lab that has succeeded is the National Ignition Facility (NIF) at the Lawrence Livermore National Laboratory (LLNL) in California. It uses 192 high-power laser beams all focused on a pencil-eraser sized gold cylinder inside of which is a hollow diamond pellet about 2 mm in diameter that contains frozen deuterium (D) and tritium (T).\* On December 13, 2022, LLNL Director **Kimberly S. Budil** announced that LLNL had achieved ignition, which means that that the fusion reaction gave off more energy than the laser beams that ignited it. This is a milestone in fusion research. See https://bit.ly/3PmUeXC. Unfortunately, the NIF laser is very inefficient and only 1% of the energy needed to make the laser beam gets to the gold cylinder. Nuclear engineer **Andrea Kritcher**, who led the team that designed the gold cylinder target, said it could take (maybe) a few decades before a practical fusion power plant could be built that would supply electricity to the electrical grid.

It turns out that the nucleus of iron is the most stable nucleus. Because of this, elements lighter than iron, especially hydrogen, can undergo nuclear fusion, as discussed above. Also, some elements heavier than iron can undergo nuclear fission, where the nucleus breaks apart when hit by a neutron. The most common nuclei that can undergo fission are uranium-235 and plutonium-239. The energy given off when one of these nuclei splits apart supplies the energy for the atomic bomb and nuclear reactors that can be used for power generation.

## APPENDIX 2-1
## THE DISTRIBUTIVE LAW

The simplest definition of the distributive law of algebra, for any variables a, b, and c, is

$$a(b + c) = ab + ac$$

---

\*The reason that D and T are used in the NIF and not the much more common and much less expensive isotope protium ($^1$H) is that the fusion reaction of D + T is *much* faster than the fusion reaction of four $^1$H atoms. You might ask: Doesn't our sun use the fusion reaction of four $^1$H atoms to produce its energy? The answer is yes. But our sun contains so much $^1$H that even though the reaction is very slow, enough of the $^1$H atoms react to give off the massive amount of energy that powers our sun.

## Appendix 2-1 The Distributive Law

where $ab = (a)(b) = a \times b$, and so on. To show that the distributive law is reasonable, let's test it with real numbers, say 3, 4, and 5: Does $3(4 + 5) = (3)(4) + (3)(5)$? Well, $3(4 + 5) = 3 \times 9 = 27$ and $(3)(4) + (3)(5) = 12 + 15 = 27$. So the answer is yes. It is true that $3(4 + 5) = (3)(4) + (3)(5)$. You can try any numbers you wish—the distributive law always works!

If there were more numbers (or symbols), such as $a$, $b$, $c$, $d$, and $e$, the distributive law would look like

$$a(b + c + d + e) = ab + ac + ad + ae$$

Suppose that we have a division of the form $(x + y)/z$ and we want to apply the distributive law. (We have changed variables to $x$, $y$, and $z$ for variety.) The first thing to recognize is that division by a number is the same as multiplication by 1 over the number ("1 over a number" is called the **reciprocal** of the number). In our case, $z$ is the number and $1/z$ is the reciprocal of the number. We can now write

$$\frac{x + y}{z} = \frac{1}{z}(x + y)$$

Let's test this out with some actual numbers. Does

$$\frac{7 + 8}{5} = \frac{1}{5}(7 + 8)?$$

Yes, since $\quad \dfrac{7 + 8}{5} = \dfrac{15}{5} = 3 \quad$ and $\quad \dfrac{1}{5}(7 + 8) = \dfrac{1}{5}(15) = 3$

Now let's apply the distributive law to

$$\frac{1}{z}(x + y)$$

We get

$$\frac{1}{z}(x + y) = \frac{x}{z} + \frac{y}{z}$$

So it must be true that applying the distributive law to $(x + y)/z$ gives

$$\frac{x + y}{z} = \frac{x}{z} + \frac{y}{z}$$

since

$$\frac{x + y}{z} = \frac{1}{z}(x + y) = \frac{x}{z} + \frac{y}{z}$$

# 36 RELATIVE ATOMIC WEIGHT

Therefore, in our discussion of the weighted average, we used the distributive law correctly, and it is true that

$$\frac{(2)(120) + (3)(130) + (1)(150)}{6} = \frac{(2)(120)}{6} + \frac{(3)(130)}{6} + \frac{(1)(150)}{6}$$

## APPENDIX 2-2
## USING A CALCULATOR

We will assume that you have a full scientific calculator that uses algebraic logic. If you have a Hewlett–Packard calculator that uses Reverse Polish Notation (RPN), please consult the instruction manual that comes with your calculator.

Some students may have an older calculator that doesn't have algebraic logic. Since the instructions in this appendix assume algebraic logic, you should test your calculator according to the following procedure.

Compute $(2)(3) + (4)(5)$ by pushing the following buttons on your keyboard: **2, ×, 3, +, 4, ×, 5, =**
You should read the following in the display.

| BUTTON PUSHED | DISPLAY READS |
|:---:|:---:|
| 2 | 2. |
| × | 2. |
| 3 | 3. |
| + | 6. |
| 4 | 4. |
| × | 4. |
| 5 | 5. |
| = | 26. |

The correct answer is 26. Doing the calculation by hand we get

$$(2)(3) + (4)(5) = 6 + 20 = 26$$

If your calculator doesn't have algebraic logic, your answer will probably be 50. The calculator doesn't follow the distributive law correctly. It is doing the problem this way:

$$(2)(3) + (4)(5) = (6 + 4)(5) = (10)(5) = 50$$

If you have one of these calculators, you must write down intermediate steps or store them in memory.

Appendix 2-2 Using a Calculator 37

Let's work out the examples in this chapter using a calculator. We won't list the display readings — just the buttons you're supposed to push. For clarity, we have written numbers less than 1 with a 0 before the decimal point, but you don't have to enter the 0 into the calculator. To a calculator, .3 is the same as 0.3. **NOTE:** It is *very important* that on paper or on a computer for you to write a zero before a "naked" decimal point. Otherwise, the decimal point could be missed leading to an error. Also, a fly might have dropped a speck on the paper, leading you to think it was a decimal point. It is highly unlikely that the fly would have also dropped a zero at the same time in exactly the right spot!

**EXAMPLE 1** Calculator entries are (we have left out the commas in the numbers for clarity)

**0.5, x, 50 000, +, 0.3, x, 38 000, +, 0.1, x, 30 000, +, 0.1, x, 25,000, =**
The answer is 41,900. ■

**EXAMPLE 2** Calculator entries are

**0.9964, x, 14.00307, +, 0.0036, x, 15.00011, =**
The answer is 14.006659 u. ■

**EXAMPLE 3** Calculator entries are

**0.9976, x, 15.99491, +, 0.0004, x, 16.99913, +, 0.002, x, 17.99916, =**
The answer is 15.99932 u. ■

**EXAMPLE 4** Calculator entries are

**0.9964, x, 14, +, 0.0036, x, 15, =**   The answer is 14.0036 u. ■

**NOTE:** Another pitfall in using a calculator:

Take $\frac{6+4}{8+2} = \frac{10}{10} = 1$. Try it on a calculator this way: **6, +, 4, ÷, 8, +, 2, =** You will probably get **8.5**. Can you figure out what the calculator is doing? Now try **(, 6, +, 4, ), ÷, (, 8, +, 2, ), =**   This gives the correct answer!

## PROBLEMS

KEYED PROBLEMS

1. Using the method of the weighted average, find the average annual salary of Ph.D. chemists in the United States in 1986 from the following data.

## 38 RELATIVE ATOMIC WEIGHT

| JOB | SALARY | THOSE MAKING THIS SALARY, % |
|---|---|---|
| College/University | $41,900 | 26 |
| Industry | $49,900 | 74 |

2. From Table 2-1, we see that chlorine has two naturally occurring isotopes: chlorine-35 and chlorine-37. Calculate the relative atomic weight of chlorine. Compare your answer with the actual value in Table 2-2.
3. From Table 2-1, we see that neon has three naturally occurring isotopes: neon-20, neon-21, and neon-22. Calculate the relative atomic weight of neon. Compare your answer with the actual value listed in Table 2-2. It should come out very close.
4. Calculate the approximate relative atomic weight of chlorine by using the mass numbers instead of the isotopic masses. Compare your answer with the actual value listed in Table 2-2.
5. From Table 2-2, what are the relative atomic weights of the following elements: bromine, helium, cobalt, silver, antimony.

**SUPPLEMENTAL PROBLEMS**

6. Define the following terms:
   a. unified atomic mass unit
   b. relative atomic weight
   c. percent abundance of an isotope
   d. fractional abundance of an isotope
   e. nucleon
   f. binding energy
   g. nuclear fusion
   h. nuclear fission
   i. distributive law of algebra

7. Calculate the relative atomic weight of the following elements listed in Table 2-1. In each case, compare your answer with the value listed in Table 2-2. It should come out very close.
   a. hydrogen
   b. boron
   c. carbon
   d. iron
   e. tin
   f. uranium

8. Calculate the approximate relative atomic weights of the elements listed in Problem 7 by using the mass numbers instead of the isotopic masses.

9. An $^{16}O$ atom has 8 protons, 8 neutrons, and 8 electrons. Using the following masses: proton = 1.0072765 u, neutron = 1.0086650 u, electron = 0.00054858 u, calculate what the mass of an $^{16}O$ atom would be if there were no such thing as binding energy. The difference between this value and the actual isotopic mass of an $^{16}O$ atom, 15.99491 u, is the mass which has been converted into binding energy. What is the size of the mass (in unified atomic mass units) which has been converted into binding energy?

---

**SPECIAL PERCENTAGE PROBLEMS**

10. Convert the following decimals to percents.

    a. 0.75   c. 1.61   e. 0.101
    b. 0.42   d. 0.031  f. 0.002

11. Convert the following percents into decimals.

    a. 38%    d. 0.24%
    b. 7%     e. 1.46%
    c. 135%   f. 11.77%

12. Find 25% of 70.
13. Find 7.2% of 250.
14. Find 0.02% of 150
15. Find 150% of 30.

16. What percent of 12 is 7?
17. What percent of 180 is 30?
18. What percent of 7 is 0.03?

---

**UPDATE: NATIONAL AVERAGE SALARY FOR A CHEMIST IN THE US**
(See Example 4 and Problem 1)

As of 2022 (first column) and 2018 (second column) the median annual salary for full-time chemistry employees was:

| | | | |
|---|---|---|---|
| Industry | $136,000 | Full Professors | $104,820 |
| Academic | $84,150 | Associate Professors | $81,274 |
| Government | $116,196 | Assistant Professors | $70,791 |
| US Average | $105,000 | Lecturers | $56,712 |

**NOTE:** The **median** means the middle salary of the group, so ½ of the chemists earn more than the median, and ½ of the chemists earn less than the median. The median annual salary for chemical engineers in 2022 was $137,000 per year. The US Average includes self-employed and other sectors.

**A Forest Fire:** Fires such as this demonstrate a vigorous chemical reaction, the burning of wood with the oxygen in the air, releasing heat, carbon dioxide, and water vapor, along with many other chemicals, some of which are irritants and some, toxic. Burning houses give off additional chemicals, many of which are also irritants and/or toxic. See https://bit.ly/2rnc9TV for an article on the health problems caused by the smoke from wildfires. *Photo is in the Public Domain.*

# MOLECULES AND THE BALANCED CHEMICAL EQUATION

So far we have discussed atoms and their composition. But that's just the beginning of the study of chemistry. Most of the things that you see around you are made up of different combinations of atoms. All these different combinations give us wood, plastics, water, metal alloys, plants, animals, people, air, and so on. Thus, in order to really get into chemistry, we must look at combinations of atoms.

## 3-1
### MOLECULES CONSIST OF TWO OR MORE ATOMS

As of 2022, there are 118 different known elements and thus 118 different kinds of atoms (not counting isotopes of the different elements).

It is rather common for some of these atoms to combine with one another to form molecules. **A molecule is two or more atoms that are bound together.**

As an example of a molecule, let's look first at two hydrogen atoms. Remember that the symbol for a hydrogen atom is H. Two of these atoms can combine to form a molecule called a hydrogen molecule. Three representations of the hydrogen molecule are shown in Figure 3-1.

In these diagrams, each H stands for a hydrogen atom. Each of the diagrams is a way of representing a hydrogen molecule. The straight line between the two H atoms represents a chemical bond. **A chemical bond is the attractive force that holds atoms together.** The spring between them also represents the chemical bond but reminds us that the bond acts something like a spring and that the H atoms vibrate back and forth. The hazy cloud indicates that the electrons of the hydrogen atoms form the chemical bond.

41

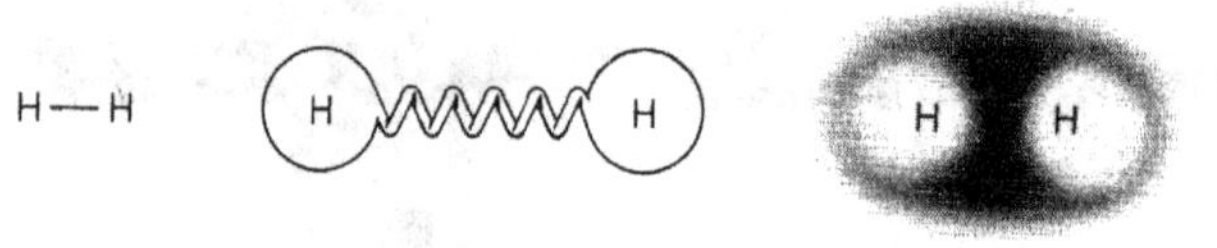

**42** MOLECULES AND THE BALANCED CHEMICAL EQUATION

**FIGURE 3-1** Three representations of the hydrogen molecule.

Of the three, the most common way of representing the hydrogen molecule is with the straight line connecting the two atoms, because it is so easy to do. You just have to remember that the line stands for a chemical bond.

Writing a hydrogen molecule as H—H can be abbreviated even further. Chemists frequently write a hydrogen molecule as $H_2$. The 2 following the H means that there are two H atoms combined in a molecule. The 2 is placed half a space down and called a subscript. $H_2$ is pronounced "H two" and is called the **chemical formula** of the hydrogen molecule.

## 3-2
## BALANCING CHEMICAL EQUATIONS CONSERVES ATOMS

Chemists have developed a notation to describe what happens when two hydrogen atoms combine to form a hydrogen molecule. This notation is called the **chemical equation** and is useful for describing *any* chemical reaction. For the hydrogen reaction the chemical equation can be written as

$$H + H \rightarrow H-H$$

and is read "One hydrogen atom plus one hydrogen atom forms (or yields) one hydrogen molecule." Another way to read it is "One hydrogen atom reacts with another hydrogen atom to form a hydrogen molecule." The hydrogen atoms on the left side of the arrow are called the reactants, and the hydrogen molecule on the right side of the arrow is called the product. In general, in a chemical equation, **reactants** are substances written to the left of the arrow, and **products** are substances written to the right of the arrow.

The arrow ($\rightarrow$) separating the reactants from the products can be read: "reacts to give" or "yields" or "forms." Chemists also use other symbols to separate the reactants and the products. We won't use them in this book, but we will mention them here because you may run across them in other books and articles. The other symbols commonly used to separate the reactants and products are the double arrow ($\leftrightarrows$) and the equals sign (=).

A shorter way of writing the chemical equation for making a hydrogen molecule is

$$H + H \rightarrow H_2$$

This is sometimes written in another, even shorter, way,

$$2H \rightarrow H_2$$

where the 2H stands for two hydrogen atoms.

There is a great difference between 2H and $H_2$. The 2H are separate hydrogen atoms that are about to combine to form a hydrogen molecule. The $H_2$ means that the atoms have been joined by a chemical bond to form a hydrogen molecule.

H atoms "want" to combine to make $H_2$ in a very vigorous way, giving off a great deal of heat. The third hottest welding torch uses this reaction to generate its heat. This torch is called the **atomic hydrogen torch** and reaches from 3400 to 4000 °C. It can melt tungsten, the element with the highest melting point at 3422 °C. See https://bit.ly/2Sb64bi.

**EXAMPLE 1** In the equation $H_2 \rightarrow 2H$, name the reactant and the product?

*Solution:* $H_2$ is the reactant and 2H are the products. Note that we have reversed our original equation on page 42, which was $2H \rightarrow H_2$; accordingly, the reactants and the products are also reversed. ■

Most atoms can combine to form molecules, and a molecule can be made up of many atoms. When there are two atoms in a molecule, such as the hydrogen molecule, we say that it is a **diatomic molecule** ("di" means two).

**EXAMPLE 2** Write the chemical equation for the formation of the diatomic oxygen molecule from two oxygen atoms.

*Solution:* $O + O \rightarrow O_2$ or $2O \rightarrow O_2$. The O atoms are the reactants, and the $O_2$ molecule is the product. ■

Let's look at some more molecules and equations. A common reaction is the burning of hydrogen molecules and oxygen molecules to make water molecules. This reaction can be written as

$$H_2 + O_2 \rightarrow H_2O$$

where $H_2$ and $O_2$ are the reactants, and $H_2O$ is the product. (What we have written here is not the final version; please read on. If you are also wondering how $O_2$ becomes O in $H_2O$, read on.) As we said before, hydrogen exists as a diatomic molecule under normal conditions, so in the equation above we have written it as $H_2$.

We write $O_2$ for the same reason. Oxygen atoms "want" to combine to form molecules. All the oxygen in the air (about 21% of the air is oxygen) exists as diatomic molecules of $O_2$.

When chemists talk about hydrogen and oxygen gases, they mean the diatomic molecules $H_2$ and $O_2$, the form in which they exist at room temperature. Below is a way of writing and naming these simple molecules even when they are made up of specific isotopes.* When chemists talk about specific atoms, they say

---

* The prefix "di", "tri", "tetra", etc. (see page 297 for more prefixes) can be used for naming elemental molecules. Examples are **dioxygen** for $O_2$, **dihydrogen** for $H_2$, **trioxygen** for $O_3$, also called ozone and tetraphosphorus for $P_4$, also called white phosphorus. **NOTE: When referring to any elements or molecules, it is assumed that all the atoms have their naturally-occurring isotopic percentages.** See Table 2-1 for examples. If they don't, it will be made very clear by showing the isotope(s) involved. For example, consider dihydrogen. If all the H atoms are replaced with deuterium atoms, we would write $D_2$. If only one H atom is replaced with a deuterium atom, we would write HD. It is called hydrogen deuteride. A dioxygen molecule containing both $^{16}O$ and $^{18}O$ atoms would be written as $^{16}O^{18}O$ or $^{34}O_2$, where 16 + 18 = 34. Heavy water has both H atoms replaced with deuterium atoms and would be written as $D_2O$ (called deuterium oxide), where it is assumed that the O atoms have their naturally-occurring isotopic percentages. There is even $D_2^{18}O$, called "doubly labeled" water. It is very useful in measuring energy expenditure in humans (https://bit.ly/3IH9qdR). Isotopically substituted molecules, both stable and radioactive, have revolutionized chemistry and biochemistry, as they allow specific atoms to be followed in a chemical reaction. **NOTE:** The term monooxygen can be used for oxygen atoms or atomic oxygen. Monooxygenases are common enzymes in biological systems.

so: "Speaking of hydrogen atoms and oxygen atoms . . . ," or "Speaking of monatomic hydrogen and monatomic oxygen . . ." (**monatomic** means one atom) is what you will hear when they are referring to H atoms and O atoms. However, chemists say that water, $H_2O$, consists of hydrogen and oxygen. This could be misleading because here they mean hydrogen and oxygen atoms.

The chemical formula $H_2O$ represents a water molecule. This formula tells us that the water molecule is made up of two hydrogen atoms and one oxygen atom. It is usually drawn

$$H\diagup O \diagdown H$$

because this is a way to show how it looks. Remember, the lines represent chemical bonds. The letter symbols H and O represent the H atom and the O atom, respectively.

From the diagram, you might think that the formula for water should be written as HOH. Many times it is, but $H_2O$ is more popular. Water is never written as $OH_2$.

Chemical formulas don't usually tell you how the atoms are connected to each other. For that information, a diagram of the molecule must be drawn as was done above for the water molecule. This diagram is called the **structural formula** of the molecule because it shows how the atoms of a molecule are connected to one another.

Now let's go back to the equation

$$H_2 + O_2 \rightarrow H_2O$$

There is something wrong with it. If you count the number of atoms of oxygen on the left side of the arrow, you will find two oxygen atoms in the oxygen molecule. But on the right side of the arrow there is only one oxygen atom in the water molecule. What happened to the other oxygen atom?

The other oxygen atom must have been used to make another water molecule. We can draw what happens, as follows:

$$\begin{array}{c} H-H \\ H-H \end{array} + O-O \rightarrow \begin{array}{c} H\diagup O\diagdown H \\ H\diagup O\diagdown H \end{array}$$

Now we have used up all the oxygen on the left of the arrow to make two water molecules. However, the second water molecule needed two hydrogen atoms, so we had to use another $H_2$ on the left side of the arrow.

We can write this structural formula equation much more easily as the chemical equation

$$2H_2 + O_2 \rightarrow 2H_2O$$

The notation 2H$_2$ means two separate H$_2$ molecules. The notation 2H$_2$O means two separate H$_2$O molecules. In words, you can read the equation as "Two hydrogen molecules react with one oxygen molecule to form two water molecules."

The numbers put in front of the molecules are called **coefficients**.

The equation for making water from hydrogen and oxygen illustrates something that is very important, namely: **The same number of atoms of each element must appear on each side of the arrow in a chemical equation**, whether they are combined in molecules or appear as free atoms. We say that in a chemical reaction the atoms of each element are **conserved**, which simply means that atoms aren't created or destroyed in a chemical reaction; they must all be accounted for.

Note that molecules are *not* usually conserved. This is reasonable, and in fact necessary, since chemical reactions make and break up molecules.

Chemical equations that have the same number of atoms of each element on each side of the arrow are said to be **balanced**. The process of figuring out the correct coefficients is called "balancing the equation."

There is one further point about chemical equations you should be aware of at this time. Chemical equations tell us what the reactants and products are. However, they don't usually tell us *how* the reacting substances collided to form the products. The how of chemical reactions is usually very complicated and not very easy to discover. You will appreciate this more when you have studied more chemistry.

**EXAMPLE 3**  The symbol for the element carbon is C. Carbon can combine with hydrogen to make, among other things, a molecule called *methane*. Methane is the main component of natural (cooking) gas and has the formula CH$_4$. How many atoms of each element are there in a CH$_4$ molecule? In two CH$_4$ molecules?

*Solution:* There are one C atom and four H atoms in each methane molecule. In two methane molecules, 2CH$_4$, there are two C atoms and 8 H atoms. If we draw the structural formula of methane twice, you can just count the atoms.

$$\begin{array}{cc} \text{H} & \text{H} \\ | & | \\ \text{H}-\text{C}-\text{H} \quad & \text{H}-\text{C}-\text{H} \quad \blacksquare \\ | & | \\ \text{H} & \text{H} \end{array}$$

**EXAMPLE 4**  The symbol for the element nitrogen is N. Nitrogen gas, which makes up about 78% of the air, contains diatomic molecules N$_2$. In the hot flame inside an automobile engine (or any other hot flame), N$_2$ and O$_2$ combine to form nitric oxide, NO, which is one of the causes of air pollution. Write the balanced equation for the reaction.

**Solution:** $N_2 + O_2 \rightarrow 2NO$. The structural formulas are

$$N\equiv N + O=O \rightarrow \begin{matrix} N=O \\ N=O \end{matrix}$$

The equation can be read "One molecule of nitrogen reacts with one molecule of oxygen to form two molecules of nitric oxide." ∎

As you can see from the structural formulas in Example 4, some molecules have more than one bond joining the atoms together. The atoms in $N_2$ are joined by a triple bond, represented by three lines; the atoms in $O_2$ and NO are joined by a double bond, represented by two lines. We will discuss bonding in Chapter 15.

**NOTE:** In our discussion of the water equation above, we drew the oxygen molecule with one bond. That was done so we wouldn't have to stop the discussion to explain that the oxygen molecule really has a double bond connecting the atoms.

A few more words about writing balanced equations are in order. We can use the water equation

$$2H_2 + O_2 \rightarrow 2H_2O$$

as an example.

We could have balanced the equation by writing, for instance,

$$4H_2 + 2O_2 \rightarrow 4H_2O$$

Atoms are conserved (there are as many of each element on the left side of the arrow as on the right side), but there are twice as many as the minimum needed to balance the equation. This is not done: **The coefficients should be the smallest possible whole numbers**.

As another possibility, you might ask if it is acceptable to write

$$H_2 + \tfrac{1}{2}O_2 \rightarrow H_2O$$

One-half of $O_2$ is just one O atom, and the equation is balanced. This method of using fractions to balance chemical equations is not incorrect, but it is not used as commonly as whole numbers.

Actually, any chemical equation usually represents a large number of atoms and molecules. So taking one-half of $O_2$ is not really dividing the molecule into atoms; we're just taking one-half of a large number of $O_2$ molecules. For example, if we have 1000 oxygen molecules, the fraction $\tfrac{1}{2}$ denotes 500 of them.

If you use fractions as coefficients to balance a chemical equation, they must be "reasonable" fractions. What are "unreasonable" fractions? If you tried to balance the water equation in the following way you would be "unreasonable":

$$\tfrac{1}{2}H_2 + \tfrac{1}{4}O_2 \rightarrow \tfrac{1}{2}H_2O \quad \textbf{unreasonable}$$

It is unreasonable because it is more complicated. Chemists like to keep things clear and simple.

There is another incorrect way that people sometimes try to balance chemical equations. For example, the water equation is sometimes incorrectly written as

$$H_2 + O_2 \rightarrow H_2O_2 \quad \textbf{!! wrong !!}$$

This equation is balanced but is wrong. The $H_2O$ has been incorrectly changed to $H_2O_2$, which is the formula for hydrogen peroxide. If you have ever put hydrogen peroxide on your hair to bleach it, you *know* that it is not water. So when you balance chemical equations, please, you must *never* change the chemical formulas of the atoms or molecules.

The following is a summary of the rules for balancing chemical equations.

1. The same number of atoms of each element must appear on each side of the equation.
2. **Never** change the chemical formula of any of the atoms or molecules appearing in the equation. This means that you should not change subscripts in a chemical formula when you balance a chemical equation.
3. You *may* take multiples or fractions of the atoms and molecules by changing the coefficients.
4. The coefficients must be the smallest possible whole numbers or reasonable fractions.
5. The object is to keep things as simple and as clear as possible.

The following examples of how to balance equations may help you to really understand how to use these rules.

**EXAMPLE 5** Methane, $CH_4$, can burn in oxygen to give carbon dioxide ($CO_2$) and water. Write the balanced equation for this reaction.

*Solution:* The unbalanced equation is

$$CH_4 + O_2 \rightarrow CO_2 + H_2O$$

# 48 MOLECULES AND THE BALANCED CHEMICAL EQUATION

Since there is one C atom on each side of the arrow, let's start by counting the H atoms. There are 4 on the left and 2 on the right. Thus we need to take 2 H$_2$O molecules to give 4 H atoms on the right side of the equation.

$$CH_4 + O_2 \rightarrow CO_2 + 2H_2O$$

Counting the O atoms, we find 2 on the left and 4 on the right. We need to take 2 O$_2$ molecules to get 4 O atoms on the left:

$$CH_4 + 2O_2 \rightarrow CO_2 + 2H_2O$$

The equation is now balanced. Writing structural formulas we have

$$\begin{array}{c} H \\ | \\ H-C-H \\ | \\ H \end{array} + \begin{array}{c} O=O \\ O=O \end{array} \rightarrow O=C=O + \begin{array}{c} H^{\nearrow O \searrow} H \\ \\ H^{\nearrow O \searrow} H \end{array}$$

Here the CO$_2$ has been drawn with two bonds to each oxygen atom to make the picture more realistic. ∎

**EXAMPLE 6**  The metal aluminum, Al, reacts with oxygen to form aluminum oxide, Al$_2$O$_3$, which is usually a white powder. However, Al$_2$O$_3$ can be "grown" into large crystals that are clear and colorless like glass. When certain metal atoms are included in the crystal, it becomes colored. Rubies and sapphires are two colored Al$_2$O$_3$ crystals. Write the balanced equation for the reaction between Al and O$_2$ to form Al$_2$O$_3$.

*Solution:* The unbalanced equation is

$$Al + O_2 \rightarrow Al_2O_3$$

To begin balancing the equation, look at the number of oxygen atoms first. There are 2 O atoms on the left and 3 on the right. If we take 3 O$_2$ molecules and 2 Al$_2$O$_3$ molecules, there are 6 O atoms on each side:

$$Al + 3O_2 \rightarrow 2Al_2O_3$$

To finish the balancing, we need 4 Al atoms on the left side to equal the 4 on the right:

$$4Al + 3O_2 \rightarrow 2Al_2O_3 \quad \blacksquare$$

**EXAMPLE 7**  Octane, C$_8$H$_{18}$, is a major component of gasoline. It burns in oxygen to give CO$_2$ and water. Write the balanced equation for the combustion of octane.

***Solution:*** The unbalanced equation is

$$C_8H_{18} + O_2 \rightarrow CO_2 + H_2O$$

The best way to start is to balance the C atoms:

$$C_8H_{18} + O_2 \rightarrow 8CO_2 + H_2O$$

Then balance the H atoms:

$$C_8H_{18} + O_2 \rightarrow 8CO_2 + 9H_2O$$

This gives us 16 O atoms from the 8 $CO_2$ molecules and gives us 9 O atoms from the 9 $H_2O$ molecules; altogether we have 25 O atoms on the right side of the equation. The best way to put 25 O atoms on the left side is to write $\frac{25}{2}O_2$. Remember that $\frac{1}{2}O_2$ is one O atom. Thus 25 times that is 25 O atoms. The balanced equation is

$$C_8H_{18} + \tfrac{25}{2}O_2 \rightarrow 8CO_2 + 9H_2O$$

This equation is a good example of the use of fractional coefficients. If you want only whole number coefficients, multiply all the coefficients by 2:

$$2C_8H_{18} + 25O_2 \rightarrow 16CO_2 + 18H_2O \quad \blacksquare$$

You might sometimes wonder just how to get started in balancing an equation. For instance, what atoms do you balance first, second, and so on. The following approach is helpful.

1. ***First, balance the atoms for elements that occur in only one substance on each side of the equation.*** For example, in the equation discussed in Example 7, namely,

    $$C_8H_{18} + O_2 \rightarrow CO_2 + H_2O$$

    we balanced the C and the H first, since they occur in only one substance on each side of the equation.

2. ***Second, balance free atoms (such as Fe or Al) or elemental molecules (such as $O_2$ or $Cl_2$) last.*** This is because you can put coefficients in front of them without changing the number of atoms of any other elements in the equation. This is why we balanced the oxygen last in Example 7.

As is usual with most things that are unfamiliar, the best way to learn how to balance equations is to balance a great number of them.

## 3-3
## BALANCE GROUPS AS A WHOLE IN EQUATIONS WITH MORE COMPLEX MOLECULAR FORMULAS

So far, the substances we have discussed have had rather simple formulas: $H_2$, $O_2$, $H_2O$, $CH_4$, $C_8H_{18}$, $Al_2O_3$, and so on. There are a large number of substances that have more complicated formulas. For example, the substance "papermaker's alum," or aluminum sulfate, has the formula

$$Al_2(SO_4)_3$$

The group of atoms represented by the $SO_4$ is called the *sulfate group*.[1] Three sulfate groups combine with two aluminum atoms to make aluminum sulfate. The parentheses allow us to take three sulfate groups:

$Al_2(SO_4)_3$ ← This 3 indicates three sulfate groups.

As far as counting atoms in the formula of aluminum sulfate, we have

2 Al atoms     3 S atoms     12 O atoms

The following two examples will give you some practice in working with chemical formulas that contain parentheses. To read the formula $Al_2(SO_4)_3$, you would say "ay el two ess oh four taken three times."

**EXAMPLE 8** How many atoms of each element are in the formula $Ca_3(PO_4)_2$?

*Solution:* There are 3 Ca atoms, 2 P atoms, and 8 O atoms. The substance is called *calcium phosphate* and $PO_4$ is the *phosphate group*. The $PO_4$ group has a $-3$ charge and is another example of a polyatomic ion. ∎

**EXAMPLE 9** Balance the following equation,

$$Ca(OH)_2 + H_3PO_4 \rightarrow Ca_3(PO_4)_2 + HOH$$

where we have written water as HOH. This will make balancing the equation easier.

*Solution:* One way to simplify the balancing is to count the groups, some of which may be in parentheses, as units, if possible. This would mean counting the OH and $PO_4$ groups as units in this equation.

---

[1] The $SO_4$ group is not considered a molecule because it always has a $-2$ charge. It is called a **polyatomic ion**. In Chapter 16, you will learn more about polyatomic ions.

### 3-3 Balance Groups as a Whole in Equations with More Complex Molecular Formulas

First balance the Ca, since it appears in only one substance on each side of the arrow:

$$3Ca(OH)_2 + H_3PO_4 \rightarrow Ca_3(PO_4)_2 + HOH$$

Next balance the $PO_4$ group, since it also appears in only one substance on each side of the arrow:

$$3Ca(OH)_2 + 2H_3PO_4 \rightarrow Ca_3(PO_4)_2 + HOH$$

Then balance the H and the OH. There are 6 OH groups and 6 H atoms on the left; we need 6 HOH molecules on the right:

$$3Ca(OH)_2 + 2H_3PO_4 \rightarrow Ca_3(PO_4)_2 + 6HOH$$

The equation is now balanced. ∎

To see how easy balancing can be when you keep the groups in parentheses together, try balancing the equation in Example 9 without keeping the groups together. Just balance it using individual atoms.

## PROBLEMS

### KEYED PROBLEMS

1. In the equation $N_2 \rightarrow 2N$, what is the reactant and what are the products?

2. Write the chemical equation for the formation of the diatomic chlorine molecule from two chlorine atoms. The symbol for chlorine is Cl.

3. The symbol for the element nitrogen is N. Nitrogen can combine with hydrogen to make a molecule called ammonia, $NH_3$. How many atoms of each element are there in an $NH_3$ molecule? In two $NH_3$ molecules?

4. Nitric oxide, NO, reacts slowly with the oxygen, $O_2$, in air to form nitrogen dioxide, $NO_2$. $NO_2$ is the primary cause of the reddish-brown haze you can see in polluted air. Write the balanced equation for the reaction between NO and $O_2$ to form $NO_2$.

5. Ethane, $C_2H_6$, can burn in oxygen to give carbon dioxide, $CO_2$, and water. Write the balanced equation for this reaction.

6. The metal iron, Fe, reacts with oxygen, $O_2$, to give the reddish rust called ferric oxide, $Fe_2O_3$. Write the balanced equation for this reaction.

7. $C_{12}H_{26}$ is a component of kerosene. It burns in oxygen to give $CO_2$ and water. Write the balanced equation for the combustion of $C_{12}H_{26}$.

8. How many atoms of each element are in the formula $Fe_2(SO_3)_3$?

9. Balance the following equation:

$$Al(OH)_3 + H_2SO_4 \rightarrow Al_2(SO_4)_3 + HOH$$

## SUPPLEMENTAL PROBLEMS

10. Define the following terms.
    a. molecule
    b. chemical formula
    c. balanced chemical equation
    d. diatomic molecule
    e. structural formula
    f. coefficient

11. How many atoms of each element are in the following compounds?
    a. KCl
    b. $KClO_4$
    c. $AgNO_3$
    d. $C_6H_{14}$
    e. $C_3H_7COOH$
    f. $H_2S$
    g. $(NH_4)_2CO_3$
    h. $Zn(NO_3)_2$
    i. $Mg(CN)_2$
    j. $Ca(HCO_3)_2$

12. Balance the following equations with the smallest possible whole numbers or reasonable fractions.
    a. $SO_2 + O_2 \rightarrow SO_3$
    b. $PCl_5 \rightarrow PCl_3 + Cl_2$
    c. $CaH_2 + H_2O \rightarrow Ca(OH)_2 + H_2$
    d. $(NH_4)_2Cr_2O_7 \rightarrow Cr_2O_3 + N_2 + H_2O$
    e. $Na + O_2 \rightarrow Na_2O$
    f. $H_2 + Cl_2 \rightarrow HCl$
    g. $P + O_2 \rightarrow P_2O_3$
    h. $NH_3 + H_2SO_4 \rightarrow (NH_4)_2SO_4$
    i. $Zn + Pb(NO_3)_2 \rightarrow Zn(NO_3)_2 + Pb$
    j. $Cu + S \rightarrow Cu_2S$
    k. $Al + H_3PO_4 \rightarrow H_2 + AlPO_4$
    l. $NaNO_3 \rightarrow NaNO_2 + O_2$
    m. $H_2O_2 \rightarrow H_2O + O_2$
    n. $BaO_2 \rightarrow BaO + O_2$
    o. $Al + Cl_2 \rightarrow AlCl_3$
    p. $P_4 + O_2 \rightarrow P_4O_{10}$
    q. $H_2 + N_2 \rightarrow NH_3$
    r. $BaCl_2 + (NH_4)_2CO_3 \rightarrow BaCO_3 + NH_4Cl$
    s. $PbO_2 \rightarrow PbO + O_2$
    t. $Al + HCl \rightarrow AlCl_3 + H_2$
    u. $Fe_2(SO_4)_3 + Ba(OH)_2 \rightarrow BaSO_4 + Fe(OH)_3$
    v. $KClO_3 \rightarrow KCl + O_2$
    w. $Mg + N_2 \rightarrow Mg_3N_2$
    x. $C_3H_7CHO + O_2 \rightarrow CO_2 + H_2O$
    y. $NaHCO_3 + HCl \rightarrow NaCl + H_2O + CO_2$
    z. $Zn(OH)_2 + H_2SO_4 \rightarrow ZnSO_4 + HOH$

aa. $C_4H_9OH + O_2 \rightarrow CO_2 + H_2O$
bb. $CaC_2 + H_2O \rightarrow C_2H_2 + Ca(OH)_2$
cc. $CaCO_3 + H_3PO_4 \rightarrow Ca_3(PO_4)_2 + CO_2 + H_2O$
dd. $C_3H_7COOH + O_2 \rightarrow CO_2 + H_2O$

13. A rocket propellant consisting of hydrazine, $N_2H_4$, and hydrogen peroxide, $H_2O_2$, reacts to form nitrogen, $N_2$, and water. Write the balanced equation for the reaction.

14. Yeast can ferment glucose, $C_6H_{12}O_6$, to give ethanol, $C_2H_5OH$, and carbon dioxide. Write the balanced equation for this reaction.

15. Carbon dioxide can be removed from the air of a spacecraft by reacting it with lithium hydroxide, LiOH, to form lithium carbonate, $Li_2CO_3$, and water. Write the balanced equation for this reaction.

16. "Quicklime," CaO, can react with water to form "slaked lime," $Ca(OH)_2$. Write the balanced equation for this reaction.

17. In a blast furnace, iron ore, $Fe_2O_3$, is burned with carbon monoxide, CO, to give iron metal, Fe, and carbon dioxide, $CO_2$. Write the balanced equation for this reaction.

---

## RUBIES, SAPPHIRES, WATCH CRYSTALS AND THE FIRST LASER

In Example 6 on page 48, the formation of **aluminum oxide** by burning aluminum in oxygen was discussed. Aluminum oxide, $Al_2O_3$, occurs in nature in large deposits of the ore bauxite (see pages 345 and 358). It also occurs as the mineral **corundum**. Pure corundum, which can be made synthetically in special furnaces by melting $Al_2O_3$ powder at 2072 °C, is clear, colorless and very hard.

Synthetic corundum is much more scratch resistant than glass, so it is used as the crystal in some watches. In this application, it is called a "sapphire crystal." Because of its hardness, $Al_2O_3$ is also used as an abrasive, say in sandpaper, where it is called **emery** (as in emery paper and emery board). Emery is a mineral of $Al_2O_3$ that has some black iron oxides mixed into it. (Another type of sandpaper contains silicon carbide, called **carborundum**, which is also black.) Fine powders of $Al_2O_3$ are used in polishing applications. See https://bit.ly/3fPeP55.

Corundum can also be made in different colors by adding various metals to it. It is called a **sapphire** if it is any color (including no color) except red. When it is red, it is called a **ruby**. To make corundum red, about 1% of the aluminum atoms are replaced by chromium atoms. The other colors are gotten by adding trace amounts of one or more metals such as iron, titanium, chromium, vanadium, and/or magnesium. The colors are so beautiful that these minerals are considered gemstones. For photos in color see https://bit.ly/37bRVRu.

The **world's first laser**, called a **ruby laser**, was made from a cylindrical rod (1 centimeter in diameter and a few centimeters long) of synthetic ruby by **Theodore H. Maiman** (1927-2007), working at the Hughes Aircraft Co., on May 16, 1960 (https://bit.ly/3qa93zM). See page 218 for more on lasers.

**A Photomicrograph of an Ice Crystal (in this case, a snowflake):** The six points of the star are due to the way water molecules join together in the solid. An interesting thing about solid water (ice) is that it floats on liquid water, whereas most, but not all, solids sink in their liquid. This is because when water molecules join together, they form an open structure, not a solid-packed one. See https://bit.ly/2EDLzfx for many diagrams of ice's molecular structure. It is a good thing that ice floats on water. If it didn't, lakes would freeze from the bottom up, eventually becoming solid blocks of ice, killing all life in them, except maybe some bacterial and fungal spores, certain viruses, tardigrades (also called "water bears" or "moss piglets") and the Patagonian dragon. Life on earth would most likely be very different.

*Photo courtesy of Charles Schmitt - Own work stellate snowflake photographed Vermont2015. https://en.wikipedia.org/wiki/Snowflake#/media/File:Snowflake_Detail.jpg CC BY-SA 4.0*

# ELEMENTS, COMPOUNDS, AND MIXTURES

Different elements exist as solids, liquids, or gases. The arrangement of the atoms in different elements also varies. In this chapter we will briefly discuss such differences.

## 4-1
## EACH ELEMENT CONSISTS OF A SINGLE KIND OF ATOM

The following is a list of some of the elements and their formulas that we have mentioned so far.

| hydrogen | $H_2$ | aluminum | Al | argon | Ar |
| oxygen | $O_2$ | radium | Ra | calcium | Ca |
| nitrogen | $N_2$ | lead | Pb | | |
| carbon | C | helium | He | | |

Some of the elements can exist as individual atoms at room temperature. Gases such as helium (He) and argon (Ar) exist as individual atoms. They are monatomic.

Some elements exist as individual molecules at room temperature. The ones like this that we have talked about are $H_2$, $O_2$, and $N_2$, which all exist as diatomic gas molecules.

Other elements that we have mentioned (C, Al, U, Ra, Fe, Pb) exist as atoms bonded together in a solid at room temperature. One example of such a solid is a metal; another is a diamond, which is a nonmetal. It is impossible to find individual molecules in these solids, so we write the chemical formulas as single atoms. This is because it's simple to do and nobody has thought of a better way of writing it.

56 ELEMENTS, COMPOUNDS, AND MIXTURES

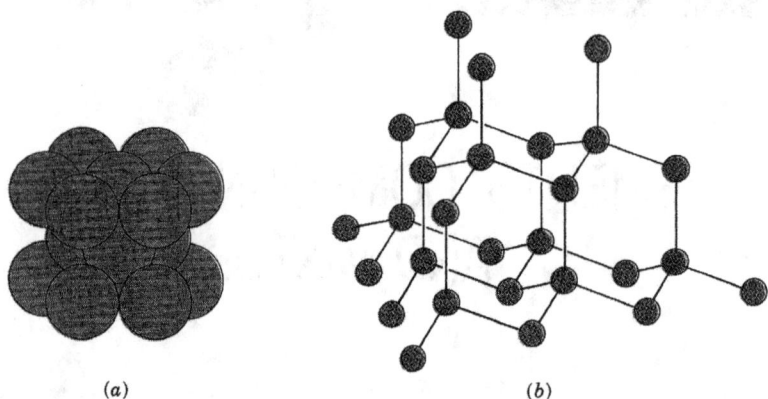

**FIGURE 4-1** Two arrangements that atoms of elements can take: (a) metal atoms; (b) carbon atoms in diamond. There are no separate molecules in either.

Figure 4-1 illustrates some of the arrangements that atoms of elements can take. Notice that in a diamond, each carbon atom is connected to several other carbon atoms. It is impossible to find individual molecules. A diamond crystal large enough to see consists of billions and billions of carbon atoms. It can be thought of as a giant molecule. We will refer to such an arrangement as a **giant array** of atoms. Even though there are billions of carbon atoms in the giant array, we still write the chemical formula of diamond as C. This may seem strange because the chemical formula makes it appear as though the diamond is made up of separate carbon atoms. However, in this case the formula simply indicates that diamonds are are made up only of carbon atoms.

Metals can also be considered as consisting of a giant array of atoms.

Two elements, mercury (Hg) and bromine ($Br_2$), are liquid at room temperature.

## 4-2
## COMPOUNDS CONSIST OF DIFFERENT KINDS OF ATOMS

Suppose that a substance contains two or more atoms of *different* atomic number connected with a chemical bond. We call this type of substance a **compound**. A quick way to spot a compound is to look at its chemical formula. If the chemical formula is made up of two or more *different* symbols (elements), the substance is a compound. Examples of compounds we have already mentioned are water, $H_2O$; calcium phosphate, $Ca_3(PO_4)_2$; carbon dioxide, $CO_2$; aluminum sulfate, $Al_2(SO_4)_3$; methane, $CH_4$; phosphoric acid, $H_3PO_4$; aluminum oxide, $Al_2O_3$; calcium hydroxide, $Ca(OH)_2$; nitric oxide, NO; octane, $C_8H_{18}$; hydrogen peroxide, $H_2O_2$.

Another way to tell whether the formula represents an element or a compound is as follows. If there is only *one* capital letter in the formula, it's an element. If there are *two or more* capital letters, it's a compound.

## 4-2 Compounds Consist of Different Kinds of Atoms

Again, as was true of elements, some of the compounds just mentioned, like water, exist as separate molecules at room temperature. Water is a liquid and a gas at room temperature. The gaseous form is usually called water vapor. There are separate molecules of water in both the gas and the liquid, but in the liquid they are attracted to each other by very weak bonds, which help keep water as a liquid. You will learn more about these bonds, called hydrogen bonds, if you study more chemistry.

In other compounds, like sodium chloride (common table salt, NaCl), there are no individual molecules because each atom is connected to several other atoms. We have discussed a similar situation with the diamond. In a crystal of NaCl large enough to see, there are over a quadrillion ions.* This type of arrangement, illustrated in Figure 4-2, is also called a giant array of atoms. However, we still write the chemical formula of sodium chloride as NaCl.

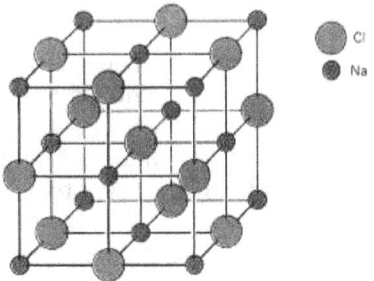

**Figure 4-2** A model of the NaCl crystal structure. The large balls represent $Cl^-$ ions, and the small balls represent $Na^+$ ions.

**Salt and Salt Crystals:** Crystals of NaCl are cubes as shown in the photo at the left (size: 6.7 × 1.9 × 1.7 cm). When salt is found as a mineral in salt deposits, it is called **halite** or **rock salt**. The name halite is derived from the Ancient Greek word for salt, ἅλς (háls). Halite occurs in vast beds that resulted from the drying up of ancient lakes and seas. Salt beds may be hundreds of meters thick. In 2017, the US alone produced 42-million tons of salt, about 15% of the world's production.
*Photo courtesy of Rob Lavinsky, iRocks.com, CC-BY-SA-3.0*
*https://en.wikipedia.org/wiki/Halite#/media/File:Halite-249324.jpg*

*An **ion** is a charged particle formed by the loss or gain of one or more electrons by a neutral atom or group of atoms. In the sodium chloride crystal, each sodium atom has a charge of +1 because it has given one electron to a chlorine atom, which then has a charge of −1. It is the attraction of the plus and minus charges for each other that holds the crystal together. The $Na^+$ ions are small because they have lost an electron. The $Cl^-$ ions are large because they have gained an electron. Because the bond between $Na^+$ and $Cl^-$ involves ions, this bond is called an **ionic bond**. Ionic bonds are very strong giving salt a high melting point. For more information about ions and ionic bonds, see Section 15-7.

Although this way of writing the chemical formula makes it appear that sodium chloride is made up of separate molecules, the formula only indicates that there is one Na atom for each Cl atom. Chemists generally can determine whether the formula of a substance represents separate atoms or molecules or a giant array of atoms. As you learn more chemistry, you will also be able to determine this.

In addition to liquids and solids, many compounds exist as gases at room temperature. Examples are $CO_2$, $CH_4$, and NO. The molecules in the gas are separate.

Please do not get the idea, from what we have said, that the type of bonding is the same in substances that are as different as NaCl, $H_2O$, diamond, and metals such as aluminum. The bonding is very different; for the full story, you will have to study more chemistry than is in this book. However, some details are given in Chapter 15.

**EXAMPLE 1** Which of the following formulas represent elements, and which represent compounds? The formulas are $Cl_2$, $N_2O_4$, $C_2H_6$, $S_8$, $P_4O_{10}$, K, $SO_2$, Fe, CO, and $I_2$.

*Solution:* It doesn't matter whether you know the names of the substances to answer this one. However, for the curious, we will include the names in the answer.

| ELEMENTS | | COMPOUNDS | |
|---|---|---|---|
| $Cl_2$ | chlorine | $N_2O_4$ | dinitrogen tetraoxide |
| $S_8$ | sulfur | $C_2H_6$ | ethane |
| K | potassium | $P_4O_{10}$ | tetraphosphorus decaoxide |
| Fe | iron | $SO_2$ | sulfur dioxide |
| $I_2$ | iodine | CO | carbon monoxide |

## 4-3
## MIXTURES CONSIST OF TWO OR MORE SUBSTANCES

A **mixture** has two or more substances (compounds, elements, or both) mixed up together. These different substances are not connected to each other by chemical bonds and can be mixed in *any* proportions.

For example, the compound sodium chloride (NaCl) and the element iron (as a powder) mixed together form a mixture. Another mixture, usually called a **solution**, is the compound sugar dissolved in the compound water. The key here is that you can separate the substances in the mixture without breaking chemical bonds. You could get the iron powder out of the salt with a magnet, and you could get the sugar out of the water by evaporating the water.

Some chemists might argue that solutions are not true mixtures because of the weak "interactions" (most chemists do not think of them as real chemical bonds) between the water molecules and the molecules of the dissolved substance. Nor can you usually dissolve as much of something in water as you might want to, since most substances have a limited solubility. However, up to the solubility limit, proportions can vary. Many chemists do think of solutions as mixtures, and in this book we shall consider solutions as mixtures.

**EXAMPLE 2** Which of the following sets of substances are mixtures, which are compounds, and which are elements?

salt and sugar
sulfur and aluminum powder
air (separate molecules of $O_2$, $N_2$, Ar, $CO_2$, $H_2O$, etc.)
liquid oxygen

potassium metal
salt
$S_8$
$PCl_3$
pure ice
ginger ale

*Solution:*

| MIXTURES | COMPOUNDS | ELEMENTS |
|---|---|---|
| salt and sugar | salt | liquid oxygen |
| sulfur and aluminum powder | $PCl_3$ | potassium metal |
| air | pure ice | $S_8$ |
| ginger ale | | |

■

## PROBLEMS

### KEYED PROBLEMS

1. Which of the following formulas represent elements, and which represent compounds? The formulas are $F_2$, $N_2O_5$, $C_3H_8$, $P_4$, $PCl_5$, Na, $SO_2$, Cr, $CO_2$, and $Br_2$.

2. Which of the following sets of substances are mixtures, which are compounds, and which are elements?

salt and pepper
iron powder and charcoal
polluted air
liquid nitrogen
rubidium metal

sugar
$P_4$
$SF_6$
cherry soda
steam

## SUPPLEMENTAL PROBLEMS

3. Define the following terms.
   a. element
   b. compound
   c. "giant array" of atoms
   d. mixture
   e. monatomic substance

4. List two examples of substances that fit into the following categories.
   a. elements that are liquid at room temperature
   b. monatomic elements
   c. diatomic elements
   d. compounds that are liquid at room temperature
   e. compounds that are gaseous at room temperature
   f. compounds that are a giant array of atoms
   g. elements that are a giant array of atoms

**ELEMENT UPDATE[1] – 104 to 118:** The *Table of Atomic Weights* (Table 2-2) on page 27 only includes elements up to atomic number 103. As of 2021, additional elements have been made in the laboratory, from element 104 to element 118. All are radioactive, have no stable isotopes, and do not occur in nature, at least on earth. The web links below contain much information about all the elements.[2]

| Atomic number | Name and Chemical Symbol | | Mass Number of Longest Lived Isotope |
|---|---|---|---|
| 104 | Rutherfordium | Rf | 267 |
| 105 | Dubnium[3] | Db | 268 |
| 106 | Seaborgium[3] | Sg | 269 |
| 107 | Bohrium | Bh | 278 |
| 108 | Hassium | Hs | 269 |
| 109 | Meitnerium | Mt | 282 |
| 110 | Darmstadtium | Ds | 281 |
| 111 | Roentgenium | Rg | 286 |
| 112 | Copernicium | Cn | 285 |
| 113 | Nihonium | Nh | 286 |
| 114 | Flerovium[3] | Fl | 290 |
| 115 | Moscovium | Mc | 290 |
| 116 | Livermorium | Lv | 293 |
| 117 | Tennessine | Ts | 294 |
| 118 | Oganesson[3] | Og | 295 |

[1] For a short article and a video entitled: "Here's how the periodic table gets new elements: From discovery to confirmation and naming, the path is rarely simple" see https://bit.ly/2RDM1BO.

[2] For a list of all the elements **sorted by alphabetical order**, with their names, symbols, and the origin of their names, see https://bit.ly/32ZgEKC.
For a list of all the elements **sorted by atomic number**, with their names, symbols, the origin of the name, and additional information about them, see https://bit.ly/35HZFxq.

[3] Elements 106 and 118 are the only elements that have been named after a person while they were still alive. Seaborgium is named after the American nuclear chemist **Glenn T. Seaborg** (1912-1999), and Oganesson is named after the Russian nuclear physicist **Yuri Oganessian** (b. 1933). It is interesting that element 114, flerovium, is named after **Georgy Flerov** (1913-1990) who was Oganessian's Ph.D. advisor. Both worked at the Joint Institute for Nuclear Research in Dubna, Moscow Oblast, Russia. Element 105, dubnium, is named after that city. That has been and still is a very talented research group!

**The Andromeda Galaxy:** It is a spiral galaxy and the nearest major galaxy to the earth, and in scientific notation, about $1.5 \times 10^{19}$ mi away. It contains about $1 \times 10^{12}$ stars, and has a diameter of about $1.3 \times 10^{18}$ mi. It will collide with our galaxy, the Milky Way, also a spiral galaxy, in about $5 \times 10^9$ yr. The likely outcome will be an oval galaxy instead of two spiral galaxies. Around this time our sun will expand into a red giant and engulf earth, totally destroying it. See https://bit.ly/3H356oj for an article and a computer simulation. After reading this chapter, you will be an expert on scientific notation (https://bit.ly/3gcF6Pf). *Photo courtesy of NASA*

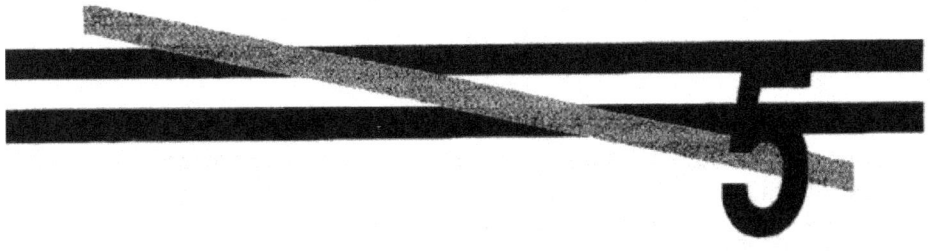

# SCIENTIFIC NOTATION

We said in Chapter 1 that atoms are very, very small. Because of this extreme smallness, a tremendous number of atoms are needed to make up an amount of matter that you could see or weigh. In the tip of your little finger, there are about this many atoms:

$$10,000,000,000,000,000,000,000$$

This number is ten sextillion.

If you took this number of atoms and laid them side by side, you could make a line of them that would reach out to the sun and back about five times. If you took this number of water molecules and mixed them with all the water in the oceans of the world, any glass of water you took would probably contain one or two of the original water molecules. For you to become as small as a water molecule, you would have to shrink your length down over ten billion times.

You can see that when we are dealing with atoms and molecules, we need to be able to use very big and very small numbers. For this reason, scientists have developed a very handy notation for large and small numbers. This notation is called **scientific notation** and is the subject of this chapter.

## 5-1
### AN EXPONENT TELLS HOW MANY TIMES TO MULTIPLY A NUMBER BY ITSELF

The key to using scientific notation is to understand numbers that have exponents attached to them. Look at the equation

$$3 \times 3 = 3^2 = 9$$

SCIENTIFIC NOTATION

The "2" that is a half-space above the number 3 is called the **exponent** or **power** of the number 3. You would read $3^2$ as "three to the second power" or, in this case, because exponents equal to 2 are so common, "three squared." Whatever you call it, the exponent is the number of times that you multiply the number by itself.

Now look at the expression $2^{10}$. This says that you multiply 2 by itself 10 times. We say that we are "raising 2 to the 10th power," or "taking 2 to the 10th." What you get when you do this is

$$2^{10} = 2 \times 2 \times 2 \times 2 \times 2 \times 2 \times 2 \times 2 \times 2 \times 2 = 1024$$

Now you can see how nice **exponential notation** is. It is much shorter to write $2^{10}$ than it is to write $2 \times 2 \times 2 \times 2 \times 2 \times 2 \times 2 \times 2 \times 2 \times 2$. But you might ask: Why not simply write $2^{10}$ as 1024? Isn't that just about as easy? Well, it is for $2^{10}$, but what about $2^{100}$? This is 100 twos multiplied together. Not only is that difficult to write out, but the actual number it multiplies out to be is about this big:

$$1,000,000,000,000,000,000,000,000,000,000$$

In this case there is no contest; $2^{100}$ beats them all as the shortest notation.

## 5-2
### A TEN WRITTEN WITH EXPONENTS ALLOWS US TO WRITE NUMBERS IN SCIENTIFIC NOTATION

Since our number system is based on the number 10, it is especially useful to write the number 10 with exponents. Some of the powers of 10 are the following:

$$10^1 = 10$$
$$10^2 = 10 \times 10 = 100$$
$$10^3 = 10 \times 10 \times 10 = 1000$$
$$10^4 = 10 \times 10 \times 10 \times 10 = 10,000$$

This list can be written in a slightly different way. We will multiply each number that has an exponent by "1":

$$10 = 1 \times 10^1$$
$$100 = 1 \times 10^2$$
$$1000 = 1 \times 10^3$$
$$10,000 = 1 \times 10^4$$

## 5-2 Scientific Notation

Let's try writing a few other numbers in a similar manner:

$$200 = 2 \times 100 = 2 \times 10^2$$
$$3000 = 3 \times 1000 = 3 \times 10^3$$
$$5{,}000{,}000 = 5 \times 1{,}000{,}000 = 5 \times 10^6$$

For the last number, multiply 10 by itself six times to get 1,000,000. To read $5 \times 10^6$, say "five times ten to the sixth."

Another example that really shows the great space saving of exponential notation is

$$6{,}000{,}000{,}000{,}000{,}000{,}000{,}000{,}000{,}000{,}000{,}000 = 6 \times 10^{33}$$

Read it "six times ten to the thirty-three" or "six times ten to the thirty-third." Notice that there are 33 zeros after the 6 and that the exponent of the 10 is 33. Thus, to figure out the exponent for this number, just count the zeros.

**EXAMPLE 1** Write the number 90,000,000,000,000,000 as a 9 multiplied by a 10 with some exponent.

*Solution:* Since there are 16 zeros after the 9, the answer is $9 \times 10^{16}$, which can be read "nine times ten to the sixteenth." ∎

This way of writing exponential notation is called **scientific notation,** probably because scientists use it a lot. Writing numbers in scientific notation has the following layout. You first write a number between 1 and 10, then a "times sign," and finally a 10 with some exponent. Let's show this as follows:

**number between 1 and 10 $\times$ $10^{\text{exponent}}$**

Actually, a "one" can also be used as the number. For instance, $1 \times 10^6$ is a perfectly good number in scientific notation. One reason that scientists like this notation is that it makes it very easy to keep track of the decimal point when multiplying and dividing numbers. And anything that makes it easy to keep track of the decimal point is welcome news indeed.

So far, we haven't written any decimal points. So let's begin. Suppose we want to write the number 362 in scientific notation. We would write the following:

$$362 = 3.62 \times 100 = 3.62 \times 10^2$$

**66** SCIENTIFIC NOTATION

Another example is to write 47,625 in scientific notation:

$$47{,}625 = 4.7625 \times 10{,}000 = 4.7625 \times 10^4$$

If it isn't clear how we did these conversions to scientific notation, the following discussion on some points of math may help.

## 5-3
## SOME DETAILS OF MULTIPLICATION AND FRACTIONS

A basic rule of our number system is that the **order of addition does not matter.** For example,

$$2 + 3 + 4 = 3 + 2 + 4 = 4 + 3 + 2 = 9$$

There are even more ways of mixing up the order than we have shown. Adding in any order would add up to nine.

Another rule of our number system is that the **order of multiplication does not matter.** For example,

$$4 \times 3 = 3 \times 4 = 12$$

**EXAMPLE 2** Write the product $3 \times 4 \times 5$ in all the six possible ways that the numbers can be mixed up. Calculate the answer of each way.

*Solution:* The six ways and the product of each way are as follows:

$$3 \times 4 \times 5 = 12 \times 5 = 60$$
$$3 \times 5 \times 4 = 15 \times 4 = 60$$
$$4 \times 3 \times 5 = 12 \times 5 = 60$$
$$4 \times 5 \times 3 = 20 \times 3 = 60$$
$$5 \times 4 \times 3 = 20 \times 3 = 60$$
$$5 \times 3 \times 4 = 15 \times 4 = 60 \quad \blacksquare$$

There is a subrule of the rule just mentioned. The subrule says that **the grouping of numbers in multiplication does not matter.** This subrule refers to using parentheses around the numbers. For example, $3 \times 4 \times 5$ can be written as

$$(3 \times 4) \times 5 \quad \text{or as} \quad 3 \times (4 \times 5)$$

## 5-3 Some Details of Multiplication and Fractions

The first notation says that you are supposed to multiply the 3 × 4 first:

$$(3 \times 4) \times 5 = 12 \times 5 = 60$$

The second notation says that you multiply the 4 × 5 first:

$$3 \times (4 \times 5) = 3 \times 20 = 60$$

Any way you do it, you always get the same answer. Parentheses are very useful because they break mathematical expressions into smaller pieces that are easier to work with.

The multiplication sign is often left out, and the parentheses serve just as well. For instance,

$$3 \times (4 \times 5) = 3(4 \times 5)$$

**EXAMPLE 3** In the product 8 × 10 × 2, insert parentheses in reasonable places and multiply.

*Solution:* You will learn from experience how to decide what is "reasonable." For instance, writing (8 × 10 × 2) is not very reasonable, because it doesn't change anything. Here are some reasonable possibilities:

$$8(10 \times 2) = 8 \times 20 = 160$$

and

$$(8 \times 10)2 = 80 \times 2 = 160$$

You can also replace all the "×" signs by parentheses if you want to:

$$(8)(10)(2) = 160$$

The order in which you multiply the numbers does not matter. ■

We can treat fractions in the same manner that we treat whole numbers. We can thus write

$$\tfrac{1}{2} \times 10 = 10 \times \tfrac{1}{2} = 5$$

**EXAMPLE 4** Write the product $\tfrac{1}{2} \times 10 \times 4$ with parentheses in reasonable places and multiply.

**Solution:**

$$(\tfrac{1}{2} \times 10)4 = 5 \times 4 = 20$$
$$\tfrac{1}{2}(10 \times 4) = \tfrac{1}{2} \times 40 = 20$$
$$(\tfrac{1}{2})(10)(4) = 20 \quad \blacksquare$$

If you look at the equation

$$\tfrac{1}{2} \times 10 = \tfrac{10}{2} = 5$$

you see that multiplication by a fraction whose numerator (top part) is 1 is the same thing as division by the denominator (bottom part) of the fraction. This, together with the rules of our number system already given, allows us to make some very useful manipulations with fractions.

Consider the multiplication $\tfrac{1}{2} \times 10$. We can easily put a 1 under the 10 so that it looks like a fraction. This is just for our convenience: When any number is divided by 1, the number doesn't change. Doing this we have

$$\tfrac{1}{2} \times 10 = \tfrac{1}{2} \times \tfrac{10}{1}$$

We can then write the two fractions as one fraction because the rule for multiplying fractions says that you "multiply the tops and multiply the bottoms":

$$\frac{1}{2} \times \frac{10}{1} = \frac{1 \times 10}{2 \times 1}$$

Now we have a product in both the numerator (top) and denominator (bottom) of the fraction. Since the order of multiplication doesn't matter, it is all right to interchange the 1 and the 2 in the denominator:

$$\frac{1 \times 10}{2 \times 1} = \frac{1 \times 10}{1 \times 2}$$

Splitting the fraction up again gives

$$\frac{1 \times 10}{1 \times 2} = \frac{1}{1} \times \frac{10}{2}$$

which of course comes out to be

$$1 \times \tfrac{10}{2} = 1 \times 5 = 5$$

Writing the whole thing out in one step gives us

$$\frac{1}{2} \times 10 = \frac{1}{2} \times \frac{10}{1} = \frac{1 \times 10}{2 \times 1} = \frac{1 \times 10}{1 \times 2} = \frac{1}{1} \times \frac{10}{2} = 1 \times 5 = 5$$

The nice thing about math is that it is so flexible. Once you understand the rules, you can make things work for you by simplifying and rearranging.

**EXAMPLE 5**  Multiply the following after rearranging it as was just demonstrated.

$$30 \times \tfrac{1}{3}$$

*Solution:* First write it as one big fraction, then rearrange and divide:

$$30 \times \frac{1}{3} = \frac{30}{1} \times \frac{1}{3} = \frac{30 \times 1}{1 \times 3} = \frac{30 \times 1}{3 \times 1} = \frac{30}{3} \times \frac{1}{1} = 10 \times 1 = 10 \quad \blacksquare$$

## 5-4
## SCIENTIFIC NOTATION USING POSITIVE EXPONENTS

Now we can go back and see how we wrote 362 in scientific notation. Certainly we can multiply 362 by 1 and it doesn't change: $362 \times 1 = 362$. We can, however, write 1 in a special way—namely, as $\tfrac{100}{100}$. Therefore,

$$\frac{100}{100} = 1$$

Then if we remember that any number can be multiplied by 1 in any form and the number doesn't change, we can write

$$362 = 362 \times 1 = 362 \times \frac{100}{100}$$

We can rearrange things a bit and turn 362 into scientific notation:

$$362 \times \frac{100}{100} = \frac{362}{1} \times \frac{100}{100} = \frac{362}{100} \times \frac{100}{1} = 3.62 \times 100 = 3.62 \times 10^2$$

Once you understand the reason for it, it is perfectly all right to just count the decimal places to convert numbers to scientific notation. Since 362 can be written as 362., to get 3.62 you count two decimal places to the left.

## 70 SCIENTIFIC NOTATION

Then, since moving the decimal point two places to the left is the same as dividing by 100 (after all, 3.62 is 100 times smaller than 362.), you have to multiply the 3.62 by 100 or $10^2$ so you don't change its value.

Moving the decimal point is illustrated below using arrows under the numbers.

$$362. = 3\,6\,2. = 3.62 \times 10^2$$

Two arrows mean we have moved the decimal point two places to the left.

**EXAMPLE 6**   Convert 47,625. to scientific notation.

*Solution:* 47625. = 4 7 6 2 5. = $4.7625 \times 10^4$. Notice again that the exponent of the 10 is the same number as the number of places (represented by arrows) that you moved the decimal point to the left. ∎

**EXAMPLE 7**   Convert 94.7 to scientific notation.

*Solution:* 94.7 = 9 4.7 = $9.47 \times 10^1$ ∎

**EXAMPLE 8**   Convert 5,000,764 to scientific notation.

*Solution:* 5,000,764. = 5 0 0 0 7 6 4. = $5.000764 \times 10^6$ ∎

To convert numbers from scientific notation to regular notation, all you have to do is to reverse the procedure shown in Examples 6 through 8. **Regular notation** is the usual everyday way of writing numbers without exponents.

**EXAMPLE 9**   Convert $6.4 \times 10^5$ to regular notation.

*Solution:* $6.4 \times 10^5$ = 6.4 0 0 0 0 = 640,000., or 640,000 without the decimal point. ∎

Notice that we have moved the decimal point five places to the right (represented by five arrows) in Example 9. That is because $10^5$ = 100,000, and to multiply a number by 100,000 you have to move the decimal point five places to the right.

**EXAMPLE 10**  Convert $4.09 \times 10^8$ to regular notation.

*Solution:* $4.09 \times 10^8 = 4.09000000 = 409,000,000.$, or $409,000,000$ without the decimal point. The decimal point has been moved eight places to the right. ∎

**EXAMPLE 11**  Convert $37.57 \times 10^7$ to regular notation.

*Solution:* Notice that $37.57 \times 10^7$ is not in scientific notation because 37.57 is not between 1 and 10. This does not matter. Just move the decimal point seven places to the right: $37.57 \times 10^7 = 37.5700000 = 375,700,000.$, or $375,700,000$ without the decimal. ∎

We can summarize the rules for converting numbers from regular notation to scientific notation and vice versa.

1. To convert from regular notation to scientific notation, move the decimal point to the *left*.
2. To convert from scientific notation to regular notation, move the decimal point to the *right*.
3. Rules 1 and 2 apply to numbers greater than 1. We will discuss numbers less than 1 later in this chapter.

You can multiply numbers in scientific notation very easily if the following rule is used: **When you multiply numbers with exponents, you add the exponents.** Of course, the numbers that have the exponents, called **base numbers** or **bases**, have to be the same. This will be explained after the following examples. To illustrate the rule just given, multiply $10^2 \times 10^3$. We know that

$$10^2 = 100 \quad \text{and} \quad 10^3 = 1000$$

Since

$$100 \times 1000 = 100,000 = 10^5$$

it must be true that

$$10^2 \times 10^3 = 10^5$$

where we have added the exponents $(2 + 3 = 5)$.

The reason for adding the exponents might be clearer from the following way of looking at this illustration. Let's write out all the tens that we are dealing with:

$$100 = 10 \times 10 \quad \text{and} \quad 1000 = 10 \times 10 \times 10$$

Multiplying $100 \times 1000 = 10 \times 10 \times 10 \times 10 \times 10 = 10^5$, which is the same as we got by adding the exponents. You can see from this that each ten in the product is, in a way, "added," so that the exponent on the $10^5$ is the "sum" of the *number of tens* that were multiplied together.

Now you can see why the base numbers have to be the same when you add the exponents. If they were not, you couldn't multiply the base numbers together the way we did. It is this multiplication of base numbers that gives us the rule for adding exponents.

**EXAMPLE 12** Multiply $10^8 \times 10^{11}$.

*Solution:* Adding exponents we get

$$10^8 \times 10^{11} = 10^{8+11} = 10^{19}$$

The $10^8$ means that there are eight tens multiplied together. The $10^{11}$ means that there are 11 tens multiplied together. So all in all there are 19 tens multiplied together to give $10^{19}$. ∎

We can now proceed to multiply more complicated numbers. For instance, multiply $3 \times 10^5$ by $2 \times 10^7$. The key to doing this is to (1) group together the "plain" numbers and (2) group together the numbers having exponents:

$$3 \times 10^5 \times 2 \times 10^7 = 3 \times 2 \times 10^5 \times 10^7 = 6 \times 10^{5+7} = 6 \times 10^{12}$$

The rearrangement is allowed because the order of multiplication doesn't matter.

**EXAMPLE 13** Multiply $6 \times 10^4$ by $9 \times 10^{27}$.

*Solution:* First regroup and then multiply:

$$6 \times 10^4 \times 9 \times 10^{27} = 6 \times 9 \times 10^4 \times 10^{27} = 54 \times 10^{4+27} = 54 \times 10^{31}$$

To get the answer into scientific notation, we must write it as $5.4 \times 10^{32}$. To get this form, we have to convert 54 to scientific notation:

$$54 \times 10^{31} = 5.4 \times 10^1 \times 10^{31} = 5.4 \times 10^{32}$$ ∎

## 5-4 Scientific Notation Using Positive Exponents

**EXAMPLE 14** Multiply the following and put the answer into scientific notation: $(9 \times 10^{15})(8 \times 10^{20})$.

*Solution:* As before, first regroup and then multiply:

$$(9 \times 10^{15})(8 \times 10^{20}) = 9 \times 10^{15} \times 8 \times 10^{20}$$
$$= 9 \times 8 \times 10^{15} \times 10^{20} = 72 \times 10^{35}$$

Converting $72 \times 10^{35}$ into scientific notation gives $7.2 \times 10^{36}$. ∎

The next thing that we have to look at is the division of numbers expressed in scientific notation. But first, let's look at the division of some numbers expressed in powers of 10. You know that

$$\frac{1000}{100} = 10$$

If we write it in a slightly different way, we have

$$\frac{1000}{100} = \frac{10 \times 10 \times 10}{10 \times 10} = \frac{10}{1} \times \frac{10}{10} \times \frac{10}{10} = \frac{10}{1} \times 1 \times 1 = 10$$

You can see that each 10 in the denominator divides a 10 in the numerator, giving 1.

In other words, each 10 on the bottom "removes" a 10 on the top. At the start there are three 10's on the top and two 10's on the bottom. After division, there is only one 10 on the top.

This method of "removing" the 10's gives us the idea that is used in dividing numbers with exponents. But don't forget that "removing" 10's is really dividing 10's. Since $1000 = 10^3$ and $100 = 10^2$, we can write

$$\frac{1000}{100} = \frac{10^3}{10^2}$$

We know that the division 1000/100 gives 10. so it must be true that

$$\frac{10^3}{10^2} = 10$$

Since we can write 10 as $10^1$, we have

$$\frac{10^3}{10^2} = 10^1$$

74    SCIENTIFIC NOTATION

And if you look carefully, you realize that $3 - 2 = 1$. So, you can **divide numbers with exponents by subtracting the exponents.** Writing our problem with the subtraction clearly indicated, we have

$$\frac{10^3}{10^2} = 10^{3-2} = 10^1 = 10$$

As we said before, we can do this subtraction of the exponents because the two 10's on the bottom "removed" two of the 10's on the top, leaving only one 10.

**EXAMPLE 15**   Perform the following division:

$$\frac{10^8}{10^3}$$

*Solution:* By subtracting exponents we get

$$\frac{10^8}{10^3} = 10^{8-3} = 10^5$$

To see it more clearly, let's write out all the 10's:

$$\frac{10^8}{10^3} = \frac{10 \times 10 \times 10 \times \cancel{10} \times \cancel{10} \times \cancel{10} \times 10 \times 10}{\cancel{10} \times \cancel{10} \times \cancel{10}}$$
$$= 10 \times 10 \times 10 \times 1 \times 1 \times 1 \times 10 \times 10$$
$$= 10 \times 10 \times 10 \times 10 \times 10$$
$$= 10^5 \quad \blacksquare$$

If there are numbers multiplying the 10's that have exponents, just rearrange. Then you can divide the "plain" numbers first; after that, you can divide the 10's that have exponents.

**EXAMPLE 16**   Compute

$$\frac{6 \times 10^7}{4 \times 10^2}$$

*Solution:* Separating the expression into two parts we have

$$\frac{6}{4} \times \frac{10^7}{10^2} = 1.5 \times 10^5 \quad \blacksquare$$

**EXAMPLE 17** Compute

$$\frac{(8 \times 10^3)(3 \times 10^7)}{(6 \times 10^2)(2 \times 10^4)}$$

*Solution:* Regroup, multiply the numerators and the denominators, and then divide:

$$\frac{8 \times 3 \times 10^3 \times 10^7}{6 \times 2 \times 10^2 \times 10^4} = \frac{24 \times 10^{10}}{12 \times 10^6} = \frac{24}{12} \times \frac{10^{10}}{10^6} = 2 \times 10^4 \quad \blacksquare$$

## 5-5
## SCIENTIFIC NOTATION USING NEGATIVE EXPONENTS

Suppose now, in doing division with exponents, you run into the following expression:

$$\frac{10^2}{10^4}$$

If you follow the usual rule and subtract exponents, you will get

$$\frac{10^2}{10^4} = 10^{2-4} = 10^{-2}$$

What does $10^{-2}$ mean? Is it reasonable? Let's see. First write out all the 10's:

$$\frac{10^2}{10^4} = \frac{10 \times 10}{10 \times 10 \times 10 \times 10} = \frac{10}{10} \times \frac{10}{10} \times \frac{1}{10} \times \frac{1}{10} = 1 \times 1 \times \frac{1}{10} \times \frac{1}{10} = \frac{1}{100}$$

It is clear, then, that

$$\frac{10^2}{10^4} = \frac{1}{100}$$

The two 10's on the top have "removed" two of the four 10's on the bottom, leaving two 10's on the bottom.

Since $10^2/10^4 = 10^{-2}$ by our rule of subtracting exponents, and since $10^2/10^4 = 1/100$ by division as was shown above, it must be true that

$$10^{-2} = \frac{1}{100}$$

Furthermore, since $100 = 10^2$, we have that $1/100 = 1/10^2$ and thus

$$10^{-2} = \frac{1}{10^2}$$

## 76 SCIENTIFIC NOTATION

What we have shown in this case, and what is true for any numbers, is that **a number with a negative exponent really represents 1 over the same number with a positive exponent.**

If we let $n$ stand for *any* number, we can say that

$$10^{-n} = \frac{1}{10^n}$$

**EXAMPLE 18** Divide

$$\frac{10^3}{10^4}$$

*Solution:* By using the subtraction rule we get

$$\frac{10^3}{10^4} = 10^{3-4} = 10^{-1} = \frac{1}{10^1} = \frac{1}{10} \quad \blacksquare$$

**EXAMPLE 19** Divide

$$\frac{10^3}{10^3}$$

*Solution:* By using the subtraction rule we get

$$\frac{10^3}{10^3} = 10^{3-3} = 10^0$$

The expression $10^0$ may look a bit strange, but if we work it out by dividing the 10's, we see that

$$\frac{10^3}{10^3} = \frac{10 \times 10 \times 10}{10 \times 10 \times 10} = 1 \quad \blacksquare$$

From Example 19, it would seem that

$$10^0 = 1$$

In fact, a basic rule of exponents is that **any number to the zero power equals one.**

Now let's look at some more properties of negative exponents. We show below that we can represent a 10 with a negative exponent as a decimal that is less than 1.

## 5-5 Scientific Notation Using Negative Exponents

Since

$$\frac{1}{10} = 0.1 \quad \text{and} \quad \frac{1}{10} = 10^{-1}$$

it must be true that

$$10^{-1} = 0.1$$

Similarly,

$$\frac{1}{100} = 0.01 = 10^{-2}$$

We can make a table of negative exponents and their decimal equivalents.

| NUMBER | DECIMAL | EXPONENT FORM |
|---|---|---|
| $\frac{1}{10}$ | 0.1 | $10^{-1}$ |
| $\frac{1}{100}$ | 0.01 | $10^{-2}$ |
| $\frac{1}{1000}$ | 0.001 | $10^{-3}$ |
| $\frac{1}{10,000}$ | 0.0001 | $10^{-4}$ |

**EXAMPLE 20** Compute

$$\frac{4 \times 10^{-3}}{2 \times 10^4}$$

*Solution:* Group and divide as we have done before:

$$\frac{4 \times 10^{-3}}{2 \times 10^4} = \frac{4}{2} \times \frac{10^{-3}}{10^4} = 2 \times 10^{-3-4} = 2 \times 10^{-7}$$

Remember that $-3$ "take away" 4 is written as $-3 - 4$. ∎

**EXAMPLE 21** Compute

$$\frac{7 \times 10^2}{3 \times 10^{-4}}$$

**78** SCIENTIFIC NOTATION

*Solution:* After grouping and dividing we have

$$\frac{7 \times 10^2}{3 \times 10^{-4}} = \frac{7}{3} \times \frac{10^2}{10^{-4}} = 2.33 \times 10^{2-(-4)}$$

$$= 2.33 \times 10^{2+4} = 2.33 \times 10^6 \quad \blacksquare$$

Example 21 illustrates *subtraction of a negative number*. Notice that $2 - (-4) = 2 + 4 = 6$, since a "minus times a minus equals a plus" and the expression $-(-4)$ is really just like $-1 \times -4$ or $(-1)(-4)$. The expression $(-1)(-4)$ can be written as $-1(-4)$, where we have left out the first set of parentheses. Now all we have to do is leave off the 1 and we have $-(-4)$, which equals $+4$.

**EXAMPLE 22** Compute

$$\frac{9 \times 10^{-2}}{4 \times 10^{-6}}$$

*Solution:* Again, regroup and then divide:

$$\frac{9 \times 10^{-2}}{4 \times 10^{-6}} = \frac{9}{4} \times \frac{10^{-2}}{10^{-6}} = 2.25 \times 10^{-2-(-6)}$$

$$= 2.25 \times 10^{-2+6} = 2.25 \times 10^4 \quad \blacksquare$$

**EXAMPLE 23** Compute

$$\frac{6.4 \times 10^{-4}}{2}$$

*Solution:* The answer is

$$\frac{6.4 \times 10^{-4}}{2} = \frac{6.4}{2} \times 10^{-4} = 3.2 \times 10^{-4} \quad \blacksquare$$

You might occasionally see someone try to solve Example 23 in the following way:

$$\frac{6.4 \times 10^{-4}}{2} = 3.2 \times 10^{-2} \quad \leftarrow \text{!! \textbf{THIS IS WRONG} !!}$$

They have forgotten one of the rules of multiplication. They have not grouped and multiplied correctly. (See Examples 3 and 4 in this chapter for a review of grouping.) Since division by 2 is like multiplication by $\frac{1}{2}$, we can write

$$\frac{6.4 \times 10^{-4}}{2} = \tfrac{1}{2} \times 6.4 \times 10^{-4}$$

and using the rules we have already learned,

$$\tfrac{1}{2} \times 6.4 \times 10^{-4} = 3.2 \times 10^{-4}$$

**EXAMPLE 24** Compute

$$\frac{9.6 \times 10^{-8}}{10^{-6}}$$

*Solution:*

$$\frac{9.6 \times 10^{-8}}{10^{-6}} = 9.6 \times \frac{10^{-8}}{10^{-6}} = 9.6 \times 10^{-8-(-6)} = 9.6 \times 10^{-2} \quad \blacksquare$$

**EXAMPLE 25** Compute

$$\frac{5}{2 \times 10^{11}}$$

*Solution:*

$$\frac{5 \times 1}{2 \times 10^{11}} = \frac{5}{2} \times \frac{1}{10^{11}} = 2.5 \times 10^{-11}$$

where we have used the rule that $1/10^n = 10^{-n}$. Another way to look at this is to remember that $1 = 10^0$:

$$\frac{1}{10^{11}} = \frac{10^0}{10^{11}} = 10^{0-11} = 10^{-11} \quad \blacksquare$$

We will now discuss how to change numbers less than 1 from regular notation to scientific notation. We will also discuss the reverse procedure, namely, changing numbers less than 1 from scientific notation to regular notation.

**EXAMPLE 26** Convert 0.34 into scientific notation.

80  SCIENTIFIC NOTATION

*Solution:* To do this, we will use the fact that $10/10 = 1$ and that multiplication by 1 doesn't change a number:

$$0.34 = 0.34 \times \tfrac{10}{10} = 0.34 \times 10 \times \tfrac{1}{10} = 0.34 \times 10 \times 10^{-1} = 3.4 \times 10^{-1}$$ ∎

Once you understand what you are doing, you can simply move the decimal point. In Example 26, you would move the decimal point one place to the right and then multiply by $10^{-1}$.

**EXAMPLE 27**  Convert 0.00076 to scientific notation.

*Solution:* We will do this problem by moving the decimal point. We need to move the decimal point four places to the right to get 7.6. At the same time we have to multiply by $10^{-4}$, since moving the decimal point four places to the right has multiplied the number by $10^4$:

$$0.0\,0\,0\,7\,6 = 7.6 \times 10^{-4}$$

Notice that the negative exponent is the same number as the number of places (arrows) that we moved the decimal point to the *right*. ∎

**EXAMPLE 28**  Convert $6.47 \times 10^{-3}$ to regular notation.

*Solution:* We will do this in two ways:

first way:    $6.47 \times 10^{-3} = 6.47 \times 0.001 = 0.00647$
second way:   $6.47 \times 10^{-3} = 0\,0\,0\,6.4\,7 = 0.00647$

In the second way, we moved the decimal point three places to the *left*. ∎

We can summarize the rules for converting numbers from regular notation to scientific notation and vice versa, when the numbers are *less* than 1.

1. To convert from regular notation to scientific notation, move the decimal point to the *right*.
2. To convert from scientific notation to regular notation, move the decimal point to the *left*.

# 5-6
## USING A CALCULATOR IN SCIENTIFIC NOTATION

Although it is important that you know how to perform calculations without a calculator, using a calculator can certainly save you a great amount of work. To do calculations in this section, you will need a full scientific calculator. We will assume that you have a calculator that uses algebraic logic, not Reverse Polish Notation (RPN, in calculators made by Hewlett–Packard). If you have one of the RPN calculators, see your instruction manual.

To enter the exponent part of a number in scientific notation, calculators have either an **EE** or an **exp** button. The following instructions will use the **EE** notation. Let's look at how you would enter some of the examples from this chapter into a calculator. Refer to previous sections of this chapter for the numerical answers. Each of the following examples refers to the same example number previously discussed in this chapter.

*Example 1.* $9 \times 10^{16}$ is entered by punching the buttons **9, EE, 16**. Notice that a "times sign" is *not* punched between the number part and the exponent part, even though there is one when you write the number. The exponent part is shown on the right side of the display, which will look something like this: $\underline{\quad 9. \quad\quad 16}$ The decimal point is entered automatically.

*Example 12.* $10^8 \times 10^{11}$. Entry: **1, EE, 8, ×, 1, EE, 11, =**. You may have to put a **1** before the **EE**. Some calculators put it in for you, whereas others consider **EE, 8** as if you had entered **0, EE, 8**, which of course equals zero. You might check yours.

*Example 13.* $6 \times 10^4 \times 9 \times 10^{27}$. Entry: **6, EE, 4, ×, 9, EE, 27, =**.

*Example 14.* $(9 \times 10^{15})(8 \times 10^{20})$. Entry: **9, EE, 15, ×, 8, EE, 20, =**.

*Example 15.* $10^8/10^3$. Entry: **1, EE, 8, ÷, 1, EE, 3, =**.

*Example 16.* $(6 \times 10^7)/(4 \times 10^2)$. Entry: **6, EE, 7, ÷, 4, EE, 2, =**.

**NOTE:** The "/" in Example 16 means the same thing as the "÷" sign. It is the line separating the numerator and the denominator.

*Example 17.*

$$\frac{(8 \times 10^3)(3 \times 10^7)}{(6 \times 10^2)(2 \times 10^4)}$$

Entry: **8, EE, 3, ×, 3, EE, 7, ÷, 6, EE, 2, ÷, 2, EE, 4, =**. Notice that in this example you need two ÷ signs. The second term in the denominator *divides* the fraction. Don't make the mistake of punching in a × sign.

*Example 18.* $10^3/10^4$. Entry: **1, EE, 3, ÷, 1, EE, 4, =**.

*Example 19.* $10^3/10^3$. Entry: **1, EE, 3, ÷, 1, EE, 3, =**.

*Example 20.* $(4 \times 10^{-3})/(2 \times 10^4)$. Entry: **4, EE, +/−, 3, ÷, 2, EE, 4, =**.

Notice that the **+/−** button is used to enter a negative exponent. On most calculators, the following entries would both give $4 \times 10^{-3}$: **4, EE,**

82  SCIENTIFIC NOTATION

$+/-$, **3** or **4**, **EE**, **3**, $+/-$. On a few calculators, only the second choice works.

*Example 21.* $(7 \times 10^2)/(3 \times 10^{-4})$. Entry: **7, EE, 2, ÷, 3, EE, +/−, 4, =**.

*Example 22.* $(9 \times 10^{-2})/(4 \times 10^{-6})$. Entry: **9, EE, +/−, 2, ÷, 4, EE, +/−, 6, =**.

*Example 23.* $(6.4 \times 10^{-4})/2$. Entry: **6.4, EE, +/−, 4, ÷, 2, =**.

*Example 24.* $(9.6 \times 10^{-8})/10^{-6}$. Entry: **9.6, EE, +/−, 8, ÷, 1, EE, +/−, 6, =**.

*Example 25.* $5/(2 \times 10^{11})$. Entry: **5, ÷, 2, EE, 11, =**.

Notice that the calculator always displays an answer in scientific notation if the number is too big or too small to fit in the display. On some calculators, an answer will always be displayed in scientific notation if the numbers are entered in scientific notation. On others, the answer will be in regular notation if it will fit in the display.

Most calculators have the ability to convert numbers in the display back and forth between regular notation and scientific notation. The following are two ways they do this (your calculator may do one or the other or neither).

| BUTTON(S) | WHAT THEY CONVERT |
|---|---|
| F ↔ E | Regular notation to scientific notation |
| F ↔ E | Scientific notation to regular notation |
| INV,EE,= | Scientific notation to regular notation |
| EE,= | Regular notation to scientific notation |

# 5-7
## POWERS AND ROOTS OF NUMBERS

The **square root** of a number is another number that, when multiplied by itself, gives the original number. Thus 2 is the square root of 4 because $2 \times 2 = 4$.

The expression $\sqrt{4}$ means the square root of 4. Another way of expressing the square root of 4 is to write

$$\sqrt{4} = 4^{1/2}$$

where the $\frac{1}{2}$ is a superscript on the 4. Since $\sqrt{4} = 2$, it must be that $4^{1/2} = 2$. You can convince yourself that $4^{1/2}$ is a sensible way to write the square root of 4 by looking at the following:

$$2 \times 2 = 4^{1/2} \times 4^{1/2} = 4^{1/2+1/2} = 4^1 = 4$$

Since $4^{1/2} \times 4^{1/2} = 4$, $4^{1/2}$ must be the square root of 4 from our definition of the square root. We have just used the rule that you add exponents when you multiply numbers.

How do we simplify an expression like $(9 \times 4)^{1/2}$? Let's multiply and see:

$$(9 \times 4)^{1/2} = 36^{1/2} = 6$$

But a rule of math says that we can also write

$$(9 \times 4)^{1/2} = 9^{1/2} \times 4^{1/2} = 3 \times 2 = 6$$

Since the answer is the same whether we multiply first and then take the square root, or take the square root of each term and then multiply, this rule seems reasonable. In general, for any numbers $a$ and $b$ and exponent $1/n$, we obtain

$$(a \times b)^{1/n} = a^{1/n} \times b^{1/n}$$

If you want to raise $a \times b$ to a power, say $m$, then

$$(a \times b)^m = a^m \times b^m$$

**EXAMPLE 29**  Calculate $(25 \times 4)^{1/2}$.

*Solution:*

$$(25 \times 4)^{1/2} = 25^{1/2} \times 4^{1/2} = 5 \times 2 = 10$$

or

$$(25 \times 4)^{1/2} = (100)^{1/2} = 10 \quad \blacksquare$$

**EXAMPLE 30**  Calculate $(4 \times 5)^2$.

*Solution:*

$$(4 \times 5)^2 = 4^2 \times 5^2 = 16 \times 25 = 400$$

or

$$(4 \times 5)^2 = (20)^2 = 20 \times 20 = 400 \quad \blacksquare$$

84   SCIENTIFIC NOTATION

**EXAMPLE 31**   Using a calculator, find $7^{1/2}$.

*Solution:* Press the following keys: **7, $\sqrt{\ }$** or **7, INV, $x^2$**. The display will read 2.6457513. Notice that the sequence **INV, $x^2$** is the same as square root. **INV** stands for "inverse." The reason the button is called "inverse" is that taking a square root and squaring are "opposite" operations. Addition and subtraction are also inverse or "opposite" operations, as are multiplication and division. ∎

Any whole number, fraction, or decimal can be used as an exponent. For instance, what does $27^{1/3}$ mean? The expression $27^{1/3}$ is a number that, when cubed (**cubed** means multiplied by itself three times), gives 27. The expression $27^{1/3}$ is read "27 to the one-third" or "the cube root of 27." The **cube root** of a number is another number that, when multiplied by itself three times, gives the original number. To understand it, look at the following:

$$27^{1/3} \times 27^{1/3} \times 27^{1/3} = 27^{1/3+1/3+1/3} = 27^1 = 27$$

What is the numerical value of $27^{1/3}$? It is 3. This is so because

$$3 \times 3 \times 3 = 27$$

**EXAMPLE 32**   What is $8^{2/3}$?

*Solution:*

$$8^{2/3} = 8^{1/3 \times 2} = (8^{1/3})^2 = 8^{1/3} \times 8^{1/3} = 2 \times 2 = 4 \quad \blacksquare$$

Only simple cases, where things "work out," can be done in your head. The more complicated cases require a calculator.

**EXAMPLE 33**   Compute $5^{2/3}$ on a calculator.

*Solution:* $2/3 = 0.66666\ldots$ . Press **5, $y^x$, 0.6666 . . . , =**. The display reads 2.92402. If your calculator has parentheses buttons, you can use them in the following way. Press **5, $y^x$, (, 2, ÷, 3, ), =**. The notation $0.6666\ldots$ means that the 6's repeat indefinitely. The author entered seven 6's into his calculator for this problem. Since you cannot enter fractions as exponents in a calculator, you must convert the fraction to its decimal equivalent or use parentheses. ∎

**NOTE:** On some calculators, the $y^x$ key may read $x^y$, $a^x$, or $a^y$. ∎

**EXAMPLE 34** Compute $8^{4.67}$ on a calculator.

*Solution:* Press **8, y$^x$, 4.67, =.** The display will read 16498. ∎

Even negative powers and roots can easily be handled on a calculator.

**EXAMPLE 35** Compute $26^{-0.024}$ on a calculator.

*Solution:* Press **26, y$^x$, 0.024, +/−, =.** The display reads 0.924785. ∎

Notice that taking roots of *negative numbers* gives an error message. Try taking $(-2)^{1/2}$ on your calculator. It will not work because the square root of a negative number, when multiplied by itself, would have to give the number. Since a "minus times a minus gives a plus," you cannot find two identical negative numbers that, when multiplied, give a negative number. For this reason, the square roots of negative numbers are called **imaginary numbers.**

You can easily take powers and roots of numbers in scientific notation. Let's look at $(8 \times 10^4)^2$. Clearing parentheses gives

$$8^2 \times (10^4)^2$$

What does $(10^4)^2$ mean? Working it out we have

$$(10^4)^2 = 10^4 \times 10^4 = 10^8$$

so

$$(8 \times 10^4)^2 = 8^2 \times (10^4)^2 = 64 \times 10^8$$

In general, for any exponents $m$ and $n$ we have

$$(10^m)^n = 10^{m \times n} \text{ or } 10^{mn}$$

and

$$(10^m)^{1/n} = 10^{m/n}$$

**EXAMPLE 36** Compute $(10^6)^3$.

*Solution:*

$$(10^6)^3 = 10^{6 \times 3} = 10^{18}$$ ∎

**EXAMPLE 37** Compute $(10^{10})^{1/2}$.

*Solution:*

$$(10^{10})^{1/2} = 10^{10/2} = 10^5 \quad \blacksquare$$

**EXAMPLE 38** Compute $(9 \times 10^8)^{1/2}$.

*Solution:*

$$(9 \times 10^8)^{1/2} = 9^{1/2} \times 10^{8 \times 1/2} = 9^{1/2} \times 10^{8/2} = 3 \times 10^4 \quad \blacksquare$$

**EXAMPLE 39** Compute $(8 \times 10^{15})^{1/3}$.

*Solution:*

$$(8 \times 10^{15})^{1/3} = 8^{1/3} \times 10^{15 \times 1/3} = 8^{1/3} \times 10^{15/3} = 2 \times 10^5 \quad \blacksquare$$

Sometimes you must "adjust" the exponent of the number so that the answer does not have a fractional exponent in it. This would only be necessary if you had a calculator that didn't have scientific notation capabilities.

**EXAMPLE 40** Compute $(4.6 \times 10^7)^{1/2}$.

*Solution:* Since 7 is not exactly divisible by 2, we can write

$$(4.6 \times 10^7)^{1/2} = (46 \times 10^6)^{1/2} = 46^{1/2} \times 10^{6 \times 1/2} = 6.8 \times 10^3$$

where we have multiplied 4.6 by 10 and divided $10^7$ by 10. The value of the number hasn't changed. By this procedure, we avoid fractional exponents in the answer. We found $46^{1/2}$ by using a calculator. $\quad \blacksquare$

**EXAMPLE 41** Compute $(9.3 \times 10^{-5})^{1/2}$.

*Solution:*

$$(9.3 \times 10^{-5})^{1/2} = (93 \times 10^{-6})^{1/2} = 93^{1/2} \times 10^{-6 \times 1/2}$$
$$= 93^{1/2} \times 10^{-6/2} = 9.6 \times 10^{-3} \quad \blacksquare$$

**EXAMPLE 42** Compute $(10^9)^{1/4}$.

*Solution:*

$$(10^9)^{1/4} = (10 \times 10^8)^{1/4} = 10^{1/4} \times 10^{8 \times 1/4}$$
$$= 10^{1/4} \times 10^{8/4} = 10^{0.25} \times 10^{8/4} = 1.8 \times 10^2 \quad\blacksquare$$

As mentioned before, if you use a scientific calculator to find powers and roots of numbers, you don't have to adjust the exponent. The calculator will do all the work for you. However, for simple cases, the procedure outlined in Examples 36 through 42 works very well. (Of course, we used a calculator to find $46^{1/2}$, $93^{1/2}$, and $10^{0.25}$.)

**EXAMPLE 43** Compute $(6.62 \times 10^{11})^{1/3}$.

*Solution:* Press **6.62**, **EE**, **11**, **y$^x$**, **0.3333** . . . , **=** or **6.62**, **EE**, **11**, **INV**, **y$^x$**, **3**, **=** or **6.62**, **EE**, **11**, **y$^x$**, **(**, **1**, **÷**, **3**, **)**, **=**. The display will read 8.7154 03, which in handwritten scientific notation would be $8.7154 \times 10^3$. The **INV**, **y$^x$** operation is the same as the **y$^{1/x}$** or **$\sqrt[x]{y}$** button found on some calculators. If you want to take $4^{1/3}$, for example, you can do it in up to four ways, depending on which calculator you have. Push: **4**, **y$^x$**, **0.333** . . . , **=** or **4**, **INV**, **y$^x$**, **3**, **=** or **4**, **y$^{1/x}$**, **3**, **=** or **4**, **y$^x$**, **(**, **1**, **÷**, **3**, **)**, **=**. The answer is 1.5874. $\quad\blacksquare$

**EXAMPLE 44** Compute $(4.58 \times 10^{-22})^{3.62}$.

*Solution:* Press **4.58**, **EE**, **+/−**, **22**, **y$^x$**, **3.62**, **=**. The display reads 5.6537 −78 or $5.65 \times 10^{-78}$, where we have rounded to three digits. $\quad\blacksquare$

One final word about using calculators. It is important that you know how to do simple calculations without a calculator. After all, your calculator may malfunction at a critical time. In addition, you should have a feeling for what a reasonable answer is when using a calculator. Otherwise, you have no chance at all to spot a wrong answer caused by pushing the wrong buttons. There is an old rule that computer programmers use that applies to calculators also. It is the **GIGO** rule:

Garbage In, Garbage Out.

In other words, don't turn off your brain when you turn on your calculator.

## PROBLEMS

### KEYED PROBLEMS

1. Write 50,000,000 as a 5 multiplied by a 10 with some exponent.
2. Write the product $2 \times 5 \times 10$ in all the six possible ways that the numbers can be mixed up. Calculate the answer for each way.
3. In the multiplication $6 \times 7 \times 10$ put parentheses in reasonable places and multiply.
4. Write the multiplication $\frac{1}{4} \times 16 \times 5$ with parentheses in reasonable places and multiply.
5. Multiply the following after rearranging as was done in Example 5:

$$20 \times \tfrac{1}{4}$$

6. Convert 38,423 to scientific notation.
7. Convert 25.6 to scientific notation.
8. Convert 7,360,000 to scientific notation.
9. Convert $8.2 \times 10^5$ to regular notation.
10. Convert $5.72 \times 10^8$ to regular notation.
11. Convert $58.85 \times 10^7$ to regular notation.
12. Multiply $10^6 \times 10^{13}$.
13. Multiply $8 \times 10^5$ by $3 \times 10^{15}$.
14. Multiply the following and put the answer into scientific notation: $(5 \times 10^{12})(7 \times 10^6)$.
15. Perform the following division:

$$\frac{10^7}{10^5}$$

16. Compute

$$\frac{4 \times 10^8}{3 \times 10^5}$$

17. Compute

$$\frac{(5 \times 10^5)(8 \times 10^8)}{(9 \times 10^2)(3 \times 10^4)}$$

Problems 89

18. Divide

    $$\frac{10^7}{10^8}$$

19. Divide

    $$\frac{10^9}{10^9}$$

20. Compute

    $$\frac{8 \times 10^{-6}}{6 \times 10^5}$$

21. Compute

    $$\frac{4 \times 10^8}{3 \times 10^{-7}}$$

22. Compute

    $$\frac{7 \times 10^{-4}}{3 \times 10^{-9}}$$

23. Compute

    $$\frac{9.3 \times 10^{-15}}{3}$$

24. Compute

    $$\frac{8.34 \times 10^{-12}}{10^{-9}}$$

25. Compute

    $$\frac{9}{4 \times 10^{15}}$$

26. Convert 0.72 into scientific notation.
27. Convert 0.00048 into scientific notation.
28. Convert $4.25 \times 10^{-3}$ into regular notation.
29. Calculate $(3 \times 27)^{1/2}$.

30. Calculate $(8 \times 2)^2$.
31. Using a calculator, find $5^{1/2}$.
32. What is $27^{2/3}$?
33. Compute $7^{2/3}$ on a calculator.
34. Compute $5^{6.71}$ on a calculator.
35. Compute $17^{-0.055}$ on a calculator.
36. Compute $(10^4)^5$.
37. Compute $(10^{16})^{1/4}$.
38. Compute $(25 \times 10^{12})^{1/2}$.
39. Compute $(64 \times 10^{21})^{1/3}$.
40. Compute $(7.2 \times 10^9)^{1/2}$.
41. Compute $(3.6 \times 10^{-7})^{1/2}$.
42. Compute $(10^{15})^{1/4}$.
43. Compute $(5.73 \times 10^{15})^{1/5}$.
44. Compute $(7.47 \times 10^{-16})^{5.41}$.

## SUPPLEMENTAL PROBLEMS

45. Convert the following numbers to scientific notation.
    a. 3264
    b. 582
    c. 0.043
    d. 0.000572
    e. 4,670,000
    f. 0.000009
    g. 6,000,000,000
    h. 7001

46. Convert the following numbers to regular notation.
    a. $3.7 \times 10^2$
    b. $4.89 \times 10^7$
    c. $5.1 \times 10^{-2}$
    d. $8.92 \times 10^{-9}$
    e. $5.117 \times 10^5$
    f. $32.4 \times 10^3$
    g. $0.01 \times 10^{-2}$
    h. $0.32 \times 10^4$

47. Compute the following. Leave the answers in scientific notation.
    a. $(4.2 \times 10^2)(3.6 \times 10^8)$
    b. $(8 \times 10^{15})(6 \times 10^{23})$
    c. $(5.3 \times 10^{-2})(6 \times 10^5)$
    d. $(3.1 \times 10^{-5})(2 \times 10^{-10})$
    e. $(4.9 \times 10^6)(8 \times 10^{-12})$
    f. $(3 \times 10^{-10})(4)$

48. Compute the following. Leave the answers in scientific notation.
    a. $\dfrac{8 \times 10^5}{4 \times 10^2}$
    b. $\dfrac{6 \times 10^{-2}}{4 \times 10^7}$
    c. $\dfrac{4.7 \times 10^{-12}}{8.2 \times 10^{-15}}$
    d. $\dfrac{7.43 \times 10^{10}}{2 \times 10^{-4}}$
    e. $\dfrac{3.2}{4 \times 10^3}$
    f. $\dfrac{2.7 \times 10^{-2}}{4}$

49. Compute

$$\dfrac{(4.07 \times 10^{-8})(3.26 \times 10^{-5})}{8.99 \times 10^{-7}}$$

50. Compute

$$\frac{5.88 \times 10^5}{(3.16 \times 10^7)(7.02 \times 10^{-6})}$$

51. Compute

$$\frac{(3.27 \times 10^4)(8.53 \times 10^7)}{(5.55 \times 10^8)(7.76 \times 10^{-5})}$$

*For Problems 52 through 54:* To work these out without a calculator, convert all terms to numbers with the same exponent. Then you can add them. Example: $3 \times 10^3 + 2 \times 10^5 = 3 \times 10^3 + 200 \times 10^3 = 203 \times 10^3 = 2.03 \times 10^5$. Another way to work this example is $3 \times 10^3 + 2 \times 10^5 = 0.03 \times 10^5 + 2 \times 10^5 = 2.03 \times 10^5$.

If one number is much smaller than the other number in addition or subtraction, you can usually ignore the smaller one. For example, $3 \times 10^5 + 2 \times 10^{-4} = 3 \times 10^5$. We have ignored the $2 \times 10^{-4}$ because it is so much smaller than $3 \times 10^5$. Let's write the numbers in regular notation to see this more clearly. Since $3 \times 10^5 = 300,000$ and $2 \times 10^{-4} = 0.0002$, when we add them we get $300,000 + 0.0002 = 300,000.0002$. Certainly we can ignore the 0002 after the decimal point. An analogy may be helpful in explaining this. If you are weighing an elephant and a flea lands on his ear, do you think the weight you get would be any different? I don't think so either!

52. Compute $7.61 \times 10^4 + 9.23 \times 10^5 + 4.61 \times 10^{-3} + 0.712$.
53. Compute $2.21 \times 10^{-5} - 8.90 \times 10^{-6}$.
54. Compute $4.88 \times 10^{-7} - 3.22 \times 10^{-2} + 5.66 \times 10^8$.
55. Use a calculator to find the following powers and roots:
    a. $(4.26)^{4/5}$
    b. $(8.99)^{1.26}$
    c. $(6.25)^{-0.011}$
    d. $(5.42 \times 10^{-11})^{-4.45}$
    e. $(4.77 \times 10^9)^{0.76}$
    f. $(3.2 \times 10^{-4})^{-0.015}$

**FUN WITH LARGE NUMBERS**

It is interesting to see how many atoms are in objects ranging in size from the human body to the entire known universe. **NOTE:** The isotope with the longest measured half-life is xenon-124 at $1.8 \times 10^{22}$ years. See https://bit.ly/3s4gStP.

| | |
|---|---|
| Atoms in the human body // Atoms in the earth | $10^{28}$ // $10^{50}$ |
| Atoms in the sun // Stars in our milky way galaxy | $10^{57}$ // $2 \times 10^{11}$ |
| Atoms in our milky way galaxy | $2.4 \times 10^{67}$ |
| Galaxies in known universe // Stars in the known universe | $3 \times 10^{11}$ // $10^{24}$ |
| Atoms in the known universe | $10^{80}$ |
| An upper estimate of the number of possible molecules* | $10^{180}$ |

*See references in the article "Artificial Intelligence leaps into chemical space," *Chemical & Engineering News*, April 6, 2020, page 30.

**Significant Figures – Accuracy and Precision:** As you go from top to bottom, the number of **significant figures** with which the time can be read increases. In other words, the **precision** with which you can read the time increases. But you don't know how **accurate** each watch is from just looking at it. You would need to **calibrate** each watch to increase its **accuracy**. You can do this by setting it to the very accurate time on your computer or tablet (if they are connected to the internet) or smartphone. After reading this chapter, you'll understand **accuracy** and **precision**.

*Photos are in the Public Domain.*

# 6

# SIGNIFICANT FIGURES

In science classes, students frequently measure objects and then divide numbers like 2.51 inches and 6.82 inches. If they use a calculator, they might get (depending on the calculator they used)

$$\frac{2.51}{6.82} = 0.3680352$$

Notice that 0.3680352 seems to have too many digits or figures to be reported as an answer, since the actual measurements only had 3 digits each. Why is it wrong to report the answer to 7 digits? This chapter discusses the way to determine how many digits you should report in your answer.

Reporting the proper number of digits after a calculation is one way that the reader knows how many digits with which you could "measure the object." In other words, this helps tell the reader how good your equipment is and how good an experimenter you are.

## 6-1
### THE PERCENT UNCERTAINTY OF A MEASUREMENT IS THE KEY TO UNDERSTANDING SIGNIFICANT FIGURES

Now back to our division. The numbers 2.51 and 6.82 were derived from an experiment—say, inches measured with a ruler graduated in tenths of an inch (see Figure 6-1). The students had to estimate the hundredths place, so 2.51 and 6.82 are not known too well, maybe to within ±1 in their last digit. (Read ±1 as "plus or minus one.") The last digit, the one in the hundredths place, is an estimated digit (see Figure 6-2). So 2.51 and 6.82 are each known to within ±0.01. In other words, they are each known to within ±1 in their last digit. The 2.51 could be anywhere between 2.52 and 2.50, and 6.82 could be between 6.83 and 6.81.

So how do we decide how many digits to report in the answer? A good way is to find out what percent the error is of the measurement. In our case, ±0.01 is the error and 2.51 and 6.82 are the measurements.

Our two measurements are thus good to within about

$$\frac{0.01}{2.51} \times 100 = 0.4\% \quad \text{and} \quad \frac{0.01}{6.82} \times 100 = 0.1\%$$

**NOTE:** The "±" can be taken as just "+" for the percent error since the negative error is the same as the positive error.

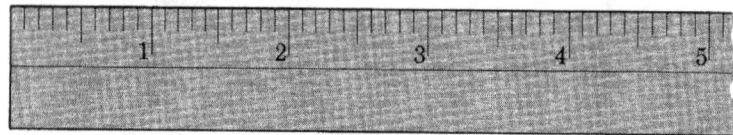

**FIGURE 6-1** A ruler graduated in tenths of an inch.

(See Chapters 2 and 10 for discussions on percent.)

In these calculations, the 0.01 is the ±1 in the hundredths place, and a number with a percent sign after it is called the **percent uncertainty**. The division of the two numbers, 2.51 and 6.82, should be reported to no more than about 0.4%, which is the largest percent uncertainty of our two ruler measurements. Taking 0.4% of 0.3680352, we get 0.3680352 × 0.004 = 0.001. This means that 0.368052 should be reported to the nearest 0.001, or 0.368. The 8 in the thousandths place is known to within ±1, and our answer is in line with our measurements. You could express the answer as 0.368 ± 0.001, showing that the 8 is uncertain by ±1. Even if you don't write a ± after a number, people will assume that your number is good to within ±1 in the last place that you have written.

A good rule of thumb when dividing or multiplying two or more numbers is as follows. The percent uncertainty of the answer should be about the same as the largest percent uncertainty in any of the numbers that make up the problem.

**EXAMPLE 1** You want to divide the measurement 6.3 feet by 2.7 feet. Each number is known to within ±1 in the tenths place. What should your answer be, to fall within the proper percent uncertainty?

*Solution:* On a calculator 6.3/2.7 = 2.3333333. The percent uncertainty of each number is 0.1/6.3 × 100 = 1.6% and 0.1/2.7 × 100 = 3.7%. The larger of the two is 3.7%. Taking 3.7% of 2.3333333, we get 2.3333333 × 0.037 = 0.086 or about 0.1. We should therefore report the answer is 2.3. If you wanted to be extra careful, you could report your answer as 2.3 ± 0.1, which specifically says that there is an uncertainty of ±1 in the tenths place. ∎

**EXAMPLE 2** You multiply 4.68 × 7.36 on a calculator and get 34.4448. How should the answer be reported if 4.68 and 7.36 are good to within ±2 in the hundredths place?

*Solution:* Take percent uncertainties of each number:

$$\frac{0.02}{4.68} \times 100 = 0.43\% \qquad \frac{0.02}{7.36} \times 100 = 0.27\%$$

Now take 0.43% (the largest percent uncertainty) of 34.4448:

$$0.0043 \times 34.4448 = 0.15$$

The answer is good to within values between $\pm 1$ and $\pm 2$ in the tenths place (or you could say the answer is good to within values between 0.10 and 0.20, since 0.15 is between these two numbers). To be safe from overreporting your answer, you should use the larger uncertainty and report your answer as 34.4 $\pm$ 0.2. If you were to write the answer as 34.4, the reader would assume you meant 34.4 $\pm$ 0.1, which is a bit misleading about the uncertainty of the answer. ■

As you can see, this procedure is a bit cumbersome. Fortunately, there is a shortcut that works most of the time if all the numbers in the problem are good to within $\pm 1$ in the last place: **When you are doing a multiplication or division, the number of digits reported in your answer should be the same as the smallest number of digits in any of the numbers of the problem.** Example 3 shows how easy the shortcut is to use.

**EXAMPLE 3**  In multiplying 4.73 × 14.47, what is a reasonable answer containing the proper number of digits?

*Solution:* 4.73 × 14.47 = 68.4431 on a calculator. The 4.73 has three digits, and the 14.47 has four. The answer should have three digits and should be reported as 68.4. ■

## 6-2
### RECOGNIZING SIGNIFICANT FIGURES IS EASIEST WHEN NUMBERS ARE EXPRESSED IN SCIENTIFIC NOTATION

The **significant figures** of a number are the digits that are known with certainty *plus* one digit that is uncertain. In Example 1 the numbers 6.3 and 2.7 each have two significant figures. The 6 and 2 are known for sure and the 3 and 7 are in doubt. That's why we said that each number is known to within $\pm 0.1$. The answer, 2.3, also has two significant figures. The 2 is known for sure, and the 3 is doubtful.

In Example 2 the numbers 4.68 and 7.36 each have three significant figures; their product, 34.4, is also expressed with three significant figures.

There is a problem in recognizing what digits are significant when the number contains zeros.

1. *Zeros in the middle of the number.* All zeros in the middle of a number are significant. An example would be 307.02. This number has five significant figures.

2. ***Zeros after a decimal point.*** An example is 32.00. The zeros are significant, and the number 32.00 has four significant figures.
3. ***Zeros before the number.*** An example is 0.00246. This number has three significant figures. The zero before the decimal point is for clarity only and tells you that a decimal point follows—we could have written the number as .00246. The two 0's between the decimal point and the 2 locate the decimal point. Zeros before a number that are used to locate the decimal point are not significant. A better way to see this is to convert 0.00246 into scientific notation:

$$0.00246 = 2.46 \times 10^{-3}$$

Now you can see that the number has only three significant figures. The best way to figure out significant figures is to convert numbers to scientific notation.

4. ***Zeros after the number but before the decimal point.*** An example would be 32,000. Now we have a problem in figuring out the significant figures. If we convert the number to scientific notation, we get four possibilities:

| number: | $3.2000 \times 10^4$ | $3.200 \times 10^4$ | $3.20 \times 10^4$ | $3.2 \times 10^4$ |
|---|---|---|---|---|
| significant figures: | 5 | 4 | 3 | 2 |

Which is correct? It depends on our original number. Were three zeros really significant? Was the measurement that obtained it good to within $\pm 1$ in 32,000? In other words, is only the last zero in doubt, and are all the others known for sure? If the answer is yes, then the percent uncertainty is

$$\frac{1}{32{,}000} \times 100 = 0.003\%$$

and in scientific notation we should write $3.2000 \times 10^4$. However, maybe the measurement was good to within only $\pm 1000$ in 32,000 or

$$\frac{1000}{32{,}000} \times 100 = 3.1\%$$

Then we should write the number as $3.2 \times 10^4$, since the 2 is doubtful. The measurement could have been good to within some other value—say, to within 100 or 10. Then we would write $3.20 \times 10^4$ or $3.200 \times 10^4$. So how we write the number depends on how good our measurement is. When you write a number in scientific notation, people assume that all the digits in the **mantissa** part (the part of the number before the

## 6-2 Recognizing Significant Figures

"×" sign) are significant. This is not always true if you write a number in regular notation. In regular notation you must somehow indicate which digits are significant when there is an ambiguity.

**EXAMPLE 4** How many significant figures are in each of the following numbers? 400.01; 376.10; 0.00003317; 7,8̄00 (the zero with the bar over it is doubtful).

*Solution:* 400.01 has five significant figures.

376.10 has five significant figures.

$0.00003317 = 3.317 \times 10^{-5}$ has four significant figures.

$7.8\bar{0}0 = 7.80 \times 10^3$ has three significant figures (remember, a doubtful digit is significant). ■

You could occasionally make a mistake in determining significant figures by using the shortcut in which you report in your answer the smallest number of significant figures that were in the problem. The example below illustrates this.

**EXAMPLE 5** In the division 11.2/9.9, how many significant figures should there be in the answer?

*Solution:* 11.2/9.9 = 1.131313 on a calculator. Using our shortcut, we would report the answer as 1.1, since 9.9 has two significant figures. Let's calculate the percent uncertainty of each number in the problem:

$$\frac{0.1}{11.2} \times 100 = 0.89\% \qquad \frac{0.1}{9.9} \times 100 = 1.01\%$$

The answer, 1.1, has a percent uncertainty of

$$\frac{0.1}{1.1} \times 100 = 9.1\%$$

The data in the problem are better than those in the answer, so we are justified in reporting the answer as having about a 1% error. To do this we can add another digit and report the answer as

$$\frac{11.2}{9.9} = 1.13$$

The percent uncertainty is

$$\frac{0.01}{1.13} \times 100 = 0.88\%$$

This is very close to the largest percent uncertainty of 1.01%, and we can safely report three significant figures in our answer. ∎

To understand what happened in Example 5, compare 99 and 100. The percent uncertainties of these numbers are

$$\frac{1}{99} \times 100 = 1.01\% \quad \text{and} \quad \frac{1}{100} \times 100 = 1.00\%$$

The percent uncertainties are about the same, but 99 has two significant figures and 100 has three. What this discussion means, then, is that numbers like 99 and 98 can be considered as having three significant figures in calculations. And numbers between 980 and 999 can be considered as having four significant figures in calculations. Can you show that this is true?

# 6-3
## ACCURACY AND PRECISION TELL US HOW GOOD OUR MEASUREMENTS ARE

At this point we will discuss the factors that affect the percent uncertainty of a measurement.

When you make a measurement, there are two main things that determine how close your reading comes to the actual or true value. They are (1) to what uncertainty you can read the instrument; (2) how close the instrument reading is to the true value.

The uncertainty with which you can read the instrument is called the **precision** of the measurement. If you make a series of measurements with the same instrument, the agreement between the numbers is an indication of the precision of the measurement. This means that if you consistently read an instrument carefully, your measurements will have a high precision.

Another way to look at precision is to consider the number of significant figures that you can read off an instrument. Assuming that you know how to read it, a ruler graduated in hundredths of an inch will give more significant figures than a ruler graduated in tenths of an inch. The rod in Figure 6-2 was measured to be 2.51 inches using a ruler graduated in tenths of an inch. If we had used a ruler graduated in hundredths of an inch, our rod might have been measured as being 2.514 inches. Clearly, 2.514 has more

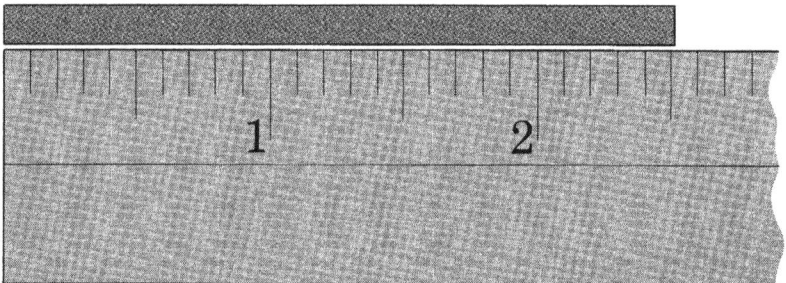

**FIGURE 6-2** A magnified view of our ruler measuring a rod that is 2.51 inches long. Notice that the hundredth place is estimated to be "1" but could be a bit more or less, depending on how good you are at reading rulers.

significant figures than does 2.51. Since 2.514 has a smaller percent uncertainty than does 2.51, the measurement 2.514 inches is more precise. Thus, the more significant figures an instrument can give, the more precise the instrument. This assumes, of course, that you know how to read the instrument and that it is working properly. We might note here that instruments capable of high precision tend to be more expensive than lower-precision instruments.

How close the reading is to the true value is called the **accuracy** of the measurement. Figure 6-3 gives two pictorial representations of accuracy and precision for a series of measurements.

Now let's look at a problem you might have if your instrument is defective. Consider again the ruler discussed at the beginning of this chapter so that you can measure the rod that is 2.51 inches long. The percent uncertainty of your reading is $0.01/2.51 \times 100 = 0.4\%$. This is the **precision** of your measurement. Now suppose your ruler is warped (bent) and the rulings on it are closer together than they should be. Then you will read more of them in measuring the rod. Possibly the actual length of the rod is 2.45 inches, and because of the warping you read 2.51 inches. Your reading is off by 2.51 inches − 2.45 inches = 0.06 inches. The percent uncertainty is $0.06/2.51 \times 100 = 2.4\%$. This is the accuracy of your measurement. Thus you can be a very good ruler reader (high precision, 0.4%) and still get bad results (low accuracy, 2.4%) if you use a bad ruler.

The "bad ruler" problem is recognized by scientists who make every effort to calibrate their equipment. To calibrate our ruler, we would get a rod of known length (say, from the National Institute of Standards and Technology) and measure it with our ruler. We could then compensate for the warp and improve the accuracy of our measurements. The precision would remain the same. A better solution, of course, is to get a good ruler.

100 SIGNIFICANT FIGURES

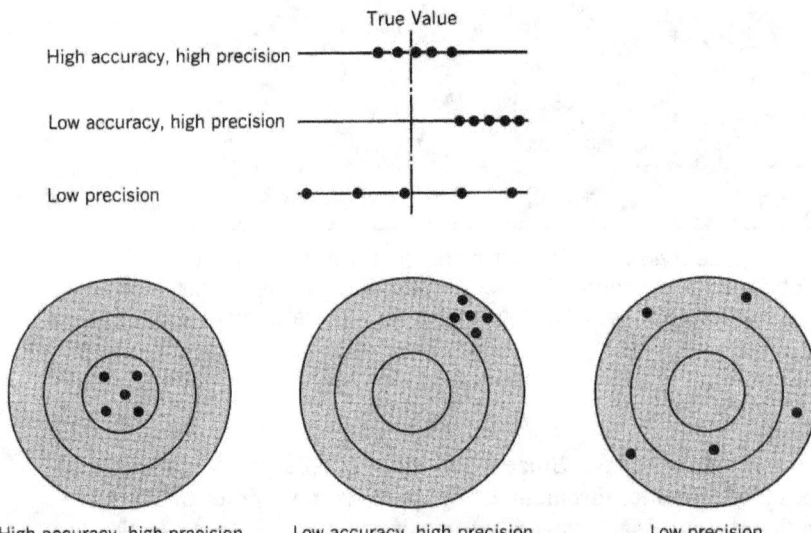

**FIGURE 6-3** Two pictorial representations of accuracy and precision. The large dots represent the measurements. In the bottom diagrams, the "true value" is in the center of the smallest circle. Notice that in the case of low precision, we don't talk about the accuracy. Although the average of the low-precision values may *by chance* give the correct answer, it is not likely. It is thus meaningless to talk about the accuracy of measurements of low precision.

**EXAMPLE 6** We will assume that the instructor in a lab class can read an instrument as well as it was designed to be read. In weighing a sample of magnesium oxide on an analytical balance that reads in grams (abbreviated "g"), the instructor gets 2.0342 g. A student, who is not yet an expert in reading the balance, weighs the same sample and gets 2.035 g. Who made the more precise weighing?

*Solution:* 2.0342 g has less percent uncertainty (and more significant figures) than does 2.035 g. The instructor's weight is more precise. ∎

**EXAMPLE 7** Refer to Example 6. An expert from the National Institute of Standards and Technology weighs the same magnesium oxide sample on a special balance known to be calibrated. The weight is 2.0349 g. Whose weight is more accurate, the instructor's or the student's? (Assume 2.0349 g is the true weight.)

*Solution:* The student's weight is more accurate because it is closer to the true value. This is probably due to luck. Since the instructor's reading was

> **Modern analytical balances** provide a **digital readout** and are very easy to use. The ones generally bought for student laboratories can be read to a precision of 0.1 mg. Suppose your experiment asks you to weigh about 5 g of NaCl. First you must weigh a container to hold the NaCl. This might be a "weighing boat." Say it has a weight of 3.0567 g. To then weigh about 5 g of NaCl, you would need to add NaCl to the weighing boat to give a reading of about 5 g + 3.0567 g = 8.0567 g. But it is very hard to get this exact amount. If the reading were 8.0632 g, this would give you 8.0632 g − 3.0567 g = 5.0065 g, which would be fine for your experiment. **NOTE:** The last digit in 5.0065 g is the 0.1 mg place. See https://binged.it/3fAvzgY for a video on how to use an analytical balance. The **tare button** shown in the video removes the weight of the weighing boat saving you the trouble of a subtraction. The video above calls a weighing boat a "weigh boat."

so far from the true value, either the instructor had a really bad day or the balance is not reliable. ∎

In general, when you make a measurement, your answer reflects a combination of three things: (1) the precision inherent (built-in) in the instrument; (2) the accuracy and precision to which you read the instrument; and (3) the inherent accuracy of the instrument. Good technical skill in reading an instrument will improve the accuracy and precision of your measurements. Careful attention to calibration (comparison to a known standard) will improve your accuracy. Using high-quality instruments will improve both your accuracy and precision.

## 6-4
### DETERMINING SIGNIFICANT FIGURES IN ADDITION AND SUBTRACTION IS DIFFERENT FROM DOING SO IN MULTIPLICATION AND DIVISION

Determining significant figures in addition and subtraction is different from doing so in multiplication and division. The following will illustrate why. Suppose you want to measure 123.46 mL of water.[1] You could do it in either (please continue reading on p 102)

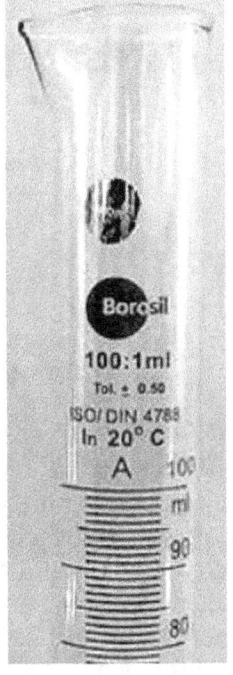

[1] Note that "mL" stands for milliliters (there are 4.93 mL in a teaspoon). A liter (L) is a measure of volume that is equal to about a quart (1 liter = 1.06 quarts). One milliliter equals one-thousandth of a liter. In other words, there are 1000 mL in 1 L.

On the next page, we will discuss using a 100-mL graduated cylinder and a 50-mL buret. At the right is an image of the top of a 100-mL graduated cylinder. Notice how close the 1-mL lines are and how careful you must be to accurately read the level of the liquid. There are 2 classes of graduated cylinders, A and B. The 100-mL graduated cylinder at the right is marked class A and has an accuracy of ±0.5 mL. Class B 100-mL graduated cylinders have an accuracy of ±1 mL. Graduated cylinders are made from glass or various plastics. Images of a 50-mL buret are on page 170.

*The image of the graduated cylinder is CC BY-SA 4.0. The link to the license is https://bit.ly/3IiAWxT. The image has been cropped, converted to grey scale, and the contrast has been enhanced.*

102  SIGNIFICANT FIGURES

of two ways using (1) a 50-mL buret accurate to within ±0.01 mL and (2) a 100-mL graduated cylinder, accurate to within ±1 mL (see Figure 6-4 for a description). The two ways are the following.

1. Use the 50-mL buret to measure two 50.00-mL portions of water (the last two zeros are significant). Then use the buret to measure 23.46 mL of water. Add all the portions together to get 123.46 mL of water.
2. Use the 100-mL graduated cylinder to measure 100 mL of water. Use the 50-mL buret to measure 23.46 mL of water. Add all the portions to get 123.46 mL of water.

Procedure 1 really does give 123.46 mL of water if you are very good at using a buret. Procedure 2 doesn't. Why? Because the 100-mL graduated

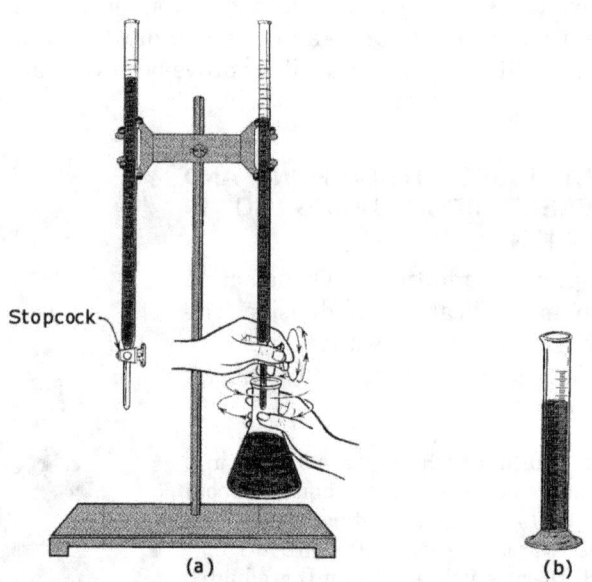

**Figure 6-4** (a) A 50-mL buret. (b) A 100-mL graduated cylinder. A solution is dripped out of the buret slowly. It is easy to control the amount of fluid coming out of the buret by using the stopcock (a valve). A good operator can let out a fraction of a drop at a time. The graduated cylinder, on the other hand, is used to add large amounts of solution. Since the solution must be poured out, here is how it is done. The amount of solution wanted is poured into the graduated cylinder. Fine adjustments of the volume can be made with a dropper. When the amount of solution you want is in the cylinder, the whole amount is poured into a beaker or other container.

## 6-4 Determining Significant Figures in Addition and Subtraction

cylinder can be read only to within ±1 mL. So if you add 23.46 mL of water to the 100 mL of water, you still have an uncertainty in your volume of ±1 mL. The best you can say is that you have 123 mL of water.

The least precise volume measure determines the uncertainty of the final volume. In procedure 2 we would add the volumes as follows:

$$
\begin{array}{r}
100. \\
+\ 23.46 \\
\hline
123.46
\end{array}
$$

The 4 and 6 in the answer are not significant because one of the numbers, here the 100, doesn't have any digits to the right of the decimal point. Thus we must write the sum as 123 mL.

Some authors use a "?" to represent the unknown digits:

$$
\begin{array}{r}
100.?? \\
+\ 23.46 \\
\hline
123.??
\end{array}
$$

The rule is that a "?" number plus a known number gives a "?" number, and "?" numbers are not significant.

**EXAMPLE 8** Add 236.1 and 3.247.

*Solution:*

$$
\begin{array}{r}
236.1 \\
+\ 3.247 \\
\hline
239.347
\end{array}
\quad \text{Using "?" we see that} \quad
\begin{array}{r}
236.1?? \\
+\ 3.247 \\
\hline
239.3??
\end{array}
$$

so we must report the sum as 239.3. ∎

**EXAMPLE 9** Compute 47.32 − 3.1.

*Solution:*

$$
\begin{array}{r}
47.32 \\
-\ 3.1? \\
\hline
44.2?
\end{array}
$$

We must report the sum as 44.2. ∎

104  SIGNIFICANT FIGURES

To see why this method of retaining significant figures is reasonable, let's look at some percent uncertainties. Take the expression 100 mL + 23.46 mL. The percent uncertainties are

$$\frac{1}{100} \times 100 = 1\% \quad \text{and} \quad \frac{0.01}{23.46} \times 100 = 0.04\%$$

For the two possible answers, we have

$$\frac{1}{123} \times 100 = 0.8\% \quad \text{and} \quad \frac{0.01}{123.46} \times 100 = 0.008\%$$

Clearly, the sum of 100 mL and 23.46 mL must be reported as 123 mL and not as 123.46 mL, because 0.8% is a bigger percent uncertainty than 0.008%. Moreover, 0.8% is close to 1%, which is the largest percent uncertainty of our two numbers, 100 mL and 23.46 mL. *Remember, it is the largest percent uncertainty of our original numbers which determines the percent uncertainty of our final answer.*

## 6-5
## ROUNDING A NUMBER PRESERVES THE PROPER NUMBER OF SIGNIFICANT FIGURES

In our discussion about measuring out 123.46 mL of water, we rounded 123.46 mL to 123 mL. How do we round numbers?

The convention for rounding numbers depends on the digits to the right of the last digit we will keep. When we rounded 123.46 mL to three significant figures, we looked at the 46. This is less than 50, so we dropped it and left the 3 alone. If the volume had been 123.58 mL, we would have rounded it to 124 mL because 58 is larger than 50.

So we can use the following rule: **If the digits to the right of the ones we want to keep are greater than 5 or 50 or 500 or 5000 or a 5 with as many zeros as we need, the last digit we keep is increased by 1. If the digits to the right are less than 5 or 500 or 5000, and so on, the last digit is left unchanged.**

**EXAMPLE 10**  Round 57.337 to three significant figures.

*Solution:* We will keep the 57.3 as is and discard the 37, which is less than 50. ∎

**EXAMPLE 11**  Round 3.986 to three significant figures.

*Solution:* Since the last digit, a 6, is larger than 5, we increase the 8 to a 9 and get 3.99. ∎

**EXAMPLE 12**  Round 15.9932 to three significant figures.

*Solution:* Since 932 is larger than 500, we increase 15.9 by 0.1 to get 16.0. ∎

If the digits to the right of the ones we want to keep are exactly 5 or 50 or 500, and so on, most scientists follow this convention: **If the last digit to be kept is even, leave it unchanged. If the last digit to be kept is odd, increase it by 1. Zero is considered an even number.** Can you figure out why? (The answer is given after Example 14.)

**EXAMPLE 13**  Round 25.550 to three significant figures.

*Solution:* In 25.5̄50 the digit with the bar is odd, so we increase it by 1 and the number becomes 25.6. ∎

**EXAMPLE 14**  Round 125.00 to two significant figures.

*Solution:* In 12̄5.00 the digit with the bar is even, so we leave it unchanged. The final answer is 120 or, even better, $1.2 \times 10^2$. ∎

The reason for the procedure shown in Examples 13 and 14 above is this: In rounding a series of numbers, the last significant digit will probably be even about half the time and odd the other half. So about half the time we increase the digit, and half the time we leave it unchanged. That's why we consider zero to be an even number. There are then five even digits (0, 2, 4, 6, 8) and five odd digits (1, 3, 5, 7, 9). In this way, our final answer hopefully won't be biased by the rounding procedure.

## 6-6
## EXACT NUMBERS HAVE AN INFINITE NUMBER OF SIGNIFICANT FIGURES

Sometimes talking about significant figures doesn't mean very much. Suppose you have three apples. You would write 3 apples. It wouldn't make much sense to write 3.0 or 3.00 apples because you have *exactly* three apples. The zeros after the decimal point do not mean anything if you're talking about whole apples. Of course, if you wanted to tell someone that you had 3.50 apples, you would mean that you had 3½ apples to three significant figures.

Thus, when we express numbers of objects that come as unique pieces, significant figures do not apply. The numbers we use are called **exact numbers** and have an infinite number of significant figures.

Numbers that are defined can also be considered exact numbers. One example is the definition that 12 in. = 1 ft. Since the "12" and the "1" are both defined, they are exact numbers. Other examples are constants that are defined, such as Avogadro's constant, $6.022\ 140\ 76 \times 10^{23}$ mol$^{-1}$ (what "mol$^{-1}$" means will be discussed in Chapter 8) and the speed of light, 299 792 458 m/s. (Spaces are used instead of commas for clarity.)

For a chemical example of exact numbers, consider the molecule $H_2O$. The subscripts denote exactly 2 atoms of H and 1 atom of O in each molecule of $H_2O$.

For an example of a chemical equation, consider the following:
$3H_2 + N_2 \rightarrow 2NH_3$. The coefficients 3, 1, and 2 are exact numbers. They denote exactly 3 molecules of $H_2$, 1 molecule of $N_2$, and 2 molecules of $NH_3$. If the balanced equation has a fractional coefficient, the fraction can also be considered an exact number. You could also clear the fraction and turn it into a whole number. See Example 7 in Chapter 3 for a chemical equation where this is done.

## 6-7
### THE MORAL OF THIS CHAPTER IS TO ALWAYS PRESENT YOUR RESULTS IN THE PROPER NUMBER OF SIGNIFICANT FIGURES

In science as in life, you want to be honest but not foolish. Presenting the proper number of significant figures in an answer is honest. Presenting more than the proper number is certainly misleading and possibly dishonest. Presenting less than the proper number of significant figures is foolish—you should always make yourself look as good as honesty allows. It is also important that readers know the number of significant figures you are reporting. Don't mislead them. See pages vi, viii, xv and 371 for quotes from famous people about the importance of honesty in science.

### PROBLEMS
KEYED PROBLEMS
1. You are dividing 8.4 feet by 3.6 feet. Each number is known to within ±1 in the tenths place. What should your answer be?
2. You are multiplying 6.84 x 8.76 on a calculator. If each number is good to within ±2 in the hundredths place, how should the answer be reported?
3. In multiplying 2.13 x 27.86, what is a reasonable answer?
4. How many significant figures are in the following: 2.001; 97.300; 0.001161?
5. In the division 1.23/0.98, what is the answer containing the proper number of significant figures?
6. Which of the following weights is more precise, 1.0342 g or 1.134 g?
7. If the "true" weight is 1.1328 g, which of the weights in Problem 6 is more accurate?
8. Add 429.6 g and 7.685 g.
9. Compute 82.47 − 6.5.
10. Round 28.941 to three significant figures.

11. Round 4.888 to three significant figures.

12. Round 33.996 to three significant figures.

13. Round 96.750 to three significant figures.

14. Round 445.00 to two significant figures.

## SUPPLEMENTAL PROBLEMS

15. Round the following numbers to three significant figures.
    a. 4.268
    b. 25.144
    c. 0.03055
    d. $3.987 \times 10^{12}$
    e. $9.231 \times 10^{-4}$
    f. 32650

16. In the following multiplications, round your answer to three significant figures.
    a. $4.23 \times 6.41$
    b. $25.2 \times 87.6$
    c. $125.1 \times 9.66$
    d. $0.342 \times 0.768$
    e. $(4.31 \times 10^6)(9.32 \times 10^4)$
    f. $(2.86 \times 10^{-4})(6.68 \times 10^{-8})$

17. In the following divisions, round your answer to three significant figures.
    a. 8.32/4.66
    b. 32.1/25.21
    c. 562/3.25
    d. 0.167/0.0876
    e. $(6.37 \times 10^{-5})/(5.42 \times 10^6)$
    f. $(2.11 \times 10^{-3})/(6.68 \times 10^{-2})$

18. In the following multiplications, round your answer to the appropriate number of significant figures.
    a. $3.2 \times 4.6$
    b. $5.11 \times 15.32$
    c. $9.86 \times 3.20$
    d. $0.030 \times 2.61$
    e. $(3.216 \times 10^3)(4.23 \times 10^5)$
    f. $(1.2 \times 10^{-2})(3.13 \times 10^{-4})$

19. In the following divisions, round your answer to the appropriate number of significant figures.
    a. 4.3/9.6
    b. 1.67/26.38
    c. 2.20/1.86
    d. 0.312/0.0026
    e. $(4.6 \times 10^3)/(2.33 \times 10^4)$
    f. $(2.671 \times 10^{-2})/(7.32 \times 10^{-6})$

20. Perform the following additions and subtractions and round your answer to the appropriate number of significant figures.
    a. $3.26 + 4.1$
    b. $5.77 + 6.00$
    c. $251.6 + 1.167$
    d. $25.44 - 13.1$
    e. $187.5 - 57.92$
    f. $200.00 - 3.00$

21. In the molecular interpretation of the chemical equation $2H_2 + O_2 \rightarrow 2H_2O$, do the coefficients have only one significant figure or are they exact numbers? Explain.

**Astronaut Stephanie Diana Wilson** (born 1966), STS-120 (Space Transportation System-120) mission specialist, with a model of Harmony (left photo), both floating freely on the middeck of the Space Shuttle Discovery while docked with the International Space Station in 2007. (Harmony is the final contribution of the US to the International Space Station.) Wilson attended Harvard University, receiving a Bachelor of Science degree in engineering science in 1988 and a Master of Science degree in aerospace engineering from the University of Texas in 1992. She flew to space onboard three Space Shuttle missions (STS-120, STS-121, STS-131) and spent a total of 42 days in space. The photo on the right is Wilson's official NASA portrait.

In 2020, NASA announced the names of the 18 astronauts (out of 48 active astronauts) who will be the astronaut corps of the Artemis Team. The list includes astronaut Wilson. See https://bit.ly/2W0eW8e. If Wilson is selected for the first moon landing, she will be the first woman to walk on the moon (https://cbsn.ws/3jS0mYn). The first uncrewed Artemis I test launch took place on November 16, 2022. The Orion crew capsule, which can hold four astronauts, returned safely to earth on December 11, 2022. It circled the moon and travelled 1.4 million miles or 2.25 million kilometers. See https://go.nasa.gov/3UUKXXL for more details.

Harmony is a "utility hub," providing additional electrical power and electronic data capability for the Space Station. It weighs approximately 14,288 kilograms or 31,500 pounds. It measures 7.2 meters or 24 feet in length and has a diameter of 4.4 meters or 14 feet. STS-120 was a 10-million kilometer or 6.2-million mile (total distance flown in outer space) Space Shuttle mission to the International Space Station (ISS). After reading this chapter, you will be able to do the unit conversions in this essay.

*Photos courtesy of NASA. Both are in the public domain.*

# 7

# UNITS AND UNIT CONVERSIONS

Now that we have covered some of the essentials of atoms, molecules, and scientific notation, we can begin a systematic study of some of the calculations necessary to "do" chemistry. This chapter will cover units and unit conversions.

## 7-1
### UNITS AND THEIR ABBREVIATIONS ARE USED TO DESCRIBE PHYSICAL QUANTITIES

*It is extremely important to keep track of and use the correct units and abbreviations in a problem, as the disasters below show when units and abbreviations are mixed up or not communicated carefully.*[1] Units indicate how a physical quantity is measured. For instance, 3 inches is the number 3 with the unit "inches" attached. The unit part sets the scale or basic multiple; thus, 3 inches is not 3 feet or 3 yards, even though the 3 is the same in each expression.[2]

---

[1] In 1999, the $125 million NASA Mars Climate Orbiter spacecraft crashed after a 286-day journey to Mars. Lockheed Martin, the contractor, was sending thruster data in English units of force to NASA, while NASA's navigation team was expecting metric units of force.

In 2004, at Tokyo Disneyland's Space Mountain, an axle broke on a roller coaster train mid-ride, causing it to derail. Fortunately, no one was hurt. The cause was new axles that were mistakenly ordered using English units instead of metric units.

In 1983, Air Canada Flight 143 ran completely out of fuel about halfway through its flight. The fuel gauge on the plane was broken, so the pilot asked for fuel by weight. Unfortunately, he used the numerical value for pounds/liter instead of the numerical value for kilograms/liter (he wanted kilograms of fuel) in a multistep unit conversion and got less than half the fuel he thought he was getting, since 1 lb is only 0.45 kg. Fortunately, the pilot made a safe emergency landing, as he was an expert glider pilot, and only 10 passengers were slightly injured during the emergency evacuation. See especially the Gimli Glider – Wikipedia section called "Miscalculation during fueling." **NOTE:** Air Canada was one of the few airlines that used both the Imperial system of units (the everyday system of units used in the United States that has pounds and feet) and the metric system, clearly a recipe for disaster. After Flight 143, they changed to using only the metric system.

In 1999, a patient received 0.5 g of phenobarbital (a sedative) instead of 0.5 gr when a surgical resident misread the label on the container. (Grams are sometimes incorrectly abbreviated "gr" instead of "g." This could have led to the error.) The patient received 15 times the intended dose since the gram is 15 times larger than the grain. Fortunately, the patient survived.

[2] As discussed on page 110, in scientific use, when a number has a unit after it, the unit is always abbreviated. Scientists would always write, say, 6.45 g, never 6.45 grams. This was not done in this paragraph for clarity.

# 110 UNITS AND UNIT CONVERSIONS

Units are always abbreviated when they are used after a number, with singular and plural forms having the same abbreviation and being used interchangeably. Abbreviations for some common units are as follows:

foot or feet = ft   gram or grams = g   inch or inches = in.   grain or grains = gr
kilogram or kilograms = kg   pound or pounds = lb   calorie or calories = cal
second or seconds = s   meter or meters = m   millimeter or millimeters = mm
centimeter or centimeter = cm   milligram or milligrams = mg   liter or liters = L
milliliter or milliliters = mL   microgram or micrograms = µg (µ → Greek letter "mu.")

**NOTE:** A period is *not* put after an abbreviation unless it comes at the end of a sentence. The exception to this rule is the abbreviation for inch or inches (in.) which might be confused with the word "in" if a period isn't used.

**WARNING:** Be careful when using the units "gram" and "grain." The grain is a very old unit and is sometimes used to measure water hardness (amount of calcium carbonate). In the medical field it is sometimes used to weigh drugs. In the ammunition area it is used to weigh bullets and the gunpowder used to propel the bullet. A gram is 15 times larger than a grain. Weighing out grams instead of grains could injure or kill a person as discussed on p. 109.

## 7-2
### "ONE" IS THE IDENTITY ELEMENT IN MULTIPLICATION

If any number is multiplied by 1, the number is not changed. We say that 1 is the **identity element** in multiplication. For example,

$$7 \times 1 = 7$$

$$\tfrac{4}{3} \times 1 = \tfrac{4}{3}$$

$$2.36 \times 1 = 2.36$$

Generally, if $n$ stands for any number, it follows that

$$n \times 1 = n$$

The number "1" can take many forms. For instance,

$$\frac{3}{3} = 1 \qquad \frac{9.46}{9.46} = 1 \qquad \frac{xy}{xy} = 1 \qquad \frac{\tfrac{7}{3}}{\tfrac{7}{3}} = 1$$

where we have used the fact that *any* number (except zero) divided by itself equals 1.

Just as was true of numbers alone, numbers with units attached don't change when they are multiplied by 1 in any form. For example, 3 ft × 1 = 3 ft. If the 1 should happen to be 9.46/9.46, say, then

$$3 \text{ ft} \times \frac{9.46}{9.46} = 3 \text{ ft}$$

## 7-3
## UNITS DIVIDE IN A WAY SIMILAR TO ANY ALGEBRAIC QUANTITY

It seems reasonable that 3 ft/3 ft = 1, since any quantity divided by itself equals 1. Thus it must be true that both 3/3 = 1 and ft/ft = 1. To convince yourself of this, think of the following example. If my lab bench is 20 ft long and your lab bench is 10 ft long, my bench is *twice* as long as your bench:

$$\frac{20 \text{ ft}}{10 \text{ ft}} = 2$$

Notice that my bench is *not* 2 ft longer than your bench—it is twice as long. It should now be clear that **units can be divided in a way similar to numbers or any algebraic quantity.** Thus in this example we have the following:

$$\frac{20}{10} = 2 \quad \text{and} \quad \frac{\text{ft}}{\text{ft}} = 1$$

**EXAMPLE 1** My experiment needs 10 mL of methanol as a solvent. Yours needs 2 mL. How many times more methanol do I need than you need? (Methanol is also called wood alcohol. It is jellied to make sterno and is very poisonous.)

*Solution:*

$$\frac{10 \text{ mL}}{2 \text{ mL}} = 5$$

Thus I need five times as much methanol as you do. ∎

## 7-4
## IDENTITIES AND THEIR RATIOS ALLOW US TO CONVERT FROM ONE UNIT TO ANOTHER

The fact that units have many of the properties of algebraic quantities and can be operated on separately from the numbers that they come with is very useful. Consider the identity [3]

$$12 \text{ in.} = 1 \text{ ft}$$

[3] The numbers 12 and 1 are exact numbers because the relationship between feet and inches is a definition. Thus the "1" and the "12" can be considered as having an infinite number of significant figures in any calculation.

# 112 UNITS AND UNIT CONVERSIONS

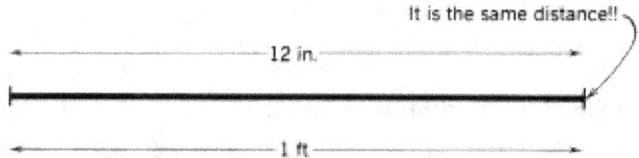

**FIGURE 7-1** A line has the same length, no matter what you call it.

This is called an **identity** because both the 12 in. and the 1 ft represent a distance that is *physically* the same. The line drawn in Figure 7-1, which represents 1 ft or 12 in., is the same length no matter what you call it.

If both sides of the identity, 12 in. = 1 ft, are divided by 12 in., we get

$$\frac{12 \text{ in.}}{12 \text{ in.}} = \frac{1 \text{ ft}}{12 \text{ in.}}$$

Since 12 in./12 in. = 1, it must be true that

$$\frac{1 \text{ ft}}{12 \text{ in.}} = 1$$

If we instead divide both sides of the identity 12 in. = 1 ft by 1 ft, we get

$$\frac{12 \text{ in.}}{1 \text{ ft}} = \frac{1 \text{ ft}}{1 \text{ ft}}$$

Since 1 ft/1 ft = 1, it must again be true that

$$\frac{12 \text{ in.}}{1 \text{ ft}} = 1$$

Two forms of "1" have been obtained:

$$\frac{12 \text{ in.}}{1 \text{ ft}} \quad \text{and} \quad \frac{1 \text{ ft}}{12 \text{ in.}}$$

These quantities are called **ratio identities** because they are ratios made from an identity. The use of ratio identities allows us to easily convert from one unit to another.

## 7-5
## CONVERTING FROM ONE UNIT TO ANOTHER USES RATIO IDENTITIES

Suppose that we want to convert 36 in. into feet. Then we simply multiply 36 in. by the ratio identity that divides the inches and leaves feet. In other words, **put the unit you want to end up with on top and the unit you want to divide on the bottom:**

$$36 \text{ in.} \times \frac{1 \text{ ft}}{12 \text{ in.}} = \frac{36 \text{ in.}}{1} \times \frac{1 \text{ ft}}{12 \text{ in.}} = \frac{36 \cancel{\text{ in.}}}{12 \cancel{\text{ in.}}} \times \frac{1 \text{ ft}}{1} = 3.0 \times 1 \text{ ft} = 3.0 \text{ ft}$$

Doing this in one step we have

$$36 \cancel{\text{ in.}} \times \frac{1 \text{ ft}}{12 \cancel{\text{ in.}}} = 3.0 \text{ ft}$$

(Notice that we are being careful about significant figures.) If the other ratio identity had been used by mistake, we would have gotten

$$36 \text{ in.} \times \frac{12 \text{ in.}}{1 \text{ ft}} = \frac{432 \text{ in.}^2}{1 \text{ ft}} = 432 \frac{\text{in.}^2}{\text{ft}}$$

where in.$^2$ = in. × in. Obviously, this is undesirable because we wanted the unit feet.

Now let's do this problem in the reverse direction. Convert 3.0 ft to inches. For this, the ratio identity 12 in./1 ft is used:

$$3.0 \cancel{\text{ ft}} \times \frac{12 \text{ in.}}{1 \cancel{\text{ ft}}} = 36 \text{ in.}$$

To help you to convert easily from one unit to another, we will outline a definite procedure that you can use. To help you spot the *given* and *asked for* in a problem, we will put a (G) after the given and an (AF) after the asked for. This procedure will also be followed for most of the problems in the next two chapters, Chapters 8 and 9. After that, you should have enough practice in solving problems to determine the given and asked for by yourself.

For instance, in the problem

How many pounds (AF) are there in 200 g (G)?

the 200 g are given and the pounds are asked for.

Let's solve this problem using the procedure given in the previous paragraph.

## 114 UNITS AND UNIT CONVERSIONS

**PROBLEM** How many pounds (AF) are there in 200 g (G)?

*Steps in the solution:*

1. Determine the given and the asked for. In this problem we have indicated them by a (G) and an (AF).
2. Write down an identity relating the *units* of the given and asked for. In this problem the identity is 1 lb = 453.6 g.
3. Write down the given, put a "x" sign after it, draw a line, and follow it with an "=" sign:

$$200 \text{ g} \times \frac{\phantom{xxx}}{\phantom{xxx}} =$$

4. From the identity in step 2, put the *asked-for unit (lb) on top of the line* and the *given unit (g) under the line*:

$$200 \text{ g} \times \frac{1 \text{ lb}}{453.6 \text{ g}} =$$

5. Compute the answer. Be *sure that the given unit divides out*, thus *leaving the asked for unit*:

$$200 \text{ g} \times \frac{1 \text{ lb}}{453.6 \text{ g}} = \frac{200}{453.6} \text{ lb} = 0.441 \text{ lb} \quad \blacksquare$$

Remember, the only thing that converting from one unit to another does is to change the *name* we give to a certain quantity. **It doesn't change that quantity physically.** Thus all ratio identities must equal 1 so that they only change the *name* of the quantity, not the size of it. **NOTE: In following chapters, we will use "identities" that do change the quantity physically. They are not strictly identities, but the mathematics of doing conversions from one unit to another works perfectly,** *if you carefully follow the rules for using them.*

Many of the identities that are used in this book are based on the metric system. Since scientists use the metric system almost exclusively, now is a good time to get familiar with it.

## 7-6
### THE METRIC SYSTEM USES PREFIXES IN FRONT OF A UNIT TO INDICATE A POWER OF 10

Due to the total chaos regarding units of measure in France, in 1791 the French Academy of Sciences presented a system of measurement to the French National Assembly — the metric system.[4] It took effect in France in 1799. In 1960, an international committee, the CGPM,[5] agreed on a particular choice of metric units. This is the modern set of metric units and is called the **International System of Units**. It is abbreviated **SI**, after the French name, Le Système International d'Unités, since the CGPM meets in France. In the SI system,

---

[4] At the time, there were around eight hundred units of measure in use in France. These had up to a quarter of a million different definitions because each unit could differ from town to town and even for different goods. Unbelievable!

[5] CGPM = **C**onférence **G**énérale des **P**oids et **M**esures or in English, the General Conference on Weights and Measures. See pages xix and 404 for more information.

## TABLE 7-1

| FIVE SI BASE UNITS | | |
|---|---|---|
| PHYSICAL QUANTITY | UNIT | SYMBOL OR ABBREVIATION |
| Length | meter | m |
| Mass | kilogram | kg |
| Time | second | s |
| Temperature | kelvin | K |
| Amount of substance | mole | mol |

a prefix, or group of letters, is placed in front of the SI unit to indicate a power of 10.

Each physical quantity has an **SI base unit.** The five SI base units that are used in this book are listed in Table 7-1.

The prefixes in the SI system are listed in Table 7-2. The prefixes that are in boldface type are the ones that are used in this book or that are so commonly used by chemists that you should be familiar with them.

To show you how to use Table 7-2, let's choose the prefix *milli*. *Milli* means $10^{-3}$ or one-thousandth. Therefore we can write the following identities:

Length: $10^{-3}$ meter = 1 millimeter or $10^{-3}$ m = 1 mm

Mass: $10^{-3}$ gram = 1 milligram or $10^{-3}$ g = 1 mg

Time: $10^{-3}$ second = 1 millisecond or $10^{-3}$ s = 1 ms

## TABLE 7-2

| SI PREFIXES | | | | | |
|---|---|---|---|---|---|
| MULTIPLE | PREFIX | SYMBOL | MULTIPLE | PREFIX | SYMBOL |
| $10^{18}$ | exa | E | $10^{-1}$ | **deci** | **d** |
| $10^{15}$ | peta | P | $10^{-2}$ | **centi** | **c** |
| $10^{12}$ | tera | T | $10^{-3}$ | **milli** | **m** |
| $10^{9}$ | giga | G | $10^{-6}$ | **micro** | **$\mu$**[b] |
| **$10^{6}$** | **mega**[a] | **M** | $10^{-9}$ | **nano** | **n** |
| **$10^{3}$** | **kilo** | **k** | $10^{-12}$ | **pico** | **p** |
| $10^{2}$ | hecto | h | $10^{-15}$ | femto | f |
| 10 | deca | da | $10^{-18}$ | atto | a |

[a] The prefixes in boldface type are used in this book or are commonly used by chemists.

[b] The letter $\mu$ is Greek "mu" and is read "micro" in this case.

116 UNITS AND UNIT CONVERSIONS

We can multiply both sides of each of these identities by $10^3$. Then we obtain the following identities, which are equivalent to those just given but look slightly different:

$$\text{Length:} \quad 1 \text{ m} = 10^3 \text{ mm}$$
$$\text{Mass:} \quad 1 \text{ g} = 10^3 \text{ mg}$$
$$\text{Time:} \quad 1 \text{ s} = 10^3 \text{ ms}$$

To show that both sets of identities are equivalent to one another, let's use both of them to convert 58 g to milligrams. Using the identity $10^{-3}$ g = 1 mg, we obtain

$$58 \text{ g} \times \frac{1 \text{ mg}}{10^{-3} \text{ g}} = 58 \times 10^3 \text{ mg} = 5.8 \times 10^4 \text{ mg}$$

Using the identity 1 g = $10^3$ mg, we get

$$58 \text{ g} \times \frac{10^3 \text{ mg}}{1 \text{ g}} = 58 \times 10^3 \text{ mg} = 5.8 \times 10^4 \text{ mg}$$

There is an exception to using the prefixes. When we measure time that is longer than a few hundred seconds, even scientists still use minutes and hours.

Table 7-3 presents some useful identities involving both the metric system and the everyday system used in the United States, which uses feet, inches, quarts, and so on.

## 7-7
### WHEN CONVERTING FROM ONE UNIT TO ANOTHER, BE SURE THAT THE UNITS DIVIDE OUT IN THE CORRECT WAY

This section presents many examples of unit conversions, most of which involve the metric system.

**EXAMPLE 2** Convert 100 g (G) into pounds (AF).

*Solution:* The identity is 453.6 g = 1 lb.

$$100 \text{ g} \times \frac{1 \text{ lb}}{453.6 \text{ g}} = \frac{100}{453.6} \text{ lb} = 0.220 \text{ lb} \quad \blacksquare$$

## TABLE 7-3

**SOME USEFUL IDENTITIES**

| QUANTITY | ABBREVIATION |
|---|---|
| Length | |
| 1 m = 100 cm | m = meter |
| 1 m = 1000 mm | mm = millimeter |
| 1 cm = 10 mm | cm = centimeter |
| 1 m = $10^9$ nm | nm = nanometer |
| 1 km = 1000 m | km = kilometer |
| 1 m = 39.37 in. | in. = inch |
| 1 in. = 2.54 cm | mi = mile |
| 1 mi = 5280 ft | ft = foot |
| 12 in. = 1 ft | |
| Volume | |
| 1 L = 1000 mL | L = liter |
| 1 L = 1.057 qt | mL = milliliter |
| 1 mL = 1 cm$^3$ | cm$^3$ = cubic centimeter or cc |
| 1 gal = 4 qt | qt = quart |
| | gal = gallon |
| Mass | |
| 1 kg = 1000 g | kg = kilogram |
| 1 lb = 453.6 g | g = gram |
| 1000 mg = 1 g | lb = pound |
| $10^6$ μg = $10^6$ mcg = 1 g | μg = mcg = microgram[a] |
| Time | |
| 1 min = 60 s | min = minute |
| 1 h = 60 min | s = second |
| | h = hour |
| Energy | |
| 1 cal = 4.184 J | cal = calorie |
| 1 kcal = 4.184 kJ | J = joule |
| 1 kJ = 1000 J | kJ = kilojoule |
| 1 kcal = 1 Cal = 1000 cal | kcal or Cal = kilocalorie |

[a] The letter μ is Greek "mu" and is read "micro" in this case. Sometimes μg is written as mcg, especially in nonscientific uses such as on the labels of nutritional supplements.

**EXAMPLE 3** Convert 2.00 lb (G) into grams (AF).

*Solution:* The identity is 453.6 g = 1 lb.

$$2.00 \text{ lb} \times \frac{453.6 \text{ g}}{1 \text{ lb}} = 907 \text{ g}$$ ∎

**EXAMPLE 4**  How many seconds (AF) are there in 10 min (G)?

*Solution:* The identity is 60 s = 1 min.

$$10 \text{ min} \times \frac{60 \text{ s}}{1 \text{ min}} = 600 \text{ s} = 6.0 \times 10^2 \text{ s} \quad \blacksquare$$

**EXAMPLE 5**  How many centimeters (AF) are there in 5.00 in. (G)?

*Solution:* The identity is 2.54 cm = 1 in.

$$5.00 \text{ in.} \times \frac{2.54 \text{ cm}}{1 \text{ in.}} = 12.7 \text{ cm} \quad \blacksquare$$

**EXAMPLE 6**  How many inches (AF) are there in 100 cm (G)?

*Solution:* The identity is 2.54 cm = 1 in.

$$100 \text{ cm} \times \frac{1 \text{ in.}}{2.54 \text{ cm}} = 39.4 \text{ in.} \quad \blacksquare$$

**EXAMPLE 7**  How many millimeters (AF) are there in 10 m (G)?

*Solution:* The identity is 1 m = 1000 mm.

$$10 \text{ m} \times \frac{1000 \text{ mm}}{1 \text{ m}} = 10{,}000 \text{ mm} = 1.0 \times 10^4 \text{ mm} \quad \blacksquare$$

**EXAMPLE 8**  How many kilograms (AF) are there in 500 g (G)?

*Solution:* The identity is 1000 g = 1 kg.

$$500 \text{ g} \times \frac{1 \text{ kg}}{1000 \text{ g}} = 0.500 \text{ kg} \quad \blacksquare$$

**EXAMPLE 9**  How many liters (AF) are there in 250 mL (G)?

*Solution:* The identity is 1000 mL = 1 L.

$$250 \text{ mL} \times \frac{1}{1000 \text{ mL}} = 0.250 \text{ L} \quad \blacksquare$$

7-7 Examples of Unit Conversions    119

Sometimes you don't have the identity needed to make a conversion. However, by using two or more identities from Table 7-3 (or some other source), you can make the conversion. Suppose you go to a gas station that sells gas by the liter. It takes 40.0 L to fill the tank of your car. How many gallons is this? Looking at Table 7-3, you don't see an identity relating liters and gallons. But you do see two identities that will do the job. They are 1 L = 1.057 qt and 4 qt = 1 gal. So write down the given (40.0 L) and draw two horizontal lines, two "×" signs, and an "=" sign. The conversion is done in two steps that are combined into one: liters → quarts → gallons.

$$40.0 \text{ L} \times \underline{\phantom{xxxxx}} \times \underline{\phantom{xxxxx}} =$$

You want gallons after the equals sign, so insert the appropriate ratio identities to divide the liters and the quarts, giving gallons:

$$40.0 \text{ L} \times \frac{1.057 \text{ qt}}{1 \text{ L}} \times \frac{1 \text{ gal}}{4 \text{ qt}} = 10.6 \text{ gal}$$

**EXAMPLE 10**  How many seconds (AF) are there in 1 h (G)?

*Solution:* The identities from Table 7.1 are 60 s = 1 min and 60 min = 1 h.

$$1 \text{ hr} \times \frac{60 \text{ min}}{1 \text{ hr}} \times \frac{60 \text{ s}}{1 \text{ min}} = 3600 \text{ s}$$

If you've studied Chapter 6, which is on significant figures, you might ponder about how many significant figures 3600 s has.  ■

**EXAMPLE 11**  A wine bottle contains 750 mL (G) of wine. How many gallons (AF) is this?

*Solution:* The three identities we need from Table 7.1 are

$$1 \text{ L} = 1000 \text{ mL} \qquad 1 \text{ L} = 1.057 \text{ qt} \qquad 1 \text{ gal} = 4 \text{ qt}$$

$$750 \text{ mL} \times \frac{1 \text{ L}}{1000 \text{ mL}} \times \frac{1.057 \text{ qt}}{1 \text{ L}} \times \frac{1 \text{ gal}}{4 \text{ qt}} = 0.198 \text{ gal} \quad ■$$

It is interesting that 0.198 gal is almost 0.200 gal or one-fifth of a gallon. Before liquor and wine manufacturers converted to the metric system, a "fifth" was exactly one-fifth of a gallon. But they rounded the "fifth" to 750 mL, thereby saving a few milliliters. If you do Problems 11 and 24 at the end of this chapter, you will see how much wine or liquor they saved, thus increasing their profits.

A final word about units. If a number has a unit attached to it and you don't write the unit, nobody except you knows what you mean. For instance, if you write only an 8 when you mean 8 kg, how is the reader supposed to know what you are talking about?

## PROBLEMS

### KEYED PROBLEMS

1. I drive 20 miles to work every day. You drive 5 miles to work every day. How many times farther is my job than your job?
2. Convert 200 g into pounds.
3. Convert 5.0 lb into grams.
4. How many seconds are there in 40 min?
5. How many centimeters are there in 8.0 in.?
6. How many inches are there in 50 cm?
7. How many millimeters are there in 200 m?
8. How many kilograms are there in 200 g?
9. How many liters are there in 725 mL?
10. How many seconds are there in a day? (*Hint:* 1 day = 24 h.)
11. How many milliliters are there in 0.200 gal?

### SUPPLEMENTAL PROBLEMS

12. The recommended dietary allowance (RDA) of vitamin C is 60 mg. How many grams is this?
13. Many people believe that high doses of vitamin C should be taken, and they take around 3 g a day. How many milligrams is this?
14. The RDA for vitamin $B_{12}$ is 3 $\mu$g. How many grams is this?
15. The distance from New York City to Seattle (Washington) is about $3.0 \times 10^3$ miles. How many kilometers is this?
16. A chemistry student weighs 150 lb. How many kilograms does the student weigh?
17. A solid rectangular object measures 38 in. × 26 in. × 34 in. on each side. How many centimeters does the object measure?

18. The New York City marathon race is 26 miles, 385 yards. How many inches is this?

19. You have been waiting on line for 1000 s to buy your textbooks. How many hours have you been waiting?

20. The label on a can of beer reads: Volume = 300 mL. How many quarts is this?

21. One milligram of natural vitamin E equals 1.49 international units (IU). How many milligrams of natural vitamin E are there in a 400-IU capsule?

22. About 0.10 lb of matter is turned into energy in the explosion of a medium-sized atomic bomb. How many grams is this?

23. How many liters are there in a half-gallon bottle of soda?

24. A bottle of bourbon whiskey contains 750 mL. This is the size that used to be called a "fifth." A "fifth" was one-fifth of a gallon. How much less bourbon are you getting now than you were getting before they switched over to metric labeling of bottles?

25. You are visiting Israel and drive into a gas station. You fill up with 48 L of gas. How many gallons is this?

26. The average male college student probably consumes about 2500 kcal/day. How many kilojoules is this? How many joules?

27. The average American diet contains about 18% of its energy in the form of refined sugar. If the average daily diet contains 9200 kJ of energy, how many kilocalories from refined sugar does the average American consume each day?

28. The yellow light from a sodium vapor lamp has a wavelength of 589 nm. How many centimeters is this?

29. Adults with a fasting serum cholesterol above 240 mg/dL are at increased risk for coronary heart disease. How many millimoles per liter is this if 1 mg/dL = 0.02586 mmol/L? (From Table 7.2 we see that dL is the abbreviation for deciliter.)

    **NOTE:** Nutrient requirements that are listed on food labels are now called the Daily Value (DV). The DV for vitamin C is 90 mg. The DV for vitamin $B_{12}$ is 2.4 µg. The current recommendations are that fasting serum cholesterol levels be less than 200 mg/dL.

**The Avogadro Project (Part 1):** The photo shows a 94-mm diameter (almost as big as a regulation softball) 1-kg single-crystal silicon sphere made for the International Avogadro Coordination's *Avogadro Project*. Diameter measurements on the sphere show an out-of-roundness of 35 nm, and the uncertainty in the mass is 3 μg. This sphere is one of the roundest objects ever made on earth. Using measurements made on this sphere, along with other measurement techniques,

**scientists have *defined* Avogadro's number as 6.022 140 76 x $10^{23}$.**

**Avogadro's number is one of the most important constants across many fields of science where it can be found in extremely important equations. That's why scientists went through so much trouble to define it. The unit conversions done with it in this chapter are only one use.**

**NOTE:** Previous experimental methods were not considered accurate or precise enough to give scientists the confidence to define Avogadro's number.

This definition of Avogadro's number officially took effect on World Metrology Day, May 20, 2019 (https://iupac.org/new-definition-mole-arrived/). **NOTE:** Since as of May 20, 2019, Avogadro's number is a definition and thus it became an exact number (see Sections 6-6 and 7-4). Part 2 of this essay is on page 157.

*Photo courtesy of the National Institute of Standards and Technology (NIST).*

# AVOGADRO'S NUMBER, THE MOLE, AND MOLECULAR WEIGHT

Consider a golf ball that weighs 100 g and a bowling ball that weighs 8000 g. The bowling ball weighs 80 times more than the golf ball.

Now consider 10 golf balls and 10 bowling balls, as is done in Figure 8-1. The 10 golf balls weigh 1000 g, and the 10 bowling balls weigh 80,000 g. But the bowling balls still weigh 80 times more than the golf balls.

What have we learned from this discussion? We have learned the following: *If one bowling ball weighs 80 times more than one golf ball, then 10 bowling balls weigh 80 times more than 10 golf balls.*

If we had a million golf balls and a million bowling balls, the bowling balls would still weigh 80 times more than the golf balls.

We could take *any number* of bowling balls and an equal number of golf balls. As long as we take the *same number* of each, the bowling balls still weigh 80 times more than the golf balls. The same principle holds when we consider atoms.

## 8-1
### ATOMIC WEIGHTS ARE RELATED TO NUMBERS OF ATOMS

In Chapter 2 we discussed the relative atomic weights of atoms. For instance, we learned that the relative atomic weight of hydrogen is about 1 u and that the relative atomic weight of oxygen is about 16 u. Oxygen atoms

# 124 AVOGADRO'S NUMBER, THE MOLE, AND MOLECULAR WEIGHT

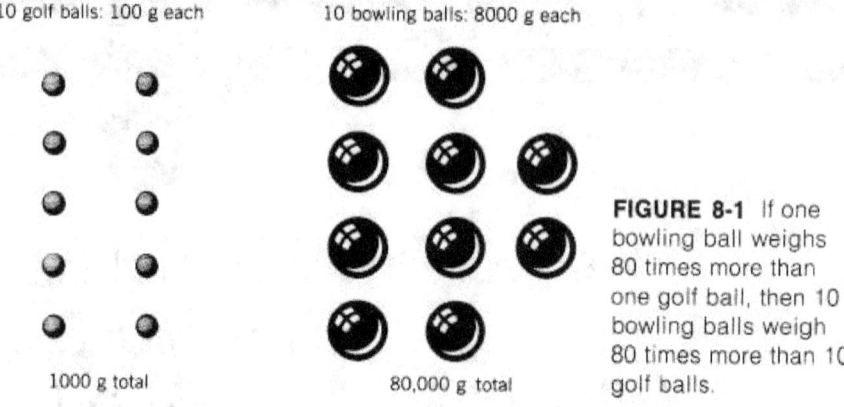

FIGURE 8-1 If one bowling ball weighs 80 times more than one golf ball, then 10 bowling balls weigh 80 times more than 10 golf balls.

weigh 16 times more than hydrogen atoms. (Ignore the different isotopes here, because they don't affect the discussion.)

If we take 10 hydrogen atoms and 10 oxygen atoms, the oxygen atoms still weigh 16 times more than the hydrogen atoms; and if we take a million of each, the oxygen atoms still weigh 16 times more than the hydrogen atoms.

Suppose now that we want to take a billion hydrogen atoms and a billion oxygen atoms. What would be the weight of each batch of atoms? Without even knowing how we are going to measure weight, we know that we need 16 times more oxygen by weight than hydrogen.

So we have discovered the following: *To take equal numbers of atoms of hydrogen and oxygen, we would weigh out 16 times more oxygen than hydrogen.*

**EXAMPLE 1** The relative atomic weight of bromine is 80 u, and the relative atomic weight of helium is 4 u. If you wanted to take an equal number of Br atoms and He atoms, what would be the ratio of their masses that you would need?

***Solution:*** Each Br atom weighs 20 times more than each He atom, because 80 u/4 u = 20. You would have to weigh out 20 times more Br than He. ∎

## 8-2
## THE GRAM IS USED AS THE UNIT OF ATOMIC WEIGHT

Now we must decide on a *unit of mass* that we will use to weigh things. The unified atomic mass unit is too small to be practical. As you know, scientists use the metric system, and one of the common units of mass they use to weigh things is the **gram**. (Since the gram is a unit of mass, you might wonder why they "weigh" samples in grams. This is just the way our language works — you will never be asked to "mass a sample." You will always be asked to "weigh a sample.")

So from now on we will speak of 1.01 g of hydrogen atoms and 16.0 g of oxygen atoms when referring to their atomic weights.

**When we attach grams to the relative atomic weight number, it is customary to drop the word "relative." So we will now speak of the "atomic weight" of an element.**

We have already found that equal numbers of H atoms and O atoms have a mass ratio of 1.01 to 16.0. Thus 1.01 g of hydrogen atoms has the same number of atoms as 16.0 g of oxygen atoms.

Since the atomic weights of the elements are in the same ratio as the weights of the individual atoms, **one atomic weight of any element contains the same number of atoms as one atomic weight of any other element.**

**EXAMPLE 2** What are the atomic weights (to 3 significant figures) of the following elements, each of which contain the same number of atoms? (Use the Table of Atomic Weights, Table 2-2.) The elements are C, Kr, U, and P.

*Solution:* The masses of each element that contain the same number of atoms are just the atomic weights in grams:

C = 12.0 g     Kr = 83.8 g     U = 238 g     P = 31.0 g     ■

## 8-3
## ONE ATOMIC WEIGHT OF ATOMS CONTAINS ONE AVOGADRO'S NUMBER OF ATOMS

**The number of atoms in one atomic weight of an element is defined as**

$$6.022\ 140\ 76 \times 10^{23}.$$

(See pp. 122 and 157 for details of how it was measured.) It is an exact number since it is defined. (Exact numbers are discussed in Section 6-6.) This number is called Avogadro's number and is named in honor of the Italian physicist **Amadeo Avogadro** (1776–1856).[1] For simplicity, we will write Avogadro's number to three significant figures as $6.02 \times 10^{23}$. So, for instance, one atomic weight of oxygen atoms, or 16.0 g of oxygen, contains $6.02 \times 10^{23}$ oxygen atoms. So we could write

**16.0 g O atoms = $6.02 \times 10^{23}$ O atoms.**

**EXAMPLE 3** How many atoms are there in 132.9 g of cesium?

*Solution:* There are $6.02 \times 10^{23}$ cesium atoms, since 132.9 g of cesium is one atomic weight of cesium.  ■

We can also find the number of grams in $6.02 \times 10^{23}$ atoms of an element.

**EXAMPLE 4** How many grams does $6.02 \times 10^{23}$ atoms of silicon weigh?

*Solution:* $6.02 \times 10^{23}$ silicon atoms correspond to one atomic weight of silicon, or 28.1 g of silicon.  ■

---

[1] His full name was Lorenzo Romano Amedeo Carlo Avogadro, conte di Quaregna e Ceretto! See page 140 for an image and a discussion of his character.

## 8-4
## A MOLE IS ONE ATOMIC WEIGHT OR ONE AVOGADRO'S NUMBER OF ATOMS

In Section 8-3, we discussed two ways of looking at a certain quantity of an element.

    1. **One atomic weight.**    2. **One Avogadro's number, $6.02 \times 10^{23}$.**

The first way refers to a certain mass of an element; the second way refers to a certain number of atoms of an element. Scientists have a name for this amount of an element. It is called a **mole**[2] of an element. **One mole of an element is <u>both</u> one atomic weight in grams of an element <u>and</u> one Avogadro's number of atoms of an element. The abbreviation for the "mole" is "mol," and it is one of the seven base units in the SI system of units. See pp 115, 136 and 403.**

**EXAMPLE 5**   How much does one mole of chlorine atoms weigh?

*Solution:* One mole of chlorine atoms weighs one atomic weight of chlorine, which is 35.453 g. ∎

**EXAMPLE 6** How many atoms are in a mole of chlorine (Cl) atoms?

*Solution:* A mole of chlorine atoms, like a mole of anything, contains $6.02 \times 10^{23}$ chlorine atoms. **NOTE:** It is customary to write "chlorine atoms" and not "atoms of chlorine" when atoms (or molecules) are used after a number. ∎

Another way to look at a mole is this. From a pile of atoms of an element, count out $6.02 \times 10^{23}$ atoms. Put them on a balance that reads in grams. **The balance will read the mass in grams of one mole of the element or one atomic weight.** The heavier the atoms the greater the atomic weight. (As shown in Example 7, you could never do this experiment.)

Since 2019, Avogadro's number is usually referred to as **Avogadro's constant**. In this text the terms Avogadro's number and Avogadro's constant are used interchangeably, although there is a slight difference. **When used without a unit, as just the number $6.022\,140\,76 \times 10^{23}$, it is called Avogadro's number. When used with its unit, the "mole," it is written as $6.022\,140\,76 \times 10^{23}$ mol$^{-1}$ and it is then called Avogadro's constant.**

What does $6.022\,140\,76 \times 10^{23}$ mol$^{-1}$ mean and where did it come from? (See Sec 5-5 to review negative exponents.) If we are talking about atoms, we could write

$$6.022\,140\,76 \times 10^{23} \text{ atoms} = 1 \text{ mol (of atoms)}.$$

This equation can be used to convert atoms to moles and moles to atoms and as discussed in the rest of this chapter, molecules to moles and moles to molecules.

Dividing both sides of the above equation by 1 mol gives

$$\frac{6.022\,140\,76 \times 10^{23} \text{ atoms}}{1 \text{ mol}} = \frac{1 \text{ mol}}{1 \text{ mol}} = 1$$

The above fraction gives us a hint as to the unit that Avogadro's number has.

---

[2]"The name *mole* is an 1897 translation of the German unit *Mol*, coined by the Baltic German chemist Wilhelm Ostwald (1853-1932) in 1894 from the German word *Molekül*," which means molecule in English. Quote from the Wikipedia article about the mole.

## 8-4 A Mole is One Atomic Weight or One Avodadro's Number of Atoms

"Atoms" is a descriptive term and not a unit, so scientists prefer to write 6.022 140 76 x $10^{23}$ $\frac{1}{\text{mol}}$ or 6.022 140 76 x $10^{23}$ mol$^{-1}$. Most of the time it is used with this unit, and then it is called Avogadro's constant. **NOTE:** Previous experiments had come up with different values for Avogadro's number which were not as accurate or precise, and scientists were not ready to treat it as a definition. See Sec 8-9 for more details about Avogadro's constant and the mole.

**EXAMPLE 7** If you could count one atom per second, how many years would it take you to count to 6.02 x $10^{23}$ atoms?

**NOTE:** This example tries to give you an idea of how large Avodadro's number is. It is more difficult than the other examples in this chapter. Also, the abbreviation for a year is different in different fields of science. It could be "y", "yr", or "a" (from the Latin *annus*). I will choose "yr" for clarity. There doesn't seem to be an official abbreviation for "day" so I chose to use "day" or "days."

*Solution:* First we must calculate the number of seconds in one year. Since 60 s = 1 min, 60 min = 1 h, 24 h = 1 day, and 365 days = 1 yr (this is not quite accurate, but close enough for our purposes), we have

$$1 \text{ yr} \times \frac{365 \text{ days}}{1 \text{ yr}} \times \frac{24 \text{ hr}}{1 \text{ day}} \times \frac{60 \text{ min}}{1 \text{ hr}} \times \frac{60 \text{ s}}{1 \text{ min}} = 365 \times 24 \times 60 \times 60 = 31{,}526{,}000 \text{ s}$$

or about 3.2 x $10^7$ s in one year. We have used four ratio identities in a row to convert from years to seconds.

Let's say we count 1 atom per second. Then we can say that 1 atom = 1 s. "Atom" or "atoms" are not units but descriptions. Here it is handy to use them as a unit. **As mentioned on page 114, this is one of the first examples where the "identity" is not really an identity, since atoms and seconds are not the same physical quantity. But the mathematics of doing this kind of conversion, where we convert one unit to another, works perfectly,** *if you carefully follow the rules for using them.* We shall do this type of unit conversion a lot in future chapters. Thus

$$1 \text{ atom} = 1 \text{ s}$$
$$3.2 \times 10^7 \text{ s} = 1 \text{ yr}$$

Now all we have to do is convert 6.02 x $10^{23}$ atoms to the equivalent number of years.

$$6.02 \times 10^{23} \text{ atoms} \times \frac{1 \text{ s}}{1 \text{ atom}} \times \frac{1 \text{ yr}}{3.2 \times 10^7 \text{ s}} = \frac{6.02 \times 10^{23}}{3.2 \times 10^7} \text{ yr} = 1.9 \times 10^{16} \text{ yr}$$

or about 2 x $10^{16}$ yr. Let's see how many billions of years this is. Since 1 billion yr = 1,000,000,000 yr = $10^9$ yr, we have

$$2 \times 10^{16} \text{ yr} \times \frac{1 \text{ billion yr}}{10^9 \text{ yr}} = 2 \times 10^7 \text{ billion yr} = 20{,}000{,}000 \text{ billion yr}$$

Because the universe is only about 13.8 billion yr old, it is hardly likely that any intelligent life form could have counted as high as Avogadro's number! ∎

## 8-5
### CALCULATING MOLES AND NUMBERS OF ATOMS USES ATOMIC WEIGHTS AND AVOGADRO'S NUMBER

A very common and useful thing to be able to do is to calculate the number of moles and number of atoms in a sample of an element. To do this kind of calculation, we will use the methods developed in Chapter 7.

Consider the element calcium (symbol Ca). The atomic weight of Ca is 40.1 g. This corresponds to one mole of Ca as well as $6.02 \times 10^{23}$ atoms of Ca. So we can write the following equality (the abbreviation of "mole" is "mol"):

$$40.1 \text{ g of Ca} = 1 \text{ mol of Ca} = 6.02 \times 10^{23} \text{ atoms of Ca}$$

It is customary to leave out the "of" in this type of expression as well as to write "Ca atoms" instead of "atoms of Ca":

$$40.1 \text{ g Ca} = 1 \text{ mol Ca} = 6.02 \times 10^{23} \text{ Ca atoms}$$

If we want to, we can write this expression as three identities:

$$40.1 \text{ g Ca} = 1 \text{ mol Ca}$$

$$40.1 \text{ g Ca} = 6.02 \times 10^{23} \text{ Ca atoms}$$

$$1 \text{ mol Ca} = 6.02 \times 10^{23} \text{ Ca atoms}$$

We can use these identities to help us solve problems like Examples 8 through 13. What we do is select the identity that has both the given (G) and asked-for (AF) units in it. Then we form a ratio identity that divides the given unit and leaves the asked-for unit. Refer to Chapter 7 for more details if necessary.

Notice that in problems involving moles and grams, units divide in the same way that they did in Chapter 7. Thus

$$\frac{\cancel{\text{mol Ca}}}{\cancel{\text{mol Ca}}} = 1 \quad \text{and} \quad \frac{\cancel{\text{g Ca}}}{\cancel{\text{g Ca}}} = 1 \quad \text{and so on.}$$

**EXAMPLE 8** How many grams of Ca (AF) are there in 2.0 mol of Ca (G)?

*Solution:* The identity is 40.1 g = 1 mol Ca. We want to convert mol Ca to g Ca. First write

$$2.0 \text{ mol Ca} \times \text{———} =$$

and then write

$$2.0 \text{ mol Ca} \times \frac{40.1 \text{ g Ca}}{1 \text{ mol Ca}} = 80 \text{ g Ca} \quad \blacksquare$$

**EXAMPLE 9**  How many moles of Ca (AF) are there in 20 g of Ca (G)?

*Solution:* The identity is 40.1 g Ca = 1 mol Ca. We want to convert g Ca to mol Ca:

$$20 \text{ g Ca} \times \frac{1 \text{ mol Ca}}{40.1 \text{ g Ca}} = 0.50 \text{ mol Ca} \quad \blacksquare$$

**EXAMPLE 10**  How many atoms of Ca (AF) are there in 20 g of Ca (G)?

*Solution:* The identity is 40.1 g Ca = $6.02 \times 10^{23}$ Ca atoms. We want to convert g Ca to Ca atoms:

$$20 \text{ g Ca} \times \frac{6.02 \times 10^{23} \text{ Ca atoms}}{40.1 \text{ g Ca}} = \frac{20}{40.1} \times 6.02 \times 10^{23} \text{ Ca atoms}$$

$$= 3.0 \times 10^{23} \text{ Ca atoms} \quad \blacksquare$$

**EXAMPLE 11**  How many atoms of Ca (AF) are there in 2.0 mol of Ca (G)?

*Solution:* The identity is 1 mol Ca = $6.02 \times 10^{23}$ Ca atoms. We want to convert mol Ca to Ca atoms:

$$2.0 \text{ mol Ca} \times \frac{6.02 \times 10^{23} \text{ Ca atoms}}{1 \text{ mol Ca}} = 12 \times 10^{23} \text{ Ca atoms}$$

$$= 1.2 \times 10^{24} \text{ Ca atoms} \quad \blacksquare$$

**EXAMPLE 12**  How many moles of Ca (AF) are there in $3 \times 10^{23}$ atoms of Ca (G)?

*Solution:* The identity is 1 mol Ca = $6.02 \times 10^{23}$ Ca atoms. We want to convert Ca atoms to mol Ca:

$$3 \times 10^{23} \text{ Ca atoms} \times \frac{1 \text{ mol Ca}}{6.02 \times 10^{23} \text{ Ca atoms}} = \frac{3 \times 10^{23}}{6.02 \times 10^{23}} \text{ mol Ca}$$

$$= 0.5 \text{ mol Ca} \quad \blacksquare$$

**EXAMPLE 13** How many grams of Ca (AF) are there in $1.2 \times 10^{24}$ atoms of Ca (G)?

*Solution:* The identity is 40.1 g Ca = $6.02 \times 10^{23}$ Ca atoms. We want to convert Ca atoms to g Ca:

$$1.2 \times 10^{24} \text{ Ca atoms} \times \frac{40.1 \text{ g Ca}}{6.02 \times 10^{23} \text{ Ca atoms}}$$

$$= \frac{1.2 \times 10^{24}}{6.02 \times 10^{23}} \times 40.1 \text{ g Ca} = 80 \text{ g Ca} \quad \blacksquare$$

## 8-6
### ONE MOLE OF MOLECULES CONTAINS AN AVOGADRO'S NUMBER OF MOLECULES WHOSE MASS IS ONE MOLECULAR WEIGHT

As indicated on page 136 in our detailed discussion of the meaning of the word "mole," you might expect that chemists would talk about moles of molecules. They do. A mole of molecules is defined as $6.022\ 140\ 76 \times 10^{23}$ molecules, or

$$1 \text{ mol of molecules} = 6.022\ 140\ 76 \times 10^{23} \text{ molecules}$$

This is reasonable, since one mole of any substance contains one Avogadro number of particles of that substance. In this case, the particles are molecules. (Remember that for calculations in this text, Avogadro's number is shortened to $6.02 \times 10^{23}$.)

The mass (in grams) of one mole of a molecule is called the **molecular weight** of that molecule, just as the mass (in grams) of one mole of the naturally-occurring mixture of isotopes of an element is called the **atomic weight** of that element. Putting this all together, we get:

$$1 \text{ mol of molecules} = 6.02 \times 10^{23} \text{ molecules} = 1 \text{ molecular weight}$$

**This equation allows us to convert between moles of molecules, the number of molecules and the weight of these molecules.**

The question now is: How do we calculate molecular weights? There are no tables of molecular weights readily available because it is easy to calculate the molecular weight of a molecule from the atomic weights of the atoms that make up that molecule. As the following examples show, you just add up the atomic weight of all the atoms in the molecule (https://bit.ly/3Ey2r7C).

**EXAMPLE 14** What is the molecular weight of the diatomic molecule $O_2$?

*Solution:* The atomic weight of oxygen is 16.0 g; and since $O_2$ has two O atoms, the molecular weight of $O_2$ is $2 \times 16.0 \text{ g} = 32.0 \text{ g}$ $\quad \blacksquare$

**EXAMPLE 15** What is the molecular weight of water, $H_2O$?

*Solution:* The atomic weights are H = 1.01 g and O $\blacksquare$ = 16.0 g. The molecular weight of water is 1.01 g + 1.01 g + 16.0 g = 18.0 g.

**EXAMPLE 16** Calculate the molecular weight of glucose, $C_6H_{12}O_6$.

*Solution:* The atomic weights are C = 12.0 g, H = 1.01 g, O = 16.0 g. Then:

$$
\begin{aligned}
&\text{For C we have } 6 \times 12.0 \text{ g} = 72.0 \text{ g} \\
&\text{For H we have } 12 \times 1.01 \text{ g} = 12.1 \text{ g} \\
&\text{For O we have } 6 \times 16.0 \text{ g} = \underline{96.0 \text{ g}} \\
&\text{The molecular weight is} \quad\quad 180.1 \text{ g} \quad \blacksquare
\end{aligned}
$$

**EXAMPLE 17** Calculate the molecular weight of $Ca_3(PO_4)_2$.

*Solution:* The atomic weights are Ca = 40.1 g, P = 31.0 g, O = 16.0 g. Then:

$$
\begin{aligned}
&\text{For Ca we have } 3 \times 40.1 \text{ g} = 120.3 \text{ g} \\
&\text{For P we have } \phantom{0}2 \times 31.0 \text{ g} = \phantom{0}62.0 \text{ g} \\
&\text{For O we have } \phantom{0}8 \times 16.0 \text{ g} = \underline{128.0 \text{ g}} \\
&\text{The molecular weight is:} \quad\quad 310.3 \text{ g} \quad \blacksquare
\end{aligned}
$$

## 8-7
### CALCULATING MOLES AND NUMBERS OF MOLECULES USES MOLECULAR WEIGHTS AND AVOGADRO'S NUMBER

As we did for atoms, we can write three identities for a molecule that will help us to convert between grams, moles, and number of molecules. For the water molecule, $H_2O$, we have

$$18.0 \text{ g } H_2O = 1 \text{ mol } H_2O = 6.02 \times 10^{23} \text{ } H_2O \text{ molecules}$$

We can also express this as three separate identities:

$$18.0 \text{ g } H_2O = 1 \text{ mol } H_2O$$
$$18.0 \text{ g } H_2O = 6.02 \times 10^{23} \text{ } H_2O \text{ molecules}$$
$$1 \text{ mol } H_2O = 6.02 \times 10^{23} \text{ } H_2O \text{ molecules}$$

**EXAMPLE 18** Using the above identities, calculate how many moles and molecules of $H_2O$ (AF) there are in 9.0 g of $H_2O$ (G).

**132** AVOGADRO'S NUMBER, THE MOLE, AND MOLECULAR WEIGHT

*Solution:* To get moles, the identity is 18.0 g H$_2$O = 1 mol H$_2$O. We want to convert g H$_2$O to mol H$_2$O:

$$9.0 \text{ g H}_2\text{O} \times \frac{1 \text{ mol H}_2\text{O}}{18.0 \text{ g H}_2\text{O}} = 0.50 \text{ mol H}_2\text{O}$$

To get molecules, the identity is 18.0 g H$_2$O = 6.02 × 10$^{23}$ H$_2$O molecules. We want to convert g H$_2$O to H$_2$O molecules:

$$9.0 \text{ g H}_2\text{O} \times \frac{6.02 \times 10^{23} \text{ H}_2\text{O molecules}}{18.0 \text{ g H}_2\text{O}} = 3.0 \times 10^2 \text{ H}_2\text{O molecules} \quad \blacksquare$$

**EXAMPLE 19** How many moles of glucose (AF) are there in 135 g of glucose (G)?

*Solution:* The molecular weight of glucose is 180 g. The identity is 180 g C$_6$H$_{12}$O$_6$ = 1 mol C$_6$H$_{12}$O$_6$. We want to convert g C$_6$H$_{12}$O$_6$ to mol C$_6$H$_{12}$O$_6$:

$$135 \text{ g C}_6\text{H}_{12}\text{O}_6 \times \frac{1 \text{ mol C}_6\text{H}_{12}\text{O}_6}{180 \text{ g C}_6\text{H}_{12}\text{O}_6} = 0.750 \text{ mol C}_6\text{H}_{12}\text{O}_6 \quad \blacksquare$$

## 8-8
## CALCULATING MOLES OF ATOMS IN MOLECULES USES THE MOLECULAR FORMULA

It is useful to know how to calculate the number of moles of atoms in a mole of molecules. This can be done by setting up the proper identities. For example, each molecule of methane, CH$_4$, has one atom of C and four atoms of H. And 1 mol of CH$_4$ contains 1 mol of C and 4 mol of H. We can write the following identities:

$$1 \text{ CH}_4 \text{ molecule} = 1 \text{ C atom} = 4 \text{ H atoms}$$
$$1 \text{ mol CH}_4 = 1 \text{ mol C} = 4 \text{ mol H}$$

Of course, 1 molecule of CH$_4$ doesn't actually equal 1 atom of C or 4 atoms of H. But we can use the equals sign here if we think of it as reading "contains." The same goes for the second identity involving moles.

**EXAMPLE 20** Using the identities just given, how many moles of C atoms and moles of H atoms (AF) are there in 5.0 mol of CH$_4$ (G)?

8-8 Calculating Moles of Atoms in Molecules Uses the Molecular Formula  133

*Solution:* The identities are

$$1 \text{ mol CH}_4 = 1 \text{ mol C} \quad \text{and} \quad 1 \text{ mol CH}_4 = 4 \text{ mol H}$$

To get moles of C, we want to convert mol $CH_4$ to mol C:

$$5.0 \text{ mol CH}_4 \times \frac{1 \text{ mol C}}{1 \text{ mol CH}_4} = 5.0 \text{ mol C}$$

To get moles of H, we want to convert mol $CH_4$ to mol H:

$$5.0 \text{ mol CH}_4 \times \frac{4 \text{ mol H}}{1 \text{ mol CH}_4} = 20 \text{ mol H} \quad \blacksquare$$

**EXAMPLE 21** How many atoms of C (AF) and atoms of H (AF) are in 5 molecules of $CH_4$ (G)?

*Solution:* The identities are

$$1 \text{ CH}_4 \text{ molecule} = 1 \text{ C atom} \quad \text{and} \quad 1 \text{ CH}_4 \text{ molecule} = 4 \text{ H atoms}$$

To get atoms of C, we want to convert $CH_4$ molecules to C atoms:

$$5 \text{ CH}_4 \text{ molecules} \times \frac{1 \text{ C atom}}{1 \text{ CH}_4 \text{ molecule}} = 5 \text{ C atoms}$$

To get atoms of H, we want to convert $CH_4$ molecules to H atoms:

$$5 \text{ CH}_4 \text{ molecules} \times \frac{4 \text{ C atoms}}{1 \text{ CH}_4 \text{ molecule}} = 20 \text{ H atoms} \quad \blacksquare$$

**EXAMPLE 22** How many atoms of H (AF) are there in 2.0 mol of pentane, $C_5H_{12}$ (G)?

There are two ways to solve this problem. See the discussion after this example for some details of what we have done.

*Solution 1:* This approach takes the following route:

$$\text{mol pentane} \rightarrow \text{mol H} \rightarrow \text{H atoms}$$

**134** AVOGADRO'S NUMBER, THE MOLE, AND MOLECULAR WEIGHT

The identities are

$$1 \text{ mol pentane} = 12 \text{ mol H} \quad \text{and} \quad 1 \text{ mol H} = 6.02 \times 10^{23} \text{ atom H}$$

$$2.0 \cancel{\text{mol pentane}} \times \frac{12 \cancel{\text{mol H}}}{1 \cancel{\text{mol pentane}}} \times \frac{6.02 \times 10^{23} \text{ H atoms}}{1 \cancel{\text{mol H}}}$$

$$= 1.4 \times 10^{25} \text{ H atoms}$$

***Solution 2:*** This approach takes the following route:

$$\text{mol pentane} \rightarrow \text{pentane molecules} \rightarrow \text{H atoms}$$

The identities are

$$1 \text{ mol pentane} = 6 \times 10^{23} \text{ pentane molecules} \quad \text{and} \quad 1 \text{ pentane molecule}$$
$$= 12 \text{ H atoms}$$

$$2.0 \cancel{\text{mol pentane}} \times \frac{6.02 \times 10^{23} \cancel{\text{pentane molecules}}}{1 \cancel{\text{mol pentane}}}$$

$$\times \frac{12 \text{ H atoms}}{1 \cancel{\text{pentane molecule}}} = 1.4 \times 10^{25} \text{ H atoms} \blacksquare$$

In Example 22, we used two identities to convert 2.0 mol of pentane into $1.4 \times 10^{25}$ H atoms. We could, of course, have worked the solution in two steps using one identity at a time. For example, referring to solution 1 in Example 22, we would have

$$2 \cancel{\text{mol pentane}} \times \frac{12 \text{ mol H}}{1 \cancel{\text{mol pentane}}} = 24 \text{ mol H}$$

and

$$24 \cancel{\text{mol H}} \times \frac{6.02 \times 10^{23} \text{ H atoms}}{1 \cancel{\text{mol H}}} = 1.4 \times 10^{25} \text{ H atoms}$$

It is clearly shorter to solve the problem in one step. Again, using solution 1 in Example 22 as the example, we do the following.

1. Write down the two identities:

$$1 \text{ mol pentane} = 12 \text{ mol H}$$
$$1 \text{ mol H} = 6.02 \times 10^{23} \text{ H atoms}$$

## 8-8 Calculating Moles of Atoms in Molecules Uses the Molecular Formula

2. Write down the given, a "×" sign, a horizontal line, a "×" sign, a horizontal line, and an "=" sign:

$$2.0 \text{ mol pentane} \times \underline{\hspace{2cm}} \times \underline{\hspace{2cm}} =$$

3. Put in one identity so that the given unit is divided:

$$2.0 \text{ mol pentane} \times \frac{12 \text{ mol H}}{1 \text{ mol pentane}} \times \underline{\hspace{2cm}} =$$

4. Put in the other identity to divide the "mol H" and give the answer:

$$2.0 \text{ mol pentane} = \frac{12 \text{ mol H}}{1 \text{ mol pentane}} \times \frac{6.02 \times 10^{23} \text{ H atoms}}{1 \text{ mol H}}$$

$$= 1.4 \times 10^{25} \text{ H atoms}$$

At this point, you should work solution 2 of Example 22 in separate steps as was done for solution 1.

In your study of the previous examples, you may have noticed that sometimes the name and sometimes the chemical formula were used to identify the unit. We could have written "mol $C_5H_{12}$" or "mol pentane." Both are correct—you just have to be sure to identify the unit. Some authors will use an abbreviation to identify the unit—such as "mol G" for "moles of glucose." In solving problems, you can make the choice as long as everyone knows what you are talking about.

**EXAMPLE 23** How many atoms of oxygen (AF) are there in 49 g of sulfuric acid, $H_2SO_4$ (G)?

*Solution:* The molecular weight of $H_2SO_4$ is 98.0 g. The identities are

$$1 \text{ mol } H_2SO_4 = 98.0 \text{ g } H_2SO_4$$

$$1 \text{ mol } H_2SO_4 = 4 \text{ mol O}$$

$$1 \text{ mol O} = 6.02 \times 10^{23} \text{ O atoms}$$

Our conversion should take the route g $H_2SO_4 \rightarrow$ mol $H_2SO_4 \rightarrow$ mol O $\rightarrow$ O atoms:

$$49 \text{ g } H_2SO_4 \times \frac{1 \text{ mol } H_2SO_4}{98.0 \text{ g } H_2SO_4} \times \frac{4 \text{ mol O}}{1 \text{ mol } H_2SO_4} \times \frac{6.02 \times 10^{23} \text{ O atoms}}{1 \text{ mol O}}$$

$$= 1.2 \times 10^{24} \text{ O atoms} \quad \blacksquare$$

At this point, it might not always be clear which identities you need to use to solve a problem. As usual, the more problems you solve the better you will get!

## 8-9
### THE MOLE IS DEFINED IN TERMS OF A CERTAIN AMOUNT OF SUBSTANCE

As we have already implied, the definition of the mole can be expanded to include things other than atoms and molecules. The *official definition of the mole*, as of May 20, 2019, is:

**The mole, symbol mol, is the SI unit for an amount of substance.**[2]

**One mole contains *exactly* 6.022 140 76 × $10^{23}$ elementary entities.**

See pp 122, 126, and https://bit.ly/3HlPwoq for additional details.

**The "amount of substance" in the case of a mole is a measure of the number of specified elementary entities. The elementary entities must be specified and may be atoms, molecules, ions, electrons, photons, or other particles, or specified groups of such particles.**

**Avogadro's constant, 6.022 140 76 × $10^{23}$ $mol^{-1}$, is one of the most important constants across many fields of science where it can be found in extremely important equations. That's why scientists went through so much trouble to define it as discussed on pages 122 and 157. The unit conversions done with it in this chapter are only one use.**

NOTE: The definition of the mole used before May 20, 2019 was: The mole is the amount of substance of a system which contains as many elementary entities as there are atoms in exactly 0.012 kg (or 12 g) of carbon-12; its symbol is "mol."

Previously in this chapter, we used the mole to do unit conversions with Avogadro's constant and atomic or molecular weights. The following two examples show how the mole and Avogadro's constant can be used with electrons and photons. Electrons have very little mass (Table 1-1) and the need to calculate the mass of a certain number of electrons rarely arises. Photons have no mass and we can only convert between a number of photons and the corresponding number of moles of photons. See Sec 13-3 for a discussion of photons. They are the "particles" associated with a light wave.

**EXAMPLE 24** How many electrons (AF) are in a mole of electrons (G)?

*Solution:* There are 6.02 x $10^{23}$ electrons in 1 mol of electrons. ∎

**EXAMPLE 25** How many photons (AF) are in 3.0 mol of photons (G)?

*Solution:* The identity is 1 mol photons = 6.02 x $10^{23}$ photons. We want to convert moles of photons to number of photons:

$$3.0 \text{ mol photons} \times \frac{6.02 \times 10^{23} \text{ photons}}{1 \text{ mol photons}} = 1.8 \times 10^{24} \text{ photons} \quad \blacksquare$$

The next and last section covers the mole interpretation of chemical equations.

---

[2]The BIPM (page xix), in the French language part of the report on the SI system of units, used the French term "quantité de matière," which most people (and Google) would translate into English as "quantity of matter." (See https://bit.ly/3qNi1o9, reference #41, page 24.) In the English language part of the report, it is translated as "amount of substance." The above link also has a brief history of this area of chemistry. Notice how confusing chemistry was until the 20th century. Think of doing chemistry without understanding or knowing about atoms, isotopes, molecules, electrons, protons, or neutrons.

# 8-10
## THE MOLE INTERPRETATION OF CHEMICAL EQUATIONS USES THE COEFFICIENTS OF THE SUBSTANCES IN THE BALANCED EQUATION

We can now interpret chemical equations in a different way from that done in Chapter 3. In Chapter 3, we said that a chemical equation such as

$$2H \rightarrow H_2$$

can be interpreted as "Two atoms of H react to give one molecule of $H_2$."

Now we can look at the equation in terms of moles. From this point of view,

$$2H \rightarrow H_2$$

is interpreted as "Two moles of H react to give one mole of $H_2$."

All we have done is increase the quantity. Instead of talking about atoms and molecules, we are now talking about moles. We have gone from talking about a few atoms and molecules to talking about many atoms and molecules (on the order of Avogadro's number, $6.02 \times 10^{23}$).

This approach is very useful because it allows us to do calculations using reasonable amounts of material. Mole-sized masses can be weighed out on ordinary balances; a few atoms or molecules weigh so little that it would be impossible to weigh them.

**EXAMPLE 26** Interpret the equation $2H_2 + O_2 \rightarrow 2H_2O$ in terms of moles.

*Solution:* This is interpreted as "Two moles of $H_2$ react with one mole of $O_2$ to form two moles of $H_2O$." ∎

**EXAMPLE 27** Interpret the equation $6CO_2 + 6H_2O \rightarrow C_6H_{12}O_6 + 6O_2$ in terms of moles.

*Solution:* This is interpreted as "Six moles of $CO_2$ react with six moles of $H_2O$ to form one mole of $C_6H_{12}O_6$ and six moles of $O_2$." ∎

## PROBLEMS

### KEYED PROBLEMS

1. The relative atomic weight of calcium is 40 u, and the relative atomic weight of carbon is 12 u. If you wanted to take an equal number of Ca atoms and C atoms, what would be the ratio of their masses that you would need?

# AVOGADRO'S NUMBER, THE MOLE, AND MOLECULAR WEIGHT

2. What are the atomic weights of the following elements, all of which contain the same number of atoms? The elements are Na, Xe, W, and Pb.

3. How many atoms are there in 55.847 g of iron?

4. How many grams does $6.02 \times 10^{23}$ atoms of palladium weigh?

5. How much does 1 mol of nickel weigh?

6. How many atoms are in 1 mol of nickel?

7. How many atoms would have been counted to today's date if Avogadro had started counting one atom per second on 12:01 A.M. of January 1, 1811? Make any reasonable assumptions you like about leap years, and so on. (This problem is more difficult than the others.)

8. How many grams of silicon are there in 2.0 mol of silicon?

9. How many moles of silicon are there in 14 g of silicon?

10. How many atoms of silicon are there in 14 g of Si?

11. How many atoms of Si are there in 2.0 mol of Si?

12. How many moles of Si are there in $3.0 \times 10^{23}$ atoms of Si?

13. How many grams of Si are there in $1.2 \times 10^{24}$ atoms of Si?

14. What is the molecular weight of the diatomic molecule $F_2$?

15. What is the molecular weight of ammonia, $NH_3$?

16. Calculate the molecular weight of ethanol, $C_2H_6O$.

17. Calculate the molecular weight of $Al_2(SO_4)_3$.

18. Using the following identities, calculate how many moles and molecules of $NH_3$ there are in 34 g of $NH_3$. First we write one identity containing all the information needed, and then we split it up into three identities. As one identity we have

$$17.0 \text{ g } NH_3 = 1 \text{ mol } NH_3 = 6.02 \times 10^{23} \text{ } NH_3 \text{ molecules}$$

As three separate identities we have

$$17.0 \text{ g } NH_3 = 1 \text{ mol } NH_3$$
$$17.0 \text{ g } NH_3 = 6.02 \times 10^{23} \text{ } NH_3 \text{ molecules}$$
$$1 \text{ mol } NH_3 = 6.02 \times 10^{23} \text{ } NH_3 \text{ molecules}$$

**NOTE:** Only two of these three identities are needed to solve this problem.

19. How many moles of ethanol are there in 23 g of ethanol, $C_2H_6O$?

20. Using the identities at the beginning of Section 8-8, how many moles of C and H atoms are there in 7 mol of $CH_4$?

21. How many atoms of C and atoms of H are there in 5 molecules of ethane, $C_2H_6$?

22. How many atoms of H are there in 8.0 mol of hexane, $C_6H_{14}$?

23. How many atoms of oxygen are there in 49 g of phosphoric acid, $H_3PO_4$?

24. How many electrons are there in 2.0 mol of electrons?

25. How many photons are there in 5.0 mol of photons?

26. Interpret the equation $2NO + O_2 \rightarrow 2NO_2$ in terms of moles.

27. Interpret the equation $Mg(OH)_2 + 2HCl \rightarrow MgCl_2 + 2H_2O$ in terms of moles.

## SUPPLEMENTAL PROBLEMS

28. How many atoms are there in 19 g of fluorine?

29. Calculate the molecular weight of calcium sulfate, $CaSO_4$.

30. How many moles are there in 25 g of manganese dioxide, $MnO_2$?

31. How many grams are there in 4 mol of sodium thiosulfate, $Na_2S_2O_3$?

32. How many atoms of sulfur and oxygen are there in 3.0 mol of sulfur trioxide, $SO_3$?

33. How many atoms of chlorine and fluorine are there in 50 g of chlorine trifluoride, $ClF_3$?

34. How many moles are there in 86 g of hydrazoic acid, $HN_3$?

35. How many grams are there in $9.50 \times 10^{24}$ molecules of $CO_2$?

36. How many grams are there in $3.20 \times 10^{21}$ molecules of diethyl ether, $C_4H_{10}O$?

37. The formula of vitamin C, ascorbic acid, is $C_6H_8O_6$. How many moles of ascorbic acid are there in 4.0 g of ascorbic acid?

38. The formula of vitamin $B_{12}$ is $C_{63}H_{90}CoN_{14}O_{14}P$. How many atoms of carbon are there in $1.0 \times 10^{-10}$ mol of vitamin $B_{12}$?

39. Thyroxine, $C_{15}H_{11}I_4NO_4$, is produced by the thyroid gland. Normal blood levels in the human being are about 5 µg in 100 mL of blood serum. How many moles of thyroxine is this?

**NOTE:** $T_4$ is the common abbreviation for thyroxine because it contains four iodine atoms per molecule.

40. A birth control pill contains 1.0 mg of progesterone, $C_{21}H_{30}O_2$. How many moles of progesterone is this?

41. Some people on the liquid protein diet have died, possibly from a potassium deficiency. Normal blood serum contains about $4 \times 10^{-3}$ mol of potassium in 1 L. How many grams of potassium is this?

42. Interpret the following equation in terms of moles (this reaction is part of the Ostwald process, which is used in making nitric acid):

$$4NH_3 + 5O_2 \rightarrow 4NO + 6H_2O$$

43. Interpret the following equation in terms of moles:

$$PBr_3 + 3HOH \rightarrow P(OH)_3 + 3HBr$$

44. What is the mass, in grams, of a single gold atom?

*Amedeo Avogadro*

Avogadro seems to have been a very modest man. He was isolated scientifically in Italy and didn't push his ideas. In fact, none of his obituaries mentioned his ideas about molecules, though they described him as "religious without intolerance, learned without pedantry, wise without ostentation, a despiser of pomp, without care for riches, not ambitious for honours; ignorant of his own worth and fame, modest, temperate, and lovable." See page 217 for how Cannizzaro told the world about Avogadro's research 50 years after he had done it.   Quote from https://rsc.li/3ulw7Oi

## REMARKS ABOUT SOME NAMES USED IN THIS BOOK

In this book, we have used the terms "atomic weight," "molecular weight" and "formula weight" (see Section 10-4). However, the International Union of Pure and Applied Chemistry (IUPAC), the organization that is responsible for, among other things, setting the standards for naming things in chemistry, has also approved some additional names for terms used in this and other chapters (https://bit.ly/3UGiHZK).

For **atomic weight**, as in the heading of Table 2-2, "Table of Atomic Weights" (what we call "relative atomic weight" in some of Chapter 2), the term **relative atomic mass** can be used. However, the Table of Atomic Weights published by the IUPAC Commission on Isotopic Abundances and Atomic Weights is called "Atomic Weight of the Elements [followed by a year]" and that is the table all scientists use (https://bit.ly/2SN4Kwv). This makes sense because "relative atomic mass" (or "atomic mass", as some authors incorrectly use) could be confused with isotopic mass.

The different uses come about because we usually use grams as the unit of atomic weight, and grams are not a weight but a unit of mass (see Section 1-1). Using the term "atomic weight" makes it very clear that it refers to the weighted average mass of all the naturally occurring isotopes of an element. For more on atomic weight, see https://bit.ly/2S40dBo. For more on relative atomic mass, see https://bit.ly/2GKb5SM.

Some textbook authors use the term **molecular mass** for **molecular weight**, using the same logic as described above. Also, some textbook authors use the term **formula mass** for **formula weight** (see Section 10-4), again using the same logic as described above. However, many chemists prefer the terms "molecular weight" and "formula weight."

Since one atomic weight, one molecular weight and one formula weight each contain one mole of atoms, molecules or formula units, the IUPAC also allows the use of the term **molar mass**, which is the mass of one atomic weight of atoms, the mass of one molecular weight of molecules and compounds, and the mass of one formula weight of the simplest formula of atoms for a giant array (see Section 10-4). The term "molar mass" is used in an increasing number of chemistry texts and the chemical literature, but many chemists prefer to use the terms "atomic weight," "molecular weight" and "formula weight." It can all be very confusing. **HINT:** *If you are taking a chemistry course, use the names that your instructor uses.*

**Space Shuttle Liftoff:** The solid rocket propellant in each solid rocket booster (the two long grey cylinders on each side of the dark center liquid-fuel tank) is commonly referred to as *Ammonium Perchlorate Composite Propellant*. The white exhaust is mostly due to white aluminum oxide coming from the combustion of the aluminum powder in the propellant. The chemical reaction is $4Al + 3O_2 \rightarrow 2Al_2O_3$. The oxygen comes from the decomposition of ammonium perchlorate.

This propellant is called a "composite" propellant to distinguish it from previous solid propellants such as gunpowder and mixtures of nitroglycerin and nitrocellulose. These types of propellants are not suitable for use in large rockets such as the Space Shuttle boosters.

The composition of this composite propellant is:
- Ammonium perchlorate ($NH_4ClO_4$, the oxidizer, 69.6% by weight).
- Atomized aluminum powder (Al, a fuel, 16% by weight).
- Iron(III) oxide, also called ferric oxide ($Fe_2O_3$, the catalyst which speeds up the combustion, 0.4% by weight).
- A synthetic polymer. It acts as a binder and fuel (12.04% by weight). A **polymer** is a very large molecule made up of repeating small molecules that are linked together with chemical bonds. These small molecules are called **monomers**. The polymer chain can be made from only one monomer, or it can be made from 2 or more different monomers.

**NOTE:** The perchlorate ion ($ClO_4^-$) can get into groundwater and food (through plastic food packaging and food processing equipment), and if ingested, can cause endocrine and reproductive problems, is a possible carcinogen, and can cause infant and fetal brain damage. The amount in groundwater is regulated by the EPA (https://bit.ly/2CjIsvw, https://nyti.ms/37Trz6H).    *Photo is in the Public Domain.*

# STOICHIOMETRY

The last item discussed in Chapter 8 was the mole interpretation of chemical equations. In this chapter, we shall use this interpretation to solve numerical problems involving balanced chemical equations. The solution of these kinds of problems is called **stoichiometry** (from the Greek *stoicheion* [element] and *metry* [measurement]). Solving these types for problems is one of the basic items in every chemist's toolkit and well worth mastering.

## 9-1 IDENTITIES FROM THE BALANCED CHEMICAL EQUATION USE THE COEFFICIENTS OF THE SUBSTANCE IN THE EQUATION

Consider again the equation $2H_2 + O_2 \rightarrow 2H_2O$. The mole interpretation of this balanced equation is: 2 mol of $H_2$ react with 1 mol of $O_2$ to give 2 mol of $H_2O$. We could also write this as

$$2 \text{ mol } H_2 + 1 \text{ mol } O_2 \rightarrow 2 \text{ mol } H_2O$$

Now if we replace the "+" and the "$\rightarrow$" signs with "=" signs, we get

$$2 \text{ mol } H_2 = 1 \text{ mol } O_2 = 2 \text{ mol } H_2O$$

This expression looks very much like the identities that we wrote in Chapters 7 and 8. However, as was true of some of the identities in Chapter 8, this mole identity must be interpreted carefully. *It is meaningless without reference to the balanced chemical equation.* After all, 2 mol of $H_2$ do not really equal 1 mol of $O_2$, but 2 mol of $H_2$ do react with 1 mol of $O_2$ according to the balanced chemical equation. The reason that we like to write an identity between the substances of a balanced chemical equation is that it allows us to easily solve problems.

One story that might make the use of the "=" sign here a bit clearer is the story of the two cavemen, Charlie and Ogg. Charlie has three clams, and he wants some of the delicious dates that Ogg has. So Ogg agrees to trade eight dates for the three clams. Thus, in the cave barter system, three clams are equivalent to eight dates, or

$$3 \text{ clams} = 8 \text{ dates}$$

But you must be aware that if clams get scarce, things could change—maybe then you could get eight dates for only one clam. The trade would then be

$$1 \text{ clam} = 8 \text{ dates}$$

This is something like the substances in the balanced chemical equation. Only if you know the balanced chemical equation that relates the substances can you write the mole identity.

**EXAMPLE 1** Write the mole identity between the carbon and the oxygen in the reaction $C + O_2 \rightarrow CO_2$. (There is no need to consider the $CO_2$ for this problem.)

*Solution:* The mole identity is 1 mol C = 1 mol $O_2$. ∎

**EXAMPLE 2** Write the identity between carbon and oxygen in the reaction $2C + O_2 \rightarrow 2CO$. (Again, ignore the CO for this problem.)

*Solution:* The identity is 2 mol C = 1 mol $O_2$. ∎

As you can see from Examples 1 and 2, the mole identity changes if the balanced chemical equation changes, even though the substances are the same. You might wonder why carbon reacts in two different ways with oxygen. Only $CO_2$ is formed when there is a lot of oxygen present, such as in open flames. Some CO is formed when the oxygen supply is limited, such as in an internal combustion engine.

**EXAMPLE 3** Write the three mole identities from the reaction $2H_2 + O_2 \rightarrow 2H_2O$.

*Solution:* The identities are 2 mol $H_2$ = 1 mol $O_2$, 2 mol $H_2$ = 2 mol $H_2O$, and 1 mol $O_2$ = 2 mol $H_2O$. ∎

## 9-2
## MOLE RELATIONSHIPS FROM THE BALANCED CHEMICAL EQUATION
## USE THE COEFFICIENTS OF THE SUBSTANCES IN THE EQUATION

We can use identities like the ones in Example 3 to relate moles of different substances in a chemical reaction.

**EXAMPLE 4**   How many moles of $O_2$ (AF) are needed to completely burn 5.0 mol of $H_2$ (G) to form water?

*Solution:* The balanced equation for the reaction is $2H_2 + O_2 \rightarrow 2H_2O$. The identity relating moles $H_2$ and moles $O_2$ is 2 mol $H_2$ = 1 mol $O_2$. Then we convert 5.0 mol $H_2$ into moles $O_2$:

$$5.0 \text{ mol } H_2 \times \frac{1 \text{ mol } O_2}{2 \text{ mol } H_2} = 2.5 \text{ mol } O_2 \quad \blacksquare$$

**EXAMPLE 5**   How many moles of ammonia, $NH_3$, (AF) can be produced from 12 mol of $H_2$ (G) in the reaction $3H_2 + N_2 \rightarrow 2NH_3$?

*Solution:* The identity relating moles $NH_3$ and moles $H_2$ is 3 mol $H_2$ = 2 mol $NH_3$. Now we convert 12 mol $H_2$ to moles of $NH_3$:

$$12 \text{ mol } H_2 \times \frac{2 \text{ mol } NH_3}{3 \text{ mol } H_2} = 8 \text{ mol } NH_3 \quad \blacksquare$$

## 9-3
## MASS RELATIONSHIPS FROM THE BALANCED CHEMICAL EQUATION
## USE BOTH THE COEFFICIENTS AND THE ATOMIC OR MOLECULAR WEIGHTS

Mass relationships between substances in a balanced equation are very important. This is so because chemicals are commonly weighed; therefore, if you want to know how much reactant to use or how much product you can get, you must know how to do mass calculations.

It is the mole identity that allows us to convert from moles of one substance to moles of another. We can then convert from grams to moles and from moles to grams by using the atomic or molecular weights.

Suppose that we are interested in two substances in a chemical reaction. We are given grams of one substance and asked to find grams of the other. The following conversions would be needed,

$$\text{grams substance 1} \xrightarrow{A} \text{moles substance 1} \xrightarrow{B} \text{moles substance 2} \xrightarrow{C} \text{grams substance 2}$$

or simply

$$\text{grams} \xrightarrow{A} \text{moles} \xrightarrow{B} \text{moles} \xrightarrow{C} \text{grams}$$

Conversions **A** and **C** use the atomic or molecular weights in the identity. Conversion **B** uses the appropriate mole relationship from the balanced equation.

Let's look at the following problem in detail. How many grams of $O_2$ (AF) are needed to burn 8.0 g of $H_2$ (G) in the reaction $2H_2 + O_2 \rightarrow 2H_2O$? We are given 8.0 g of $H_2$. We want to find grams of $O_2$. We can do the following.

**A.** *Convert g $H_2$ to moles $H_2$:* The identity is 1 mol $H_2$ = 2.02 g $H_2$. Therefore, we have

$$8.0 \text{ g } H_2 \times \frac{1 \text{ mol } H_2}{2.02 \text{ g } H_2} = 4.0 \text{ mol } H_2$$

**B.** *Convert the 4.0 mol $H_2$ to mol $O_2$.* The identity is 2 mol $H_2$ = 1 mol $O_2$. Therefore, we have

$$4.0 \text{ mol } H_2 \times \frac{1 \text{ mol } O_2}{2 \text{ mol } H_2} = 2.0 \text{ mol } O_2$$

**C.** *Convert the 2.0 mol $O_2$ into grams $O_2$.* The identity is 1 mol $O_2$ = 32.0 g $O_2$. Therefore, we have

$$2.0 \text{ mol } O_2 \times \frac{32.0 \text{ g } O_2}{1 \text{ mol } O_2} = 64 \text{ g } O_2$$

Thus it takes 64 g of oxygen to burn 8.0 g of hydrogen.

This problem could have been done in one step using all three identities at once. The three identities are

**A.** 1 mol $H_2$ = 2.02 g $H_2$.
**B.** 2 mol $H_2$ = 1 mol $O_2$.
**C.** 1 mol $O_2$ = 32.0 g $O_2$.

The solution, in one step, is

$$8.0 \text{ g } H_2 \times \frac{1 \text{ mol } H_2}{2.02 \text{ g } H_2} \times \frac{1 \text{ mol } O_2}{2 \text{ mol } H_2} \times \frac{32.0 \text{ g } O_2}{1 \text{ mol } O_2} = 64 \text{ g } O_2$$

$$\phantom{8.0 \text{ g } H_2 \times \ } \uparrow \phantom{\frac{1 \text{ mol } H_2}{2.02 \text{ g } H_2}} \uparrow \phantom{\frac{1 \text{ mol } O_2}{2 \text{ mol } H_2}} \uparrow$$

*Conversion:*   **A**   **B**   **C**

Let's look closely at the procedure used to solve this problem in one step.

1. Write down the given and the asked for (G) = 8.0 g $H_2$, (AF) = ? g $O_2$.
2. Write down the three identities:
   
   **A.** *To convert grams given to moles given:* 1 mol $H_2$ = 2.02 g $H_2$.
   
   **B.** *To convert moles given to moles asked for:* 2 mol $H_2$ = 1 mol $O_2$.
   
   **C.** *To convert moles asked for to grams asked for:* 1 mol $O_2$ = 32.0 g $O_2$.

3. Write down the given, a "×", a horizontal line, a "×", a horizontal line, a "×", a horizontal line, and an "=":

$$8.0 \text{ g } H_2 \times \underline{\hspace{2cm}} \times \underline{\hspace{2cm}} \times \underline{\hspace{2cm}} = ? \text{ g } O_2$$

4. Fill the lines with the appropriate ratio identities and compute the answer:

$$8.0 \text{ g } H_2 \times \frac{1 \text{ mol } H_2}{2.02 \text{ g } H_2} \times \frac{1 \text{ mol } O_2}{2 \text{ mol } H_2} \times \frac{32.0 \text{ g } O_2}{1 \text{ mol } O_2} = 64 \text{ g } O_2$$

Conversion:     A      B      C

The following examples should illustrate the method of solving these kinds of problems.

**EXAMPLE 6** How many grams of $O_2$ (AF) are needed to burn completely 81 g of Al (G) in the following reaction:

$$4Al + 3O_2 \rightarrow 2Al_2O_3$$

*Solution:* The identities needed are **(A)** 1 mol Al = 27.0 g Al, **(B)** 4 mol Al = 3 mol $O_2$, and **(C)** 1 mol $O_2$ = 32.0 g $O_2$. Our conversion should take the route g Al → mol Al → mol $O_2$ → g $O_2$:

$$81 \text{ g Al} \times \frac{1 \text{ mol Al}}{27.0 \text{ g Al}} \times \frac{3 \text{ mol } O_2}{4 \text{ mol Al}} \times \frac{32.0 \text{ g } O_2}{1 \text{ mol } O_2} = 72 \text{ g } O_2 \quad \blacksquare$$

Conversion:     A      B      C

**EXAMPLE 7** How many grams of water (AF) are produced from the combustion of 0.50 mol of methane, $CH_4$ (G)? The reaction is

$$CH_4 + 2O_2 \rightarrow CO_2 + 2H_2O$$

**Solution:** This problem is a bit shorter than Example 6 because we are given the moles of methane. Therefore, conversion **A**, where grams are converted to moles, can be eliminated.

The identities are **(B)** 1 mol $CH_4$ = 2 mol $H_2O$ and **(C)** 1 mol $H_2O$ = 18.0 g $H_2O$. Our conversion should take the route mol $CH_4 \to$ mol $H_2O \to$ g $H_2O$:

$$0.50 \text{ mol CH}_4 \times \frac{2 \text{ mol H}_2\text{O}}{1 \text{ mol CH}_4} \times \frac{18.0 \text{ g H}_2\text{O}}{1 \text{ mol H}_2\text{O}} = 18 \text{ g H}_2\text{O} \quad \blacksquare$$

Conversion:             **B**         **C**

**EXAMPLE 8** How many moles of $CO_2$ (AF) are produced by the oxidation of 90 g of glucose, $C_6H_{12}O_6$ (G)? The reaction is

$$C_6H_{12}O_6 + 6O_2 \to 6CO_2 + 6H_2O$$

**Solution:** Again, this problem only uses two identities. They are **(A)** 1 mol glucose = 180 g glucose and **(B)** 1 mol glucose = 6 mol $CO_2$. Our conversion should take the route g glucose $\to$ mole glucose $\to$ mole $CO_2$:

$$90 \text{ g glucose} \times \frac{1 \text{ mol glucose}}{180 \text{ g glucose}} \times \frac{6 \text{ mol CO}_2}{1 \text{ mol glucose}} = 3.0 \text{ mol CO}_2 \quad \blacksquare$$

Conversion:             **A**         **B**

**EXAMPLE 9** How many grams of calcium carbonate, $CaCO_3$ (AF), are needed to completely neutralize (react with) 109.5 g of hydrochloric acid, HCl (G)? The reaction is $CaCO_3 + 2HCl \to CaCl_2 + H_2O + CO_2$. (Note: Hydrochloric acid is a water solution of hydrogen chloride, HCl. It is a strong acid. $CaCO_3$ acts as a base. The reaction between an acid and a base is called **neutralization.** $CaCO_3$ is the main ingredient in TUMS antacid tablets, and hydrochloric acid is "stomach acid.")

**Solution:** This solution requires all three conversions. The identities are **(A)** 1 mol HCl = 36.46 g HCl, **(B)** 2 mol HCl = 1 mol $CaCO_3$, and **(C)** 1 mol $CaCO_3$ = 100.1 g $CaCO_3$. Our conversion should take the route g HCl $\to$ mol HCl $\to$ mol $CaCO_3 \to$ g $CaCO_3$:

$$109.5 \text{ g HCl} \times \frac{1 \text{ mol HCl}}{36.46 \text{ g HCl}} \times \frac{1 \text{ mol CaCO}_3}{2 \text{ mol HCl}} \times \frac{100.1 \text{ g CaCO}_3}{1 \text{ mol CaCO}_3}$$

$$= 150.3 \text{ g CaCO}_3 \quad \blacksquare$$

Conversion:             **A**         **B**         **C**

# 9-4
## THE LIMITING REAGENT IS THE REACTANT THAT IS USED UP FIRST IN A CHEMICAL REACTION

So far, the stoichiometry problems that we have solved assumed that all the reactants were used up. Frequently, one of the reactants will be in excess.

An example from everyday life should illustrate what we mean. Suppose you are in charge of making hero sandwiches (subs, torpedoes, grinders, hoagies, po' boys) in the school cafeteria. For each hero, you need one piece of hero bread, one slice of bologna, and one slice of cheese. Assume that mustard, mayonnaise, lettuce, and tomato are in great excess. You have 100 pieces of hero bread, 80 slices of bologna, and 60 slices of cheese. How many heros can you make? Only 60—the cheese is the limiting ingredient. It runs out first. The bread and the bologna are in excess and will be left over.

Consider again the reaction $2H_2 + O_2 \rightarrow 2H_2O$. This says that 2 mol of $H_2$ react with 1 mol of $O_2$. But suppose you are given a flask with 2 mol of $H_2$ and 2 mol of $O_2$ in it. After the reaction, what would be left in the flask? Well, the 2 mol of $H_2$ would react with 1 mol of $O_2$ to form 2 mol of $H_2O$. *And 1 mol of $O_2$ would remain unreacted.* So after the reaction there would be 1 mol of $O_2$ and 2 mol of $H_2O$ in the flask. The $H_2$ would have reacted completely. The $H_2$ is the **limiting reagent.** The limiting reagent is completely used up in a chemical reaction. It is the reactant that "limits" the reaction because when it is used up, the reaction stops. Thus, the limiting reagent determines how much of the products are made. Let's make a table to illustrate this:

| balanced equation: | $2H_2$ | + $O_2$ | $\rightarrow 2H_2O$ |
|---|---|---|---|
| moles before reaction: | 2 | 2 | 0 |
| moles that react: | −2 | −1 | +2 |
| moles after reaction: | 0 | 1 | 2 |

In the row "moles that react," a minus sign indicates that a substance is being used up, and a plus sign indicates that a substance is being formed. $H_2$ and $O_2$ are being used up; $H_2O$ is being formed. The last row, "moles after reaction," is just the sum of the first two rows.

**EXAMPLE 10** For the reaction $C + O_2 \rightarrow CO_2$, you are given the reactants 2.0 mol of C and 1.5 mol of $O_2$. What substances are left after reaction, and how many moles of each are there?

***Solution:*** From the balanced equation we know that 1 mol of C reacts with 1 mol of $O_2$. Therefore, 1.5 mol of $O_2$ must react with 1.5 mol of C. Let's make a table:

150   STOICHIOMETRY

| balanced equation: | C | + | $O_2$ | → | $CO_2$ |
|---|---|---|---|---|---|
| moles before reaction: | 2.0 | | 1.5 | | 0.0 |
| moles that react: | −1.5 | | −1.5 | | +1.5 |
| moles after reaction: | 0.5 | | 0.0 | | 1.5 |

Therefore, after reaction, 0.5 mol of C and 1.5 mol of $CO_2$ remain. The $O_2$ is the limiting reagent, and there is no $O_2$ left. ∎

There is a simple way to determine the limiting reagent. Consider the following imaginary reaction: $3A + 7B \rightarrow 2C + D$, where A, B, C, and D represent different substances. The balanced equation indicates that

$$3 \text{ mol A} = 7 \text{ mol B} = 2 \text{ mol C} = 1 \text{ mol D}$$

The "3 mol A", "7 mol B", "2 mol C", and "1 mol D" *can each be considered as a unit,* and each will be called a **reaction equivalent** (abbreviated REQ). Thus we have that

$$3 \text{ mol A} = 1 \text{ REQ A} \qquad 7 \text{ mol B} = 1 \text{ REQ B}$$
$$2 \text{ mol C} = 1 \text{ REQ C} \qquad 1 \text{ mol D} = 1 \text{ REQ D}$$

*only* for the reaction $3A + 7B \rightarrow 2C + D$.

You can see that 1 REQ of any reactant will completely react with 1 REQ of any other reactant (to give, of course, 1 REQ of each product). Thus 1 REQ A reacts completely with 1 REQ B. Therefore, if the moles of each reacting substance are converted to REQs, *the substance with the smallest REQ is the limiting reagent.*

To convert from moles to REQs, we just use the appropriate identity.

**EXAMPLE 11** In the reaction $3A + 7B \rightarrow 2C + D$, if 5 mol of A and 8 mol of B are mixed, which reactant will be completely used up after reaction (i.e., which reactant is the limiting reagent)?

*Solution:* The identities needed are 3 mol A = 1 REQ A and 7 mol B = 1 REQ B. We have

$$5 \text{ mol A} \times \frac{1 \text{ REQ A}}{3 \text{ mol A}} = 1.67 \text{ REQ A}$$

and

$$8 \text{ mol B} \times \frac{1 \text{ REQ B}}{7 \text{ mol B}} = 1.14 \text{ REQ B}$$

Since there are fewer REQs of B, reactant B will be completely used up. Reactant B is the limiting reagent. ∎

If you examine Example 11 closely, you will see that all we really did to find the REQs was to divide the given moles of A and B by their respective coefficients from the balanced chemical equation:

$$\frac{5 \text{ mol A}}{3} = 1.67 \qquad \frac{8 \text{ mol B}}{7} = 1.14$$

(We have left off most units because in this one case they serve no useful purpose.)

So we have a "shortcut" to determine the limiting reagent. **To determine the limiting reagent, divide the given moles of each reactant by its respective coefficient from the balanced chemical equation; the reactant with the smallest quotient is the limiting reagent.** We don't have to mention the REQ anymore—it was just introduced to derive the preceding rule.

**EXAMPLE 12** You have 10 mol of Al powder (G) and 10 mol of ferric oxide, $Fe_2O_3$ (G). When these react according to the equation $2Al + Fe_2O_3 \rightarrow Al_2O_3 + 2Fe$, which reactant will be completely used up (i.e., which reactant will be the limiting reagent (AF))?

**NOTE:** This reaction is called the **thermite reaction** and produces molten iron and a great shower of sparks.

*Solution:* Just divide the given moles by the respective coefficients:

For Al: $10/2 = 5$     For $Fe_2O_3$: $10/1 = 10$

We have left off all the units this time. Since Al gives the smallest quotient, 5, the Al will be all used up. Al is the limiting reagent. ∎

Usually you will be given grams of reactants and asked to determine the limiting reagent. Then you must first convert all grams of reactants to moles of reactants.

**EXAMPLE 13** In the presence of plenty of oxygen, methane burns to give carbon dioxide and water. However, if methane burns with a shortage of oxygen, some carbon monoxide, CO, will be produced according to the reaction $2CH_4 + 3O_2 \rightarrow 2CO + 4H_2O$. If 80 g $CH_4$ and 96 g $O_2$ (G) are ignited, which reactant is the limiting reagent (AF)?

*Solution:* Convert all gram quantities to moles. The identities needed are 1 mol $CH_4$ = 16.0 g $CH_4$ and 1 mol $O_2$ = 32.0 g $O_2$. Thus we have

$$80 \text{ g CH}_4 \times \frac{1 \text{ mol CH}_4}{16.0 \text{ g CH}_4} = 5.0 \text{ mol CH}_4$$

$$96 \text{ g O}_2 \times \frac{1 \text{ mol O}_2}{32.0 \text{ g O}_2} = 3.0 \text{ mol O}_2$$

To find the limiting reagent, divide the moles by the respective coefficients from the balanced equation.

For $CH_4$: 5.0/2 = 2.5    For $O_2$: 3.0/3 = 1

Oxygen is the limiting reagent. Methane is in excess and cannot burn efficiently, thus giving carbon monoxide instead of carbon dioxide. ∎

Sometimes the problem will ask you to calculate not only the limiting reagent but also how much of each reactant was left over and how much of each product was formed. To solve this kind of problem, you must use all the techniques presented in this chapter.

**EXAMPLE 14**    Referring to Example 13, calculate the number of grams of $CH_4$ (AF) left after the reaction; also calculate the number of grams of CO and $H_2O$ (AF) formed.

**NOTE:** 80 g $CH_4$ and 96 g $O_2$ are the (G) in this problem.

*Solution:* We found in Example 13 that $O_2$ is the limiting reagent. Therefore, $O_2$ determines how much $CH_4$ is burned and how much CO and $H_2O$ are formed. Thus the problem is similar to the following one. How many grams of $CH_4$ (AF) are needed to react with 96 g of $O_2$ (G), and how many grams of CO and $H_2O$ (AF) are formed?

We will break the solution into five parts.

1. Calculate moles of $CH_4$. The identities needed are 1 mol $O_2$ = 32.0 g $O_2$ and 3 mol $O_2$ = 2 mol $CH_4$:

$$96 \text{ g O}_2 \times \frac{1 \text{ mol O}_2}{32.0 \text{ g O}_2} \times \frac{2 \text{ mol CH}_4}{3 \text{ mol O}_2} = 2.0 \text{ mol CH}_4$$

2. Calculate moles of CO formed. The identities needed are 1 mol $O_2$ = 32.0 g $O_2$ and 3 mol $O_2$ = 2 mol CO:

$$96 \text{ g O}_2 \times \frac{1 \text{ mol O}_2}{32.0 \text{ g O}_2} \times \frac{2 \text{ mol CO}}{3 \text{ mol O}_2} = 2.0 \text{ mol CO}$$

3. Calculate moles of H$_2$O formed. The identities needed are 1 mol O$_2$ = 32.0 g O$_2$ and 3 mol O$_2$ = 4 mol H$_2$O:

$$96 \text{ g O}_2 \times \frac{1 \text{ mol O}_2}{32.0 \text{ g O}_2} \times \frac{4 \text{ mol H}_2\text{O}}{3 \text{ mol O}_2} = 4.0 \text{ mol H}_2\text{O}$$

4. Calculate moles of all substances present after the reaction. A table is useful:

| balanced equation: | 2CH$_4$ | + 3O$_2$ | → 2CO | + 4H$_2$O |
|---|---|---|---|---|
| moles before reaction: | 5.0 | 3.0 | 0.0 | 0.0 |
| moles that react: | −2.0 | −3.0 | +2.0 | +4.0 |
| moles after reaction: | 3.0 | 0.0 | 2.0 | 4.0 |

**NOTE:** This table gives the number of moles of CH$_4$ and O$_2$ before reaction (from 80 g CH$_4$ and 98 g O$_2$), as calculated in Example 13.

5. Calculate the number of grams of all substances present after the reaction. The table in step 4 (in the row labeled "moles after reaction") gives the number of moles of each substance after the reaction. The identities needed are 1 mol CH$_4$ = 16.0 g CH$_4$, 1 mol CO = 28.0 g CO, and 1 mol H$_2$O = 18.0 g H$_2$O:

$$3.0 \text{ mol CH}_4 \times \frac{16.0 \text{ g CH}_4}{1 \text{ mol CH}_4} = 48 \text{ g CH}_4$$

$$2.0 \text{ mol CO} \times \frac{28.0 \text{ g CO}}{1 \text{ mol CO}} = 56 \text{ g CO}$$

$$4.0 \text{ mol H}_2\text{O} \times \frac{18.0 \text{ g H}_2\text{O}}{1 \text{ mol H}_2\text{O}} = 72 \text{ g H}_2\text{O}$$

There is no O$_2$ present after the reaction because it was the limiting reagent. ∎

If you are ever in doubt whether a problem involves a limiting reagent, remember this. If the mass of more than one reactant is given in the problem, you can be pretty sure it's a limiting reagent problem.

## PROBLEMS

### KEYED PROBLEMS

1. Write the mole identity between sulfur and oxygen in the reaction $S + O_2 \rightarrow SO_2$.

2. Write the mole identity between sulfur dioxide and oxygen in the reaction $2SO_2 + O_2 \rightarrow 2SO_3$.

3. Write the three mole identities from the reaction $4Al + 3O_2 \rightarrow 2Al_2O_3$.

4. How many moles of $O_2$ are needed to burn 5.0 mol of Al in the reaction of Example 3 above?

5. How many moles of HCl can be produced from 10 mol of $Cl_2$ by the reaction $H_2 + Cl_2 \rightarrow 2HCl$?

6. How many grams of $O_2$ are needed to react completely with 128 g of $SO_2$ in the reaction $2SO_2 + O_2 \rightarrow 2SO_3$?

7. How many grams of water are produced from the combustion of 0.50 mol of octane, $C_8H_{18}$? The reaction is $C_8H_{18} + \frac{25}{2}O_2 \rightarrow 8CO_2 + 9H_2O$.

8. How many moles of $CO_2$ are produced by the oxidation of 23 g of ethanol, $C_2H_5OH$? The reaction is $C_2H_5OH + 3O_2 \rightarrow 2CO_2 + 3H_2O$.

9. How many grams of magnesium hydroxide (milk of magnesia), $Mg(OH)_2$, are needed to neutralize (or react with) 40.0 g of HCl? The reaction is $Mg(OH)_2 + 2HCl \rightarrow MgCl_2 + 2HOH$.

10. For the reaction $S + O_2 \rightarrow SO_2$, you are given the reactants 4 mol of S and 3 mol of $O_2$. What substances are left after the reaction, and how many moles of each are there?

11. In the reaction $3A + 7B \rightarrow 2C + D$, if 10 mol of A and 8 mol of B are mixed, which reactant will be completely used up after the reaction (i.e., which reactant is the limiting reagent)?

12. You have 40 mol of Al powder and 15 mol of ferric oxide, $Fe_2O_3$. When these react according to the equation $2Al + Fe_2O_3 \rightarrow Al_2O_3 + Fe$, which reactant will be completely used up—that is, which reactant will be the limiting reagent?

13. When methane burns with excess oxygen, carbon dioxide and water are produced: $CH_4 + 2O_2 \rightarrow CO_2 + 2H_2O$. If 8.0 g of $CH_4$ and 64 g of $O_2$ are ignited, which reactant is the limiting reagent?

14. Referring to Problem 13, calculate the number of grams of $O_2$ left after the reaction; also calculate the number of grams of $CO_2$ and $H_2O$ formed.

## SUPPLEMENTAL PROBLEMS

15. Certain recipes use a mixture of baking soda (sodium bicarbonate, $NaHCO_3$) and vinegar (a 5% solution of acetic acid, $HC_2H_3O_2$) as a substitute for baking powder. How many grams of $NaHCO_3$ should be added to react completely with 20 g of $HC_2H_3O_2$? The reaction is $NaHCO_3 + HC_2H_3O_2 \rightarrow NaC_2H_3O_2 + CO_2 + H_2O$.

16. Aspirin, $C_9H_8O_4$, is synthesized from salicylic acid, $C_7H_6O_3$, and acetic anhydride, $C_4H_6O_3$, in the reaction $C_7H_6O_3 + C_4H_6O_3 \rightarrow C_9H_8O_4 + HC_2H_3O_2$. How many grams of aspirin can be made from 100 g of salicylic acid?

17. The Haber process for the synthesis of ammonia uses the reaction $N_2 + 3H_2 \rightarrow 2NH_3$. How many grams of hydrogen, $H_2$, are required to react completely with 7.0 g of $N_2$?

18. The antacid tablet TUMS contains $CaCO_3$ as the active ingredient. How many grams of hydrochloric acid, HCl, can be neutralized by 1.0 g $CaCO_3$ (about two TUMS tablets)? The reaction is $2HCl + CaCO_3 \rightarrow CaCl_2 + CO_2 + H_2O$.

19. The modern Hall process for producing aluminum was invented in the 1880s; prior to that, aluminum was more expensive than gold. Aluminum was made by the reduction of an aluminum salt with metallic sodium; $3Na + AlCl_3 \rightarrow Al + 3NaCl$. If 10 g Na and 10 g $AlCl_3$ are mixed together, how many grams of Al metal will be produced?

20. The "fixer" in photography (also known as "hypo") is sodium thiosulfate, $Na_2S_2O_3$. It removes undeveloped silver bromide (AgBr) from the film by the reaction $AgBr + 2Na_2S_2O_3 \rightarrow Na_3[Ag(S_2O_3)_2] + NaBr$. How many grams of $Na_2S_2O_3$ is required to remove 0.50 g of AgBr?

    (**NOTE:** Ignore the square brackets in the formula to the right of the arrow.)

21. Calcium oxide (CaO, quicklime) reacts with water to form calcium hydroxide ($Ca(OH)_2$, slaked lime) by the reaction $CaO + H_2O \rightarrow Ca(OH)_2$. If 5.0 g of CaO are reacted with 10 g of $H_2O$, what is the limiting reagent? How many grams of each substance will be left over after the reaction is completed?

22. Propyne, $C_3H_4$, can react with $H_2$ in the presence of a nickel catalyst to produce propane, $C_3H_8$: $C_3H_4 + 2H_2 \rightarrow C_3H_8$. (This is an example of a process called **hydrogenation**.) If 25 g $C_3H_4$ and 20 g $H_2$ are mixed together, how many grams of $C_3H_8$ will be produced?

23. Acetylene, $C_2H_2$, can be made by the reaction $CaC_2 + 2H_2O \rightarrow C_2H_2 + Ca(OH)_2$. $CaC_2$ is called calcium carbide and is used in the "carbide cannon" (which makes a loud boom used to start sailing races). How many grams of acetylene can be made from 50 g of $CaC_2$?

# STOICHIOMETRY

24. Ethyl alcohol, $C_2H_5OH$, can be made by fermenting sugar (such as glucose, $C_6H_{12}O_6$) according to the reaction $C_6H_{12}O_6 \rightarrow 2C_2H_5OH + 2CO_2$. This process can be used to make alcohol for drinking as well as for use in gasohol. How many grams of glucose are needed to make 100 g of ethyl alcohol?

25. Aluminum reacts with bromine to form aluminum bromide according to the reaction $2Al + 3Br_2 \rightarrow 2AlBr_3$. How many grams of bromine are needed to react completely with 250 g Al?

## MOLE DAY and Pi DAY

```
MOLE DAY
OCTOBER 23
6:02 AM TO 6.02 PM
6.02 X 10 E 23

10 E 23 is the same as 10²³
```

**Mole Day** is an unofficial holiday celebrated by many chemists, chemistry students and chemistry enthusiasts every October 23, between 6:02 a.m. and 6:02 p.m. There is a National Mole Day Foundation (NMDF) which was established on May 15, 1991.

The American Chemical Society (ACS) sponsors National Chemistry Week, which occurs from the Sunday through Saturday during the week on which October 23 falls. This makes Mole Day an integral part of National Chemistry Week.

Many high schools celebrate Mole Day as a way to get their students interested in chemistry, with various activities related to chemistry and moles. If you were born on October 23, you can have a double celebration! **NOTE:** Remember that one mole of elementary entities (see footnote on page 136) contains one Avogadro number (6.02 x $10^{23}$) of elementary entities. It seems reasonable that this day could also be called Avogadro's Day but calling it Mole Day is much more fun. There are Mole Day t-shirts (https://bit.ly/2WcTcpQ), mugs (https://bit.ly/3EXBoAw) and all sorts of clever and funny images (https://bit.ly/39BXaLY).

A living mole. *Photo courtesy of Kenneth Catania, Vanderbilt University,* https://bit.ly/3ETpXtE

There is another day celebrated by scientists and mathematicians. It is called **Pi ($\pi$) Day**, and is celebrated every March 14, which is 3/14 because the first 3 digits of pi are 3.14. You might remember that pi is the relation between the radius of a circle and the circumference of a circle, as in $C = 2\pi r$. It is also the relation between the radius of a circle and the area of a circle as in $A = \pi r^2$. Pi is one of the most important numbers in science and mathematics. It is used so frequently by scientists and mathematicians that it is one of only two dedicated "special" number buttons on almost all scientific calculators, the other one being "e" (see page 353).

# The Avogadro Project (Part 2)

NIST scientists **Bob Vocke** (left) and **Savelas Rabb** (right) in front of their Thermo Scientific™ Neptune™ Series high resolution mass spectrometer. *Photo courtesy of J. Stoughton/NIST*

On page 122, we discussed the silicon sphere used in defining Avogadro's constant. The story of how numerous international groups made and analyzed the sphere is interesting.

The element silicon consists of three naturally-occurring isotopes, whose isotopic mass and percent abundance are: $^{28}$Si, 27.98 u, 92.2%; $^{29}$Si, 28.98 u, 4.7%; $^{30}$Si, 29.97 u, 3.1%. To make accurate measurements of the sphere, it had to be as close to 100% pure silicon–28 as possible.

A Russian group separated the isotopes so that the sphere was 99.995% silicon–28. A German group then made the sphere and did the preliminary analysis of its isotopic composition. Because the analysis was so challenging, they asked researchers in other countries to also work on the problem.

In the US, engineer and geochemist Bob Vocke and chemist Savelas Rabb at NIST (National Institute of Standards and Technology) took up the challenge (https://bit.ly/2TA2UMn). The German group had been using sodium hydroxide (NaOH) to dissolve the silicon so it could be vaporized and put into their **mass spectrometer** (a device that separates atoms and molecules on the basis of their mass, called a **mass spec** for short). But Vocke and Rabb quickly found that the sodium was clogging up the mass spec and reducing the silicon signal. Rabb, being a superb chemist, realized that using **tetramethylammonium hydroxide, TMAH,**\* would dissolve the silicon and then completely vaporize in the mass spec, leaving no residue. This allowed a strong silicon signal and quickly all the other groups started using TMAH. It was now possible to get highly accurate and precise measurements to define Avogadro's constant. A YouTube video that has Vocke and Rabb discussing all this is at https://bit.ly/2pjJFwh.

\*TMAH = $(CH_3)_4NOH$. $(CH_3)_4N^+$ is similar to $NH_4^+$, shown in Table 16-1, where $CH_3$ groups, called methyl groups, replace the H atoms.

A **superconductor** has essentially no resistance to electricity, thus not generating any waste heat when electricity is flowing through it. Superconductors were discovered in 1911 by the Dutch physicist **Heike Kamerlingh Onnes** (1853-1926). He cooled liquid mercury (mercury is a liquid at room temperature) with liquid helium, whose temperature is −269 °C or 4 K, and the now solid mercury became superconducting.

In the photo, a magnet is levitated above a superconducting disk that has been cooled with liquid nitrogen, whose temperature is −195.79 °C or 77 K (see Section 12-5). Substances that become superconducting at a temperature above that of liquid helium, such as the disc in the photo, are called **high-temperature superconductors (HTS)**.[1] Many high-temperature superconductors have been made. The chemical formulas for two of them are **$YBa_2Cu_3O_7$** and **$HgBa_2Ca_2Cu_3O_8$**. The second one is superconducting at 134 K or −139 °C, one of the highest temperatures so far recorded for an HTS.

Superconducting magnets are used in MRI imaging machines.[2] They use liquid helium to cool their magnets that produce a magnetic field up to 60,000 times that of earth's magnetic field. The noise that MRIs make is due to the vibration of the coils of wire used to make the magnetic field (these are called electromagnets).[3] This vibration is caused by the pulsing of electricity through the wires (and thus rapidly changing the magnetic field) needed to obtain the images.

As of 2021, no practical room-temperature superconductors have been found. These would make the use of liquid nitrogen or liquid helium unnecessary and would greatly reduce the amount of electricity that is used.

*Photo courtesy of Peter Nussbaumer, CC BY-SA 3.0, https://bit.ly/2EaMiEv.*

[1] The first HTS was discovered in 1986 by IBM researchers **Georg Bednorz** and **K. Alex Müller**. They shared the 1987 Nobel Prize in Physics for this discovery.

[2] MRI stands for **m**agnetic **r**esonance **i**maging. In chemistry, it is called NMR, for **n**uclear **m**agnetic **r**esonance, and is used to help in the determination of molecular structures. The name was changed for clinical use because it was thought that people would be afraid of the word "nuclear."

[3] The superconducting wires are made from fibers of a niobium-titanium alloy (the superconductor) embedded in Cu or Al. As of 2021, wires that are superconducting at liquid nitrogen temperatures don't seem to be practical for current MRI machines.

# PERCENT COMPOSITION, EMPIRICAL FORMULAS, AND MOLECULAR FORMULAS

This chapter discusses certain mass and percentage topics involving compounds. We begin with a review of percent.

The techniques discussed in this chapter are commonly used by chemists when they are calculating the percent composition of a compound. And if given the percent composition of a compound, chemists can calculate the empirical formula of the compound.

## 10-1
### PERCENT IS BASED ON 100

If you get 95% on an exam, this means that out of 100 possible points, you got 95 points. Percent (abbreviated "%") means parts per hundred and is used mostly for convenience. It seems easier to say that you got 95% (meaning 95 out of a possible 100) on an exam than to say that you got 0.95 (out of a possible 1).[1]

**EXAMPLE 1** A student gets 82 questions out of 100 correct on an exam. What is the fraction correct and the percent correct?

*Solution:* The fraction correct is 82/100 = 0.82. The percent correct is:

$$\frac{82}{100} \times 100 = 82\%. \blacksquare$$

---

[1] A baseball player's batting average is another example of a decimal number. It is based on 1000 instead of 100. **Joan Joyce** (1940-2022), the most dominant pitcher in the history of women's fast-pitch softball, was also an exceptional batter with a lifetime batting average (LBA) of 0.327. In exhibition play, she struck out two of the greatest major league batters, **Hank Aaron** (1934-2021), who had a major league LBA of 0.305, and **Ted Williams** (1918-2002) who had a major league LBA of 0.344. Joyce's LBA means she averaged 327 hits per 1000 times at bat. The decimal is calculated from 327/1000 = 0.327. Her percent of hits would be:

$$\frac{327}{1000} \times 100 = 0.327 \times 100 = 32.7\%.$$

160   PERCENT COMPOSITION, EMPIRICAL FORMULAS, AND MOLECULAR FORMULAS

The difference between the fraction correct and the percent correct is simply a factor of 100. Multiplying the fraction correct by 100 gives the percent correct.

Sometimes an instructor will give an exam that doesn't have 100 points. A percent correct can still be calculated, as Example 2 shows.

**EXAMPLE 2**   A student takes an exam consisting of 80 short-answer questions. The student gets 67 questions correct. What is the percent correct on this exam?

*Solution:* The percent correct is 67/80 × 100 = 84%.  ∎

A formula for calculating the percent correct (or the grade) on an exam is

$$\frac{\text{number of questions correct}}{\text{total number of questions}} \times 100 = \% \text{ correct}$$

This formula assumes that all the questions have the same value. If questions have different values, you must calculate your grade from the following formula:

$$\frac{\text{number of points you received}}{\text{total number of points on exam}} \times 100 = \% \text{ of points correct}$$

Another thing you should realize is that

$$\% \text{ points correct} + \% \text{ points incorrect} = 100\%$$

**EXAMPLE 3**   On an exam, a student lost 7 points out of a possible 30. What percentage did the student score?

*Solution:* The number of points the student received was 30 − 7 = 23. The grade on the exam was 23/30 × 100 = 77%.  ∎

In Example 3 the percent points incorrect on the exam was 7/30 × 100 = 23%. The sum of the % points correct + % points incorrect = 77% + 23% = 100%.

## 10-2
### PERCENT COMPOSITION IS THE MASS PERCENT OF THE ELEMENTS IN A COMPOUND

The percent by mass of all the elements in a compound is called the **percent composition.** Each element's percent by mass is found in a way that is similar to calculating exam scores. Take the mass of an element in the compound, divide by the total mass of the compound, and multiply by 100:

$$\frac{\text{element's mass}}{\text{total mass of compound}} \times 100 = \% \text{ by mass of the element}$$

In practice, the atomic and molecular weights are used in this calculation.

**EXAMPLE 4**  What is the percent by mass of carbon in $CO_2$?

*Solution:* Atomic and molecular weights are easily available numbers that we can use. The atomic weight of carbon is 12.0 g, and the molecular weight of $CO_2$ is 44.0 g. Therefore,

$$\frac{12.0 \text{ g C}}{44.0 \text{ g } CO_2} \times 100 = 27.3\% \text{ carbon} \quad \blacksquare$$

**EXAMPLE 5**  What is the percent by mass of oxygen in $CO_2$?

*Solution:* The atomic weight of oxygen is 16.0 g. There are two oxygen atoms in a molecule of $CO_2$, so the total weight of oxygen in 1 mol (44.0 g) of $CO_2$ is 32.0 g. Thus

$$\frac{32.0 \text{ g O}}{44.0 \text{ g } CO_2} \times 100 = 72.7\% \text{ oxygen} \quad \blacksquare$$

Notice that when we combine the percent by mass from Example 4 with that from Example 5, our total is 100%:

$$27.3\% \text{ carbon} + 72.7\% \text{ oxygen} = 100\% \text{ total}$$

*A listing of all the percents by mass of the elements in a compound is called the percent composition of that compound.*

162  PERCENT COMPOSITION, EMPIRICAL FORMULAS, AND MOLECULAR FORMULAS

**EXAMPLE 6**  What is the percent composition of $CO_2$?

*Solution:* From the two previous examples, the percent composition of $CO_2$ is 27.3% carbon and 72.7% oxygen.  ∎

**EXAMPLE 7**  What is the percent composition of water, $H_2O$?

*Solution:* The atomic weights are H = 1.01 g and O = 16.0 g. The molecular weight of $H_2O$ is 18.0 g. The percent composition is

$$\frac{2 \times 1.01 \text{ g H}}{18.0 \text{ g H}_2\text{O}} \times 100 = 11.1\% \text{ hydrogen}$$

$$\frac{1 \times 16.0 \text{ g O}}{18.0 \text{ g H}_2\text{O}} \times 100 = 88.9\% \text{ oxygen}$$

The "2" in 2 × 1.01 g H is needed because of the mole identity 2 mol H = 1 mol $H_2O$. The "1" in 1 × 16.0 g O shows that 1 mol O = 1 mol $H_2O$.  ∎

**EXAMPLE 8**  What is the percent composition of glucose, $C_6H_{12}O_6$?

*Solution:* The atomic weights are C = 12.0 g, H = 1.01 g, and O = 16.0 g. The molecular weight of glucose is 180 g. The percent composition is

$$\frac{6 \times 12.0 \text{ g C}}{180 \text{ g glucose}} \times 100 = 40.0\% \text{ carbon}$$

$$\frac{12 \times 1.01 \text{ g H}}{180 \text{ g glucose}} \times 100 = 6.73\% \text{ hydrogen}$$

$$\frac{6 \times 16.0 \text{ g O}}{180 \text{ g glucose}} \times 100 = 53.3\% \text{ oxygen}$$  ∎

If you are given a quantity of a substance and you know the percent composition, you can calculate the mass of each element in the sample.

**EXAMPLE 9**  What is the mass of each element in 30.0 g of glucose?

*Solution:* From Example 8 we know the percent composition of glucose. Thus the mass of each element in 30.0 g of glucose is (remember to convert percents into the equivalent decimal form):

10-3 Calculating Empirical Formulas and Molecular Formulas    163

mass of carbon:       30.0 g × 0.400 = 12.0 g carbon
mass of hydrogen:     30.0 g × 0.0673 = 2.02 g hydrogen
mass of oxygen:       30.0 g × 0.533 = 16.0 g oxygen  ■

## 10-3
## AN EMPIRICAL FORMULA AND A MOLECULAR FORMULA CAN BE CALCULATED FROM THE PERCENT COMPOSITION AND MOLECULAR WEIGHT OF THE COMPOUND

A chemical formula that gives the relative number of each type of atom in a molecule of a substance is called the **empirical formula** of the substance. It is the simplest formula of a substance. Notice the words "relative number" in this definition. The empirical formula is not necessarily the same as the actual molecular formula. Take the case of glucose, whose actual molecular formula is $C_6H_{12}O_6$. The empirical formula of glucose is $CH_2O$. This is the simplest formula that has the correct ratio of carbon, hydrogen, and oxygen atoms. The empirical weight of $CH_2O$ is 30.0 g. **Empirical weight** is the term we shall use for the mass represented by the empirical formula. If we take six empirical formulas, we have $6 \times CH_2O$ or $C_6H_{12}O_6$ with a molecular weight of $6 \times 30.0$ g = 180 g.

**EXAMPLE 10**   The empirical formula of ethylene is $CH_2$. The molecular weight of ethylene is 28.0 g. What is the molecular formula of ethylene?

*Solution:* The empirical weight of $CH_2$ is 14.0 g. If we divide 28.0 g by 14.0 g, we find the number of empirical formula units in the molecular formula. Since 28.0 g/14.0 g = 2, the molecular formula of ethylene is $2 \times CH_2 = C_2H_4$.  ■

**EXAMPLE 11**   The empirical formula of hydrogen peroxide is HO. The molecular weight is 34.0 g. What is the molecular formula of hydrogen peroxide?

*Solution:* The empirical weight of HO is 17.0 g. Since 34.0 g/17.0 g = 2, the molecular formula of hydrogen peroxide is $2 \times HO = H_2O_2$.  ■

Sometimes the empirical formula and the molecular formula are the same. An example would be the empirical formula of sulfur dioxide, $SO_2$, whose empirical weight is 64.1 g. The molecular weight of sulfur dioxide is also 64.1 g, and thus the molecular formula of sulfur dioxide is $SO_2$.

# PERCENT COMPOSITION, EMPIRICAL FORMULAS, AND MOLECULAR FORMULAS

It is important for chemists to know the molecular formulas of substances. Since the percent composition and the molecular weights can be determined experimentally, we can use these data to determine molecular formulas.

The simplest way to approach the problem is to assume that we have 100 g of the substance. This is just a convenience—any amount would do—but taking percents of 100 is very easy, and everybody does it this way.

**EXAMPLE 12** The percent composition of a hydrocarbon is 75.0% carbon and 25.0% hydrogen. The molecular weight is 16.0 g. What is the molecular formula?

*Solution:* To calculate the empirical formula, we use percent composition. Taking 100 g of hydrocarbon, we see that it has 75.0 g carbon and 25.0 g hydrogen. Since the ratio of atoms in the molecule is the same as the ratio of moles of atoms in the substance, we now calculate the number of moles of each element in these masses:

$$75.0 \text{ g C} \times \frac{1 \text{ mol C}}{12.0 \text{ g C}} = 6.25 \text{ mol C}$$

$$25.0 \text{ g H} \times \frac{1 \text{ mol H}}{1.01 \text{ g H}} = 24.8 \text{ mol H}$$

Writing the empirical formula as $C_{6.25}H_{24.8}$ is not very simple—we must simplify it. We can do this by dividing both subscripts by the smallest one. This will give a "1" as the smallest subscript:

$$6.25/6.25 = 1 \qquad 24.8/6.25 = 3.97$$

We round 3.97 to 4 because we need a whole number of atoms in our empirical formula. The difference between 3.97 and 4 is due to experimental error and/or rounding error. The empirical formula is thus $C_1H_4$ or $CH_4$. The empirical weight is 16.0 g. Since the molecular weight is also 16.0 g, and 16.0 g/16.0 g = 1, the molecular formula is $CH_4$. ∎

**EXAMPLE 13** The percent composition of a hydrocarbon is found to be 82.7% carbon and 17.4% hydrogen. The molecular weight is 58.1 g. What is the empirical formula and the molecular formula?

*Solution:* Taking 100 g of hydrocarbon, we see that it has 82.7 g carbon and 17.4 g hydrogen. The number of moles of each element is

## 10-3 Calculating Empirical Formulas and Molecular Formulas

$$82.7 \text{ g C} \times \frac{1 \text{ mol C}}{12.0 \text{ g C}} = 6.89 \text{ mol C}$$

$$17.4 \text{ g H} \times \frac{1 \text{ mol H}}{1.01 \text{ g H}} = 17.2 \text{ mol H}$$

Again, writing the empirical formula as $C_{6.89}H_{17.2}$ is not the simplest way, and we must divide both subscripts by the smallest one:

$$6.89/6.89 = 1 \qquad 17.2/6.89 = 2.5$$

An empirical formula of $CH_{2.5}$ is still not the simplest, since we want whole number subscripts. If we double both subscripts, we get $C_2H_5$. This is a fine empirical formula whose empirical weight is 29.1 g. Since the molecular weight of our hydrocarbon is 58.1 g, we divide 58.1 g by 29.1 g and get 58.1 g/29.1 g = 2; the molecular formula of our hydrocarbon is $2 \times C_2H_5 = C_4H_{10}$. ■

Example 13 illustrates an interesting point. If the empirical formula turns out to have noninteger (noninteger means not a whole number) subscripts, try multiplying the subscripts by 2 or 3 or some other small whole number. This will usually turn the subscripts into whole numbers.

**EXAMPLE 14** In Example 8 we calculated the percent composition of glucose from the formula $C_6H_{12}O_6$. Let's work backward from the percent composition and the molecular weight to arrive at the molecular formula of $C_6H_{12}O_6$. Glucose is 40.0% carbon, 6.73% hydrogen, and 53.3% oxygen and has a molecular weight of 180 g.

*Solution:* Assuming we are given a sample of 100 g glucose, we have 40.0 g carbon, 6.73 g hydrogen, and 53.3 g oxygen. The moles of each element are

$$40.0 \text{ g C} \times \frac{1 \text{ mol C}}{12.0 \text{ g C}} = 3.33 \text{ mol C}$$

$$6.73 \text{ g H} \times \frac{1 \text{ mol H}}{1.01 \text{ g H}} = 6.66 \text{ mol H}$$

$$53.3 \text{ g O} \times \frac{1 \text{ mol O}}{16.0 \text{ g O}} = 3.33 \text{ mol O}$$

Dividing each number by the smallest, 3.33, we get

$$3.33/3.33 = 1 \qquad 6.66/3.33 = 2 \qquad 3.33/3.33 = 1$$

**166**  PERCENT COMPOSITION, EMPIRICAL FORMULAS, AND MOLECULAR FORMULAS

The empirical formula is $C_1H_2O_1$ or $CH_2O$, which has an empirical weight of 30.0 g. Since 180 g/30.0 g = 6, the molecular formula is 6 × $CH_2O$ = $C_6H_{12}O_6$. ∎

**EXAMPLE 15**  A gaseous compound has the following percent composition: 5.05% hydrogen and 95.0% fluorine. Its molecular weight is 120 g. What is the molecular formula of the compound?

*Solution:* Assuming we are given 100 g of gas, we have 5.05 g hydrogen and 95.0 g fluorine. The number of moles of each element is

$$5.05 \text{ g H} \times \frac{1 \text{ mol H}}{1.01 \text{ g H}} = 5.00 \text{ mol H}$$

$$95.0 \text{ g F} \times \frac{1 \text{ mol F}}{19.0 \text{ g F}} = 5.00 \text{ mol F}$$

Dividing by the smallest number we get

$$5.00/5.00 = 1 \qquad 5.00/5.00 = 1$$

The empirical formula is HF, and the empirical weight is 20.0 g. Since 120 g/20.0 g = 6, the molecular formula is 6 × HF = $H_6F_6$.

**NOTE:** The formula for hydrogen fluoride is commonly written as HF; however, in the gas phase, HF molecules stick together and on the average form "clumps" of six. Thus, in the gas phase, we can write the molecular formula of hydrogen fluoride as $H_6F_6$. Another way to write it would be $(HF)_6$. ∎

## 10-4
### THE FORMULA WEIGHT OF A GIANT ARRAY IS THE MASS OF THE SIMPLEST RATIO OF ATOMS IN THE ARRAY

If a substance is a "giant array" of atoms (see Chapter 4) such as NaCl, the empirical formula would be NaCl. It wouldn't make much sense to talk about the molecular formula of NaCl, since there are no separate molecules in the NaCl crystal. The empirical weight of NaCl is 58.4 g and is the mass of the simplest ratio of atoms in the NaCl crystal. Many authors call the mass of the simplest ratio of atoms in the giant array the **formula weight.** Thus the formula weight of NaCl is 58.4 g.

## PROBLEMS

### KEYED PROBLEMS

1. A student gets 88 questions out of 100 correct on an exam. What is the fraction correct and percent correct?

2. A student takes an exam consisting of 89 short-answer questions. The student gets 72 questions correct. What is the percent correct on this exam?

3. On an exam, you lost 12 points out of a possible 40 points. What was your grade?

4. What is the percent by mass of sulfur in sulfur dioxide, $SO_2$?

5. What is the percent by mass of oxygen in sulfur dioxide, $SO_2$?

6. What is the percent composition of $SO_2$?

7. What is the percent composition of hydrogen peroxide, $H_2O_2$?

8. What is the percent composition of diethyl ether, $C_4H_{10}O$?

9. What is the mass of each element in 50.0 g of diethyl ether, $C_4H_{10}O$?

10. The empirical formula of benzene is CH. The molecular weight of benzene is 78.1 g. What is the molecular formula of benzene?

11. The empirical formula of an oxide of nitrogen is $NO_2$. The molecular weight is 92.0 g. What is the molecular formula of the oxide of nitrogen?

12. A smelly liquid is 15.78% carbon and 84.22% sulfur. The molecular weight is 76.13 g. What is the molecular formula of the liquid?

13. The percent composition of a hydrocarbon is found to be 85.6% carbon and 14.4% hydrogen. The molecular weight is 56.1 g. What is the empirical formula and the molecular formula?

14. The percent composition of butyric acid is 54.5% carbon, 36.3% oxygen, and 9.15% hydrogen. The molecular weight of butyric acid is 88.1 g. What is the molecular formula of butyric acid?

15. A sulfur–fluorine compound has the following percent composition: 25.2% sulfur and 74.8% fluorine. Its molecular weight is 254 g. What is the molecular formula of the compound?

### SUPPLEMENTAL PROBLEMS

16. Calculate the percent composition of the following interhalogen compounds: $ClF_3$, $BrF_5$, $IF_7$. (An interhalogen compound consists only of a combination of halogens, which are the elements F, Cl, Br, I, and At.)

17. Calculate the percent composition of plaster of Paris, $(CaSO_4)_2 \cdot H_2O$. What is the percent of $H_2O$ in plaster of Paris? The "·" in the formula $(CaSO_4)_2 \cdot H_2O$ is read "dot" and signifies that the $H_2O$ is "water of hydration." This water of hydration is loosely held to the $CaSO_4$ and can be removed by heating.

18. Calculate the percent composition of copper sulfate pentahydrate, which appears as deep blue crystals. The formula is $CuSO_4 \cdot 5H_2O$, where the $5H_2O$ is the water of hydration (see Problem 17). If the water is removed by heating, the crystals turn white.

19. Calculate the percent composition of calcium nitride, $Ca_3N_2$.

20. The percent composition of ascorbic acid (vitamin C) is 40.9% carbon, 54.5% oxygen, and 4.55% hydrogen. The molecular weight is 176 g. What is the molecular formula of ascorbic acid?

21. The percent composition of a strong acid is 2.04% hydrogen, 32.65% sulfur, and 65.31% oxygen. What is the molecular formula of the strong acid if the molecular weight is 98.1 g?

22. An unpleasant gaseous compound is 22.54% phosphorus and 77.46% chlorine. The molecular weight is 137.3 g. What is the molecular formula of the compound?

23. Upon analysis, the black coating found on tarnished silver contains 87.1% silver and 12.9% sulfur. The formula weight is 248 g. What is the chemical formula?

24. This problem is more difficult: When 5.000 g of a certain hydrocarbon is burned, 13.72 g of $CO_2$ and 11.23 g of $H_2O$ are produced. What is the empirical formula of the hydrocarbon? (Hint: First calculate the percent of C in $CO_2$ and the percent of H in $H_2O$. Second, calculate the number of grams of C and H in the hydrocarbon. Finally, calculate the percent of C and H in the hydrocarbon and then determine the empirical formula.)

25. Calculate the formula weight of the following substances that consist of giant arrays of atoms: KBr, SiC.

26. This is a question about the differences and similarities of the terms "molecular weight" and "formula weight." Are all molecular weights also formula weights? Are all formula weights also molecular weights?

## PROOF THAT LEAD HAS AT LEAST TWO ISOTOPES: AN IMPORTANT STOICHIOMETRY PROBLEM IN THE HISTORY OF CHEMISTRY[1]

In Chapter 1, we discussed the fact that most elements have more than one isotope, based on the research of **Frederick Soddy** (p. 8). At the time, scientists knew that both uranium-238 and thorium-232 were radioactive and that lead was in all the ores of both uranium (U) and thorium (Th). But they didn't know the isotopic composition of the lead in the ores. Was all the lead the same isotope, or was the lead in each of the ores different?

In 1914, Prof. **Theodore Williams Richards** (1868-1928) of Harvard University decided to find out by measuring the atomic weight of lead from each of these ores (he was in reality measuring the isotopic mass, as the actual atomic weight of all the isotopes of naturally-occurring lead is 207.2 g/mol). At the time, Prof. Richards was the world's leading expert in determining the atomic weight of elements and was the first American to win a Nobel Prize in Chemistry (in 1914) for this research. The lead in both ores is all in the compound lead sulfide (PbS). Here is the conceptual way Prof. Richards solved the problem.

He prepared 100.00 g of purified lead sulfide from each ore. The lead sulfide from the **uranium ore gave 86.54 g of metallic lead** after removal of the sulfur. The lead sulfide from the **thorium ore gave 86.64 g of metallic lead** after removal of the sulfur. It was known at the time that the atomic weight of sulfur was 32.06 g.

The amount of sulfur removed from the uranium ore's lead sulfide was 100.00 g – 86.54 g = **13.46 g S**, and the amount removed from the thorium ore's lead sulfide was 100 g – 86.64 g = **13.36 g S**. There is 1 mol Pb and 1 mol S in each formula unit of PbS, so **1 mol Pb = 1 mol S**.

The moles of sulfur and lead in each ore are:

U ore: 13.46 g S x 1 mol S/32.06 g S = 0.4198 mol S = **0.4198 mol Pb**.
Th ore: 13.36 g S x 1 mol S/32.06 g S = 0.4167 mol S = **0.4167 mol Pb**.

For a sample of an element that has only one isotope, the mass number and the isotopic mass will be the same to 3 significant figures.[2] Here are the isotopic masses computed as the (number of grams/number of moles) in the sample:

**For the uranium ore's Pb: 86.54 g Pb/0.4198 mol Pb = 206.1 g Pb/mol Pb**
**For the thorium ore's Pb: 86.64 g Pb/0.4167 mol Pb = 207.9 g Pb/mol Pb**

Rounding to three figures, the mass numbers of the isotopes of lead are:

**From the uranium ore: lead-206 (comes from the decay of $^{238}$U)**
**From the thorium ore: lead-208 (comes from the decay of $^{232}$Th)**

Both lead isotopes are stable and their discovery was an important confirmation of Prof. Soddy's deduction that an element could consist of atoms of different isotopes. These results accelerated the search for the natural isotopic composition of all the elements. **The results of all this decades-long research by hundreds of talented scientists has given us Table 1-2, which lists the natural isotopes of the elements.**

[1] This discussion was inspired by Example 4-6 on page 100 of the text *General Chemistry* by **Linus Pauling**, 3rd edition, 1970.

[2] You can check this by comparing Tables 1-2 and 2-2.

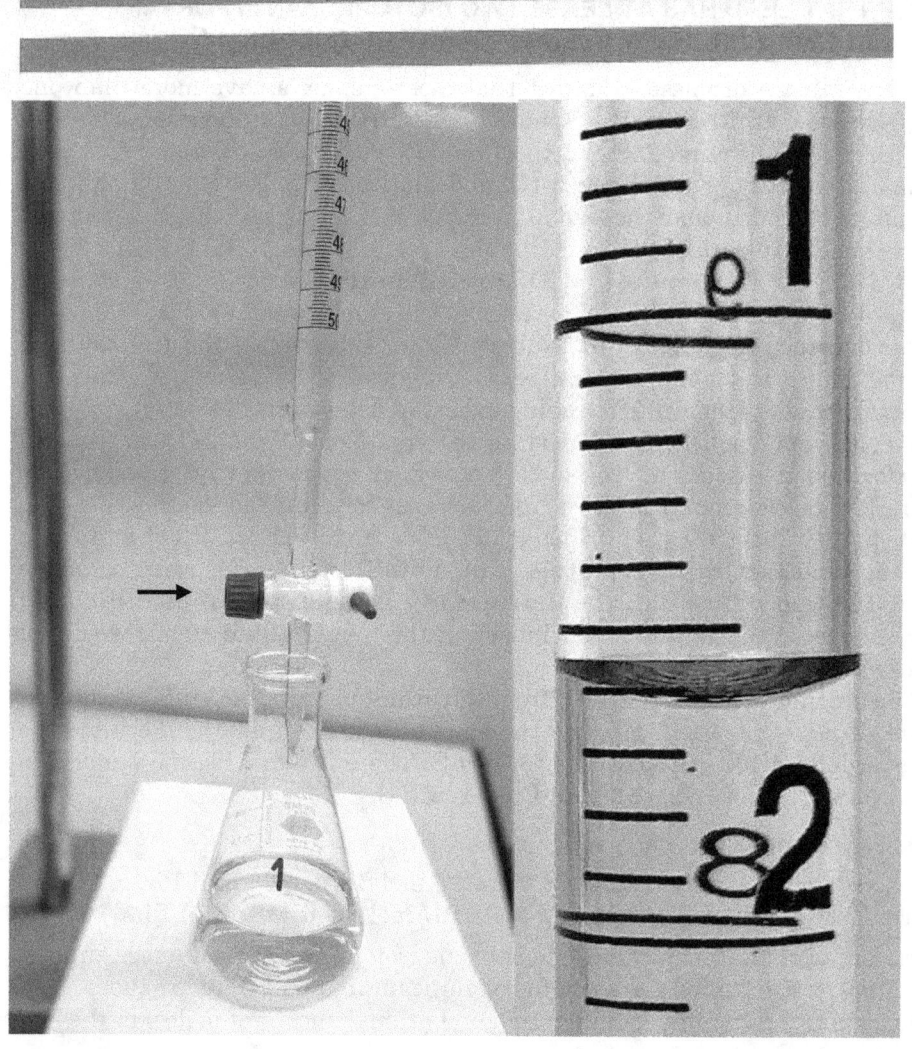

**A Buret (some spell it burette) Performing a Titration:** The left photo shows the bottom part of the 50-mL buret (markings go up to 0 mL at the top of the buret) in an Erlenmeyer flask (with a "1" on it) ready to titrate (the neck of the flask is narrowed to reduce splashing). The Teflon stopcock (arrow) is a rotating valve used to control the flow of liquid. The right photo is a magnification of part of the buret. It is more reliable to read the bottom of the dark curve (called the meniscus). It can easily be read to the tenths place, say 1.6 mL (read going down). The next place can be estimated, thus reading, say, 1.61 mL. So a buret can be read to two decimal places, with the last one estimated (reading between the lines). This means that for volumes less than 10 mL, we can read a buret to 3 significant figures. (But see Example 5, page 97 for exceptions.) At 10 mL and above, we can read to 4 significant figures. The accuracy of the last digit depends upon the skill of the titrator. *Left photo is in the Public Domain. Right photo courtesy USGS.*

# MOLARITY AND SOLUTION STOICHIOMETRY

In this chapter we will discuss the kinds of solutions in which a solid or a liquid is dissolved in a liquid. An example of a solid dissolved in a liquid is a salt–water solution. A solid, salt, is dissolved in a liquid, water. An example of a liquid dissolved in a liquid is grain alcohol dissolved in water. Vodka is an example of such a solution.

In a solution in which a solid has been dissolved in a liquid, the solid is called the **solute.** The liquid is called the **solvent.** In a solution of a liquid in a liquid, the liquid present in smaller amounts is usually called the solute; however, the distinction is sometimes not clear.

**EXAMPLE 1** If you prepare a solution by mixing water and sugar, which component is the solute and which is the solvent?

*Solution:* Water, a liquid, is the solvent; sugar, a solid, is the solute. ∎

## 11-1
### MOLARITY IS A UNIT OF CONCENTRATION INVOLVING MOLES AND LITERS

From your study of stoichiometry in Chapter 9, you might realize that we can do stoichiometric calculations with solutions. To perform such calculations, we must have a concentration unit that involves moles. One such concentration unit that is very useful is called **molarity.** A solution with a molarity of one is defined as follows:

> If one mole of a solute is dissolved in enough solvent to make one liter of solution, the resulting solution has a **molarity of one.** The abbreviation of molarity is M, so this solution would be referred to as 1 M and would read "one molar."

Of course, we don't always have to make up solutions that are 1 M. We can make solutions of any molarity, provided the solute can be dissolved in the solvent. If two moles of solute are dissolved in enough solvent to make one liter of solution, we would have a 2 molar solution (2 M). Figure 11-1 shows the procedure for making up one liter of a 0.500 M water solution of

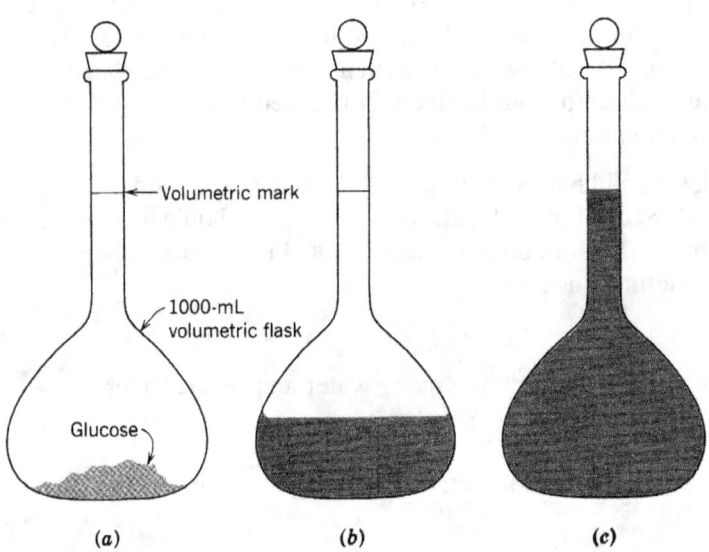

**FIGURE 11-1** The procedure for making 1 L of 0.500 M glucose solution. (a) Weigh out 0.500 mol of glucose. This would be 90.1 g of glucose. Place the glucose in a 1000-mL volumetric flask. (b) Fill the flask about half full with water. Shake the flask to dissolve the glucose. If the flask is filled up to the volumetric mark before the glucose is dissolved, it will be difficult to dissolve the glucose. The neck is too narrow to allow good mixing, and the volume after mixing might turn out to be more than 1000 mL. However, the narrow neck has a purpose. If an error is made in filling the flask to the volumetric mark, the error in the volume will be small. (c) Fill the flask to the volumetric mark and mix completely. Notice that a volumetric flask has only one volumetric mark on it—in this case 1000 mL. It can only be used to make 1000 mL of solution. If you want to make a different volume of solution, you must use a volumetric flask of a different size.

glucose, $C_6H_{12}O_6$. Notice that in making up the solution, we add enough water to make up one liter of solution. We don't, as a rule, add one liter of water.

It seems reasonable that we could use *any quantity of solute and solvent* to make our solutions. Since molarity is the number of moles of solute in a liter of solution, molarity is calculated by dividing the number of moles of solute by the number of liters (or fraction of a liter) of solution:

$$\text{molarity} = \frac{\text{number of moles of solute}}{\text{number of liters of solution}}$$

$$= \text{number of moles of solute per liter of solution}$$

Since the number of moles refers to "moles of solute," and the liters always refers to "liters of solution," we can simplify things by writing (remember that the abbreviation of mole is mol)

$$\text{molarity} = \frac{\text{moles}}{\text{liter}} = \frac{\text{mol}}{\text{L}} = M$$

So to calculate molarity, just divide moles of solute by liters of solution.

**EXAMPLE 2**  A solution contains 2.0 mol of NaOH dissolved in 8.0 L of solution. What is the molarity of the solution?

*Solution:* Since molarity = mol/L, we have

$$\text{molarity} = \frac{2.0 \text{ mol NaOH}}{8.0 \text{ L}} = 0.25 \frac{\text{mol NaOH}}{\text{L}} = 0.25 \text{ M NaOH}$$

Notice that a 0.25 molar NaOH solution is abbreviated 0.25 M NaOH.  ■

**EXAMPLE 3**  A solution contains 0.10 mol of $H_2SO_4$ dissolved in a total volume of 0.20 L. What is the molarity of the solution?

*Solution:*

$$\text{molarity} = \frac{\text{mol}}{\text{L}} = \frac{0.10 \text{ mol } H_2SO_4}{0.20 \text{ L}} = 0.50 \text{ M } H_2SO_4$$  ■

**EXAMPLE 4**  A solution contains 0.300 mol of glucose in 250 mL of solution. What is the molarity of the solution?

174 MOLARITY AND SOLUTION STOICHIOMETRY

*Solution:* First we must convert 250 mL to liters:

$$250 \text{ mL} \times \frac{1 \text{ L}}{1000 \text{ mL}} = 0.250 \text{ L}$$

$$\text{molarity} = \frac{\text{mol}}{\text{L}} = \frac{0.300 \text{ mol glucose}}{0.250 \text{ L}} = 1.20 \text{ M glucose} \quad \blacksquare$$

Most of the time, the amount of solute is given in grams, since that's the unit of mass that we read on our balances. Then you must convert from grams to moles in order to calculate the molarity of the solution.

**EXAMPLE 5** A solution contains 15 g of $FeCl_3$ in 500 mL of solution. What is the molarity of the solution?

*Solution:* First calculate the number of moles in 15 g of $FeCl_3$. The formula weight of $FeCl_3$ is 162 g, so 1 mol weighs 162 g. (Remember from Section 10-4 that the term "formula weight" is used instead of "molecular weight" when we want to refer to the weight of the simplest formula of a giant array.)

$$15 \text{ g FeCl}_3 \times \frac{1 \text{ mol FeCl}_3}{162 \text{ g FeCl}_3} = 0.092 \text{ mol FeCl}_3$$

Then convert 500 mL to liters:

$$500 \text{ mL} \times \frac{1 \text{ L}}{1000 \text{ mL}} = 0.500 \text{ L}$$

Now calculate the molarity:

$$\text{molarity} = \frac{\text{mol}}{\text{L}} = \frac{0.092 \text{ mol FeCl}_3}{0.500 \text{ L}} = 0.18 \text{ M FeCl}_3$$

We can do this problem in one step:

$$15 \text{ g FeCl}_3 \times \frac{1 \text{ mol FeCl}_3}{162 \text{ g FeCl}_3} \times \frac{1}{500 \text{ mL}} \times \frac{1000 \text{ mL}}{1 \text{ L}} = 0.18 \text{ M FeCl}_3$$

Notice that the product of the first two terms is 0.092 mol Fe, whereas the product of the last two terms is 1/0.500 L. Thus the product of all four terms is 0.092 mol Fe/0.500 L, just as when we use separate steps. $\blacksquare$

11-1 Molarity Is a Unit of Concentration Involving Moles and Liters 175

It is useful to devise a general formula to help us solve problems involving molarity. Since

$$M = \frac{mol}{L}$$

we can write

$$M \times L = mol$$

which is the same as

$$\frac{mol}{L} \times L = mol$$

where you must remember that the "mol/L" term stands for molarity.

**EXAMPLE 6** How many grams of NaCl must be taken to make 700 mL of an 0.20 M NaCl solution?

*Solution:* The number of liters is 700 mL × 1 L/1000 mL = 0.700 L. Using the equation

$$\frac{mol}{L} \times L = mol$$

we can calculate the number of moles of NaCl needed:

$$0.20 \frac{mol\ NaCl}{L} \times 0.700\ L = 0.14\ mol\ NaCl$$

Remember that the first term in this equation is the same as 0.20 M NaCl. To figure out the number of grams of NaCl, use the formula weight of NaCl:

$$0.14\ mol\ NaCl \times \frac{58.5\ g\ NaCl}{1\ mol\ NaCl} = 8.2\ g\ NaCl\ \blacksquare$$

Examples 5 and 6 show the relationship between grams, moles, liters, and molarity, which we can write out as

**molarity × liters = moles**

**moles × molecular (or formula) weight = grams**

176    MOLARITY AND SOLUTION STOICHIOMETRY

**EXAMPLE 7**   Using 7.3 g of $AgNO_3$, how many liters of solution are needed to make a 0.12 M solution?

*Solution:* The number of moles of $AgNO_3$ is

$$7.3 \text{ g AgNO}_3 \times \frac{1 \text{ mol AgNO}_3}{169.9 \text{ g AgNO}_3} = 0.043 \text{ mol AgNO}_3$$

Using the formula mol/L $\times$ L = mol and remembering that 0.12 M = 0.12 mol/L,

We will solve for this "L" which represents
the number of liters of solution we need.
↓

$$0.12 \, \frac{\text{mol AgNO}_3}{\text{L}} \times L = 0.043 \text{ mol AgNO}_3$$

$$L = \frac{0.043 \text{ mol AgNO}_3}{0.12 \, \frac{\text{mol AgNO}_3}{\text{L}}} = 0.36 \text{ L} \quad \blacksquare$$

Notice that in a complex fraction like

$$\frac{\text{mol}}{\frac{\text{mol}}{\text{L}}}$$

we can clear the layers by multiplying numerator and denominator by "L" (which is the same as multiplying the three-layered fraction by L/L = 1):

$$\frac{\cancel{\text{mol}} \times L}{\frac{\cancel{\text{mol}}}{\cancel{L}} \times \cancel{L}} = L$$

## 11-2
## SOLUTION STOICHIOMETRY USES MOLARITY

A good way to visualize what is involved in solution stoichiometry is to understand a laboratory procedure called **titration**. It is performed using the apparatus shown in Figure 11-2. A buret is filled to a known level with one solution, say 0.1 M NaOH (0.1 M sodium hydroxide). (A buret was described in Figure 6-4.) In a flask placed below the buret, a known quantity (say 25 mL) of another solution (e.g., hydrochloric acid, HCl) is added. The

11-2 Solution Stoichiometry Uses Molarity  177

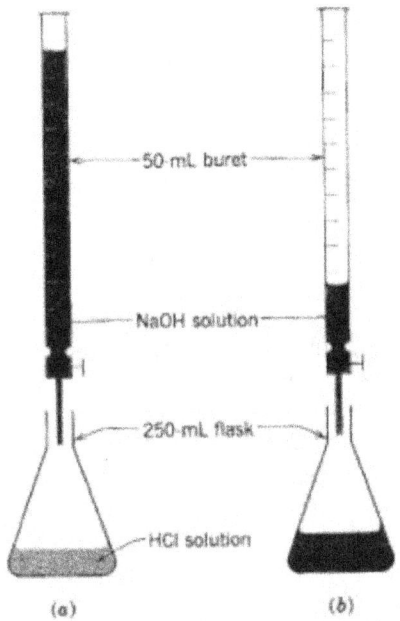

**FIGURE 11-2** An acid-base titration using a 50-mL buret. (a) The setup before any NaOH solution (the base) has been added to the HCl solution (the acid) in the flask. (b) The NaOH solution is slowly added to the flask until the titration is completed. See page 170 for photos of a buret.

A full discussion about how to do a titration of the kind mentioned above begins on page 176 (Section 11-2) and continues to page 179. Below we discuss three of the errors that can be made in such a titration. Some of the first error below can be removed by calibrating the buret, which is time consuming and not needed in a general chemistry class. The third error can be reduced by adding a smaller drop.

This discussion is for a class A 50-mL buret that can deliver up to 50 mL of solution with an error of 0.1%. This percentage is the maximum allowed variation from each other that the burets have when they come from the factory (called "tolerance" in the following link). See https://bit.ly/3rXO4BR for the error limits of burets, volumetric flasks and pipets. The textbook *Analytical Chemistry* by Gary D. Christian and coauthors has a complete discussion of titration errors and how to calibrate a buret in Sec.2.4 pp. 30-4, available for free on Amazon at https://amzn.to/3ojXbf4.

A 50-mL buret can deliver up to 50 mL of solution with an error in each reading of about ±0.01 mL (this means an error of 0.02 mL could be made). A typical titration in general chemistry is designed to deliver about 25 mL of solution from the buret. Assuming that 25 mL of NaOH is delivered during the titration, the percent error for each reading of the buret would be about 0.02/25 x 100 = 0.08 %. When we carry out a titration using a buret, we make two readings — one reading before the solution is delivered and one after the titration is finished. Thus the total error could be 0.04 mL if we were unlucky and the errors added up. The percent uncertainty for the **titration error** due to reading the buret is 0.04/25 x 100 = 0.16%.

In Section 11-2 it is mentioned that in a titration you must add about one drop or less after the **equivalence point** for the indicator to change color. This is called the **end point** of the titration. One drop is about equal to 0.05 mL. This introduces an additional **titration error** of 0.05/25 x 100 = 0.2 % if you have added about 25 mL of base. The **total titration error** could be as high as 0.1% + 0.16% + 0.2% = 0.46 %. If you can add less than 1 extra drop, then the error would be reduced. For ½ an extra drop, the total titration error would be reduced by 0.1% to 0.36%.

There are many videos online that discuss how to do a titration. Two that seem easy to understand are at https://bit.ly/3FQdwhM and https://bit.ly/3FNoCnv.*

*In these videos, the right-handed titrator uses the left hand to turn the stopcock. This is to prevent the titrator from accidently pulling out the stopcock. If you take a course that has a titration lab, listen carefully to your instructor for all the pitfalls and do your best to avoid them.

molarity of this solution is not known. We will find out what it is by doing the titration. The NaOH solution is added dropwise (a drop at a time) into the flask until an indicator that has been previously added to the HCl tells us that all the HCl has been neutralized (completely reacted with) by the NaOH. The **indicator** is a dye that changes color when the HCl is neutralized. The reaction is

$$HCl + NaOH \rightarrow NaCl + HOH$$

By reading the level of NaOH in the buret at this point, we know what volume of NaOH has been added to the HCl. From this we can calculate the molarity of the HCl solution. The following discussion will show you how to do this type of calculation.

Let's look at an actual titration of HCl with 0.100 M NaOH. We will take 25.00 mL of the HCl solution, mainly because an accurate pipet called a **volumetric pipet** (commonly available in chemistry labs) delivers 25.00 mL of solution (see Figure 11-3). The HCl solution is put into a 250-mL flask (again, it's convenient and commonly available), and an indicator called **phenolphthalein** is added. Phenolphthalein is colorless in acidic solutions and pink in basic solutions. A 50-mL buret is filled with 0.100 M NaOH solution, and the titration is performed. We find that 23.20 mL of 0.100 M NaOH is needed to turn the phenolphthalein pink. We have reached the **end**

Volumetric mark

**Figure 11-3** A 25-mL class A volumetric pipet. It has an error of no more than 0.03 mL. A volume of 25 mL is large enough so that an error of 0.03 mL gives a percent error of 0.03/25 x 100 = 0.12%. This error is acceptable for most lab work. Notice that a 25-mL pipet has only one volumetric mark, namely 25 mL. It can only be used to add 25 mL of solution. If you need to add a different amount of solution, you will need to use a pipet of a different size. Common sizes are 1 mL, 10 mL, 25 mL and 50 mL. If you take a lab course, listen carefully to any tips your instructor may give you to minimize errors. See https://bit.ly/3rXO4BR.

**point** of the titration as described in Figure 11-2. In actual practice, a very small amount of NaOH (a drop or less) over the amount needed to neutralize the HCl is added to make the solution basic. Then the phenolphthalein turns a pale pink color. The volume of NaOH needed to neutralize (or completely react with) all the HCl is called the **equivalence point**. Notice that the equivalence point volume is usually *less* than the end-point volume. The difference is called the **titration error**. It is usually small and we will ignore it.

The number of moles of NaOH used in the titration is

$$0.100 \text{ M} \times 0.0232 \text{ L} = 0.00232 \text{ mol NaOH}$$

Since the chemical equation for this reaction is HCl + NaOH → NaCl + HOH, we see that 1 mol HCl = 1 mol NaOH. Thus the number of moles of HCl neutralized is

$$0.00232 \text{ mol NaOH} \times \frac{1 \text{ mol HCl}}{1 \text{ mol NaOH}} = 0.00232 \text{ mol HCl}$$

The molarity of the HCl solution is (remember we took 25.00 mL of HCl)

$$M = \frac{\text{mol}}{L} = \frac{0.00232 \text{ mol HCl}}{0.02500 \text{ L}} = 0.0928 \text{ M HCl}$$

**EXAMPLE 8** What is the molarity of 25.0 mL of an $H_2SO_4$ (sulfuric acid) solution that requires 29.3 mL of 0.250 M NaOH (sodium hydroxide) to neutralize it?

***Solution:*** The number of moles of NaOH is

$$0.250 \frac{\text{mol NaOH}}{L} \times 0.0293 \text{ L} = 0.00733 \text{ mol NaOH}$$

The balanced chemical equation for the reaction is

$$H_2SO_4 + 2NaOH \rightarrow Na_2SO_4 + 2HOH$$

Thus

$$1 \text{ mol } H_2SO_4 = 2 \text{ mol NaOH}$$

The number of moles of $H_2SO_4$ reacting with the NaOH is

$$0.00733 \text{ mol NaOH} \times \frac{1 \text{ mol } H_2SO_4}{2 \text{ mol NaOH}} = 0.00367 \text{ mol } H_2SO_4$$

## 180 MOLARITY AND SOLUTION STOICHIOMETRY

The molarity of the $H_2SO_4$ solution is

$$M = \frac{mol}{L} = \frac{0.00367 \text{ mol } H_2SO_4}{0.0250 \text{ L}} = 0.147 \text{ M } H_2SO_4 \quad \blacksquare$$

**EXAMPLE 9** How many milliliters of 0.15 M KOH (potassium hydroxide) solution is needed to completely neutralize (react with) 100 mL of 0.20 M $H_3PO_4$ (phosphoric acid)?

*Solution:* The number of moles of $H_3PO_4$ is

$$0.20 \frac{\text{mol } H_3PO_4}{L} \times 0.10 \text{ L} = 0.020 \text{ mol } H_3PO_4$$

The balanced equation is

$$3KOH + H_3PO_4 \rightarrow K_3PO_4 + 3HOH$$

Thus

$$3 \text{ mol KOH} = 1 \text{ mol } H_3PO_4$$

The number of moles of KOH needed is

$$0.020 \text{ mol } H_3PO_4 \times \frac{3 \text{ mol KOH}}{1 \text{ mol } H_3PO_4} = 0.060 \text{ mol KOH}$$

The number of milliliters of 0.15 M KOH solution needed is

$$M = \frac{mol}{L}$$

$$0.15 \text{ M KOH} = \frac{0.060 \text{ mol KOH}}{L}$$

$$L = \frac{0.060 \text{ mol KOH}}{0.15 \frac{\text{mol KOH}}{L}} = 0.40 \text{ L}$$

Converting liters to milliliters, we get

$$0.40 \text{ L} \times \frac{1000 \text{ mL}}{1 \text{ L}} = 400 \text{ mL} = 4.0 \times 10^2 \text{ mL}$$

which is expressed to two significant figures.

We can also solve this problem in one step:

$$0.20 \text{ M H}_3\text{PO}_4 \times 0.10 \text{ L} \times \frac{3 \text{ mol KOH}}{1 \text{ mol H}_3\text{PO}_4} \times \frac{1}{0.15 \text{ M KOH}} \times \frac{1000 \text{ mL}}{1 \text{ L}}$$

$$= 4.0 \times 10^2 \text{ mL} \quad \blacksquare$$

**EXAMPLE 10** The titration in acid solution of 25.0 mL of an oxalic acid ($H_2C_2O_4$) solution required 22.5 mL of 0.097 M $KMnO_4$ (potassium permanganate) for all the oxalic acid to be used up. The balanced equation for the reaction is

$$2KMnO_4 + 5H_2C_2O_4 + 6HCl \rightarrow 10CO_2 + 2MnCl_2 + 8H_2O + 2KCl$$

What is the molarity of the oxalic acid solution? In this problem you can ignore the HCl because it is not involved in the calculation.

*Solution:* The number of moles of $KMnO_4$ used is

$$0.097 \text{ M KMnO}_4 \times 0.0225 \text{ L} = 0.00218 \text{ mol KMnO}_4$$

where we converted 22.5 mL to 0.0225 L. From the balanced equation we see that 2 mol $KMnO_4$ = 5 mol $H_2C_2O_4$. The number of moles of $H_2C_2O_4$ that reacted is

$$0.00218 \text{ mol KMnO}_4 \times \frac{5 \text{ mol H}_2\text{C}_2\text{O}_4}{2 \text{ mol KMnO}_4} = 0.00545 \text{ mol H}_2\text{C}_2\text{O}_4$$

The molarity of the oxalic acid solution is

$$M = \frac{\text{mol}}{\text{L}} = \frac{0.00545 \text{ mol H}_2\text{C}_2\text{O}_4}{0.0250 \text{ L}} = 0.218 \text{ M H}_2\text{C}_2\text{O}_4$$

The problem can also be solved in one step:

$$0.097 \text{ M KMnO}_4 \times 0.0225 \text{ L} \times \frac{5 \text{ mol H}_2\text{C}_2\text{O}_4}{2 \text{ mol KMnO}_4} \times \frac{1}{0.0250 \text{ L}}$$

$$= 0.218 \text{ M H}_2\text{C}_2\text{O}_4$$

Based on the material in Chapter 6, which was on significant figures, you might think about the percent uncertainty in 0.097 M $KMnO_4$ as compared to the rest of the numbers in the problem. $\blacksquare$

182 MOLARITY AND SOLUTION STOICHIOMETRY

Notice that in these examples, the solution with the known molarity was in the buret. One can just as well put a solution of unknown molarity in the buret. Then, of course, the solution of known molarity would have to be in the flask. The method of calculating the unknown molarity is similar in either case.

## 11-3
### WHEN A SOLUTION IS DILUTED, THE NUMBER OF MOLES OF SOLUTE DOESN'T CHANGE

When a solution is diluted with pure solvent (such as water), the number of moles of solute originally present doesn't change. All that has been added is solvent. For instance, if we add water to a sugar solution, the number of moles of sugar in the solution, before and after the water is added, remains the same.

Let's use the subscript "$i$" (standing for "initial") for the undiluted solution, and the subscript "$f$" (standing for "final") for the diluted solution. Then, since the number of moles of solute doesn't change, we can state that

**moles of solute in undiluted solution = moles of solute in diluted solution**

Let

$$\text{moles of solute in undiluted solution} = \text{moles}_i = \text{mol}_i$$

and

$$\text{moles of solute in diluted solution} = \text{moles}_f = \text{mol}_f$$

Since initial and final moles are equal, we can say that

$$\text{mol}_i = \text{mol}_f$$

Moreover, since molarity × liters = moles, we have

$$M_i \times L_i = \text{mol}_i \quad \text{and} \quad M_f \times L_f = \text{mol}_f$$

Thus

$$M_i \times L_i = M_f \times L_f$$

which is the same as saying $\text{mol}_i = \text{mol}_f$.

## 11-3 When a Solution Is Diluted, the Number of Moles of Solute Doesn't Change

Remember that $L_i$ and $L_f$ refer to the initial and final volumes of the solution. $L_f$ does *not* refer to the amount of solvent added when making the dilution.

**EXAMPLE 11** You have 200 mL of a 0.5 M NaCl solution. If you add water to bring the final volume to 500 mL, what is the new molarity of the solution?

*Solution:* We use the formula $M_i L_i = M_f L_f$. Substituting values from the problem, we have

$$(0.5 \text{ M})(200 \text{ mL}) = M_f(500 \text{ mL})$$

Notice that we didn't bother to convert milliliters to liters. It is not necessary because the milliliter units divide out anyway. Continuing, we solve for $M_f$:

$$M_f = \frac{(0.5 \text{ M})(200 \text{ mL})}{500 \text{ mL}} = 0.2 \text{ M} \quad \blacksquare$$

Chemists frequently prepare a dilute solution from a concentrated stock solution. The next example shows how to do this.

**EXAMPLE 12** A chemist needs to prepare 900 mL of 0.15 M $H_2SO_4$ (sulfuric acid) solution. Concentrated sulfuric acid is 18 M. How many milliliters of concentrated sulfuric acid and about how many milliliters of water do you need to make the diluted solution?

*Solution:* **CAUTION:** *Concentrated sulfuric acid is very dangerous. If you get it on your skin or clothing, it will cause burns or holes. You should never add water to the concentrated acid. Always pour the acid slowly into a large amount of water and stir. The reason is that concentrated sulfuric acid reacts with water to produce a great deal of heat. If a small amount of water is added to the concentrated sulfuric acid, the heat released may boil the water and spatter the acid.* Now, getting back to our example, we use the formula

$$M_i L_i = M_f L_f$$
$$(18 \text{ M})(L_i) = (0.15 \text{ M})(900 \text{ mL})$$
$$L_i = \frac{(0.15 \text{ M})(900 \text{ mL})}{18 \text{ M}} = 7.5 \text{ mL}$$

Therefore, you need 7.5 mL of concentrated sulfuric acid and about 900 mL − 7.5 mL = 892.5 mL of water.

Notice that the amount of water added may not be exactly 892.5 mL. It is possible that the volume is changed when two substances are mixed. This solution will also be warm after the mixing and would have a slightly greater volume than the room temperature solution. So for the greatest accuracy, the dilute $H_2SO_4$ solution is prepared as follows: 7.5 mL of 18 M $H_2SO_4$ is added to about 400 mL water in a 1000-mL graduated cylinder. This solution is allowed to cool to room temperature. Then additional water is added, with stirring, to bring the volume up to 900 mL. ∎

## PROBLEMS

## KEYED PROBLEMS

1. In a salt solution of water and salt, which component is the solute and which is the solvent?

2. A solution contains 1.5 mol of sucrose dissolved in 5.0 L of solution. What is the molarity of the solution?

3. A solution contains 0.250 mol of $KMnO_4$ dissolved in 0.350 L of solution. What is the molarity of the solution?

4. A solution contains 0.400 mol of fructose dissolved in 400 mL of solution. What is the molarity of the solution?

5. A solution contains 8.00 g of $CaSO_4$ dissolved in 800 mL of solution. What is the molarity of the solution?

6. How many grams of potassium bromide, KBr, must be used to make 75 mL of a 0.12 M KBr solution?

7. How many liters of solution are needed to make a 0.050 M solution from 5.2 g of $MgCl_2$?

8. What is the molarity of 25.0 mL of an oxalic acid ($H_2C_2O_4$) solution that required 23.5 mL of 0.0950 M KOH to neutralize (or react completely with) it? The reaction is

$$H_2C_2O_4 + 2KOH \rightarrow K_2C_2O_4 + 2HOH$$

9. How many milliliters of a 0.250 M NaOH solution are needed to completely neutralize 175 mL of 0.110 M $H_3PO_4$?

10. The titration, in acid solution, of 25.0 mL of oxalic acid solution required 27.3 mL of 0.125 M $KMnO_4$ for all the oxalic acid to be used up. What is the molarity of the oxalic acid?

11. You have 100 mL of a 0.15 M HCl solution. If you add water to bring the final volume to 500 mL, what is the new molarity of the solution?

12. You want to prepare 600 mL of 0.23 M $HNO_3$ (nitric acid). You will have to dilute concentrated (16 M) $HNO_3$. How many milliliters of concentrated nitric acid and about how many milliliters of water do you need to make the diluted solution?

   **CAUTION:** *Concentrated nitric acid is very corrosive. Don't get it on you. If you do, wash it off quickly with lots and lots of water.*

## SUPPLEMENTAL PROBLEMS

13. How many grams of $CuSO_4 \cdot 5H_2O$ (copper(II) sulfate pentahydrate) are needed to prepare 250 mL of a 0.075 M $CuSO_4$ solution? The centered dot in $CuSO_4 \cdot 5H_2O$ indicates that the 5 $H_2O$ molecules are in the crystal and must be counted in the formula weight. The water is called **water of hydration.**

14. How many milliliters of 0.1400 M $Ce(SO_4)_2$ (Ce is the element cerium, used in analytical chemistry) are needed to react with 50.00 mL of a 0.07362 M $FeSO_4$ solution? The reaction is

    $$2Ce(SO_4)_2 + 2FeSO_4 \rightarrow Ce_2(SO_4)_3 + Fe_2(SO_4)_3$$

15. How many milliliters of 0.5125 M HCl are needed to react completely with 35.00 mL of 0.1018 M $KMnO_4$? The reaction is

    $$16HCl + 2KMnO_4 \rightarrow 5Cl_2 + 2MnCl_2 + 8H_2O + KCl$$

16. How many milliliters of 0.102 M NaOH are needed to neutralize 9.34 mL of 0.0987 M $H_2SO_4$? The reaction is

    $$2NaOH + H_2SO_4 \rightarrow Na_2SO_4 + 2HOH$$

17. How many milliliters of solution do you get when you prepare a 0.0850 M solution from 15.2 g $ZnCl_2$? About how many milliliters of water are used?

18. How would you make 1.0 L of 0.50 M HCl from concentrated (12 M) HCl?

19. You are given 300 mL of a 1.50 M glucose solution. What final volume (in milliliters) do you need in order to dilute the glucose to 0.350 M? About how many milliliters of water are added?

20. Intravenous saline solution (NaCl solution) contains 0.85% NaCl in water. This solution is isotonic, meaning that it can be injected into veins without causing damage to blood cells or vein walls. What is the molarity of a 0.85% NaCl solution?

    (**NOTE:** A 0.85% NaCl solution contains 0.85 g NaCl in 100 mL of solution.)

21. What is the molarity of 10.00 mL of vinegar [a dilute aqueous (water) solution of acetic acid, $HC_2H_3O_2$] that is neutralized by 17.02 mL of 0.5103 M NaOH? The reaction is

$$HC_2H_3O_2 + NaOH \rightarrow NaC_2H_3O_2 + HOH$$

22. Sodium hypochlorite, NaClO, in dilute solution is a bleach and disinfectant. A common brand name is Clorox. It can be prepared by the following reaction:

$$Cl_2 + 2NaOH \rightarrow NaClO + NaCl + H_2O$$

How many grams of $Cl_2$ are needed to react with 253 mL of 0.600 M NaOH?

23. Zinc reacts with hydrochloric acid according to the following equation:

$$Zn + 2HCl \rightarrow ZnCl_2 + H_2$$

How many milliliters of 0.500 M HCl are needed to react with 1.32 g of Zn?

24. One way to determine the concentration of chloride ion, $Cl^-$, in seawater is to perform a Mohr titration according to the following reaction:

$$AgNO_3 + Cl^- \rightarrow AgCl(s) + NO_3^-$$

(**NOTE:** AgCl(s) means solid silver chloride. As the silver nitrate solution is added to the seawater during the titration, a white material appears in the flask with the seawater. This white material is the solid silver chloride. It is called a precipitate.) What is the molarity of $Cl^-$ in a sample of seawater if 25.00 mL of seawater required 26.32 mL of 0.4216 M $AgNO_3$?

25. If we need 42.8 mL of 0.296 M $H_3PO_4$ (phosphoric acid) to just neutralize 24.9 mL of KOH (potassium hydroxide), what is the molarity of the potassium hydroxide solution? The reaction is

$$H_3PO_4 + 3KOH \rightarrow K_3PO_4 + 3HOH$$

# ACIDS, BASES AND SALTS

**Svante Arrhenius** ((1859 –1927), a Swedish chemist who won the 1903 Nobel Prize in Chemistry, was the first, in 1884, to define acids and bases as:

**ACIDS ionize in water to produce hydrogen ions ($H^+$).**

**BASES ionize in water to produce hydroxide ions ($OH^-$).**

**NOTE:** In 1896, Arrhenius estimated how increases in atmospheric carbon dioxide ($CO_2$) would increase Earth's surface temperature through the **greenhouse gas effect**. See footnote 2 on page 305 and https://bit.ly/3zRQ1T4 for more details.

**NOTE:** There are other definitions of acids and bases that are more general. You will learn about these in general chemistry. The Arrhenius definition is sufficient to understand the acid-base reactions in this book.

Examples of acids are hydrochloric acid (HCl), nitric acid ($HNO_3$) and sulfuric acid ($H_2SO_4$). In water these acids ionize according to the following three equations giving $H^+$ ions as a product of the reaction:

$HCl \rightarrow H^+ + Cl^-$     $HNO_3 \rightarrow H^+ + NO_3^-$     $H_2SO_4 \rightarrow 2H^+ + SO_4^{2-}$

Examples of bases are sodium hydroxide (NaOH), potassium hydroxide (KOH) and calcium hydroxide ($Ca(OH)_2$). In water these bases ionize according to the following three equations giving $OH^-$ ions as a product of the reaction:

$NaOH \rightarrow Na^+ + OH^-$     $KOH \rightarrow K^+ + OH^-$     $Ca(OH)_2 \rightarrow Ca^{2+} + 2OH^-$

When an acid and a base react, the products of the reaction are a salt (NaCl is a salt whose everyday name happens to be "salt") and water. **The $H^+$ ion of the acid reacts with the $OH^-$ ion of the base to give water, HOH.**[1] **This process is called neutralization**, because if all the H+ and OH− react with each other, in say a titration (see page 177), the solution is said to be **neutral**.[2] Equations from this chapter that show the process of neutralization are:[3]

$HCl + NaOH \rightarrow NaCl + HOH$   (p 178 – NaCl, sodium chloride, is the salt)

$H_2SO_4 + 2NaOH \rightarrow Na_2SO_4 + 2HOH$   (p 179 example 8, p 185 problem 16 – $Na_2SO_4$, sodium sulfate, is the salt)

$3KOH + H_3PO_4 \rightarrow K_3PO_4 + 3HOH$   (p 180 example 9, p 186 problem 25 – $K_3PO_4$, potassium phosphate, is the salt)

$H_2C_2O_4 + 2KOH \rightarrow K_2C_2O_4 + 2HOH$   (p 184 problem 8 – $K_2C_2O_4$, potassium oxalate, is the salt)

$HC_2H_3O_2 + NaOH \rightarrow NaC_2H_3O_2 + HOH$   (p 186 problem 21 – $NaC_2H_3O_2$, sodium acetate, is the salt)

[1]The formula for water is usually written as $H_2O$, but in acid-base reactions it is common to write the formula as HOH to clearly show that water is formed from the reaction of $H^+$ and $OH^-$. The equation for this reaction is: $H^+ + OH^- \rightarrow HOH$.

[2]This statement will be modified in a general chemistry class where you will learn about strong acids, strong bases, weak acids and weak bases and the details of how they react. The essence of a neutralization reaction using the Arrhenius definition of acids and bases is the equation: $H^+ + OH^- \rightarrow HOH$.

[3]These equations are written without showing the ionization of the acids and bases.

*Top:* **Hot Air Balloons in Flight.** *Bottom:* **Putting Hot Air into the Balloon Using the Flame from a Propane Tank.** The balloon previously had been inflated with a large air blower. Notice the wicker basket in the bottom photo just above the word "currents." This is where the passengers stand. It's scary standing in a chest-high wicker basket when you are a mile above the ground and a hot flame, a few feet above your head, is firing into a flimsy nylon balloon. If the pilot wants to go up, the flame is turned on. To go down, a flap at the top of the balloon is opened letting out some hot air, and the flame is turned off. To go sideways, the pilot searches for wind currents by going up or down. (Yes, the author has flown in a hot air balloon.)

Why does the balloon rise? The balloon and passengers have a certain weight. If their weight is less than the weight of the air they displace, there will be a net upward force on the balloon and the balloon rises. So in this situation, the upward force on the balloon is greater than the downward gravity force on the balloon. The original air in this space that the balloon displaced must have had the same upward force on it, keeping it from moving up or down. So for the original air, the upward force (called the buoyant force) = the downward gravity force. This is called **Archimedes' principle**, named after **Archimedes of Syracuse** (287-212 BCE), a Greek mathematician, physicist, engineer, inventor, and astronomer (see https://bit.ly/2u2HSyw).   *Photos are in the Public Domain.*

# GASES AND THE IDEAL GAS LAW

Sulfur dioxide is used to make sulfuric acid, gaseous ammonia to make fertilizer. Freon gases are used in refrigerators and air conditioners, helium and argon in scientific experiments and in some industrial processes such as welding. The air around us is a gas. Because of the importance of air and other gases, it is useful to study the properties of gases.

## 12-1
## MOLECULES OF A GAS COLLIDE AND CHANGE THEIR SPEED AND DIRECTION

A gas such as air consists of molecules that are constantly moving around. At room temperature and pressure, these molecules are usually far apart, so that if you were the size of a molecule, another molecule would seem very far away most of the time. But every so often, another molecule would collide with you and bounce off. And if the gas you were a part of was in a container, occasionally you would collide with a wall of the container. A picture of what gas molecules might look like as they fly around is given in Figure 12-1. Notice that the molecules are traveling in different directions and that they have different speeds. This is reasonable, since speeds and directions can change in a collision. Figure 12-2 shows how this can happen in the case of a straight-line collision.

# 190 GASES AND THE IDEAL GAS LAW

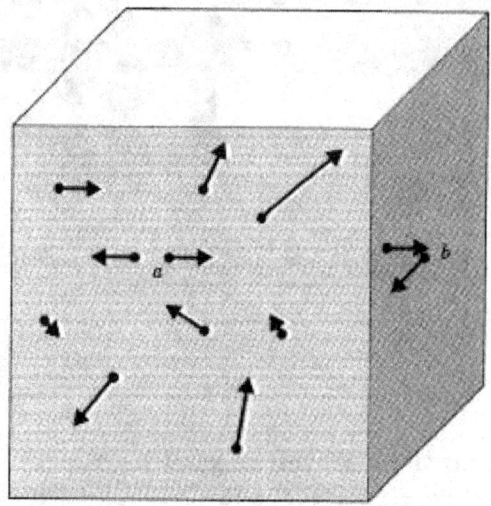

**FIGURE 12.1** The dots with arrows represent gas molecules. The arrows show the direction of motion of the molecules. The length of an arrow is proportional to the speed of the molecule. Point *a* shows two molecules bouncing apart after colliding. Point *b* shows a molecule hitting a wall and bouncing off.

## 12-2
## PRESSURE IS THE FORCE PER UNIT AREA

When gas molecules strike the walls of a container, they exert a force on the walls. This force is responsible for the pressure of the gas in the container. **Pressure** is defined as force per unit area. For example, at sea level, the air of the atmosphere exerts a pressure of 14.7 lb/in.$^2$. Read lb/in.$^2$ as "pounds per square inch." This means that each square inch of surface has a force of 14.7 lb on it. See Figure 12-3 for a good way to understand atmospheric pressure.

Other units of pressure in addition to pounds per square inch are used. Figure 12-4 shows how 14.7 lb/in.$^2$ of air pressure can support a column of mercury 760 mm high. The weight of the column of mercury is 14.7 lb for each square inch of cross section. The width of the column doesn't matter because a wider column will have more air pushing on it. Thus pressures are often reported in heights of a mercury column. A pressure of 760 mm of mercury (Hg) has been selected as "standard pressure" and is called one **atmosphere.** The unit "millimeters of mercury" is abbreviated **mmHg,** and the unit "atmosphere" is abbreviated **atm.**

**FIGURE 12-2** Speed change during collision. (*a*) Before collision: The fast molecule is catching up to the slow molecule. (*b*) After collision: The fast molecule has transferred some of its energy to the slow molecule. The fast molecule from part *a* is now the slow one, and the slow molecule from part *a* is now the fast one.

## 12-2 Pressure is the Force per Unit Area

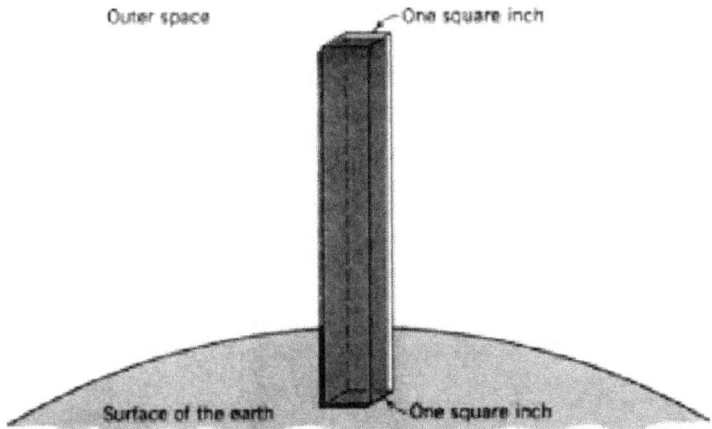

**Figure 12-3** Weight is the force resulting from the gravitational attraction between a mass and the earth. For example, if you weigh 150 lb, this means that there is a force of 150 lb attracting your mass to the earth. When we say that each square inch of the earth has a force of 14.7 lb on it because of the air in the atmosphere, we mean the following: All the air in a 1-in.$^2$ column extending from the surface of the earth up to outer space weighs 14.7 lb.

From this discussion, we can write the following equality between different pressure units:

**1 atm = 760 mmHg = 14.7 lb/in.$^2$**

The unit mmHg is often replaced by the unit **torr**. (**NOTE:** The above 3 units, and the torr, are *not* official SI units, but are widely used.) Thus **1 Torr = 1 mmHg**. It is named after the Italian scientist **Evangelista Torricelli** (1608-1647) who invented the mercury barometer (See Figure 12-4). There is no abbreviation for torr. When it is used without a number, torr is written in lower case. But as a unit after a number, it is capitalized (as in 760 Torr). This follows the rule for units named after people (with two exceptions — Celsius and Fahrenheit degrees — discussed in Section 12-3).

Relating atm, torr, and lb/in.$^2$, we can write

**1 atm = 760 Torr = 14.7 lb/in.$^2$**

Different fields of science (and medicine) sometimes use different units. For example, blood pressure is measured in **millimeters of mercury (mmHg)**. If your blood pressure is reported as 115/65, this means 115 mmHg/65 mmHg. In aviation and on TV weather reports, air pressure is reported in **inches of mercury (inHg)**, where 29.92 inHg = 1 atm. Meteorologists use **millibars (mbar)**, or the SI unit of pressure, **hectopascals (hPa)**, on their weather maps. (See page 192 for the definition of the SI unit of pressure, the **pascal**.) The conversion is

**1013.25 mbar = 1013.25 hPa = 1 atm.**

**EXAMPLE 1** How many atmospheres are there in 9.00 x 10$^2$ Torr?

*Solution:*          1 atm = 760 Torr

9.00 x 10$^2$ ~~Torr~~ x 1 atm/760 ~~Torr~~ = 1.18 atm ∎

# 192 GASES AND THE IDEAL GAS LAW

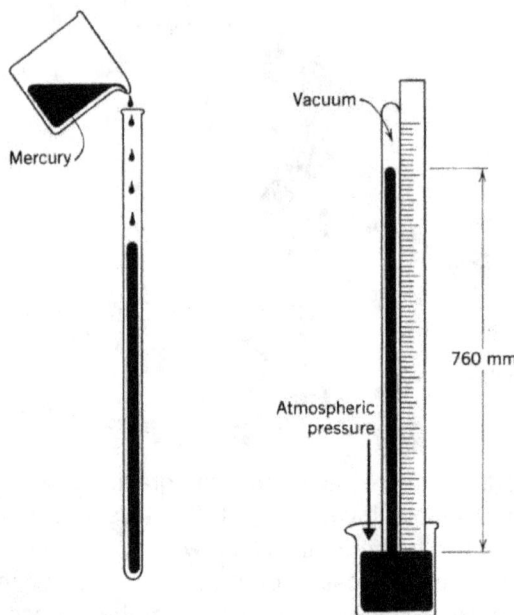

**Figure 12-4** Making a Mercury Barometer: A piece of glass tubing, about 800 mm long and closed at one end, is filled with mercury. Then it is inverted (turned upside down) and put into a beaker filled with mercury. Atmospheric pressure, acting downward on the surface of the mercury in the beaker, prevents the mercury in the tube from draining out.

The mercury level drops somewhat until the weight of the mercury in the tube is equal to the air pressure pushing on the surface of the mercury. If the length of the column of mercury is 760 mm, the air pressure is exactly 1 atm. The weight of such a column of mercury is 14.7 lb for a column whose inside cross section is one square inch. (In practice, the column's cross section is less than 1 in.². **NOTE:** The diameter of the column makes no difference because pressure only depends on the force per unit area, not just on the force. In this case, the force comes from the weight of air.)

**EXAMPLE 2** How many torr are there in 5.00 lb/in.²?

*Solution:* 760 Torr = 14.7 lb/in.²

5.00 l̶b̶/̶i̶n̶.̶² x 760 Torr/14.7 l̶b̶/̶i̶n̶.̶² = 259 Torr ∎

As mentioned on page 191, another unit of pressure, the **pascal (Pa)**, is being used more frequently because it is the pressure unit of the SI system. It is named after the French scientist **Blaise Pascal** (1623-1662) and

$$1 \text{ atm} = 101{,}325 \text{ Pa}$$

**NOTE:** Another pressure unit, used in medicine and physiology, is the **centimeter of water, cmH$_2$O**, where **1033 cmH$_2$O = 1 atm**. It is used, for example, to measure the pressure in a mechanical ventilator that breathes for a person who can't breathe on their own if needed during and/or after surgery, or who has a lung infection such as pneumonia or Covid-19 whose lungs are filling with fluid, https://bit.ly/3aQlhVa. The maximum pressure used is 35 cmH$_2$O.

Since the pascal is rather small, it is usually more convenient to use the **kilopascal**, abbreviated **kPa**. We have

**1000 Pa = 1 kPa**, and to three significant figures, **1 atm = 101 kPa**.

The tire pressure of automobiles is reported in both pounds per square inch and kilopascals. A sticker, usually placed on the driver's side door jamb, lists recommended pressures. For example, the front tire pressure on the author's 2013 Subaru Outback is listed as "220 KPA, 32 PSI." (These would properly be written 220 kPa and 32 psi. **NOTE:** "psi" is the common abbreviation for pounds per square inch, or lb/in².)

**EXAMPLE 3**  How many kilopascals are there in 5.00 atm?

*Solution:* Since 1 atm = 101 kPa, we have

5.00 ~~atm~~ x 101 kPa/1 ~~atm~~ = 505 kPa ∎

## 12-3
### DEGREES CELSIUS AND DEGREES FAHRENHEIT ARE TWO UNITS OF TEMPERATURE

The everyday unit of temperature that we use is the Fahrenheit degree, named after **Daniel Gabriel Fahrenheit** (1686–1736), a German–Dutch scientific instrument maker, who invented the mercury thermometer. After many experiments, he assigned 32 degrees Fahrenheit (abbreviated °F) as the freezing point of water. He seems to have then arbitrarily chosen 212 °F as the boiling point of water.

In science, we don't use the Fahrenheit degree. We use the Celsius degree (abbreviated °C), which is named after **Anders Celsius** (1701–1744), a Swedish astronomer. The Celsius degree used to be called the centigrade degree, but the official name after 1948 is the Celsius degree. For the Celsius system, 0 °C was chosen as the freezing point of water and 100 °C was chosen as the boiling point of water.[1] Both the Celsius and Fahrenheit degrees are mentioned in weather reports.

In degrees Fahrenheit, the temperature difference between the freezing point and boiling point of water is 180 °F. In the degrees Celsius system, the difference is 100 °C. Thus one degree Celsius is larger than one degree Fahrenheit, as is shown in Figure 12-5.

A temperature in degrees Celsius can be converted to the equivalent temperature in degrees Fahrenheit by the formula

$$°F = 9/5 \ °C + 32$$

[1] The "normal" body temperature of 37 °C (or 98.6 °F) was obtained by the German physician **Carl Wunderlich** in 1851. He measured the temperature of 25,000 people living in Leipzig, Germany. Recent research (https://elifesciences.org/articles/49555) has determined that currently, in high-income countries, normal body temperature has decreased to about 36.4 °C (97.5 °F). In 1851, the average life expectancy was about 38 y, and there was a much higher rate of chronic infection and inflammation than there is now. These probably caused the higher body temperature compared to what we see now.

# 194 GASES AND THE IDEAL GAS LAW

**FIGURE 12-5** The relative sizes of the Celsius and Fahrenheit degrees. The Celsius degree is 1.8 times larger than the Fahrenheit degree.

To derive this formula, we draw the graph of a straight line in Figure 12-6.

The equation of any straight line can be expressed as $y = mx + b$, where $m$ is the slope of the line and $b$ is the $y$ intercept. For our straight line we can write the equation as $°F = m \, °C + b$. To find out what $m$ and $b$ are, we simply substitute the values of the two known points from our graph in Figure 12-6:

| | |
|---|---|
| the equation: | $°F = m \, °C + b$ |
| substituting the first point: | $32 = m(0) + b$ |
| solving for $b$: | $b = 32$ |
| substituting the second point and the value of $b$: | $212 = m(100) + 32$ |
| solving for $m$: | $m = \dfrac{212 - 32}{100} = \dfrac{180}{100} = 1.8$ |
| the final equation: | $°F = 1.8 \, °C + 32$ |

To get the equation to look like the one you are familiar with, notice that $1.8 = 18/10$ and that $18/10 = 9/5$. Thus $°F = \frac{9}{5} °C + 32$. Maybe the form $°F = 1.8 \, °C + 32$ is more convenient if you are using a calculator.

**EXAMPLE 4** How many degrees Fahrenheit are equal to 20 °C?

*Solution:*

$$°F = 1.8 \, °C + 32 = 1.8(20) + 32 = 36 + 32 = 68 \, °F \quad \blacksquare$$

## 12-3 Degrees Celsius and Degrees Fahrenheit Are Two Units of Temperature 195

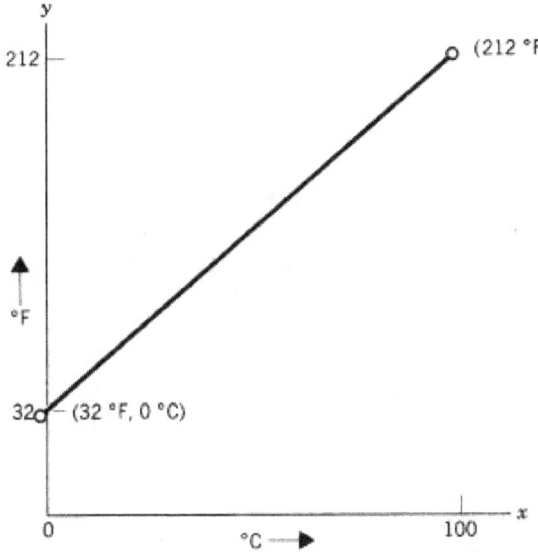

**FIGURE 12-6** The x-axis is the degrees Celsius axis. The y-axis is the degrees Fahrenheit axis. The two points whose coordinates we know are 0 °C = 32 °F and 100 °C = 212 °F. These two points determine a straight line.

It is useful to solve the equation °F = 1.8 °C + 32 for °C so that we can easily convert from degrees Fahrenheit to degrees Celsius:

$$°F = 1.8\ °C + 32$$
$$°F - 32 = 1.8\ °C$$
$$\frac{°F - 32}{1.8} = °C$$

The form you may be familiar with is $\frac{5}{9}(°F - 32) = °C$. The two forms are the same since 5/9 = 10/18 = 1/1.8. To convert 10/18 to 1/1.8, divide numerator and denominator by 10:

$$\frac{10}{18} = \frac{\frac{10}{10}}{\frac{18}{10}} = \frac{1}{1.8}, \quad \text{where } \frac{10}{10} = 1 \text{ and } \frac{18}{10} = 1.8$$

**EXAMPLE 5** Normal body temperature is 98.6 °F. What is this in degrees Celsius?

*Solution:*

$$°C = \frac{°F - 32}{1.8} = \frac{98.6 - 32}{1.8} = \frac{66.6}{1.8} = 37.0\ °C \quad \blacksquare$$

## 12-4
## AS THE TEMPERATURE OF A GAS DECREASES, THE VOLUME OF THE GAS DECREASES

To study the effect of temperature on the volume (abbreviated $V$) of a gas, we will use the piston and cylinder arrangement shown in Figure 12-7. The piston has a seal around its edge so that no gas can get in or out of the cylinder. Since the piston is very light, it easily moves up and down. This keeps the pressure inside the cylinder equal to the pressure outside. Since the outside pressure equals atmospheric pressure, we can assume that the pressure is held constant. If we place a candle under the cylinder, it will become warmer. This causes the gas inside the cylinder to expand, pushing up the piston. The volume inside the cylinder will increase.

If we now surround the cylinder with ice (Figure 12-8), the piston will go down, decreasing the volume of the gas inside the cylinder.

Now we are going to do an experiment in which we keep cooling the cylinder, well below the temperature of ice. As the gas inside the cylinder gets colder, its volume decreases. Let's plot what happens to the volume of the gas as the temperature decreases. As you can see in Figure 12-9, as the gas gets colder and colder, its volume gets smaller and smaller. But at a certain point, the gas liquefies. All gases will liquefy if they get cold enough.

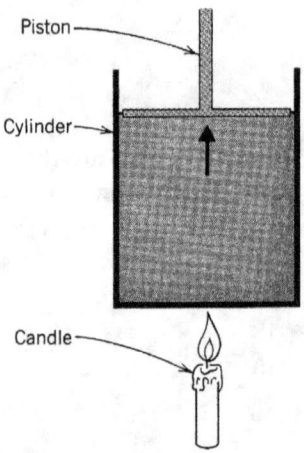

**FIGURE 12-7** A cylinder with a lightweight piston that can move up and down without friction is heated by a candle. The walls of the cylinder are very strong, and no gas can escape through the piston seals. Before the candle is put under the cylinder, the pressure inside is equal to the pressure outside. When the cylinder is heated, the piston moves up because the pressure inside the cylinder is increasing. The volume of the gas inside the cylinder increases as the piston moves up. If we assume that the heating is done very slowly, the pressure inside the cylinder is only a tiny bit greater than the pressure outside. We can assume that inside and outside pressures are "equal."

## 12-4 As the Temperature of a Gas Decreases, the Volume of the Gas Decreases

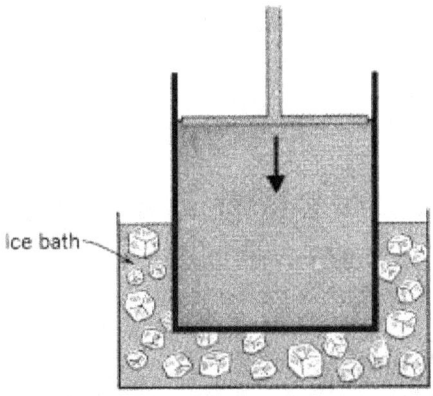

**FIGURE 12-8** The cylinder is surrounded by ice, which cools it and causes the piston to go down. The volume of gas inside the cylinder decreases as the piston falls.

Since the volume change of a liquid is extremely small as the liquid is cooled, our graph shows a leveling off of the volume as the temperature decreases. If we continue drawing the original line (the dotted part of the line in Figure 12-9), however, we find that it crosses the temperature axis at −273.15 °C. This is the temperature that would be needed to reduce the gas volume to **zero** if the gas had not liquefied. However, since real gases liquefy long

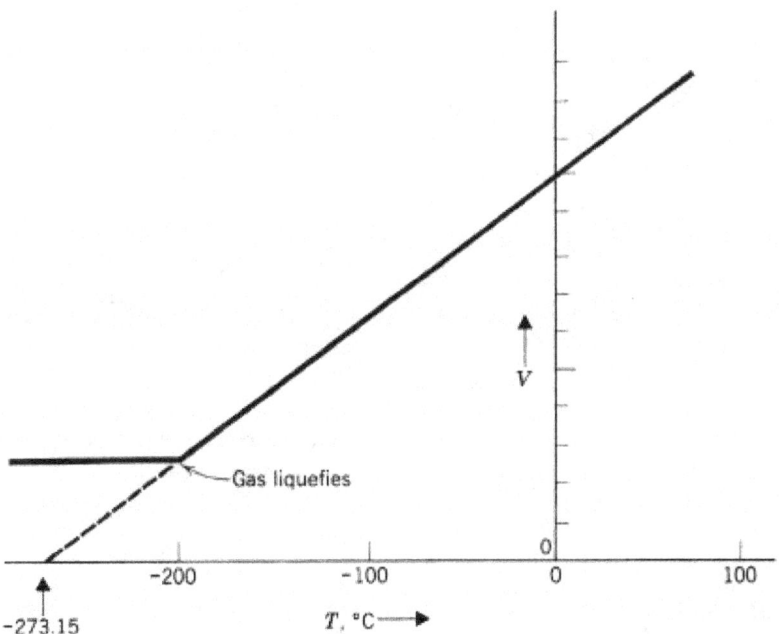

**FIGURE 12-9** A graph showing the volume of a gas versus the temperature in degrees Celsius. As the gas cools, the volume decreases. When the gas becomes sufficiently cold, it liquefies; the volume remains fairly constant upon further cooling. The dashed line shows what would happen if the gas didn't liquefy but kept decreasing in volume. Its volume would become zero at −273.15 °C.

## 198  GASES AND THE IDEAL GAS LAW

before they reach -273.15 ºC, volumes never do become zero. But -273.15 ºC is the lowest temperature possible, absolute zero. Some sources say that at absolute zero, all motion stops. This is not true because of the uncertainty principle and other laws of quantum mechanics (see Chapter 13). A detailed explanation is beyond the scope of this book, but there really is some motion (https://bit.ly/3AiFvXu).

## 12-5
### THE LOWEST POSSIBLE TEMPERATURE IS ZERO KELVINS

The lowest possible temperature is so important that it is the basis of a temperature scale. This is the absolute temperature scale and the unit is **kelvins** (abbreviated K). It is named after **Lord Kelvin (William Thomson**, 1824–1907), the British scientist who first suggested that –273.15 ºC is the lowest temperature that can exist. Some authors and organizations use the singular unit "kelvin." Both "kelvins" and "kelvin" seem to be correct and are widely used. NIST recommends "kelvins" and that will be used in this book as it follows the same grammar as "degrees." An exception is if you are writing, say, "one kelvin is a very low temperature." The plural would sound funny. *In a course, use the form your instructor uses.*

**NOTE:** We do not say "degrees kelvin"—just "kelvins." The word kelvin starts with a lower-case letter (unless it is used as a proper noun, as in the "Kelvin scale" or at the beginning of a sentence), but the abbreviation is a capital K. This is the convention scientists use for units named after people. Remember our use of pascal and Pa. (The exceptions are degrees Celsius and degrees Fahrenheit.)

The new temperature scale starts at –273.15 ºC. This temperature is called zero kelvins or absolute zero and is written as 0 K. There is much evidence that absolute zero cannot be attained. The closest that scientists (at NIST in 2017) have gotten to absolute zero for a solid object, a very small and very thin aluminum disc, is 360 µK or 0.000360 K. See https://bit.ly/3h1GlxC. The coldest that scientists have gotten with a gas is 38 pK or 0.000000000038 K (at the University of Bremen in 2021). See https://bit.ly/3BCU9Z1. The gas they used was not the kind of gas discussed in this chapter. It has *very* different properties and remains a gas at this low temperature, contrary to what we said on pp 197-198 above.

Since the Kelvin scale starts at –273 ºC (let's round it off a bit), we can write 0 K = –273 ºC. Adding 273 to each side of this equation, we have 273 K = 0 ºC. Rearranging, we get

$$0 \text{ ºC} = 273 \text{ K.}$$

Although kelvins are the same size as Celsius degrees, the Kelvin scale starts 273 degrees below the Celsius scale. Thus

$$\text{ºC} + 273 = \text{K}$$

**EXAMPLE 6**   Convert 37 ºC to kelvins.
   *Solution:* Substitute in ºC + 273 = K to give
   37 ºC + 273 = 310 K.  ∎

**EXAMPLE 7**   Convert 200 K to degrees Celsius.
   *Solution:* Since ºC + 273 = K, we can write ºC = K – 273.
   Therefore, ºC = 200 K – 273 = –73 ºC   ∎

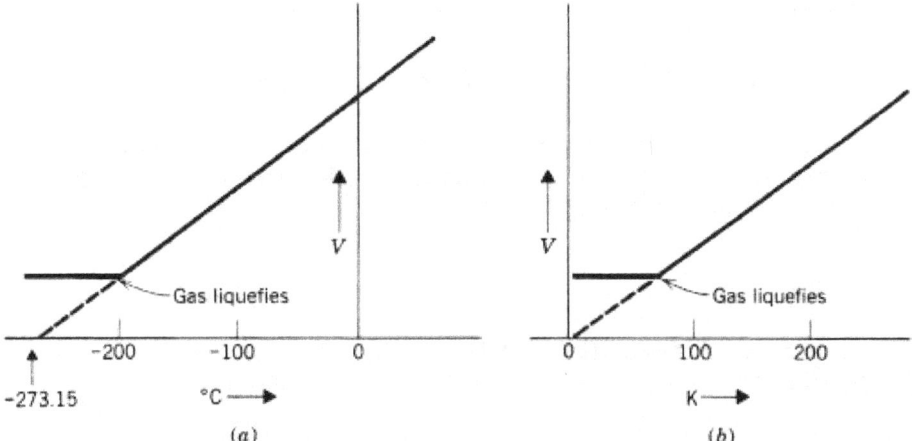

**FIGURE 12-10** (a) A redrawing of Figure 12.9. (b) The vertical axis (the volume or V axis) has been moved over to −273.15 °C. The temperature scale is now in kelvins (K), since the vertical axis is now at zero kelvins.

If we redraw Figure 12-9 and move the vertical axis (the volume axis) left to −273.15 °C, we see that the volume of a gas is directly proportional to the absolute temperature ($T$) of the gas. This means that as $T$ increases, $V$ increases, and when $T = 0$, $V = 0$ (see Figure 12-10). This is why degrees Celsius is not proportional to $V$—when degrees Celsius equals zero, $V$ does not equal zero.

## 12-6
### A NOTE ABOUT PROPORTIONALITY AND PROPORTIONALITY CONSTANTS

If you go to the store to buy 10 apples, the price you pay depends on the price per apple. Thus we have

¢ you pay for 10 apples = (¢ per apple)(10 apples)

To be a bit more mathematical, we can write (the "per" in "¢ per apple" means divided by)

$$¢ = \left(\frac{¢}{\text{apple}}\right)(\text{number of apples})$$

Suppose each apple costs 25¢. Then for the price of 10 apples we have

$$¢ = \left(\frac{25¢}{\text{apple}}\right)(10 \text{ apples}) = 250¢ = \$2.50$$

# GASES AND THE IDEAL GAS LAW

The cents you pay is **directly proportional** to the number of apples you buy. The actual price paid for your apples is determined by the price per apple—in this case 25¢. The term 25¢/apple is called the **constant of proportionality**. It converts the unit apples to the unit cents; it also sets the scale of the conversion with the number 25.

## 12-7
### THE VOLUME OF A GAS IS DIRECTLY PROPORTIONAL TO THE TEMPERATURE IN KELVINS

Just as cents is directly proportional to apples, volume ($V$) is directly proportional to absolute temperature ($T$) in kelvin (K). We can say that (the word "directly" is usually left out for simplicity)

$$V \text{ is proportional to } T$$

The symbol for "is proportional to" is "$\propto$." Thus

$$V \propto T$$

We will not discuss the actual value of the proportionality constant relating $V$ and $T$. You will see why later.

**EXAMPLE 8** If $x$ is proportional to $y$, and $k$ is the proportionality constant, write an equation relating $x$ and $y$.

*Solution:* $x = ky$. ∎

## 12-8
### THE PRESSURE OF A GAS IS INVERSELY PROPORTIONAL TO THE VOLUME OF THE GAS

Let's look at our piston–cylinder arrangement again (Figure 12-11). This time we will hold the temperature constant. The greater the pressure ($P$) on the piston, the more the piston sinks, compressing the gas in the cylinder. The volume is **inversely proportional** to the pressure. This means that as the pressure gets larger, the volume gets smaller. We can write the relationship mathematically as

$$P \propto 1/V$$

Let's see how this works. If $V$ goes from 2 to 4, $P$ goes from $\frac{1}{2}$ to $\frac{1}{4}$. We see that when $V$ doubles, $P$ becomes one-half of its original value.

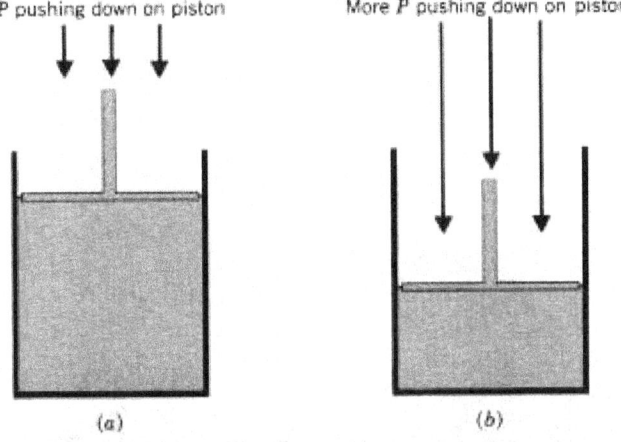

**FIGURE 12-11** The piston–cylinder arrangement at constant temperature. As the pressure increases from part a to part b, the piston sinks and the volume of the gas in the cylinder decreases.

When you breathe, you illustrate the pressure–volume relationship of a gas. Try taking a breath, but at the same time block your throat by putting your tongue toward the back of your mouth so no air can rush into your lungs. What have you done? You've lowered your diaphragm (the muscle at the bottom of your chest cavity) and expanded your rib cage, thus increasing the volume of your chest cavity. You have also decreased the air pressure inside your chest. Now, if you relax your throat, air rushes into your lungs. The higher air pressure outside your chest forces air into the lower-pressure region inside your chest. When you exhale, you do the opposite, decreasing chest volume and forcing air out of your lungs.

**EXAMPLE 9** If $x$ is inversely proportional to $y$, and $c$ is the proportionality constant, what is the equation relating $x$ and $y$?

*Solution:* $x = c(1/y) = c/y$. ∎

## 12-9
## THE PRESSURE OF A GAS IS DIRECTLY PROPORTIONAL TO THE TEMPERATURE IN KELVINS

Instead of a piston–cylinder arrangement, we will now use a closed box as shown in Figure 12-12. In this system, the volume is constant because the walls of the box are rigid and cannot move. As the temperature of the gas

**A SCUBA Diver:** The regulator at the top right reduces the high pressure in the air tanks on the diver's back so it equals the water pressure, allowing the diver to breath easily and not cause damage to the diver's lungs. **SCUBA** stands for **S**elf-**C**ontained **U**nderwater **B**reathing **A**pparatus. It was invented by French engineer **Émile Gagnan** (1900-1979) and French naval officer **Jacques-Yves Cousteau** (1910-1997) in 1943 and commercialized under the name **Aqua Lung**. Cousteau later become a famous explorer, conservationist, filmmaker, scientist, photographer, author and researcher who studied the sea and all forms of life in water.

*Photo is in the Public Domain.*

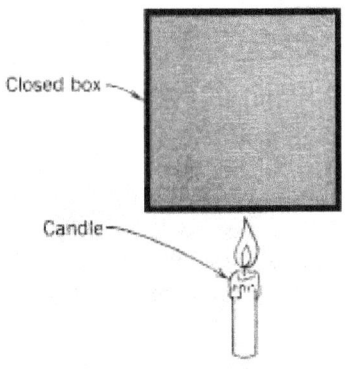

**FIGURE 12-12** A closed box being heated by a candle. The walls of the box are very strong, and no gas can pass through them. As the flame heats the gas in the box, the pressure inside increases. The volume cannot change because of the strong walls—the volume is constant.

increases, its pressure increases. Experiments show that the pressure ($P$) is proportional to the absolute temperature ($T$):

$$P \propto T$$

## 12-10
### THE PRESSURE OF A GAS IS DIRECTLY PROPORTIONAL TO THE AMOUNT OF GAS

Again we will need a closed box that is held at constant temperature. In addition, we need a gas syringe that we will use to inject gas into the box (see Figure 12-13). As gas is injected, the pressure increases. The pressure is found to be directly proportional to the amount of gas,

$$P \propto n$$

where $n$ is the number of moles of gas in the box.

**EXAMPLE 10**  If the pressure is 2 atm when 3 mol of gas are in a box, what is the pressure when 6 mol of gas are in the box?

**Solution:** Since $P \propto n$, and $n$ has doubled, $P$ must double. Thus, when there are 6 mol of gas in the box, the pressure is 4 atm.  ∎

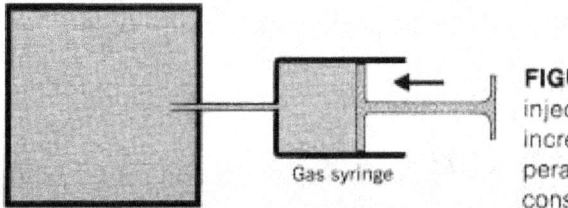

**FIGURE 12-13** As the gas is injected into the box, the pressure increases. The volume and temperature of the gas in the box are constant.

## 12-11
### THE VOLUME OF A GAS IS DIRECTLY PROPORTIONAL TO THE AMOUNT OF GAS

Here we will use our piston–cylinder arrangement and a gas syringe. Both the temperature and the pressure are held constant (see Figure 12-14). As gas is injected into the cylinder, the piston rises and the volume of the gas in the cylinder increases. The volume is found to be directly proportional to the amount of gas. We can write

$$V \propto n$$

**EXAMPLE 11** If the volume is 20 L when 1 mol of gas is in the piston–cylinder arrangement, what is the volume when 2 mol of gas are in the piston–cylinder arrangement? Assume that both the temperature and pressure are held constant.

*Solution:* Since $V \propto n$, when the number of moles doubles (from 1 to 2), the volume doubles. The final volume is 40 L. ∎

## 12-12
### THE IDEAL GAS LAW RELATES PRESSURE, VOLUME, MOLES, AND TEMPERATURE

In previous sections, we derived five proportionalities that describe the behavior of gases. We list them together here.

$V \propto T$  (at constant $P$ and $n$)

$P \propto 1/V$  (at constant $n$ and $T$)

$P \propto T$  (at constant $n$ and $V$)

$P \propto n$  (at constant $T$ and $V$)

$V \propto n$  (at constant $P$ and $T$)

Let's rewrite the second proportionality so that $V$ is on the left side of the proportionality sign. We get $V \propto 1/P$. The proportionalities involving $V$ are

$V \propto T$

$V \propto 1/P$

$V \propto n$

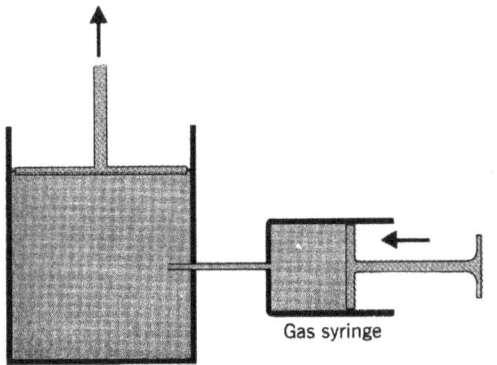

**FIGURE 12-14** As gas is injected into the cylinder, the volume increases. The pressure of the gas in the cylinder is constant because the piston can move until both the inside and outside pressures are equal.

Since $V$ is proportional to $T$, $1/P$, and $n$ individually, it must be proportional to all three at the same time. So we can write

$$V \propto nT/P$$

Multiplying both sides of the expression by $P$, we get

$$PV \propto nT$$

Now let's look at the proportionalities involving $P$. These are

$$P \propto 1/V$$
$$P \propto T$$
$$P \propto n$$

Since $P$ is proportional to $1/V$, $T$, and $n$ individually, it must be proportional to all three at the same time, and as we did before, we can write

$$P \propto nT/V$$

Multiplying both sides by $V$, we get

$$PV \propto nT$$

This is the same expression that we got above. Let's make this proportionality an equality by putting in a constant of proportionality. It is customary to use the symbol $R$ for the constant of proportionality and to insert it between the $n$ and the $T$. Doing this we get

$$\boxed{PV = nRT}$$

This equation is called the **ideal gas law**, and $R$ is called the **ideal gas law constant** or the **universal gas constant**. As you study more chemistry, you will learn more about ideal gases and why real gases, such as air, do not exactly obey the ideal gas law. The ideal gas law is called *ideal* because it is a *model* of the way gases behave. It is pretty good but not perfect. Under normal conditions, air "obeys" the ideal gas law with an error of less than 1%.

To use the ideal gas law equation, we must decide on the units we will use. The following units are very common.

Pressure ($P$) is measured in atmospheres (atm).

Volume ($V$) is measured in liters (L).

Amount of gas ($n$) is measured in moles (mol).

Temperature ($T$) is measured in kelvins (K).

To determine the units of $R$, we will substitute these units into the ideal gas law:

$$PV = nRT$$

$$\text{atm} \times \text{L} = \text{mol} \times R \times \text{K}$$

Solving for $R$ we get

$$R = \frac{\text{atm} \times \text{L}}{\text{mol} \times \text{K}}$$

Now we must determine the numerical value of $R$. To do this it is necessary to perform an experiment with a gas. In Figure 12-15, which shows our piston–cylinder arrangement, we take 1.00 mol of a gas at 1.00 atm pressure and 273 K. Putting the gas into the piston–cylinder apparatus,

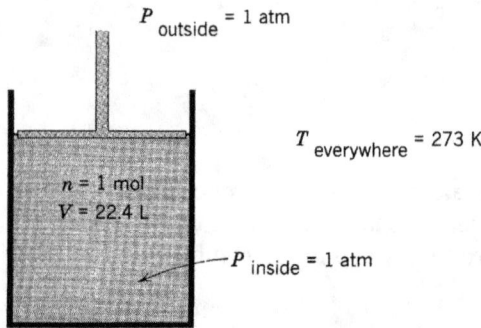

**FIGURE 12-15** An experiment to determine the numerical value of $R$. Any values of $n$, $T$, and $P$ could have been used; the ones in the diagram are convenient.

## 12-12 The Ideal Gas Law Relates P, V, n, and T

we allow the movable piston to come to its natural resting place and then we measure the volume. It turns out to be 22.4 L. Now we substitute the numerical values of $P$, $V$, $n$, and $T$ into the ideal gas law:

$$PV = nRT$$

$$(1.00 \text{ atm})(22.4 \text{ L}) = (1.00 \text{ mol})(R)(273 \text{ K})$$

Solving for $R$, we get

$$R = \frac{(1.00 \text{ atm})(22.4 \text{ L})}{(1.00 \text{ mol})(273 \text{ K})} = 0.0821 \frac{\text{atm} \cdot \text{L}}{\text{mol} \cdot \text{K}}$$

Any combination of $P$, $V$, $n$, and $T$ values will give the same numerical value for $R$ if these units are used. Of course, the experiment with the movable piston described above must be redone if the values of $P$, $V$, $n$, and $T$ are changed. Also, using a different set of units for $P$, $V$, $n$, and $T$ will give different numerical value for $R$.

**NOTE:** The calculator steps for the above calculation for "$R$" are **1, x, 22.4, ÷, 1, ÷, 273, =.** See Example 17 on page 81 for another example of this type of calculation and why you must divide twice.

Many times the units atmospheres, liters, moles, and kelvins will be used in solving problems with $PV = nRT$. Be sure to convert all units to atmospheres, liters, moles, and kelvins and use $0.0821 \frac{\text{atm} \cdot \text{L}}{\text{mol} \cdot \text{K}}$ as the value for $R$. For example, pressure might be reported in torr or kilopascals. Remember that

**1 atm = 760 Torr = 101 kPa.**

Temperature is commonly measured in degrees Celsius. Remember that °C + 273 = K. Volume is sometimes measured in milliliters (mL). Remember that **1000 mL = 1 L**.

**EXAMPLE 12** What pressure (in atmospheres) would be exerted by 3.00 mol of oxygen gas in a 5.00-L container at 300 K?

**Solution:** The given values are $V = 5.00$ L, $n = 3.00$ mol, and $T = 300$ K. Substituting these values into $PV = nRT$, we get

$$(P)(5.00 \text{ L}) = (3.00 \text{ mol})(0.0821 \frac{\text{atm} \cdot \text{L}}{\text{mol} \cdot \text{K}})(300 \text{ K})$$

$$P = \frac{(3.00 \text{ mol})(0.0821 \frac{\text{atm} \cdot \cancel{\text{L}}}{\cancel{\text{mol}} \cdot \cancel{\text{K}}})(300 \cancel{\text{K}})}{5.00 \cancel{\text{L}}}$$

$$P = \frac{(3.00)(0.0821)(300)}{5.00} \text{ atm} = 14.8 \text{ atm}$$

**NOTE:** $0.0821 \frac{\text{atm} \cdot \text{L}}{\text{mol} \cdot \text{K}}$ can also be written as $0.0821 \text{ (atm} \cdot \text{L)}/(\text{mol} \cdot \text{K})$ or $0.0821 \text{ (atm} \cdot \text{L)}(\text{mol} \cdot \text{K})^{-1}$ where $(\text{mol} \cdot \text{K})^{-1}$ is the same as $\frac{1}{\text{mol} \cdot \text{K}}$.

Notice that $(\text{mol} \cdot \text{K})^{-1} = \text{mol}^{-1} \cdot \text{K}^{-1}$. See page 83 for the rule allowing this.

How do the units in Example 12 divide out? Let's write the expression for P using only units and then put slash marks on "mol" and "K":

$$\frac{(\text{mol})(\frac{\text{atm} \cdot \text{L}}{\text{mol} \cdot \text{K}})(\text{K})}{\text{L}} \rightarrow \frac{(\cancel{\text{mol}})(\frac{\text{atm} \cdot \text{L}}{\cancel{\text{mol}} \cdot \cancel{\text{K}}})(\cancel{\text{K}})}{\text{L}}$$

Mol divides mol and K divides K, giving

$$\frac{\text{atm} \cdot \cancel{\text{L}}}{\cancel{\text{L}}} \rightarrow \text{atm}$$

L divides L, leaving only atm.

**EXAMPLE 13** What volume (in liters) will 0.500 mol of $N_2$ occupy at a pressure of 300 torr and a temperature of 37.0 °C?

*Solution:* First convert torr to atmospheres and degrees Celsius to kelvins:

$$300 \text{ Torr} \times \frac{1 \text{ atm}}{760 \text{ Torr}} = 0.395 \text{ atm}$$

$$37.0 \text{ °C} + 273 = 310 \text{ K}$$

Then substitute into $PV = nRT$. Since we want volume, let's solve this equation for $V$ first and then substitute numerical values:

$$V = \frac{nRT}{P} = \frac{(0.500 \text{ mol})(0.0821 \frac{\text{atm} \cdot \text{L}}{\text{mol} \cdot \text{K}})(310 \text{ K})}{0.395 \text{ atm}} = 32.2 \text{ L} \quad \blacksquare$$

Be sure that you understand how the units divide out.

**EXAMPLE 14** How many moles of methane gas are in a container that has the following conditions: $P = 75.0$ kPa, $V = 600$ mL, $T = 25.0$ °C?

*Solution:* First convert kilopascals to atmospheres, milliliters to liters, and degrees Celsius to kelvins:

$$75.0 \text{ kPa} \times \frac{1 \text{ atm}}{101 \text{ kPa}} = 0.743 \text{ atm}$$

$$600 \text{ mL} \times \frac{1 \text{ L}}{1000 \text{ mL}} = 0.600 \text{ L}$$

$$25.0 \text{ °C} + 273 = 298 \text{ K}$$

Solving $PV = nRT$ for $n$, as well as substituting numerical values, we get

$$n = \frac{PV}{RT} = \frac{(0.743 \text{ atm})(0.600 \text{ L})}{0.0821 \text{ (atm} \cdot \text{L)(mol} \cdot \text{K})^{-1}(298 \text{ K})} = 0.0182 \text{ mol methane} \quad \blacksquare$$

The above equation and a similar one in Example 15 below has been written using one of the notations described in the NOTE at the bottom of page 208. The different notations are common and it would be ideal if you were familiar with all of them. Again, be sure you understand how the units divide out.

A useful application of the ideal gas law is the determination of the molecular weight of an unknown gas.

**EXAMPLE 15** A sample of gas at 27.1 °C is contained in a 1.00-L glass bulb. The pressure of the gas is 1.00 atm, and the gas weighs 0.650 g. What is the molecular weight of the gas? (Be sure to convert 27.1 °C to kelvins.)

*Solution:* First calculate the number of moles of gas in the sample. To do this solve $PV = nRT$ for $n$ and substitute numerical values from the problem:

$$n = \frac{PV}{RT} = \frac{(1.00 \text{ atm})(1.00 \text{ L})}{0.0821 \text{ (atm} \cdot \text{L)(mol} \cdot \text{K})^{-1}(300 \text{ K})} = 0.0406 \text{ mol}$$

Since the statement of the problem gave us the mass of the gas, we have

given mass of gas → 0.650 g ↔ 0.0406 mol ← calculated moles of gas

The molecular weight is the number of grams per mole of a substance. Writing this statement as an equation we have:

$$\text{molecular weight} = \frac{\text{number of grams of gas}}{\text{number of moles of gas}}$$

Substituting 0.650 g of gas and 0.0406 mol of gas, we have

$$\text{molecular weight} = \frac{0.650 \text{ g}}{0.0406 \text{ mol}} = 16.0 \, \frac{\text{g}}{\text{mol}} \quad \blacksquare$$

## 12-13
### CHANGING P, V, AND T WHILE KEEPING n CONSTANT: THE COMBINED GAS LAW

Sometimes the pressure, temperature, and volume of a certain number of moles of gas are given. (Moles of gas are not changed in this discussion.) Then the values of one or two of these quantities are changed and we are asked to find the value of the third quantity. This is easily done if the ideal gas law equation is used. Let the initial values of $P$, $V$, and $T$ be denoted by the subscript "1". These values are written as $P_1$, $V_1$, and $T_1$. The ideal gas law for these initial values is $P_1V_1 = nRT_1$. Notice that $n$ and $R$ do not have subscripts. $R$ is always constant; $n$ is also constant for problems of this kind, since the number of moles of gas doesn't change.

We will denote the final condition of the gas with a subscript "2". The ideal gas law for these values is $P_2V_2 = nRT_2$.

We can solve each of these two equations for $nR$ by dividing the first equation by $T_1$ and the second one by $T_2$.

$$nR = \frac{P_1V_1}{T_1} \quad \text{and} \quad nR = \frac{P_2V_2}{T_2}$$

Since $nR$ is equal to $nR$, we have

$$\boxed{\frac{P_1V_1}{T_1} = \frac{P_2V_2}{T_2}}$$

This equation is called the **combined gas law.**

**EXAMPLE 16** You are given the following initial conditions: $P_1 = 3.00$ atm, $V_1 = 10.0$ L, and $T_1 = 298$ K. If the final conditions are $P_2 = 5.00$ atm and $T_2 = 100$ K, what is $V_2$ (in liters)?

**Solution:** Write $P_1V_1/T_1 = P_2V_2/T_2$ and solve for $V_2$:

$$V_2 = \frac{P_1V_1T_2}{P_2T_1}$$

It is convenient to rearrange this equation so that all the $P$ and $T$ terms are together:

$$V_2 = \frac{P_1}{P_2} \times \frac{T_2}{T_1} \times V_1$$

Now substitute the given numerical values:

$$V_2 = \frac{3.00 \text{ atm}}{5.00 \text{ atm}} \times \frac{100 \text{ K}}{298 \text{ K}} \times 10.0 \text{ L} = 2.01 \text{ L}$$

To check that the answer is reasonable (Should the volume really decrease?), look at the terms $P_1/P_2$ and $T_2/T_1$. As you see, the pressure increases from 3.00 to 5.00 atm. We know that an increase in pressure will decrease the volume. The temperature decreases from 298 to 100 K. This cooling will also decrease the volume. Since both the pressure and the temperature changes decrease the volume, a volume change from 10.0 to 2.01 L is certainly in the right direction. (Sometimes both changes are not in the same direction. Then you cannot tell what happens without doing the calculation.) ∎

**EXAMPLE 17**  You are given the following initial conditions: $P_1 = 0.500$ atm, $V_1 = 5.00$ L, and $T_1 = 300$ K. If the final conditions are $V_2 = 2.00$ L and $T_2 = 400$ K, what is $P_2$ (in atmospheres)?

***Solution:*** Write $P_1V_1/T_1 = P_2V_2/T_2$ and solve for $P_2$:

$$P_2 = \frac{P_1V_1T_2}{T_1V_2}$$

Rearrange to get

$$P_2 = P_1 \times \frac{V_1}{V_2} \times \frac{T_2}{T_1}$$

Substituting numerical values gives

$$P_2 = 0.500 \text{ atm} \times \frac{5.00 \text{ L}}{2.00 \text{ L}} \times \frac{400 \text{ K}}{300 \text{ K}} = 1.67 \text{ atm}$$

Is it reasonable that the pressure increases? ∎

## 12-14
### DALTON'S LAW OF PARTIAL PRESSURES SAYS THAT THE SUM OF THE PRESSURES OF THE INDIVIDUAL GASES EQUALS THE TOTAL GAS PRESSURE

If we have a mixture of three gases in a container, the total number of gas molecules in the container is equal to the sum of the number of gas molecules of each of the three gases. It follows also that the total number of moles is equal to the sum of the number of moles of each gas. Thus

$$n_T = n_1 + n_2 + n_3$$

where $n_T$ is the total number of moles and $n_1$, $n_2$, and $n_3$ are the number of moles of each component gas.

In a given container at a certain temperature, different kinds of gas all have the same volume and temperature. Thus, using the ideal gas law, $PV = nRT$, we can write (where $n_T$ is the total number of moles in the gas mixture and $P_T$ is the total pressure of the gas mixture):

$n_T = P_T V/RT$     for the entire gas mixture

$n_1 = P_1 V/RT$     for the first component in the gas mixture

$n_2 = P_2 V/RT$     for the second component in the gas mixture

$n_3 = P_3 V/RT$     for the third component in the gas mixture

Since $V$ and $T$ are constant, only $P$ can change when n changes. $R$ is always a constant.

Adding all the moles together, we get

$$n_T = n_1 + n_2 + n_3$$

Substituting the appropriate $PV/RT$ expression for $n_T$, $n_1$, $n_2$ and $n_3$, we get

$$\frac{P_T V}{RT} = \frac{P_1 V}{RT} + \frac{P_2 V}{RT} + \frac{P_3 V}{RT}$$

Now we divide both sides of the equation by $\frac{V}{RT}$. Since $\frac{\frac{V}{RT}}{\frac{V}{RT}} = 1$

$$\frac{P_T \cancel{V}}{\cancel{RT}} = \frac{P_1 \cancel{V}}{\cancel{RT}} + \frac{P_2 \cancel{V}}{\cancel{RT}} + \frac{P_3 \cancel{V}}{\cancel{RT}}$$

After dividing out all the $\frac{V}{RT}$ terms, we get    $\boxed{P_T = P_1 + P_2 + P_3}$

This is called **Dalton's law of partial pressures** (named after the British scientist **John Dalton**, 1766-1844), which says that **the total pressure of a mixture of gases is equal to the sum of the partial pressures of all the individual kinds of gas in the mixture.** The **partial pressure** of a gas is simply the pressure exerted by the gas when it is mixed with other gases. This is the same pressure the gas would have if it were in the container by itself. **NOTE:** Since Dalton's Law was derived from the ideal gas law, it is only true for ideal gases. As mentioned on page 206, real gases don't obey the ideal gas law. Thus, Dalton's law is only approximately true for real gases, but the deviation is less than 1% at room temperature and pressure, depending on the gas. Air has a deviation of less than 0.1% at room temperature and pressure. See https://bit.ly/3NPP83u for details.

**EXAMPLE 18** A gas mixture consists of the following partial pressures: 20 Torr $O_2$, 50 Torr $N_2$, and 10 Torr Ar. What is the total pressure (in torr) of the mixture of gases?

## 12-14 Dalton's Law of Partial Pressures

*Solution:* Let's use the following notation:

$$P_{O_2} = \text{partial pressure of } O_2 = 20 \text{ Torr}$$

$$P_{N_2} = \text{partial pressure of } N_2 = 50 \text{ Torr}$$

$$P_{Ar} = \text{partial pressure of Ar} = 10 \text{ Torr}$$

Then using Dalton's Law of Partial Pressures, we can write

$$P_T = P_1 + P_2 + P_3 = P_{O_2} + P_{N_2} + P_{Ar}$$

$$P_T = 20 \text{ Torr} + 50 \text{ Torr} + 10 \text{ Torr} = 80 \text{ Torr} \quad \blacksquare$$

**EXAMPLE 19** For this example, assume that a commercial deep-sea diver is working at 192 ft below sea level and is breathing a gas mixture of 3.0% $O_2$ and 97% He. (This mixture is called **Heliox**.) If the total gas pressure she breathes is 7.0 atm, what are the partial pressures of $O_2$ and He in torr? (For every 32 ft below sea level, the pressure increases by 1 atm.[2] So the total pressure at 32 ft below sea level is 2 atm. That's because there is 1 atm at the surface that comes from atmospheric pressure, and an additional 1 atm at 32 ft deep that comes from water pressure.)

---

**Sealab II and Gas Mixtures for Deep Diving**

The Sealab II project, started in 1965, successfully demonstrated that underwater workers (called **aquanauts**) could safely live and work at a 210 ft depth on the ocean floor for weeks, if they could be provided with a pressurized, livable shelter near their work. (https://bit.ly/2S8uBfP)

Inside this shelter a gas mixture of 4% oxygen, 16% nitrogen, and 80% helium, maintained at a pressure equal to the water pressure at a 210 ft depth, was used. This gas mixture is called **Trimix**. When the aquanauts came to the surface, they had to decompress for 35 hr so that the high-pressure gas dissolved in their bodies would come out slowly. Otherwise, the gas would come out of solution quickly, forming gas bubbles that cause a painful and potentially deadly condition called "the bends." (Think of what happens when you shake a warm can of soda and then open it.) For more details see the PBS show about the extensive Sealab program sponsored by the US Navy at https://to.pbs.org/2EGuzGz. In the show, there is short sequence about **Jacques-Yves Cousteau** (mentioned on page 202) and his research vessel, the Calypso.

Trimix and Heliox are the most common gas mixtures used for very deep diving. Whichever gas mixture a diver uses, it will be made specifically for the pressure the diver will be at when under the water. There are advantages and disadvantages to each mixture. See https://bit.ly/2vCOp3j for details.

---

[2]The reason that the water pressure increases by one atmosphere for every 32 ft down a diver goes is as follows. A column of water that is 1 in.² and 32 ft long weighs 14.7 lb. At the bottom of the column, the pressure is 14.7 lb/in.², which is 1 atm. Remember that in Figure 12-3, a column of air that is 1 in.² goes from ground level to outer space for the air in the column to weigh 14.7 lb.

## 214   GASES AND THE IDEAL GAS LAW

*Solution:* The partial pressure of $O_2$ is 3.0% × 7.0 atm = 0.030 × 7.0 atm = 0.21 atm or 160 Torr. The partial pressure of He is 97% × 7.0 atm = 0.97 × 7.0 atm = 6.8 atm or 5200 Torr. It is interesting that the partial pressure of $O_2$ in the air we breathe is 20.9% × 760 Torr = 0.209 × 760 Torr = 159 Torr, about the same as the partial pressure of the oxygen the diver breathed. If she had breathed a 20.9% $O_2$–79.1% He mixture, she would have gotten 1112 Torr of $O_2$, which could have caused cell damage and/or cell death, especially to the central nervous system, lungs and eyes. If she had breathed air, the $N_2$ would become toxic at this high pressure (it becomes toxic below about 100 ft) causing nitrogen narcosis (a drunk-like behavior that Jacques Cousteau called "Rapture of the Deep"). Below a depth of about 500 ft, even helium is toxic, causing high-pressure nervous syndrome. Deep diving is quite dangerous, requiring much expertise. ■

## PROBLEMS
### KEYED PROBLEMS

1. How many atmospheres are there in 1500 Torr?

2. How many torr are there in 12.00 lb/in$^2$?

3. How many kilopascals are there in 3 atm?

4. How many degrees Fahrenheit are equal to 25 °C?

5. How many degrees Celsius are equal to 86 °F?

6. Convert 100 °C to kelvins.

7. Convert 77 K to degrees Celsius.

8. If $a$ is directly proportional to $b$, and $q$ is the proportionality constant, write an equation relating $a$ and $b$.

9. If $a$ is inversely proportional to $b$, and $t$ is the proportionality constant, what is the equation relating $a$ and $b$?

10. If the pressure is 5 atm when 10 mol of gas are in a box, what is the pressure (in atmospheres) when 20 mol of gas are in the box?

11. If the volume is 15 L when 3 mol of gas are in the cylinder, what is the volume (in liters) when 6 mol of gas are in the cylinder?

12. What pressure (in atmospheres) would 2.0 mol of nitrogen gas exert if it were in a 7.0-L container at 400 K?

13. What volume (in liters) will 0.25 mol $Cl_2$ occupy at a pressure of 500 Torr and a temperature of 80 °C?

14. How many moles of $CO_2$ are there in a sample of gas that is at the following conditions: $P = 80.0$ kPa, $V = 200$ mL, and $T = 30$ °C.

15. A sample of gas at 25 °C is contained in a 1.50-L glass bulb. The pressure of the gas is 1.30 atm, and the gas weighs 4.40 g. What is the molecular weight of the gas?

16. The initial conditions of a gas are $P_1 = 1.5$ atm, $V_1 = 5.0$ L, and $T_1 = 273$ K. If the final conditions are $P_2 = 4.0$ atm and $T_2 = 200$ K, what is $V_2$ (in liters)?

17. The initial conditions of a gas are $P_1 = 0.20$ atm, $V_1 = 5.0$ L, and $T_1 = 210$ K. If the final conditions are $V_2 = 0.70$ L and $T_2 = 310$ K, what is $P_2$ (in atmospheres)?

18. A gas mixture consists of 150 Torr $CH_4$, 150 Torr $O_2$, and 500 Torr Ar. What is the total pressure of the gas (in torr)?

19. Calculate the partial pressure (in torr) of the oxygen in a 20% $O_2$–80% He mixture at 7 atm.

## SUPPLEMENTAL PROBLEMS

20. How many atmospheres are there in 500 kPa?

21. Convert 0.50 atm to torr.

22. A high fever is 104 °F. What is this temperature in degrees Celsius?

23. A child has a fever of 41 °C. What is this temperature in degrees Fahrenheit?

24. Convert 68 °F to kelvins. (Hint: First convert °F to °C.)

25. The boiling point of sodium is 892 °C. What is this in kelvins?

26. The boiling point of nitrogen is –195.8 °C. What is this in kelvins?

27. The temperature of the glowing tungsten filament of an incandescent light bulb is about 2800 K. What is this in degrees Celsius? In degrees Fahrenheit?

28. At what temperature are degrees Fahrenheit and degrees Celsius equal? (Hint: In the equation °F = 1.8 °C + 32, let both °F and °C equal $T$, substitute $T$ into the equation, and solve for $T$.)

29. What volume (in liters) will 8.0 g $O_2$ occupy at a pressure of 400 Torr and a temperature of 100 °F? (Hint: Convert grams to moles, convert degrees Fahrenheit to kelvins, and then use $PV = nRT$.)

30. What pressure (in torr) would 15 g $F_2$ have if it were in a 2.0-L container at 25 °C?

31. How many grams of $N_2O$ are there in a sample of gas that is at the following conditions: $P = 200$ kPa, $V = 700$ mL, and $T = 40.0$ °C?

## 216 GASES AND THE IDEAL GAS LAW

32. What is the molecular weight of 3.00 g of gas that is in a 400-mL container at 300 Torr and 22 °C?

33. The total pressure of a gas mixture is 1 atm. The gas consists of three components: $CH_3NC$, $O_2$, and $H_2$. The partial pressure of the $CH_3NC$ is 50 Torr, and that of the $O_2$ is 150 Torr. What is the partial pressure of the $H_2$ (in torr)? (**HINT:** Convert atmospheres to torr and use Dalton's law of partial pressures.) **NOTE:** From the author's research with $CH_3NC$, called methyl isocyanide, (https://bit.ly/38YQAzT), he can tell you that it is a *really* foul-smelling molecule.

34. A sample of gas in a rigid closed 200-mL container has a $P$ of 150 kPa and a $T$ of 50.0 °C. The container is heated to 100 °C. What is the new pressure (in kilopascals)? Assume that the volume is constant.

35. A gas in a piston–cylinder arrangement has a $P$ of 800 Torr and a $T$ of 20.0 °C. The piston is allowed to expand from a volume of 2.00 L so that the gas pressure is 1.00 atm. What is the final volume (in liters)? (Assume that $T$ doesn't change.)

36. The given conditions of a gas are $P = 20.0$ kPa, $V = 800$ mL, and $T = 23.0$ °C. Conditions are changed so that $V = 400$ mL and $T = 30.0$ °C. What is the new pressure (in kilopascals)?

37. What pressure (in atmospheres) will be exerted by $3.0 \times 10^{10}$ molecules of a gas that is in a 3.0-mL container at 15 °C? (Hint: Find the number of moles of gas by using Avogadro's constant.)

38. What is the molecular weight of a gas from the following data: $m$(mass) = 7.2 g, $P = 95$ kPa, $V = 2.5$ L, and $T = 100$ °C?

39. What pressure (in atmospheres) will be exerted by 50.0 g $C_2H_4$ in a 100-mL container at 35.0 °C?

40. How many grams of benzene, $C_6H_6$, are in a sample of gas that is in a 5.0-L container at 50 Torr and 75 °C?

41. Substitute $n = m/MW$ into $PV = nRT$ and solve for $MW$.
    **NOTE:** $MW$ is the abbreviation for molecular weight and $m$ is a symbol for mass in grams.

42. Starting with the combined gas law, $P_1V_1/T_1 = P_2V_2/T_2$, derive Boyle's law, $P_1V_1 = P_2V_2$ (named after the British chemist and physicist **Robert Boyle**, 1627—1691). This is done by holding the temperature constant (in other words, $T_1 = T_2$) and then multiplying both sides of the equation by the temperature.

43. Starting with the combined gas law, $P_1V_1/T_1 = P_2V_2/T_2$, derive Charles's law, $V_1/T_1 = V_2/T_2$ (named after the French physicist **Jacques Charles**, 1746–1823). This is done by holding the pressure constant (in other words, $P_1 = P_2$) and then dividing both sides of the equation by the pressure.

44. Because of the combined effects of nitrogen narcosis and oxygen toxicity, the maximum depth that highly trained divers breathing compressed air can go to is about 192 ft, but only for a few minutes. For advanced recreational divers, 130 ft is the limit. What is the partial pressure (in torr) of $N_2$ in the air a diver would breathe at 192 ft? (See Example 19 for data. Air is 78.1% $N_2$.)

---

## AVOGADRO'S LAW AND THE IDEAL GAS LAW

In 1811, the Italian physicist **Amedeo Avogadro** (1776–1856) realized something that was so far ahead of its time that it was not until about 50 yr later (see below) that other scientists agreed that he was right. **Avogadro's constant** is named after him and is discussed in Chapter 8.

Avogadro's idea was that, if the pressure and temperature are the same for two gases, equal volumes of each gas contain equal numbers of molecules. This statement can be easily derived from the ideal gas law, $PV = nRT$, by holding $P$ and $T$ constant.[3] For 2 gases, we get $V_1/V_2 = n_1/n_2$.

If $V_1 = V_2$, then $V_1/V_2 = 1$, and therefore $n_1/n_2 = 1$. This means that $n_1 = n_2$. Since the number of moles is proportional to the number of molecules (the proportionality constant is Avogadro's constant — see Section 8-5), then **molecules$_1$ = molecules$_2$**. For gases such as He and Ar where there are individual atoms in the gas, He atoms = Ar atoms, or in general, **atoms$_1$ = atoms$_2$**. This is also true if one gas has molecules and the other atoms, or if the gases are a mixture of both.

Avogadro's idea is now called **Avogadro's Law** (Section 12-11). It is *extremely* important because **if the volumes of 2 gases are equal at the same temperature and pressure, then, as shown in Section 8-1, the mass of each gas is proportional to the molecular weight (or atomic weight for atoms) of the molecules in the gas**.

In 1860, the Italian chemist **Stanislao Cannizzaro** (1826–1910) learned of Avogadro's idea and *for the first time constructed a logical table of atomic weights. This in turn made it possible to write correct chemical formulas for many compounds*. Upon hearing of Cannizzaro's work, the great German chemist **Julius Lothar von Meyer** (1830–1895) said "Doubt vanished and was replaced by a feeling of peaceful clarity." (https://bit.ly/2WvAmKK) **Some say that this was when modern chemistry began!**

Using **Boyle's Law** (Section 12-8 and Problem 42), **Charles's Law** (Section 12-11 and Problem 43) and **Gay-Lussac's Law**[4] (Section 12-9), *Avogadro's Law allowed the ideal gas law to be derived* (Section 12-12). Above, we did the reverse and derived Avogadro's Law *starting* with the ideal gas law.

---

[3] To do this write: $PV_1 = n_1RT$ and $PV_2 = n_2RT$. Then divide the first equation by the second equation and divide out common terms: $\cancel{P}V_1/\cancel{P}V_2 = n_1\cancel{RT}/n_2\cancel{RT}$.

[4] Named after the French chemist **Joseph Louis Gay-Lussac** (1778–1850).

**Lasers and Some Recent Nobel Prize Winners:** A laser (**l**ight **a**mplification by **s**timulated **e**mission of **r**adiation) beam can be focused into a very small spot, thus concentrating its energy. This makes lasers useful in many applications. Two are discussed below.

In the photo, a high-power laser beam is being used to weld thick sheet metal. The arrow points to where the beam hits the metal to weld pieces together.

In LASIK eye surgery, an excimer laser uses ultraviolet light to reshape the cornea. This laser doesn't heat the surrounding part of the cornea, so healing can be more rapid. See https://bit.ly/2WewI1U for more details.

The development of the field of lasers with high-intensity ultra-short pulses of laser light won the 2018 Nobel Prize in Physics for **Donna T. Strickland** (b. 1959), a physics professor at the University of Waterloo in Canada.

Two thousand eighteen, 2020 and 2022 were good years for women scientists. **Frances H. Arnold** (b. 1956), https://nyti.ms/2WtCssz, the Linus Pauling Professor of Chemical Engineering, Bioengineering, and Biochemistry at CalTech, won the 2018 Nobel Prize in Chemistry for research on directed evolution of enzymes. In 2020, **Andrea Ghez** (b. 1965), Professor of Physics and Astronomy at UCLA, shared a Nobel Prize in Physics for the discovery of black holes. **Jennifer Doudna** (b. 1964), Li Ka Shing Chancellor Chair Professor in the Department of Chemistry and the Department of Molecular and Cell Biology at UC Berkeley, and **Emmanuelle Charpentier** (b. 1968), founding director of the Max Planck Unit for the Science of Pathogens in Berlin, shared the Nobel Prize in Chemistry for their discovery of CRISPR-CAS9, a gene editing technique. See https://nyti.ms/3Iz2IjU for an article about the implications of this technique. **Carolyn Bertozzi** (b. 1966), Professor of Chemistry at Stanford University, shared the 2022 Nobel Prize in Chemistry for her discovery of a method of following chemical reactions in live animal cells without damaging the cells. See https://stanford.io/3VboXt7 for more information about Prof. Bertozzi.

Eight women have won the Nobel Prize in Chemistry. The other four are **Marie Curie** in 1911, her daughter **Irène Joliot-Curie** (1897-1956) in 1935, **Dorothy Crowfoot Hodgkin** (1910-1994) in 1964, and **Ada Yonath** (born 1939) in 2009. Four women have won a Nobel Prize in Physics. The other two are **Marie Curie** (1867-1934) in 1903 and **Maria Goeppert-Mayer** (1906-1972) in 1963.

*Photo courtesy of TRUMPF GmbH + Co. KG, CC BY-SA 3.0*
*https://en.wikipedia.org/wiki/Laser#/media/File:Lasertechnik06.jpg*

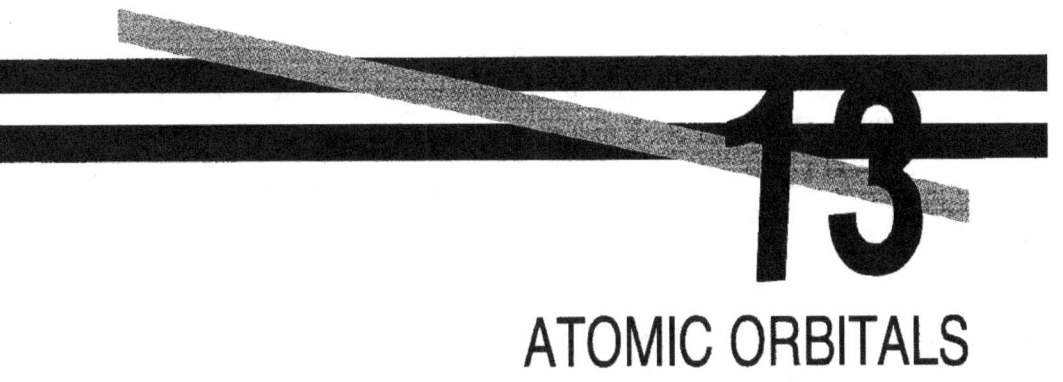

# ATOMIC ORBITALS

In Chapter 1 we promised to discuss the way electrons exist in the space surrounding the nucleus. In this chapter we shall try to give you an idea of how chemists "visualize" these electrons.

## 13-1
## THE PROBLEM OF SEEING THINGS IS RELATED TO THEIR SIZE AND HOW YOU LOOK AT THEM

Electrons have very little mass. Remember that they are almost 2000 times lighter than a proton. Because of the extremely small mass of the electron, it is impossible to clearly "see" where the electron is in the space surrounding the nucleus. It is also impossible to "see" what the electron looks like. This is because almost anything that collides with the electron, even light beams, causes it to be moved around.

It will help you to understand this moving around if you think of the following example. Suppose you wanted to know what a book looks like. The way that you see any object is to shine a light on it. The light is reflected by the object, in this case the book, and is focused into an image on the retina of your eye. The brain then interprets the image as that of a book. The light seems to have no effect on the book except to allow you to see it.

Now let's "look" at the book, not with light but with machine gun bullets. Assume that the book is made of stainless steel so that the bullets bounce off. The way that you can tell what the book looks like is to form, in some way, an image from the bullets that bounce off the book. But there is a problem—the book moves around from the force of the impact of the bullets. The image of the book that you see does not come from a stationary book — it comes from a book that is moving around. The image of the book that you will get is not clear — you will see a blurry image of a book. In addition to not being able to see the book clearly, you cannot even be sure where the book is in space. It keeps moving around from the impact of the bullets.

It is more difficult to understand why we cannot "see" an electron than to understand why we couldn't "see" the book that was hit with machine gun bullets. But the basic idea is the same. The next section will try to give you some insight into the problem of "seeing" electrons.

## 13-2
### THE UNCERTAINTY PRINCIPLE LIMITS OUR ABILITY TO "SEE" AN ELECTRON

The law of physics that explains why we cannot "see" an electron clearly is called the uncertainty principle. This principle, first stated by **Werner Heisenberg** (1901-1976) in 1927, is one of the most important principles in science. This law governs the behavior of electrons and other very small particles as far as our ability to "see" them is concerned. **The uncertainty principle can be stated in the following way: There is a limit to the accuracy of any measurement, and this limit becomes important when we try to measure (or "see") extremely small particles such as electrons.**[1]

To help you to understand why the uncertainty principle is necessary, we must first discuss a few properties of light.

## 13-3
### THE PROPERTIES OF LIGHT DEPEND ON ITS WAVELENGTH

Light can be considered as being both a particle and a wave. This "duality" of light is hard to understand, but you can get some feeling for it if you think in the following way. Light consists of particles called photons. Photons have energy but no mass, and they travel at the speed of light. There is also a wave associated with the photon. It is this wave that gives light many of its optical properties. For instance, light can be focused by the lenses of a camera, a telescope, or a microscope.

We can describe a light wave by its wavelength. The wavelength is the distance between two nearest peaks of a wave and has the symbol $\lambda$ (Greek lambda), as shown in Figure 13-1. The wavelength of visible light determines its color. Red light has a long wavelength and violet light has a short wavelength.

---

[1]It may seem strange to you that the uncertainty principle as described in this chapter depends on human observation. And that would be strange indeed! A law of nature should not depend on humans (or other intelligent life forms) interacting with nature for the law to work. There is a lot of evidence, all too esoteric to discuss here, that the uncertainty principle is inherent in the properties of all very small particles like electrons. Thus, *the uncertainty principle actually states a fundamental property of all very small particles like electrons and does not depend on the outcome of scientific experiments.* See pages 412-416 for more information. A more advanced discussion by physics Nobel Laureate **Steven Weinberg** (1933-2021) of the University of Texas/Austin, that originally appeared in the 1/19/2017 issue of *The New York Review of Books*, "The Trouble with Quantum Mechanics," is at https://bit.ly/3HLMJoq.

## 13.3 The Properties of Light Depend on Its Wavelength

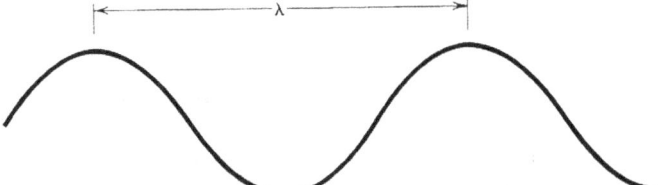

**FIGURE 13-1** The wavelength λ is the distance between two closest peaks of the wave.

The other colors have wavelengths shorter than that of red light but longer than that of violet light. The colors of the rainbow are

**red, orange, yellow, green, blue, indigo, violet**

decreasing wavelength →

In addition to visible light, there are other waves that are in many ways similar to light waves. These have wavelengths that are longer and shorter than visible light. The following is a list of these waves in order of decreasing wavelength:

**radio, radar, microwave, infrared, visible, ultraviolet, x-rays, gamma rays**

decreasing wavelength →

The photon associated with the light wave has more energy when the wavelength is short and less energy when the wavelength is long:

| **wavelength:** | **short** | **long** |
|---|---|---|
| **photon energy:** | **high** | **low** |

The radio-wave photon has very little energy. The gamma-ray photon has a great deal of energy. The photon energy of the other kinds of light are in between these two.

You become aware of the different photon energies of different kinds of light every time you get a suntan. Ultraviolet-light photons have enough energy to cause the chemical reaction that tans the skin. Visible light will not cause a tan, no matter how bright the light is. Each visible-light photon has insufficient energy, no matter how bright it is. The brightness is related to the number of photons, not the energy of the photons. However, even a few ultraviolet-light photons can cause tanning, because their energy is high enough.

Another property of light that has to do with our ability to see an object

222  ATOMIC ORBITALS

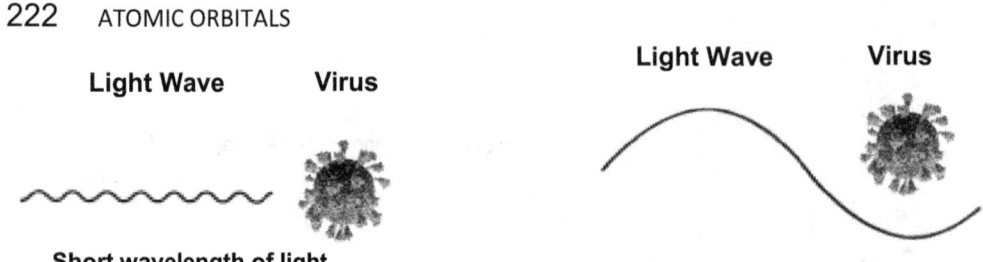

**Figure 13-2** In order to see an object clearly, the wavelength of the light must be much shorter than the object. Image of coronavirus is CC by-SA 4.0. Attribution is at https://bit.ly/3NdpLc4. Image has been converted to grey scale.

is related to the wavelength of the light. **It is an optical principle that to be able to see an object clearly, the object must be larger than the wavelength of the light used to look at it.** This is shown in Figure 13-2.

Objects that need to be magnified more than about 2000 times to be seen are smaller than the wavelength of visible light. That is why a visible-light microscope has a maximum magnification of about 2000 times.

## 13-4
### WE CANNOT SEE AN ELECTRON BECAUSE OF THE UNCERTAINLY PRINCIPLE

Now to return to our discussion of electrons. Indirect experiments and theoretical arguments indicated that electrons are *very* small. The mass has been measured (Table 1-1) and it is also *very* small. Taking these facts into consideration, what conditions would you need to see an electron clearly? You would need light with very low energy photons (because of the small mass of the electron). These very low energy photons won't bounce the electron around. But these very low energy photons have a very long wavelength and won't allow you to focus the image of the electron clearly (see Figure 13-2 for this effect in trying to see a coronavirus). So to focus the image of the electron clearly, a very short wavelength, shorter than the size of the electron, must be used.

Now we have a real dilemma. To focus the image of an electron clearly, we would have to use very short wavelength photons. Very short wavelength gamma rays may be able to do the job, but these gamma rays would have *very* high energy photons. These high energy photons would bounce the electrons all over the place, leading to a very blurry image. And as stated above, we cannot lower the energy of the photons because then the wavelength would be too long to get a clear picture of the electron.

This is a dilemma that cannot be solved. **There is no way to arrange the wavelength and photon energy so that we can see an electron clearly.**

You might ask, "Even if we cannot see an electron clearly, doesn't it really have a definite shape and isn't it really in some definite place?" The answer is that we don't know, and seemingly can never know, because of the uncertainly principle.

The **uncertainty principle** and the **wave-particle duality of matter** (see page 257) are central concepts in **quantum mechanics** (mentioned below) and physicists have been trying to understand its exact meaning since the 1920s. See https://bit.ly/2BKm6Dm and https://bit.ly/3i8qXPc for details. The essay beginning on page 412 about the double slit experiment may give you some insight into the weirdness of quantum mechanics. With all its weirdness, quantum mechanics is the most successful physical theory ever developed, which means it agrees more closely with experiments than any other theory.

## 13-5
### THE ELECTRON IS DESCRIBED WITH FUZZY PICTURES CALLED ORBITALS

The interesting thing is that any theory describing the electron bound to a nucleus that does not take into account the uncertainty principle and the wave-particle duality of matter gives results that do not agree with experiments. Another way of putting this is to say that any theory that assumes that we can see an electron clearly and that matter has only a particle nature will give results that do not agree with experiments.

**Niels Bohr**'s great theory, mentioned in Appendix 1-1 in Chapter 1, assumed that the orbit of the electron around the nucleus was completely known. His theory was great in the sense that it was the first theory that could explain many of the experiments involving the electron of the hydrogen atom. Even though Bohr's theory agreed with most, but not all, experiments done on hydrogen atoms, it failed when applied to atoms that had more than one electron.

Between 1924 and 1928, physicists developed a theory that did include the uncertainty principle and the wave-particle duality of matter. It agreed with the experimental findings for all atoms. This theory is called **quantum theory** or **quantum mechanics** and gives us a blurry picture of the electron.

The mathematics of quantum mechanics is very involved and will not be discussed in this book. But the results of solving the equation of quantum mechanics for the hydrogen atom will be diagrammed.[2] It is these diagrams that chemists use to describe the shape of the space that the electrons are in when they surround the nucleus. Be warned, however, that the diagrams chemists draw are to help them understand the space the electrons occupy in atoms—they are not pictures of the electrons themselves. Due to the uncertainty principle, the actual electrons cannot be visualized.

The diagrams that chemists draw of the electrons in the space surrounding the nucleus are called **orbitals**. Orbitals are a description of the shape of the space in which the electrons exist. Due to the uncertainly principle and the wave nature of the electrons, the orbitals are fuzzy and don't have clear boundaries, but more details must be left for a general chemistry course.

[2]There are many solutions to this equation, called Schrödinger's equation, and it is these solutions that give us the formulas used to diagram the fuzzy shapes of the various orbitals. This equation is named after **Erwin Schrödinger** (1887-1961). He received the Nobel Prize in Physics in 1933 for "formulating" (to quote the Nobel Prize organization) this equation. See https://bit.ly/31YPxLE for a very interesting story of his life. **NOTE:** Schrödinger's equation cannot be derived in the usual sense that you might be familiar with from some math classes. It is an exceptionally brilliant "formulation" that incorporates the uncertainly principle and the wave-particle duality of matter.

# 224 ATOMIC ORBITALS

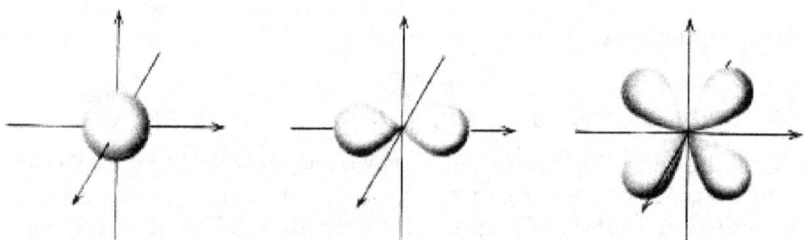

**Figure 13-3**   Examples of a few types of orbitals

There are many kinds of orbitals as shown in Figure 13-3. Some are shaped like spheres, some like dumbbells, and some like four-leaf clovers. Others are even more complicated.

## 13-6
## THE RULES FOR PLACING ELECTRONS IN ORBITALS DEPEND ON THE PROPERTIES OF THE ELECTRON

As you know, there are 118 different known elements. The number of electrons in an atom of each of these elements depends on the atomic number of the element and ranges from one electron in hydrogen to 118 electrons in element 118, which is called Oganesson, after Yuri Oganessian (born 1933), a Russian nuclear physicist who is a leading researcher in super-heavy elements. The following questions then arise:

1. What kinds of orbitals are there in these different elements?

2. Since the electrons are in orbitals, how many orbitals of each kind are there in the different atoms?

We will spend the rest of this chapter answering the first question; the second question will be answered in Chapter 14.

To begin answering these questions, we must first look at some rules that electrons follow as they are arranged in orbitals. These rules come from quantum mechanics, and it has been found from many experiments that electrons really do obey them.

### RULE 1
Electrons surrounding a nucleus are arranged in orbitals.

### RULE 2
An orbital can contain a maximum of two electrons.

### RULE 3
Electrons like to be as far apart from one another as possible because they have negative charges and objects with the same charge repel each other.

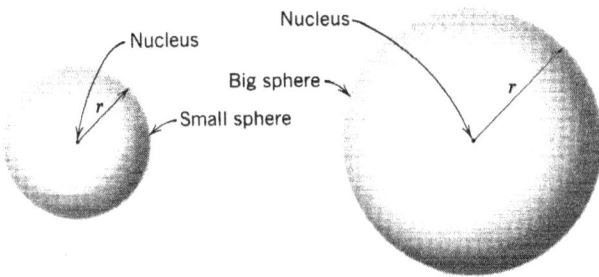

**FIGURE 13-4** The larger the radius of the sphere, the more space there is for electrons.

### RULE 4
Electrons are "lazy." This means that they like to be as close to the nucleus as possible, because being close to the nucleus means that they need less energy. Since the negative charge of the electron attracts the positive charge of the nucleus, the electron has to do work to get farther from the nucleus, just as you have to do work in climbing a flight of stairs. Being "lazy," the electron prefers not to do this work, and thus it likes to be close to the nucleus.

### RULE 5
The farther away the electrons are from the nucleus, the more room there is for them. Or, the closer the electrons are to the nucleus, the less room there is for electrons. (See Figure 13-4.)

## 13-7
## ORBITALS ARE ARRANGED IN ENERGY LEVELS

The orbitals are not spaced at equal distances from the nucleus. They tend to be bunched together in groups or levels. All the orbitals in a group or level are at about the same average distance from the nucleus and therefore have about the same energy. (This statement will be modified in Chapter 14.) The usual way to refer to orbitals having about the same energy is to say that they are in the same **energy level.**

The energy levels are numbered, beginning closest to the nucleus with energy level 1 and working outward to higher levels. Each energy level has a different number of orbitals in it, which is reasonable in light of Rule 5. The higher the energy level, the greater the number of orbitals it contains. There is a simple formula that tells us how many orbitals there are in each energy level. The formula is

$$\text{number of orbitals} = n^2$$

where $n$ is the number of an energy level; $n$ needs only to take on the values 1

## TABLE 13-1

VALUES OF $n$, $n^2$, and $2n^2$

| ENERGY LEVEL, $n$ | NUMBER OF ORBITALS, $n^2$ | NUMBER OF ELECTRONS, $2n^2$ |
|---|---|---|
| 1 | 1 | 2 |
| 2 | 4 | 8 |
| 3 | 9 | 18 |
| 4 | 16 | 32 |
| 5 | 25 | 50 |
| 6 | 36 | 72 |
| 7 | 49 | 98 |

through 7 to get all the energy levels needed to contain enough electrons for the 118 elements.

Naturally in atoms, the lower energy levels fill up with electrons first. This is a consequence of Rule 4.

We also need to know the maximum number of electrons in each energy level. Because each orbital can contain up to two electrons (Rule 2), the formula

**number of electrons = 2n²**

gives the maximum number of electrons in a given energy level.

Table 13-1 lists values of $n^2$ and $2n^2$ for the first seven energy levels.

If we add up all the numbers in the last column of Table 13-1, we get 280. But since we only need 118 electrons to make element 118, why take 7 energy levels? The reason is that the last three energy levels, 5, 6 and 7, never get completely filled. We will see how this works in Chapter 14.

## 13-8
### SPIN IS A PROPERTY OF ELECTRONS

Before we proceed, we should talk about one more property of electrons — their **spin**. The word "spin" may suggest that electrons spin like a top, but because of the uncertainty principle, we cannot say that they do. The spin of an electron is a property that seems to be like a top's spin in only one respect: It has what's called **spin angular momentum**, as do tops, which is why this property of the electron was called **spin**. (See page 231 for a discussion of momentum and spin angular momentum.) Remember that the uncertainty principle prevents us from visualizing the spin of an electron.

As we have done before for orbitals, we will talk of spin as if we actually could visualize it. We say that electrons can have a spin of either +1/2 or -1/2. (For why these numbers are fractions, see page 227.) Each of these two spin

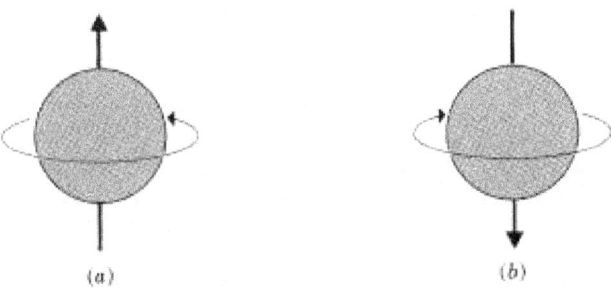

**FIGURE 13-5** The two spin states of an electron: (a) "spin-up" electron; (b) "spin-down" electron.

conditions is called a **spin state.** Chemists call the $+\frac{1}{2}$ spin state "spin up" and the $-\frac{1}{2}$ spin state "spin down." The numbers $+\frac{1}{2}$ and $-\frac{1}{2}$ come out of the mathematics of quantum mechanics and are related to the spin angular momentum. But these matters need not concern you in this course. You can think of the "spin-up" condition as the electron spinning with its top "up" and the "spin-down" condition as the electron having flipped over and having its top pointing "down." "Up" and "down" are our descriptions. The electron only "knows" that it has two spin states, which are illustrated in Figure 13-5.

If it bothers you that you are continually told "We cannot possibly visualize this or that property of electrons, but here is the way to look at it," please try not to be upset. That's the way nature has made electrons and human beings—"indescribable" and "describers," respectively. Since we do have to talk about electrons, we use the next best method, a model. So you will continually be given pictures of the unpicturable and told to imagine the unimaginable.

This discussion about spin was included here to bring out one very important point. We said before that only two electrons can fit into an orbital (Rule 2). These two electrons *must* have different spin states. One electron has to have spin *up*, and the other electron has to have spin *down*. We can symbolize a spin-up electron with an up arrow, ↑, and a spin-down electron with a down arrow, ↓. So in each orbital that is filled, the two electrons are ↑ and ↓.

To summarize: **Two electrons can be in the same orbital only if their spins are different.**

## 13-9
### THE FOUR BASIC TYPES OF ORBITALS ARE THE s, p, d, AND f ORBITALS

Earlier in this chapter, we raised the question, "What kinds of orbitals are there in the different elements?" We are now in a position to answer this question.

All the elements have their electrons arranged in four basic types of

228  ATOMIC ORBITALS

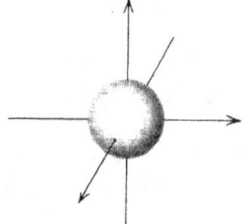

**FIGURE 13-6**  The s orbital in the first energy level.

orbitals. These orbitals are referred to as "s," "p," "d," and "f." Unfortunately, this system is a carryover from the old days, before scientists knew the shapes of the orbitals. The names aren't related to the shapes.

***The s orbital.*** The s orbital has the shape of a fuzzy sphere, as shown in Figure 13-6. There is only *one* s orbital possible in each energy level. Like all orbitals, the s orbital can hold up to two electrons. In each energy level, electrons will go into an s orbital first. The lowest energy level, $n = 1$, has *only* an s orbital.

***The p orbitals.*** The p orbital has the shape of a fuzzy dumbbell. There are *three* p orbitals in each energy level except for the first level. Those in the second energy level are shown in Figure 13-7. The three p orbitals are at right angles to one another and have names that tell us the axis they are on. The p orbital that lies on the x-axis is called $p_x$, the p

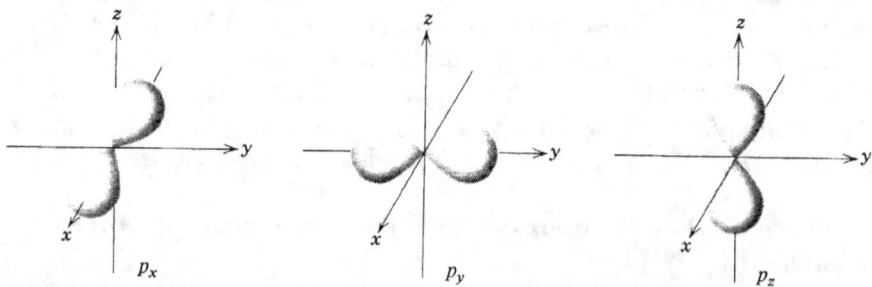

**FIGURE 13-7**  The three p orbitals in the second energy level.

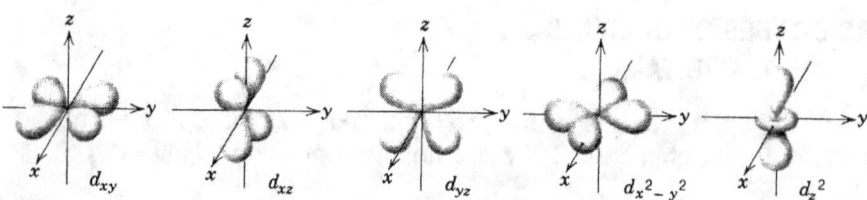

**FIGURE 13-8**  The five d orbitals in the third energy level.

### 13-9 The Four Basic Types of Orbitals are the s, p, d, and f Orbitals

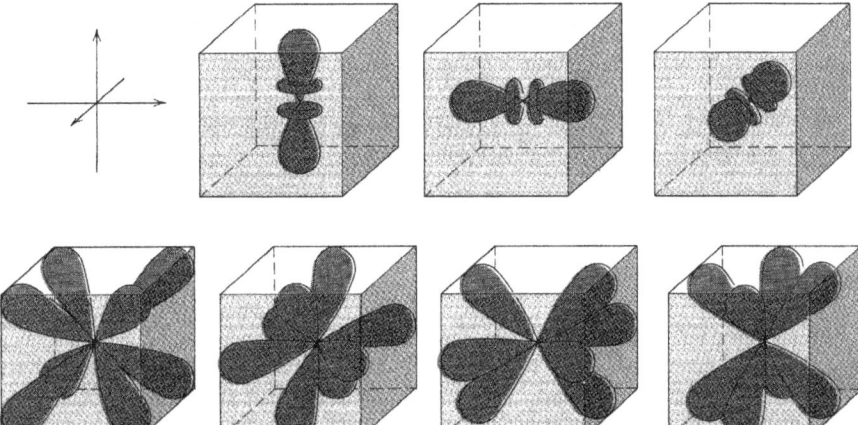

**FIGURE 13-9** The seven f orbitals in the fourth energy level.

orbital that lies on the y-axis is called $p_y$, and the p orbital that lies on the z-axis is called $p_z$. Since each p orbital can hold up to two electrons, all three p orbitals can hold up to six electrons. In each energy level, electrons will go into the p orbitals only after the s orbital has been filled.

*The d orbitals.* The d orbitals have shapes that can best be described with the help of the pictures in Figure 13-8. There are *five* d orbitals in each energy level starting with the third level ($n = 3$). Since each d orbital can hold a maximum of two electrons, the five d orbitals can hold up to 10 electrons. Each of the d orbitals has a name (see Figure 13-8). However, it is not necessary to learn these names at this time.

*The f orbitals.* The f orbitals have shapes that can best be described with the help of the pictures in Figure 13-9. They are more complicated than the d orbitals. There are *seven* f orbitals in each energy level starting with the fourth level ($n = 4$). Since each f orbital can hold a

### TABLE 13-2

ORBITALS AND ELECTRONS

| ORBITAL | NUMBER OF ORBITALS OF A GIVEN TYPE IN AN ENERGY LEVEL | MAXIMUM NUMBER OF ELECTRONS IN ALL THE ORBITALS OF A GIVEN TYPE |
| --- | --- | --- |
| s | 1 | 2 |
| p | 3 | 6 |
| d | 5 | 10 |
| f | 7 | 14 |

## 230 ATOMIC ORBITALS

maximum of two electrons, the seven $f$ orbitals can hold up to 14 electrons. There is no need for you to learn the names, and they will not be listed in the figure.

Table 13-2 gives (a) the number of orbitals of each type in an energy level and (b) the maximum number of electrons in all the orbitals of a given type.

## PROBLEMS[1]

1. Define the uncertainty principle.
2. Define wavelength.
3. What is the relationship between the photon energy and the wavelength of light?
4. Discuss the two reasons why the uncertainty principle exists.
5. Why are the effects of the uncertainty principle not noticeable in everyday living?
6. What do we mean by the term "orbital"?
7. List five rules that electrons follow as they are arranged in orbitals.
8. How many orbitals and electrons can there be in the eighth energy level?
9. What are the "quantum numbers" of the two spin states that an electron can have?
10. Complete the following: Two electrons can be in the same orbital *only* if . . . .
11. Sketch the $s$ orbital in the first energy level.
12. Sketch the $p_x$, $p_y$, and $p_z$ orbitals in the second energy level.
13. a. How many $g$ orbitals can exist in one of the higher energy levels? (The $g$ orbitals come right after the $f$ orbitals.)
    b. How many electrons does this correspond to?

---

[1] There are no keyed problems in this chapter.

## MOMENTUM AND ANGULAR MOMENTUM

In Section 13-8, we mentioned that electrons have **spin angular momentum**.

To understand spin angular momentum we must discuss linear or straight-line momentum, which is usually just called momentum. **Momentum** is defined as the mass ($m$) times the velocity ($v$) of an object. As an equation: **momentum = $m$ x $v$**. Momentum is a very important concept in physics because it describes one of the ways our universe works, as shown by the German mathematician **Emmy Noether** (1882-1935).

The above equation shows that momentum is directly proportional to both the mass and the velocity. This means that if the mass increases, the momentum increases. And, if the velocity increases, the momentum also increases. **NOTE:** For our discussion, the terms velocity and speed can be used interchangeably.

**Momentum** is related to the **force** an object can exert on another object. Let's say you lie across a bowling lane in front of the pins (*!Don't ever do this!*). If a bowler throws a 16 lb (or 7258 g) bowling ball down the lane at high speed, the ball will do serious damage to you — it might even kill you. (A large force is transferred to your body.) But if the bowler just lets the ball go so it is barely rolling down the lane, the ball will probably not hurt you much when it hits. (A small force transferred to your body.) Clearly, the greater the momentum → the more force transferred → the more damage is done in a collision.

Suppose the bowler now throws a golf ball (mass is 45.9 g — 158 times less than the bowling ball) at the same high speed. When the golf ball hits you, it will hurt, but might not do much serious damage. That's because its mass is 158 times less than the bowling ball. This shows the effect of mass on momentum. And if the bowler just lets the golf ball barely roll down the lane, you might not even feel it when it hits you. This shows the effect of velocity on momentum.

Now we will discuss **angular momentum** of which there are two types:
**1)** When an object is spinning (or rotating) on its axis, it has **spin angular momentum**. Spin angular momentum is defined as the mass times the velocity times the radius ($r$) of an object. As an equation: **spin angular momentum = $m$ x $v$ x $r$**. Here, the "$v$" is related to how fast the object is rotating or spinning. An example is a spinning ice skater. The spin angular momentum of the skater must stay the same (if no outside force is acting on her — one of the ways the universe works). If the skater is trying to spin as fast as possible, she first spins as fast as she can with her arms stretched out. This makes her effective radius large. Then she tucks her arms in close to her body. This makes her effective radius small, and since her mass doesn't change, her rotational speed must increase since the spin angular momentum must stay the same. Spin angular momentum is also related to force. Just think how hard it would be (or how much force it would take) to stop a rapidly spinning bowling ball, as compared to stopping a slowly spinning one.
**2)** When an object is rotating around a central point (as shown in Figure 1-3) it has **orbital angular momentum**, described by the same equation as above, but here $r$ is the distance between the hand and the rock. Another example would be the earth going around the sun, giving the earth orbital angular momentum. In this case, $r$ is the distance between the sun and the earth. Of course, the earth also has spin angular momentum because it is spinning on its axis. **The electron going around a nucleus has orbital angular momentum because it is going around the nucleus. It also has spin angular momentum because it behaves as if it were spinning on an axis** as discussed in Section 13-8.

| Reihen | Gruppe I.<br>—<br>R²O | Gruppe II.<br>—<br>RO | Gruppe III.<br>—<br>R²O³ | Gruppe IV.<br>RH⁴<br>RO² | Gruppe V.<br>RH³<br>R²O⁵ | Gruppe VI.<br>RH²<br>RO³ | Gruppe VII.<br>RH<br>R²O⁷ | Gruppe VIII.<br>—<br>RO⁴ |
|---|---|---|---|---|---|---|---|---|
| 1 | H=1 | | | | | | | |
| 2 | Li=7 | Be=9.4 | B=11 | C=12 | N=14 | O=16 | F=19 | |
| 3 | Na=23 | Mg=24 | Al=27.3 | Si=28 | P=31 | S=32 | Cl=35.5 | |
| 4 | K=39 | Ca=40 | —=44 | Ti=48 | V=51 | Cr=52 | Mn=55 | Fe=56, Co=59, Ni=59, Cu=63. |
| 5 | (Cu=63) | Zn=65 | —=68 | —=72 | As=75 | Se=78 | Br=80 | |
| 6 | Rb=85 | Sr=87 | ?Yt=88 | Zr=90 | Nb=94 | Mo=96 | —=100 | Ru=104, Rh=104, Pd=106, Ag=108. |
| 7 | (Ag=108) | Cd=112 | In=113 | Sn=118 | Sb=122 | Te=125 | J=127 | |
| 8 | Cs=133 | Ba=137 | ?Di=138 | ?Ce=140 | — | — | — | — |
| 9 | (—) | | | | | | | |
| 10 | — | — | ?Er=178 | ?La=180 | Ta=182 | W=184 | — | Os=195, Ir=197, Pt=198, Au=199. |
| 11 | (Au=199) | Hg=200 | Tl=204 | Pb=207 | Bi=208 | — | — | |
| 12 | — | — | — | Th=231 | — | U=240 | — | — — — |

## Dmitri Mendeleev's 1871 Periodic Table

**Dimitri Mendeleev** (1834-1907) is considered the father of the modern periodic table. **Compare this table with the one on page 246 to see how far we have come since 1871.** Note that the above table contains only the 66 elements known at the time. Now we know 118 elements.

Mendeleev ordered his table by chemical properties and atomic weight. The two main chemical properties he used were the formulas for the hydride and oxide of the element. $RH^4$, $RH^3$, etc. represent the hydrides. So oxygen in Gruppe VI (Gruppe is German for Group) has a hydride form of $RH^2$, which we now write as $H_2O$. The oxide form is labeled $RO^2$. The oxide of carbon in the column $RO^2$ would now be written as $CO_2$.

Now we order the periodic table by atomic number, but atomic numbers were unknown in 1871. In fact, it was only after **Henry Moseley**'s 1913 research that the concept of atomic number was understood. See page 256 for more about his life.

Mendeleev's great genius was that by ordering the elements according to their chemical properties and atomic weight, he realized that some were missing. For example, in Gruppe IV under Si = 28 is a —=72, a guess at the atomic weight of a missing element. He predicted the chemical and physical properties of this element, and it was discovered in 1886 with very similar properties. It was given the name germanium (Ge) and has an atomic weight of 72.59, close to Mendeleev's guess.

His other great concept was that if the atomic weight of an element put it in a column with the wrong chemical properties, he chose the chemical properties as being more important, assuming that the atomic weight was incorrect. He even attempted to correct some of these atomic weights, https://bit.ly/1Q3ZNVc.

*Table is in the Public Domain.*

# ORBITALS IN ATOMS AND THE PERIODIC TABLE

In Chapter 13 we raised two questions: "What kinds of orbitals are in the different elements?" and "How many orbitals of each kind are there in each different kind of atom?" We answered the first question in Chapter 13. We will answer the second question in this chapter.

## 14-1
### ORBITALS ARE ARRANGED IN ENERGY LEVELS IN THE ORDER s, p, d, and f

In each energy level, the *approximate* order in which the orbitals are filled with electrons is as follows:[1]

***s* first, *p* second, *d* third, *f* fourth**

- For the **first energy level** (n = 1), which has only one orbital, we see that this orbital must be an *s* orbital.

- For the **second energy level** (n = 2), which has four orbitals, we can have one *s* orbital and three *p* orbitals.

- For the **third energy level** (n = 3), which has nine orbitals, we can have one *s* orbital, three *p* orbitals, and five *d* orbitals.

- For the **fourth energy level** (n = 4), which has 16 orbitals, we can have one *s* orbital, three *p* orbitals, five *d* orbitals, and seven *f* orbitals.

Higher energy levels (n greater than 4) have room for additional types of orbitals. However, these additional orbitals are not needed to discuss the known elements. Table 14-1 summarizes the information just given on orbitals and energy levels.

[1]After element 20 (Ca), this order is not followed for *d* and *f* orbitals. See Figure 14-2 for details.

234  ORBITALS IN ATOMS AND THE PERIODIC TABLE

**TABLE 14-1  ORBITAL TYPES IN EACH ENERGY LEVEL**

| ENERGY LEVEL, n | MAXIMUM NUMBER OF ORBITALS, $n^2$ | ORBITAL TYPES IN ENERGY LEVEL | NUMBER OF ORBITALS OF EACH TYPE |
|---|---|---|---|
| | | (Only orbitals needed to build all 118 elements are shown.) | |
| 1 | 1 | s | one s, |
| 2 | 4 | s, p | one s, three p |
| 3 | 9 | s, p, d | one s, three p, five d |
| 4 | 16 | s, p, d, f | one s, three p, five d, seven f |
| 5 | 25 | s, p, d, f[a] | one s, three p, five d, seven f |
| 6 | 36 | s, p, d[a] | one s, three p, five d |
| 7 | 49 | s, p[a] | one s, three p |

**NOTE:** It is presumed that elements 113 to 118 are filling 7p orbitals, but this has not been determined experimentally.

[a] There is room in these energy levels for more orbital types. But these orbitals do not have electrons in them. We can describe all 118 known elements with only the orbitals indicated. Be sure to read the "NOTE" at the bottom of page 235 for details about the spacing of the energy levels in atoms.

## 14-2
### ATOMS OF DIFFERENT ELEMENTS HAVE THEIR ORBITALS FILLED IN A SPECIFIC ORDER

Before we can discuss the order in which electrons fill up the orbitals to make the different elements, a bit of notation is necessary. An s orbital in the first energy level (n = 1) is called a 1s orbital. An s orbital in the second energy level (n = 2) is called a 2s orbital. A p orbital in the second energy level is called a 2p orbital. And so on. As you can see, the number of the energy level of an orbital precedes the name of the orbital.

**EXAMPLE 1** What would a p orbital in the fourth energy level be called?

*Solution:* It would be called a 4p orbital. ■

**EXAMPLE 2** What would a d orbital in the fifth energy level be called?

*Solution:* It would be called a 5d orbital. ■

As was discussed in Rule 4 on page 255, electrons are "lazy." If you look at Figure 14-1, you can see that the electrons will always fill lower energy orbitals before they fill higher energy orbitals. That's why they fill the 1s orbital first, then the 2s orbital, then the three 2p orbitals, and so on up the "energy ladder."

Figures 14-1 and 14-2 show the order in which the orbitals are filled by electrons. **NOTE:** Where these figures disagree in the higher energy levels, Figure 14-2 is the correct one. **Also, the order of filling of the orbitals may not be their final configuration above element 20 (Ca).** Each small-shaded circle in Figure 14-1 represents one orbital. The shading of the circles is different for different energy levels. This is so you can easily tell which energy level a circle, and thus an orbital, is in.

14-2 Atoms of Different Elements Have Their Orbitals Filled in a Specific Order 235

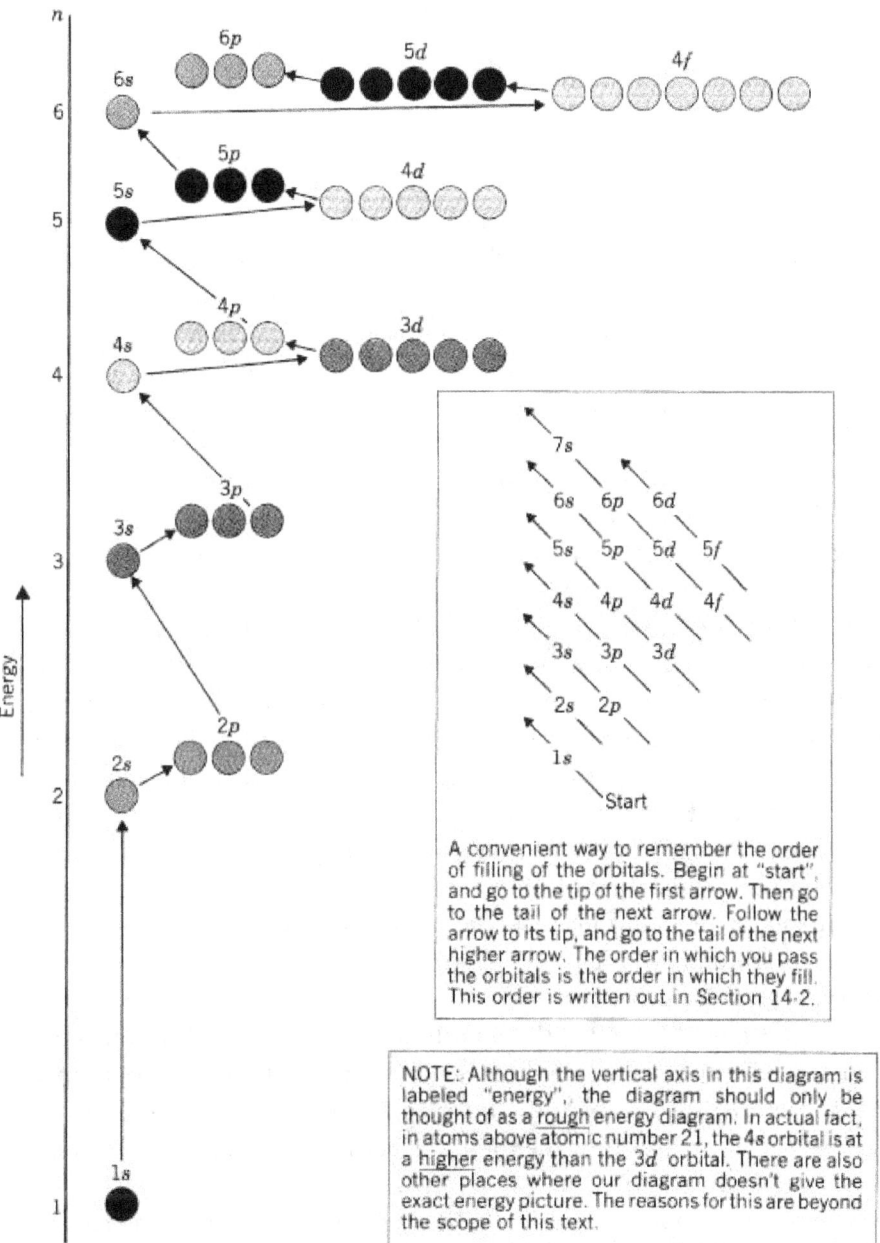

**FIGURE 14-1** Orbitals and their approximate energy. The order of filling of the orbitals is indicated by the arrows. In the upper boxed area, we present a convenient way to remember the order of filling of the orbitals. In the lower boxed area, we briefly discuss the actual energy of some orbitals.

NOTE: The energy level spacing is not the same as the energy increases. The distance between levels 2 and 3 is *less* than the distance between levels 1 and 2, and so on up the "energy ladder." Since the energy levels get closer together, there are actually an *infinite number* of them ("n" approaches infinity). But there is a highest energy level for each atom and an infinite number of energy levels is squeezed into this very small finite space. See https://bit.ly/3NjnehQ for a diagram of this. If an electron has an energy higher than this highest energy level, it can leave the atom. This process is called ionization and is discussed in Section 15-7. More about this topic will be discussed in a general chemistry course.

The arrows show the order of the filling of the orbitals with electrons.

The lowest energy level is the first one (n = 1). There is one *s* orbital in it, and this orbital is called the 1*s* orbital. An *s* orbital can contain a maximum of two electrons.

The second energy level has four orbitals: one *s* orbital, called the 2*s* orbital, and three *p* orbitals, known as the 2*p* orbitals. Notice that the three 2*p* orbitals have a slightly higher energy than the single 2*s* orbital. See Appendix 14-1 at the end of this chapter for an explanation. From Rule 4 in Chapter 13, the 2*s* orbital will fill up first and then the electrons will begin to fill the 2*p* orbitals. Again, each orbital can contain a maximum of two electrons (Rule 2 in Chapter 13).

The third energy level has nine orbitals: one *s*, three *p*, and five *d*. The 3*s* orbital will fill up first and then the electrons will go into the 3*p* orbitals. However, now we have a problem. You would normally think that after the 3*p* orbitals are filled, the 3*d* orbitals would fill up with electrons. But nature doesn't work this way. Because the 3*d* orbitals are at a higher energy than the 4*s* orbital, it is the 4*s* orbital that now begins to fill with electrons. It is only after the 4*s* orbital is filled that the 3*d* orbitals begin to fill. After the 3*d* orbitals are filled, the 4*p* orbitals start to fill, since they are at a higher energy than the 3*d* orbitals.

Notice that in Figure 14-1, the 3*d* orbitals are at about the same energy as the 4*s* and 4*p* orbitals. However, they are still called 3*d* orbitals, which means that they formally belong to the third energy level. The *d* orbitals fill one energy level later than you would think. This may seem strange, but it all has to do with the mathematics of quantum mechanics and many laboratory experiments.

After the 4*p* orbitals are filled, the 5*s* orbital fills up. Then the 4*d* orbitals fill up, followed by the 5*p* orbitals.

As shown in Figure 14-2, after the 5*p* orbitals are filled the 6*s* orbitals are filled. After the 6*s* orbitals are filled, one electron goes into a 5*d* orbital after which the 4*f* orbitals are filled. After they are filled, the 5*d* orbitals resume filling. Figure 14-2 illustrates the order of filling of the remaining orbitals which is 7*s*, 6*d*, 5*f*, 6*d* (again) and then 7*p*.

A summary of the order in which the orbitals fill with electrons is as follows (but see Figure 14-2 for some exceptions):

**1*s* 2*s* 2*p* 3*s* 3*p* 4*s* 3*d* 4*p* 5*s* 4*d* 5*p* 6*s* 4*f* 5*d* 6*p* 7*s* 5*f* 6*d* 7*p***

Now look again at Figure 14-2. This diagram shows the order in which the elements are formed by the filling of the various orbitals. Looking at Figure 14-2, we see that if there is one electron in the 1*s* orbital, we have hydrogen, H. Two electrons in the 1*s* orbital gives us helium, He.

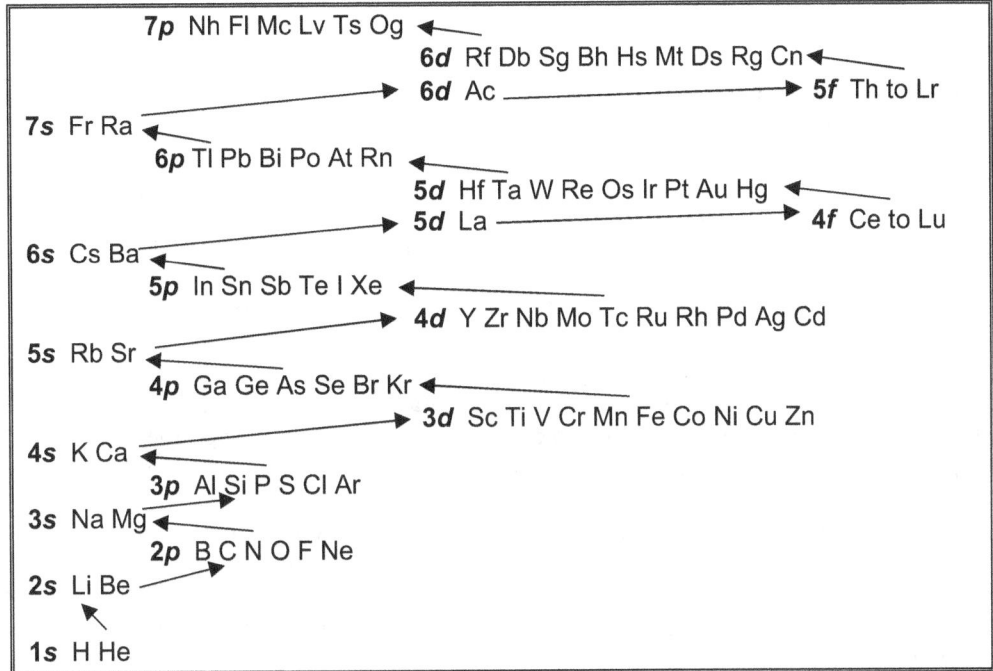

**Figure 14-2** The order of filling of the orbitals of the elements. This diagram follows the periodic table shown in Figure 14-10, and you can use this diagram to solve problems like Example 3 below. **NOTE: The order of the filling of the orbitals is not always the same as the final orbital configuration after element 20 (Ca), and there are exceptions to the orbital filling order as shown above that are beyond the scope of this book.** If you are interested, see https://rsc.li/2h49UA2 for details. **ALSO:** The energy level spacing in this figure is not correct. Figure 14-1 shows a more accurate depiction of the actual energy level spacing.

Adding an electron to the 2s orbital gives lithium, Li. Two electrons in the 2s orbital give beryllium, Be. Now we have a total of four electrons, corresponding to the atomic number of Be, which is 4.

The three 2p orbitals can contain a total of 6 electrons. These orbitals fill one electron at a time to give the elements B, C, N, O, F and Ne.

**EXAMPLE 3** Using Figure 14-2 and Figure 14-10, list all the elements in which the 5s orbital, the 4d orbitals, and the 4f orbitals are being filled. **NOTE:** As discussed on page 247, La is also considered a lanthanoid element but it does not have any 4f electrons. The 4f orbitals begin to fill with Ce.

*Solution:* 5s orbitals: Rb, Sr.
4d orbitals: Y, Zr, Nb, Mo, Tc, Ru, Rh, Pd, Ag, Cd.
4f orbitals: Ce, Pr, Nb, Pm, Sm, Eu, Gd, Tb, Dy, Ho, Er, Tm, Yb, Lu. ∎

238  ORBITALS IN ATOMS AND THE PERIODIC TABLE

## 14-3
## ORBITAL DIAGRAMS AND ELECTRON CONFIGURATIONS ARE WAYS OF REPRESENTING THE ARRANGEMENT OF ELECTRONS IN ATOMS

In Figure 14-1, we showed the order of the filling of the orbitals. If we put one electron at a time into the diagram, we will build up the **orbital diagram** (or arrangement) of the different elements.

Each electron will be represented by an arrow, either ↑ or ↓, to correspond with the two spin states, "spin up" or "spin down."

In the following discussion, Figure 14-1 has been reduced in size to save space. Notice that the circles for all the orbitals are still present, but only some of them will be filled with electrons. That's all right, since the space for the electrons is always there, even though there are no electrons in that space.

The first element is hydrogen, H, which has one electron in the $1s$ orbital. We can write this electron as $1s^1$, where the superscript "1" tells us the number of electrons that are in the $1s$ orbital. This is shown in Figure 14-3. This way of writing the electron description is called the **electron configuration** of an element.

Figure 14-4 has one arrow drawn in the circle that represents the $1s$ orbital. This is the orbital diagram of hydrogen. Below the diagram we have written the electron configuration of hydrogen.

The second electron also goes into the $1s$ orbital to give helium, He. We can write the electron configuration of He as $1s^2$, since there are two electrons in the $1s$ orbital. Figure 14-5 shows the orbital diagram and electron configuration for He.

Since each orbital can have a maximum of two electrons, the superscript on the orbital letter can only take on the values of 1 or 2.

The third electron goes into the $2s$ orbital to give the element lithium, Li. The orbital diagram and electron configuration for Li is $1s^2 2s^1$, as shown in Figure 14-6.

The fourth electron also goes into the $2s$ orbital to give beryllium, Be. The orbital diagram and electron configuration for Be is shown in Figure 14-7.

Next the $2p$ orbitals fill up to give the elements B, C, N, O, F, and Ne. The electron configurations and orbital diagrams are shown in Figure 14-8.

We will now discuss some very important points in the way electrons fill the $2p$ orbitals. Remember that there are three $p$ orbitals—$p_x$, $p_y$, and $p_z$. In studying Figure 14-8, you may wonder why we put the electrons into the $p$ orbitals in the order $p_x$, $p_y$, $p_z$. We do this to make our diagrams look

Energy level → $1s^1$ ← Number of electrons in orbital
       ↑ Type of orbital

**FIGURE 14-3** The electron configuration of hydrogen.

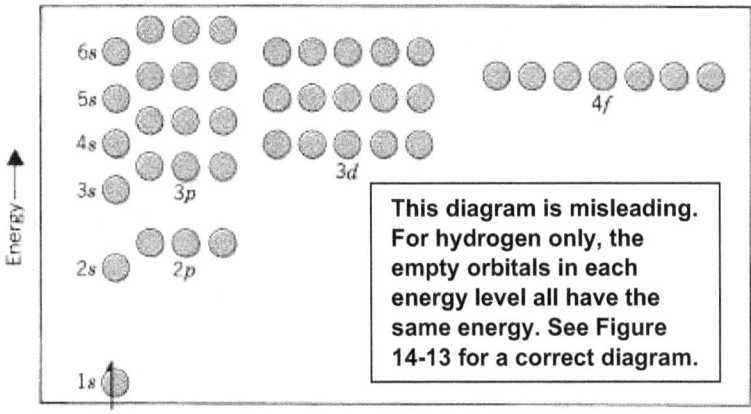

**FIGURE 14-4** The orbital diagram and electron configuration for hydrogen.

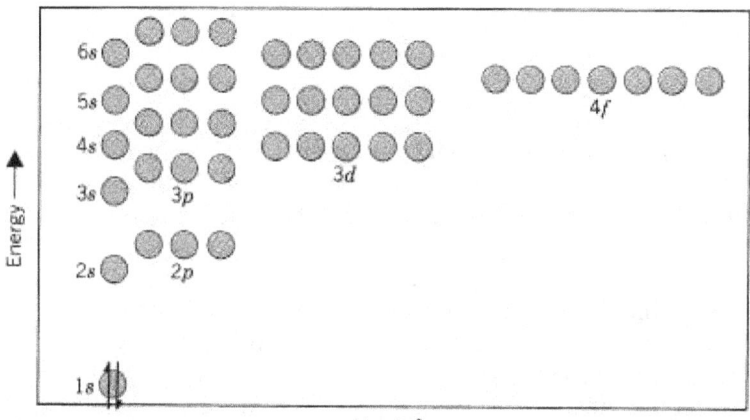

**FIGURE 14-5** The orbital diagram and electron configuration for helium.

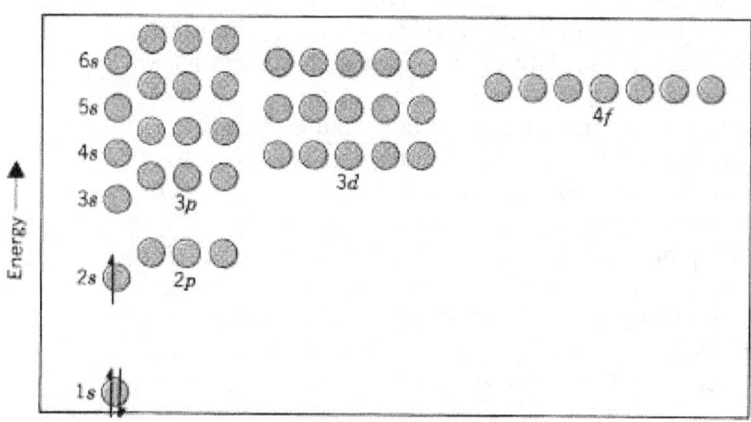

**FIGURE 14-6** The orbital diagram and electron configuration for lithium.

**240** ORBITALS IN ATOMS AND THE PERIODIC TABLE

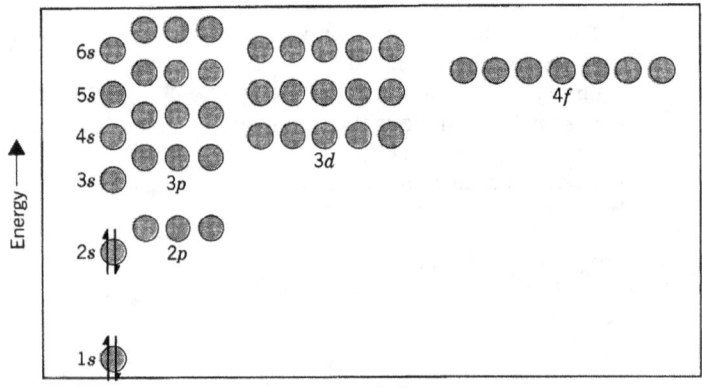

**FIGURE 14-7** The orbital diagram and electron configuration for beryllium.

consistent. We don't actually know which orbital is occupied first. Notice, however, that when electrons go into the $p_x$, $p_y$, and $p_z$ orbitals, the first three always go in with their spins in the same direction. This is the lower-energy arrangement.

The orbital diagrams demonstrate a very important point. According to Rule 3 in Chapter 13, electrons like to be as far apart from each other as possible. But this must be consistent with Rule 4, which states that they are "lazy" and like to be at the lowest energy. Following these rules, look at the orbital diagram and electron configuration for carbon. The sixth electron *doesn't* join the fifth one in the $2p_x$ orbital, which would fill up the $2p_x$ orbital. The sixth electron can get farther from the $p_x$ electron by going into the $2p_y$ orbital with *no* increase in energy. So it does just that.

The same thing happens with nitrogen: The seventh electron goes into the $2p_z$ orbital. This way the electron is as far from the other electrons as possible with no increase in energy. But after nitrogen, the electrons will start doubling up in the $2p$ orbitals, because it would take some extra energy for the electrons to go into the $3s$ and higher orbitals. And electrons are so "lazy" that they would rather "tolerate" each other than do extra work. Refer to the diagrams of oxygen, fluorine, and neon for illustrations of this phenomenon.

After neon, the eleventh electron will go into the $3s$ orbital to form sodium, Na. We could continue describing the orbital diagrams and electron configurations for all the elements, but now you can figure them out for yourself. Just use Figures 14-1 and 14-2 as guides, and remember the rules (see Chapter 13) that electrons follow when arranging themselves in orbitals.

**EXAMPLE 4** Using a diagram like Figure 14-1, draw the orbital diagram for the element silicon, atomic number 14. Also write the electron configuration for silicon.

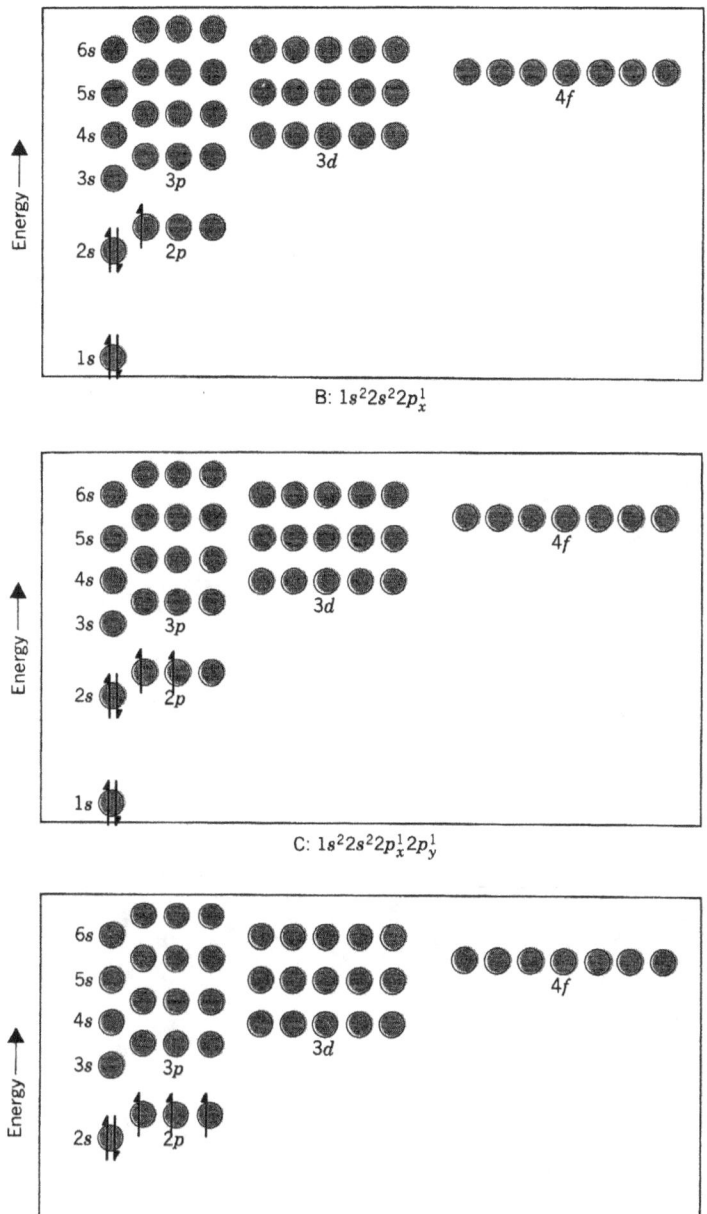

**FIGURE 14-8** The orbital diagrams and electron configurations for boron, carbon, nitrogen, oxygen, fluorine, and neon.

242 ORBITALS IN ATOMS AND THE PERIODIC TABLE

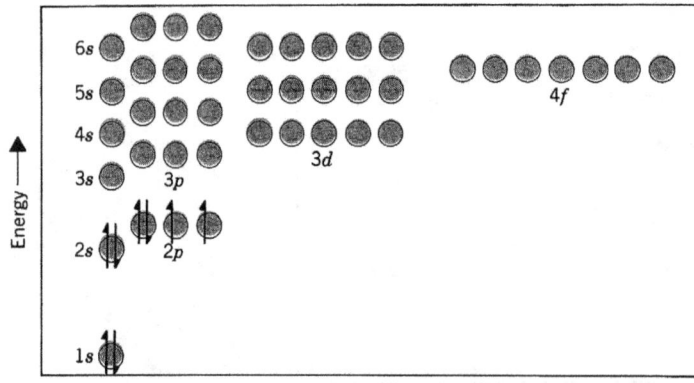

O: $1s^2 2s^2 2p_x^2 2p_y^1 2p_z^1$

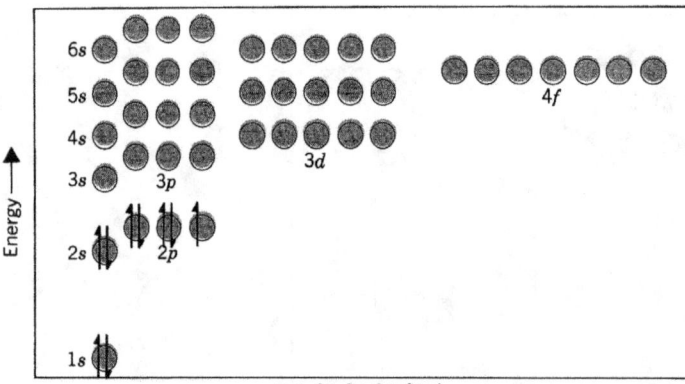

F: $1s^2 2s^2 2p_x^2 2p_y^2 2p_z^1$

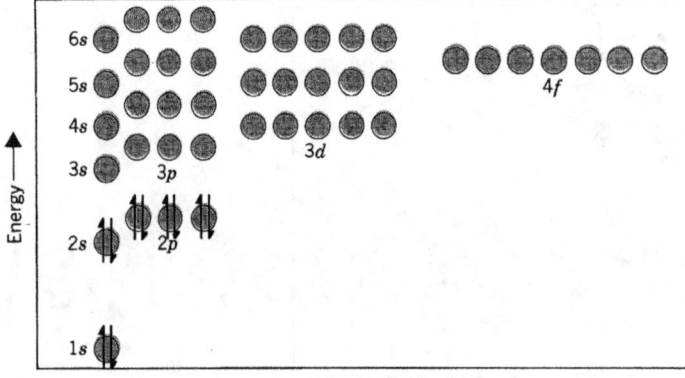

Ne: $1s^2 2s^2 2p_x^2 2p_y^2 2p_z^2$

**FIGURE 14-8** continued.

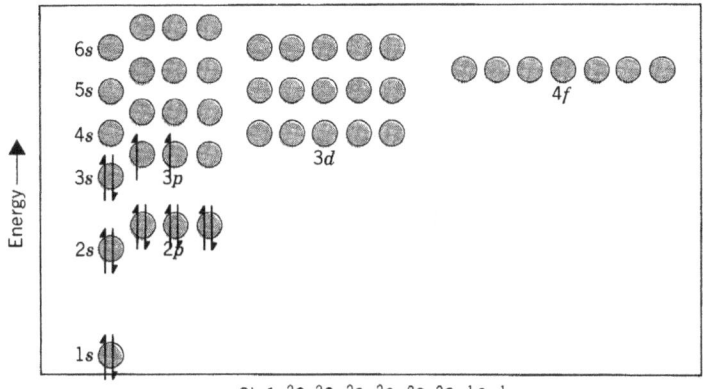

Si: $1s^2 2s^2 2p_x^2 2p_y^2 2p_z^2 3s^2 3p_x^1 3p_y^1$

**FIGURE 14-9** The orbital diagram and electron configuration for silicon.

*Solution:* See Figure 14-9. ∎

As you can see, it takes considerable space to draw the orbital diagrams as we have done. So chemists have developed a way of writing them across the page. Instead of circles to represent orbitals, they use a short line like this: ____. The electrons are drawn in as arrows. An orbital with one electron in it is drawn as $\underline{\uparrow}$, and an orbital with two electrons in it is drawn as $\underline{\uparrow\downarrow}$. (Again, we draw in the "spin-up" electrons first. Of course the electron doesn't know up from down, but chemists want all their diagrams to look the same.) The names of the orbitals are put under the short lines in the same way as the electron configurations are written under the orbital diagrams. These **horizontal orbital diagrams** for the first 10 elements are written as follows:

H: $\underline{\uparrow}\atop 1s^1$

He: $\underline{\uparrow\downarrow}\atop 1s^2$

Li: $\underline{\uparrow\downarrow}\atop 1s^2$ $\underline{\uparrow}\atop 2s^1$

Be: $\underline{\uparrow\downarrow}\atop 1s^2$ $\underline{\uparrow\downarrow}\atop 2s^2$

B: $\underline{\uparrow\downarrow}\atop 1s^2$ $\underline{\uparrow\downarrow}\atop 2s^2$ $\underline{\uparrow}\atop 2p_x^1$

C: $\underline{\uparrow\downarrow}\atop 1s^2$ $\underline{\uparrow\downarrow}\atop 2s^2$ $\underline{\uparrow}\atop 2p_x^1$ $\underline{\uparrow}\atop 2p_y^1$

## 244 ORBITALS IN ATOMS AND THE PERIODIC TABLE

N: $\underset{1s^2}{\uparrow\downarrow}$ $\underset{2s^2}{\uparrow\downarrow}$ $\underset{2p_x^1}{\uparrow}$ $\underset{2p_y^1}{\uparrow}$ $\underset{2p_z^1}{\uparrow}$

O: $\underset{1s^2}{\uparrow\downarrow}$ $\underset{2s^2}{\uparrow\downarrow}$ $\underset{2p_x^2}{\uparrow\downarrow}$ $\underset{2p_y^1}{\uparrow}$ $\underset{2p_z^1}{\uparrow}$

F: $\underset{1s^2}{\uparrow\downarrow}$ $\underset{2s^2}{\uparrow\downarrow}$ $\underset{2p_x^2}{\uparrow\downarrow}$ $\underset{2p_y^2}{\uparrow\downarrow}$ $\underset{2p_z^1}{\uparrow}$

Ne: $\underset{1s^2}{\uparrow\downarrow}$ $\underset{2s^2}{\uparrow\downarrow}$ $\underset{2p_x^2}{\uparrow\downarrow}$ $\underset{2p_y^2}{\uparrow\downarrow}$ $\underset{2p_z^2}{\uparrow\downarrow}$

**EXAMPLE 5** Draw a horizontal orbital diagram for phosphorus.

*Solution:* The atomic number of phosphorus is 15. This means that we will have to put 15 electrons into orbitals.

P: $\underset{1s^2}{\uparrow\downarrow}$ $\underset{2s^2}{\uparrow\downarrow}$ $\underset{2p_x^2}{\uparrow\downarrow}$ $\underset{2p_y^2}{\uparrow\downarrow}$ $\underset{2p_z^2}{\uparrow\downarrow}$ $\underset{3s^2}{\uparrow\downarrow}$ $\underset{3p_x^1}{\uparrow}$ $\underset{3p_y^1}{\uparrow}$ $\underset{3p_z^1}{\uparrow}$

The sum of the superscripts equals 15. Notice that the electrons in the $3p$ orbitals are as far from one another as possible with no increase in energy. Each is in a different $p$ orbital. ∎

**EXAMPLE 6** Which of the following horizontal orbital diagrams is correct?

a. $\underset{1s^2}{\uparrow\downarrow}$ $\underset{2s^2}{\uparrow\downarrow}$ $\underset{2p_x^1}{\uparrow}$ $\underset{2p_y^1}{\downarrow}$ $\underset{2p_z^1}{\uparrow}$

b. $\underset{1s^2}{\uparrow\downarrow}$ $\underset{2s^2}{\uparrow\downarrow}$ $\underset{2p_x^1}{\downarrow}$ $\underset{2p_y^1}{\downarrow}$ $\underset{2p_z^1}{\downarrow}$

c. $\underset{1s^2}{\uparrow\downarrow}$ $\underset{2s^2}{\uparrow\downarrow}$ $\underset{2p_x^2}{\uparrow\downarrow}$ $\underset{2p_y^1}{\uparrow}$ $\underset{2p_z}{\phantom{\uparrow}}$

d. $\underset{1s^2}{\uparrow\downarrow}$ $\underset{2s^2}{\uparrow\downarrow}$ $\underset{2p_x^1}{\uparrow}$ $\underset{2p_y}{\phantom{\uparrow}}$ $\underset{2p_z^1}{\uparrow}$

*Solution:*

a. The orbital diagram is incorrect because the three unpaired $p$ electrons do not have the same spin.

b. The orbital diagram is correct because the unpaired $p$ electrons all have the same spin. However, we should always draw this diagram with the spins in the up direction so that all our diagrams will look alike.

c. The orbital diagram is incorrect. One of the $2p_x^2$ electrons should be in the $2p_z$ orbital.

d. The orbital configuration is correct. However, we should always draw this diagram with the $2p_y$ orbital, not the $2p_z$ orbital, containing the second unpaired electron so that all our diagrams look alike. ∎

## 14-4
### THE PERIODIC TABLE ARRANGES THE ELEMENTS ACCORDING TO THEIR ORBITAL CONFIGURATION AND CHEMICAL PROPERTIES

Chemists have developed a table of the elements that is organized according to the way orbitals are filled; it is called the **periodic table**.[2] Figure 14-10 is an example of a periodic table. The periodic table is divided into groups and periods. The **groups** (also called **chemical families**) are in vertical columns. Each group or chemical family is identified by a number from 1-18 or a Roman numeral at the top of each column. The Roman numerals will be used in this book. The elements in each group have similar chemical properties, which is why they are also called a chemical family. The **periods** are in horizontal rows that correspond to the filling of the energy levels. The numbers at the left of each period show the energy level that is being filled for the A group elements.

Even though hydrogen, H, is listed in group IA, it has different properties than the other elements in group IA, which are all silvery metals. Hydrogen is a gas and really belongs in a group by itself as shown in Figure 14-11.

The two groups on the left side of the table (columns IA and IIA) represent elements that have electrons in the s orbitals of their highest occupied energy level. Helium, which is listed in group VIIIA, also has an s orbital in its highest energy level, and it could belong with hydrogen as shown in Figure 14-11.

The six groups to the right side of the table (columns IIIA, IVA, VA, VIA, VIIA, and VIIIA) represent elements whose p orbitals are being filled. (Remember that helium, which is in group VIIIA, is the exception.)

The eight groups toward the center represent elements whose d orbitals are being filled (columns IIIB, IVB, VB, VIB, VIIB, VIII, IB, and IIB). These elements are called the **transition metals**. The numbering system of these groups is related to the chemical properties of the elements and won't be discussed further in this book.

The rows without column numbers, which contain elements 57 (La) through 71 (Lu) and 89 (Ac) through 103 (Lr), represent elements, with the exceptions of La and Ac, whose f orbitals are being filled. See page 247 for a discussion of what orbitals are being filled in La and Ac.

[2]Before orbitals and atomic numbers were known, chemists had devised the periodic table based on the chemical properties and atomic weights of the elements. Upon **Henry Moseley's** (see p. 256) understanding of atomic number in 1913, it was clear that the periodic table should be arranged according to atomic number, even though there are 4 places in the modern periodic table where the atomic weight does not increase with atomic number. This is due to the many isotopes that are used in calculating the atomic weight of most elements. Another major advance in understanding the periodic table was made in 1922 by **Niels Bohr** (see pp 16, 223). He proposed the concept of filled "shells" (which we call the lower filled energy levels) and that the shell with the highest energy level is the valence "shell" (see p. 250). This lead to labeling areas of the periodic table as s, p, d, and f areas. Finally, in 1945, **Glenn Seaborg** (see pp 18, 61) suggested that the actinoid elements from Th to Lr were filling 5f orbitals. Ac has no 5f orbitals as shown in Fig. 14-2. This completed the arrangement of the modern periodic table by orbital type. **Dmitri Mendeleev** (1834-1907) is considered the founder of the pre-modern periodic table, mainly because he predicted the existence of 3 elements that were not known at the time and were later discovered with the properties he predicted. Element 101 is named in his honor. More about Mendeleev's periodic table can be found on page 232.

**Figure 14-10** A modern periodic table arranged approximately according to the order of filling of the orbitals. Groups or chemical families are the vertical columns. The horizontal rows are called periods. See the bottom of page 247 and Figure 14-2 for more details. **NOTE:** La and Ac are placed in the *d* orbital section in some periodic tables. The IUPAC recommends the configuration shown below. The electron shell configurations for elements 109-118 are predicted. *Periodic Table courtesy of* **Anne Helmenstine**, *Ph.D. www.sciencenotes.org*. This site has free printable periodic tables as both images and pdf files, including this one — many are in color. Sir **Martyn Poliakoff** and colleagues have made a series of short videos about the elements at www.periodicvideos.com/. Click on an element to see the video.

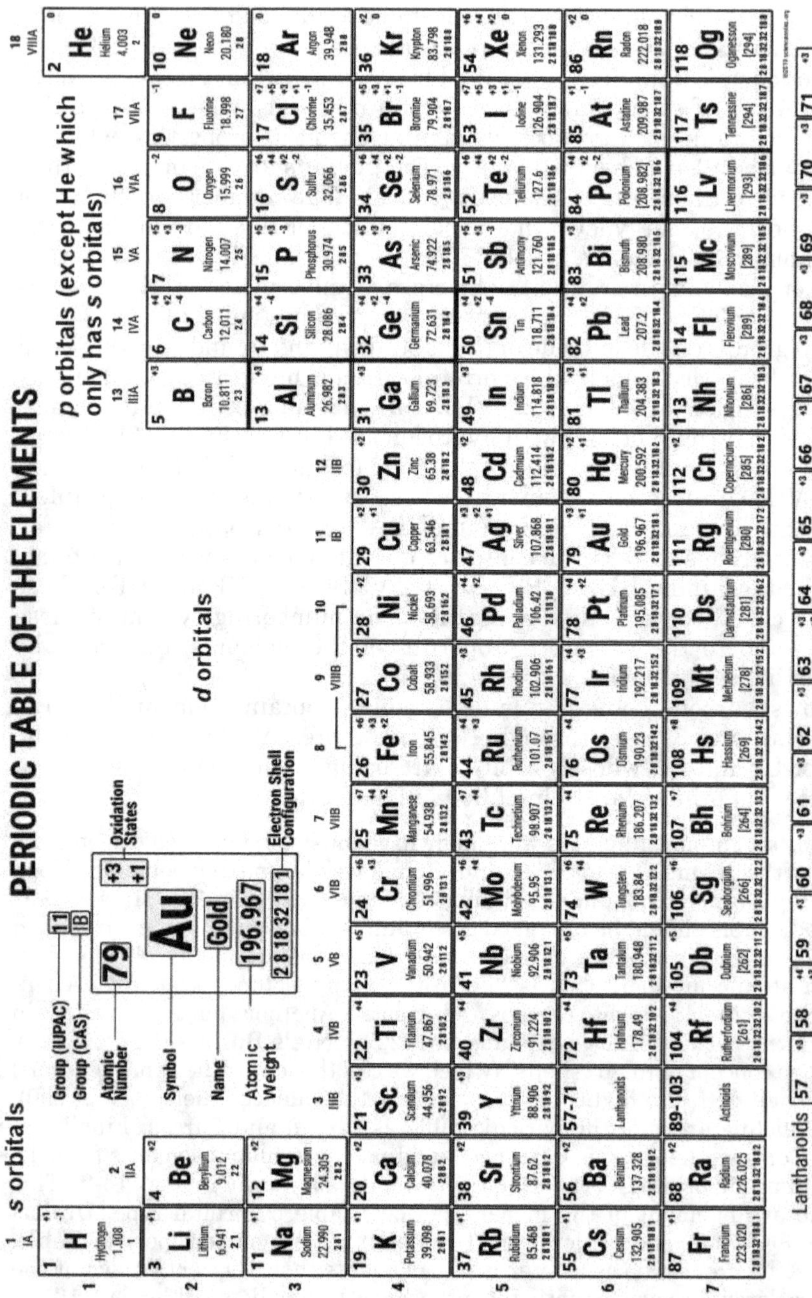

These rows have special names: The row containing elements 57 through 71 is called the **lanthanoid** or **rare earth series**, and the row containing elements 89 through 103 is called the **actinoid series**. (**NOTE:** Older but still very widely used names for *lanthanoid* and *actinoid* are *lanthanide* and *actinide*.)

Element 57, lanthanum (La), is usually placed in the lanthanoid or rare earth series of elements. It has many of the properties of elements 58 through 71 but its orbital diagram does not have any $4f$ electrons in it. The highest occupied orbitals in its orbital diagram have one $d$ electron and two $s$ electrons in this order: $5d^1 6s^2$. Note the order of filling in Figure 14-2 is different. The reason why is beyond the scope of this book. The first element in the lanthanoid series that has $4f$ electrons is element 58, cerium (Ce). A similar arrangement is made for the first element in the actinoid series, actinium (Ac), which does not have any $5f$ electrons in its orbital diagram. There is disagreement among chemists about where La and Ac should be placed in the periodic table. See https://bit.ly/2P4tFfa. The IUPAC uses the arrangement shown in Figure 14-10.

For a famous and really funny song about the periodic table called "The Elements" by **Tom Lehrer** that uses animation as it follows the lyrics to construct a modern periodic table see https://bit.ly/2WYVOoe.

**EXAMPLE 7** List the elements that are in the following groups in the periodic table (refer to Figure 14-10): Groups IIA, VIIB, VIA.

*Solution:* From Figure 14-10, we have, reading down each group:

Group IIA: Be, Mg, Ca, Sr, Ba, Ra    Group VIIB: Mn, Tc, Re, Bh

Group VIA: O, S, Se, Te, Po, Lv ∎

The first row (period 1) across the table lists elements in which the highest-energy orbital that has electrons in it is the first energy level (n = 1). The second row (period 2) lists elements in which the highest-energy orbitals that have electrons in them are the second energy level (n = 2). This pattern is similar for the third row (period 3) elements.

**EXAMPLE 8** List the elements that are in periods 1, 2, and 3 of the periodic table (refer to Figure 14-10).

*Solution:* From Figure 14-10, we have, reading across each period,

Period 1: H, He
Period 2: Li, Be, B, C, N, O, F, Ne
Period 3: Na, Mg, Al, Si, P, S, Cl, Ar ∎

The fourth row (period 4) introduces some complications. Elements 19, 20, and 31 through 36 all have their highest-energy occupied orbitals in the fourth energy level (n = 4), but elements 21 through 30 have their highest-energy occupied orbitals in the third energy level. (It is actually more complicated. A complete discussion is beyond the scope of this book.) The $d$ orbitals in the (please continue reading on page 248)

---

(Opposite). The top set of numbers (groups), just above the element boxes, goes across from 1 to 18. It is the latest system proposed by the IUPAC (International Union of Pure and Applied Chemistry). The set of numbers below these, going across from IA to VIIIA, is an older system used by the Chemical Abstract Service (CAS) of the American Chemical Society. **The CAS system is used in this book.** Periodic tables in color can also be printed from the NIST and IUPAC web sites. See https://bit.ly/2v8gkZi and https://bit.ly/2O4Zi73. Note that these tables use the lanthanide and actinide terminology.

## 248 ORBITALS IN ATOMS AND THE PERIODIC TABLE

fourth row are 3*d* orbitals. After the 4*s* orbital fills, the 3*d* orbitals begin to fill (see Figure 14-2). The *d* orbitals in periods 5 through 7 are also in one energy level lower than that of the period (row) in which they are shown.

The lanthanoid series (La, Ce through Lu) is in row 6 of the periodic table. For 14 elements in this series (Ce to Lu), the 4*f* orbitals begin to fill after the 6*s* orbital has been filled and one 5*d* electron (for La) has been added as shown in Figure 14-2. **NOTE:** As shown in Figure 14-10, La is sometimes included in the lanthanoid series as recommended by the IUPAC.

The arrangement shown in Figure 14-10 is concise so that the periodic table can fit on a page. See https://bit.ly/36ZM2KK for alternative representations of the periodic table, some showing the "long form" where the lanthanoid and actinoid series are placed in the body of the table.

## 14-5
### CHEMICAL FAMILIES HAVE SIMILAR CHEMICAL PROPERTIES

In addition to the transition metals, the lanthanoid series, and the actinoid series, there are other areas of the periodic table that have special names. Some of them are shown in Figure 14-11. As we have said, these areas are called chemical families because they have similar chemical properties.

1. The **alkali metals** (group IA): Li, Na, K, Rb, Cs, and Fr all have one electron in their highest-energy occupied orbital, which is an *s* orbital.
2. The **alkaline earth metals** (group IIA): Be, Mg, Ca, Sr, Ba, and Ra all have two electrons in their highest-energy occupied orbital, which is an *s* orbital. These electrons have opposite spins.

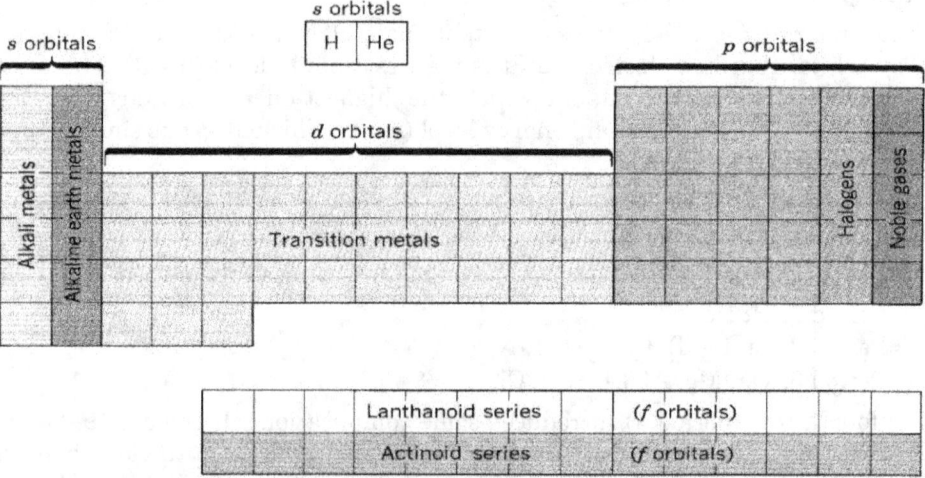

**Figure 14-11** Chemical families and regions in the periodic table up to elements 106. H and He are placed in the center. These two elements are special because they are filling the s orbital in the first energy level. H is usually placed above Li in the alkali metals family. H is normally a gas and a nonmetal, but theory suggests it could become a metal at the extremely high pressure of 500 GPa or about *five million times* normal atmospheric pressure (see Sec 12-2). Helium is a noble gas and is usually placed above Ne in the noble gas family. He has some very unusual properties close to absolute zero that the other elements don't have. One example is that it does not solidify at atmospheric pressure as do all the other elements. It does solidify at 0.95 K at 2.5 MPa or about *25 times* normal atmospheric pressure. This can only be explained using quantum mechanics (see pp 219-223). Also, as shown in Figure 14-10, but not in the above table, La and Ac are sometimes placed with the lanthanoid and actinoid series, respectively, as recommended by the IUPAC.

## 14-5 Chemical Families Have Similar Chemical Properties

3. The **halogens** (group VIIA): F, Cl, Br, I, and At all have seven electrons (two $s$ and five $p$) in their last or highest occupied energy level. We can say that they all have one electron in their last $p_z$ orbital. (As we said in section 14-3, we really don't know which $p$ orbital the lone electron is in.)

4. The **noble gases** (group VIIIA): Helium has two $s$ electrons in its highest occupied energy level, which is the first energy level. Ne, Ar, Kr, Xe, and Ra all have eight electrons (two $s$ and six $p$) in their highest occupied energy level.[2] All their $p$ orbitals are filled.

**EXAMPLE 9** Using the periodic table (Figure 14-10), list all the elements in which (a) the $3d$ orbitals are being filled and (b) the $4d$ orbitals are being filled.

*Solution:* The elements in which the $3d$ orbitals are filling are found in period 4, and those in which the $4d$ orbitals are filling are in period 5:

a. $3d$: Sc, Ti, V, Cr, Mn, Fe, Co, Ni, Cu, Zn
b. $4d$: Y, Zr, Nb, Mo, Tc, Ru, Rh, Pd, Ag, Cd ■

**EXAMPLE 10** Using the periodic table, list all the elements in which the $4f$ orbitals are being filled. **NOTE**: As discussed on page 247, lanthanum (La) is also considered a lanthanoid element as it has many properties similar to those of elements 58-71 even though it does not have any $4f$ electrons. Some periodic tables place lanthanum in the $d$-orbital section. The IUPAC recommends that it be placed as shown in Figure 14-10.

*Solution:* These are the lanthanoid elements (without La) that are filling $4f$ orbitals (there are seven $4f$ orbitals for a total of 14 electrons):

$4f$: Ce, Pr, Nd, Pm, Sm, Eu, Gd, Tb, Dy, Ho, Er, Tm, Yb, Lu ■

The periodic table is very useful to chemists because it allows them to correlate the properties of the elements.

In Figure 14-12, a portion of Figure 14-10 is presented. The nonmetals and semimetals have been copied from Figure 14-10 and shown in greater detail. The darkest-shaded boxes represent the semimetals. Also, hydrogen has been added as it is a nonmetal. All the other elements in Figure 14-10 are metals except for elements 108-118, whose chemistry is not known well enough to place them in their proper category.

**Metals** have a characteristic luster (they shine when polished), conduct electricity very well, and aren't brittle (they can be bent somewhat without breaking). **Nonmetals** usually don't conduct electricity very well (an exception is carbon in the form of graphite and graphene), do not shine like metals, and are generally rather brittle as solids. Many nonmetals are gases. **Semimetals**[3] have properties that are intermediate between those of metals and nonmetals.

Another useful feature of the periodic table is that we can very easily tell the number of electrons in the highest occupied energy level for the "A" group

---

[2] The orbital configuration of element 118, oganesson, is predicted. In fact, all the orbital configurations of the elements after element 108 (Hs) are predicted. They have not been determined by experiment. These elements are all *very* radioactive and have *very* short half-lives. Also, they have only been made in *very* small amounts. All these things make it *very* difficult to do experiments.

[3] Also called **metalloids**.

## 250 ORBITALS IN ATOMS AND THE PERIODIC TABLE

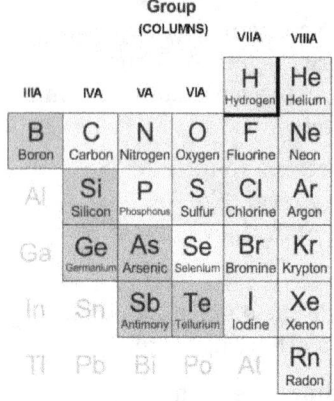

**Figure 14-12 Nonmetals and Semimetals.** This shows the portion of Figure 14-10 that contains the nonmetals and the semimetals. H has been moved over to group VIIA from its position in group IA in Figure 14-10 because it is also a nonmetal. The darkest shaded elements are semimetals (B, Si, Ge, As, Sb, Te). The lighter-shaded elements are nonmental. All the remaining elements in the periodic table shown in Figure 14-10 are metals. Elements 109-118 are metals, but few or no experiments have been done to determine their chemical properties (Element 112, Cn, is an exception and some chemistry has been done with it.) These elements are placed in Figure 14-10 based on their atomic number.

The image is CC BY-SA 4.0 for commercial purposes. The image has been converted to grey scale and cropped. Numerical group numbers are replaced by the CAS system (page 247).
https://en.wikipedia.org/wiki/Nonmetal#/media/File:Nonmetals_in_the_periodic_table.png.

elements (IA and IIA on the left, IIIA through VIIIA on the right). The Roman numeral of the group tells us the number of electrons in the highest occupied energy level. Group VIIIA elements have eight electrons in their highest energy level; the only exception is helium, which has two electrons in its highest energy level which is the first energy level.

Information about the number of electrons in the highest occupied energy level is useful because it is these "outer" electrons that are responsible for chemical bonding. These "outer" electrons are called **valence electrons**, and the highest occupied energy level is usually called the **valence shell**. See the Glossary for a definition of "valence."

**EXAMPLE 11** How many valence electrons are there in the valence shell of the following groups in the periodic table: IA, IVA, VIIA, VIIIA?

*Solution:* IA one valence electron     IVA four valence electrons
VIIA seven valence electrons     VIIIA eight valence electrons[3]

NOTE: The elements in group VIIIA all have filled *p* orbitals in their valence shell (except for helium, which has a filled 1*s* orbital in its valence shell, the first energy level). These filled orbitals, with all electrons paired, are very stable. Helium, neon, and argon don't form compounds that are stable at room temperature and pressure. In 1962, **Neil Bartlett** (1932-2008), then at the University of British Columbia, made the first noble gas compound, xenon hexafluoroplatinate (XePtF$_6$). Since then, many compounds of krypton, xenon, and two of radon (radioactive) have been prepared. These are all stable at room temperature and pressure. See https://bit.ly/393z1Mt.

[3]Except for helium, which has two valence electrons.

---

**Periodic Table Day and Two Interactive Periodic Tables.** The following link has some amusing stories about the periodic table including a discussion of Periodic Table Day, February 7 (www.periodictableday.org). It also contains short discussions about some scientists who were instrumental in developing the modern periodic table. The picture that is supposed to be **Avogadro** is actually of **John Dalton** (see page 140 for the real Avogadro). Two really good interactive periodic tables are at https://bit.ly/3JesFLO and www.inl.gov/periodic-table/. You can buy periodic table mugs, blankets, shirts and who knows what else online.

The number of valence electrons in the transition metals and the lanthanoid and actinoid series cannot always be determined by looking at the periodic table and will not be discussed in this book.

## APPENDIX 14-1
## ORBITAL SPACING IN MULTIELECTRON ATOMS

### THE ENERGY RELATIONSHIP OF ORBITALS IN A ONE-ELECTRON ATOM IS DIFFERENT FROM THAT IN MULTIELECTRON ATOMS

To understand better the arrangement of the orbitals in a many-electron atom, we must first describe the orbital arrangement in a one-electron atom such as the hydrogen atom. The hydrogen atom has only one electron, but the space for the orbitals is still around the nucleus. It is like an apartment house with only one occupied apartment on the first floor.

The orbital arrangement in the hydrogen atom is shown in Figure 14-13. All the orbitals in each energy level have the same energy. In contrast, as we saw Figure 14-1, in a multielectron atom, different types of orbitals (i.e., the $s$, $p$, $d$, and $f$ orbitals) in an energy level have different energies. To explain this difference, we will use an analogy with gravity.

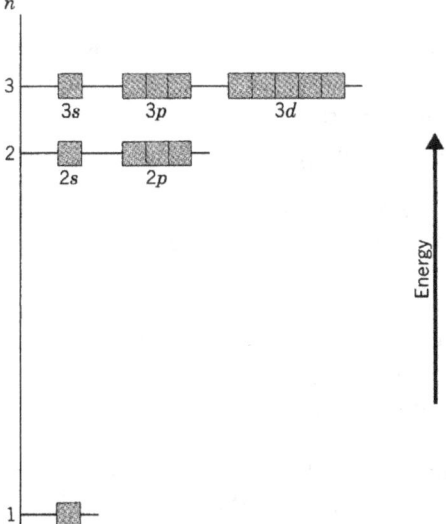

**FIGURE 14-13** The orbital diagram of the hydrogen atom.

## AS YOU LIFT AN OBJECT IN A GRAVITATIONAL FIELD, IT GAINS ENERGY

If you lift an object above the surface of the earth, it gains energy. The higher you lift it, the more energy it has. How can you tell that it has more energy? Just drop it. The higher it has been lifted, the more damage it can do when it hits the ground. So we have our first energy principle: *The farther away from the surface an object is lifted, the more energy it has.*

Now let's shift to the moon. The gravitational field on the moon is one-sixth that on earth. (This means that if you weigh 180 lb on earth, you weigh only 30 lb on the moon.) Now lift the same object we used on earth above the surface of the moon. Again, the higher you lift the object, the more energy it has. But since the moon has only one-sixth the gravity force of the earth, the object lifted 60 ft on the moon will have the same energy that it would have if lifted only 10 ft on the earth. So we have our second energy principle: *The stronger the gravitational field, the more energy an object has when lifted in that gravitational field.*

## HOW DO GRAVITATIONAL FIELDS APPLY TO ELECTRONS IN ATOMS?

In an atom, we have a positive nucleus and negative electrons. The gravitational field of the nucleus is extremely small and can be ignored. But the positive electrical charge of the nucleus is large and attracts the negative electrons. So, even though gravity isn't important in the atom, we can use the *analogy* of gravity on both the earth and the moon to help us understand energy relationships in an atom.

## THE ENERGY OF AN ELECTRON DEPENDS ON THE NUCLEAR CHARGE AND HOW FAR THE ELECTRON IS FROM THE NUCLEUS

Since the positive nucleus attracts the negative electrons, our two energy principles listed above apply. We will restate them in terms of electrical attraction.

1. The farther away an electron is from the nucleus, the higher its energy.
2. The stronger the positive nuclear charge that an electron "feels," the higher the energy of that electron around the nucleus.

We can use these principles to help explain the fact that in a multielectron atom, $2p$ orbitals have a higher energy than $2s$ orbitals. But first we must introduce the concept of shielding.

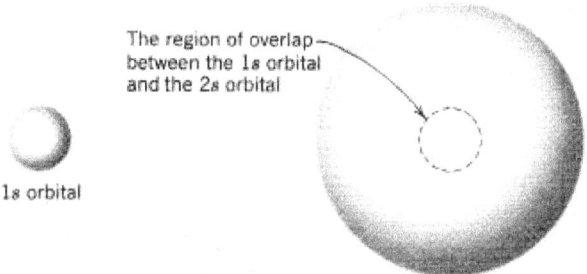

FIGURE 14-14 The overlap between the 1s orbital and the 2s orbital.

## INNER ELECTRONS CAN SHIELD OUTER ELECTRONS FROM THE NUCLEUS

Suppose that you are sitting on a 2s electron in a boron atom. Remember that a boron atom has the following electron configuration: $1s^2 2s^2 2p^1$. The 2s electrons are at a higher energy than the 1s electrons because they are, on the average, farther away from the nucleus than the 1s electrons. Therefore, the 2s electrons don't feel the full positive charge of the boron nucleus. The two 1s electrons get in the way and neutralize some of the positive nuclear charge. They partially *shield* the 2s electrons from the nucleus. The 2s electrons don't "feel" the full +5 nuclear charge of boron; they feel a somewhat smaller nuclear charge. But they *don't* "feel" only a +3 charge as you might expect if the two 1s electrons neutralized two positive nuclear charges. The reason is that the 2s electrons aren't always outside the 1s orbital. They penetrate the 1s orbital and spend some of their time close to the nucleus. There is a region of *overlap* between the 1s and the 2s orbitals. But on the *average* the 2s electrons are farther away from the nucleus than the 1s electrons. This is shown in Figure 14-14. So the shielding by the two 1s electrons is not complete—it is only a partial shielding.

The one 2p electron in boron is also only partially shielded from the nucleus by the two 1s electrons. The 2p electron can also penetrate the 1s orbital and spend some of its time close to the nucleus. This is shown in Figure 14-15.

However, the average distance between the nucleus and the 2p electron is greater than the average distance between the nucleus and the 2s

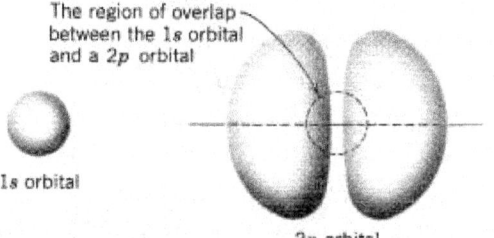

FIGURE 14-15 The overlap between the 1s orbital and a 2p orbital.

## 254 ORBITALS IN ATOMS AND THE PERIODIC TABLE

electrons, and so the penetration of the $2p$ electron is less than the penetration of the $2s$ electrons. This is reasonable because their shapes are so different. Thus the nuclear charge that the $2p$ electron "feels" is less than the charge that the $2s$ electrons "feel." So, according to our energy principles, the energy of the two kinds of electrons is different.

## THERE ARE ALSO ELECTRON-ELECTRON INTERACTIONS THAT AFFECT THE ENERGY OF THE ELECTRONS

There is an additional effect that can change the energy of the electrons. It is the repulsion that the electrons have for each other. Remember that electrons are negative, and negative charges repel each other. The situation is similar to the effect that a planet in our solar system has on the planet nearest to it. Their gravitational attraction for each other actually disturbs their orbits and changes their energy. With electrons it is not an attraction, of course, but a repulsion. The $2s$ and the $2p$ electrons repel one another, and both are repelled by the 1s electrons.

## THE 2P ELECTRON HAS A HIGHER ENERGY THAN THE 2S ELECTRONS

It turns out that when all these effects are taken into account, the $2p$ electron in boron has a slightly higher energy than the $2s$ electrons. We can generalize this and say that all three $2p$ orbitals (which have the same energy because their shapes are the same) have a slightly higher energy than the $2s$ orbital.

The same kinds of arguments also apply to higher energy levels and different kinds of orbitals.

## PROBLEMS
## KEYED PROBLEMS

1. What would a $p$ orbital in the fifth energy level be called?
2. What would a $d$ orbital in the sixth energy level be called?
3. Using Figure 14.2, list all the elements that are formed by the filling of the $4s$ orbital, the $3d$ orbitals, and the $5f$ orbitals (See Fig. 14-10).
4. Using an orbital diagram like Figure 14-1, draw in the electrons for the element phosphorus, P, atomic number 15. Also, write the electron configuration for phosphorus.

5. Draw a horizontal orbital diagram for silicon.

6. Which of the following horizontal orbital diagrams is correct? Explain what is wrong with the incorrect ones.

a. $\frac{\uparrow\downarrow}{1s^2}$ $\frac{\uparrow\downarrow}{2s^2}$ $\frac{\uparrow}{2p_x^1}$ $\frac{\downarrow}{2p_y^1}$ $\frac{\downarrow}{2p_z^1}$

b. $\frac{\uparrow\downarrow}{1s^2}$ $\frac{\uparrow\downarrow}{2s^2}$ $\frac{\uparrow}{2p_x^1}$ $\frac{\uparrow}{2p_y^1}$ $\frac{\uparrow}{2p_z^1}$

c. $\frac{\uparrow\downarrow}{1s^2}$ $\frac{\uparrow\downarrow}{2s^2}$ $\frac{\uparrow}{2p_x^1}$ $\frac{\uparrow\downarrow}{2p_y^2}$ $\frac{}{2p_z}$

d. $\frac{\uparrow\downarrow}{1s^2}$ $\frac{\uparrow\downarrow}{2s^2}$ $\frac{}{2p_x}$ $\frac{\uparrow}{2p_y^1}$ $\frac{\uparrow}{2p_z^1}$

7. List the elements that are in the following groups in the periodic table: groups IIIA, VB, and VIA.

8. List the elements that are in periods 4 and 5 of the periodic table.

9. Using the periodic table, Figure 14-10, list all the elements that are formed by having their $5d$ orbitals filled.

10. Using the periodic table, Figure 14-10, list all the elements that are formed by having their $5f$ orbitals filled.

11. How many valence electrons are there in the elements in the following groups in the periodic table: IIA, VA, and VIA?

## SUPPLEMENTAL PROBLEMS

12. Explain what the following symbols mean.

    a. $3p_x^2$
    b. $4f^7$
    c. $2s^1$
    d. $4d^{10}$

13. Name the elements that correspond to the following electron configurations.

    a. $1s^2 2s^1$
    b. $1s^2 2s^2 2p_x^1 2p_y^1$
    c. $1s^2 2s^2 2p^6 3s^1$
    d. $1s^2 2s^2 2p^6 3s^2 3p_x^2 3p_y^1 3p_z^1$
    e. $1s^2 2s^2 2p^6 3s^2 3p^6 4s^1$
    f. $1s^2 2s^2 2p^6 3s^2 3p^6 4s^2 3d^7$

14. Write a complete horizontal orbital diagram for each of the following elements.

    a. C
    b. O
    c. Mg
    d. P
    e. Ar
    f. V

## 256  ORBITALS IN ATOMS AND THE PERIODIC TABLE

15. List the chemical symbols of all the elements in the following chemical families or regions of the periodic table.

   a. alkali metals
   b. alkaline earth metals
   c. halogens
   d. noble gases
   e. lanthanoid series
   f. actinoid series

16. Using Figures 14-12 and 14-10, list the names and chemical symbols of all the elements indicated as semimetals.

17. Write out the names and chemical symbols of all the transition elements from atomic number 21 to atomic number 30.

18. Write out the names and chemical symbols of all the elements in group IVA of the periodic table.

19. The following are some horizontal orbital diagrams. Which ones are correct and which are incorrect? Explain the incorrect ones.

   a. $1s^2$ ↑↓   $2s^1$ ↑   $2p_x^1$ ↑   $2p_y^1$ ↑   $2p_z^1$ ↑

   b. $1s^2$ ↑↓   $2s^2$ ↑↓   $2p_x^2$ ↑↓   $2p_y$ __   $2p_z$ __

   c. $1s^2$ ↑↓   $2s^2$ ↑↓   $2p_x^3$ ↑↓↑   $2p_y^1$ ↑   $2p_z$ __

   d. $1s^2$ ↑↓   $2s^2$ ↑↓   $2p_x^1$ ↓   $2p_y^1$ ↓   $2p_z^1$ ↓

   e. $1s^2$ ↑↓   $2s^2$ ↑↓   $2p_x$ __   $2p_y^1$ ↑   $2p_z$ __

20. Fill in the chemical symbols for all the elements in Figure 14-11. Then extend the list to element 118, drawing in boxes as needed. Use Fig. 14-10 as a guide. Be careful when filling in La and Ac. They are in a different place in Fig. 14-11 than they are in Fig.14-10. No solution is given for this problem as it is a comparison of two tables.

---

The British physicist **Henry Gwyn Jeffreys Moseley** (1887-1915) was one of the most brilliant scientists of his time. Using x-rays, he established that the **atomic number** is the number of protons in the nucleus of an atom, thus determining which element it was. This discovery led to arranging the periodic table according to the atomic number and the modern periodic table was born. His concept of using x-rays to identify elements is now used by astronomers to identify some elements in outer space and in supernova remnants like the one on the cover of this book (see pp 400-401). Because the earth's atmosphere absorbs x-rays, these observations must be made with earth satellites (https://bit.ly/3gHedQb). Against the advice of friends and colleagues, Moseley enlisted in the British Army during WW I and was killed at the age of 27 years by a Turkish sniper in the Battle of Gallipoli. It was widely thought that he would have won the Nobel Prize in Physics in 1916 had he lived. (Nobel Prizes are only awarded to living persons.) After his death, Great Britain did not allow its best scientists to be put in harm's way. See https://www.famousscientists.org/henry-moseley/

## WAVES AND PARTICLES: LIGHT, ELECTRONS, AND BASEBALLS

On pages 219-223 we discussed the fact that light sometimes behaves as a wave and sometimes as a particle. In the 1600s, the British physicist **Isaac Newton** (1643-1727) thought that light consisted of particles. The Dutch physicist **Christiaan Huygens** (1629-1695) thought that light had wave-like properties. It turns out that both were right, but Newton was more famous, and most scientists believed him. In 1802, the British physician **Thomas Young** (1773-1829) did some experiments that proved that light really did have wave-like properties, and scientists were finally convinced.

Experiments done in the late 1800s, similar in principle to the tanning by ultraviolet light we discussed on page 221, were explained by the German-born American physicist **Albert Einstein** (1879-1955) in 1905 by assuming that light was made out of particles called photons. His 1921 Nobel Prize in Physics was for this research. In 1923, the American physicist and Nobel Prize winner **Arthur Holly Compton** (1892-1962) did experiments using gamma rays that proved that light consists of photons.

**So light sometimes behaves as if it were a wave and sometimes behaves as discrete particles called photons.** This dual nature seems impossible to understand at a "gut" level, but that is the way nature works.

Even stranger, in 1924, the French physicist **Louis de Broglie** (1892-1987) proposed that objects with mass (photons have no mass) might also have wave-like properties when they are moving. The wavelength is inversely proportional (see pages 200-201) to both the mass and the velocity, and de Broglie predicted that these waves could be detected in the behavior of very small particles like electrons. (For a derivation of his equation, see https://bit.ly/31EoVjn.) He won the 1929 Nobel Prize in Physics after this prediction was confirmed. In 1927, the American physicists **Clinton Davisson** (1881-1958) and **Lester Germer** (1896-1971), and the British physicist **George P. Thomson** (1892-1975), found by experiment that **electrons really did have a wave nature**. Davisson and Thomson shared the 1937 Nobel Prize in Physics for this discovery.

What is the wavelength of an electron? That depends on how fast it is moving. In an electron microscope that uses 200,000 volts to accelerate the electrons, the electron's wavelength is $2.51 \times 10^{-3}$ nm (see https://bit.ly/2MUjP9Y for a calculator), about 56 times smaller than the diameter of a carbon atom. See Figure 13-2 for why it is important that the wavelength to be smaller than the object being visualized. Specialized electron microscopes (see pages 285, 322 and 323) are now being used to determine molecular structures.

Since all objects that are moving have an associated wave, a baseball moving at the speed a fast pitcher could throw, about 90 mi/h, would have a wavelength of about $1 \times 10^{-25}$ nm, about one septillion ($10^{24}$) times smaller than an atom. No experiment has been able to detect such a short wavelength! To calculate your wavelength, see https://bit.ly/2ABSIfa. Enter your weight and try a jogging speed of 5 mi/h.

**Very Large-Scale Integrated Circuits (also known as "chips" or "microchips"):** The large grey rectangular chip in the center of the photo is an integrated circuit. As of 2022, the Apple M1 Ultra chip contains 114 billion transistors (these control the flow of electricity). Each transistor in these chips is 10,000 times smaller than the width of a human hair. A Dutch company, ASML, produces the giant machines used to make the latest integrated circuit chips (https://bit.ly/3XzUnui).

By comparison, the first commercial integrated circuit in 1971, the Intel 4004, had 2,300 transistors. Integrated circuits are used in computers, smartphones, tablets, TVs, game boxes, automobiles, digital medical devices like MRIs and CT scanners, and just about any device that uses electronic circuits. Due to these chips, an iPhone is millions of times more powerful than the guidance computer on Apollo 11, the first moon lander in 1969. But that guidance computer had one advantage — it never crashed — a good thing if you are landing on the moon!

**Margaret Hamilton** (b. 1936), an accomplished mathematician and computer science pioneer, is credited with having coined the term *software engineering* while developing the guidance and navigation system for the Apollo spacecraft as head of the Software Engineering Division of the MIT Instrumentation Laboratory. In 2016, President Barack Obama awarded Hamilton the Medal of Freedom See https://bit.ly/2T5tNqw for how her software saved the Apollo 11 moon lander.

The production of these large integrated circuits involves many scientific disciplines. Among them are chemistry, electrical engineering, material science, computer engineering and computer science. It is also necessary to master quantum mechanics to understand the behavior of the semiconductors that are in integrated circuits. A **semiconductor** has an electrical conductivity between that of an insulator and that of most metals. Silicon, a semimetal and a semiconductor, is the main element in most integrated circuits. It is also the 2$^{nd}$ most abundant element in the earth's crust after oxygen. See https://bit.ly/3kHtLqq for more information.

*Photo is in the Public Domain.*

# THE CHEMICAL BOND

In Chapters 13 and 14 we described atomic orbitals as the space around the atom in which electrons exist. In this chapter we will briefly discuss the space around a molecule in which electrons exist. We will also discuss chemical bonds. The discussion will be limited to atoms of the first three periods of the periodic table because the bonding of these atoms is easier to explain. This chapter, then, will try to answer the question "How and why do atoms combine using chemical bonds to form molecules?"

## 15-1
## THE COVALENT BOND OF THE HYDROGEN MOLECULE IS THE SIMPLEST COVALENT BOND

The hydrogen molecule is the simplest molecule, and thus we will use it to introduce the basic concepts of chemical bonding. A hydrogen molecule is formed, as we mentioned in Chapter 3, when two hydrogen atoms combine to form a hydrogen molecule according to the equation

$$H + H \rightarrow H_2$$

We also said that this reaction gives off a lot of heat, so the $H_2$ molecule has less energy than the two separate H atoms. (Remember from Chapter 2 that all substances in nature like to be at the lowest possible energy. This is true of electrons, as we saw in Chapters 13 and 14, and it is also true of atoms and molecules.) Since the energy of the newly formed $H_2$ molecule is less than the energy of the two separate H atoms, the H atoms "want" to combine. They are at a lower energy state as a molecule.

260 THE CHEMICAL BOND

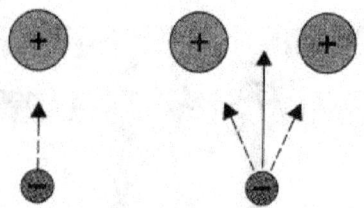

**FIGURE 15-1** The dashed arrows represent the force of attraction of the electron to one nucleus. The full arrow represents the total force of attraction of the electron to two nuclei. The longer the arrow, the greater the force.

Why should the $H_2$ molecule be at a lower energy than the two separate H atoms? In Chapter 13, we said that electrons like to be as close as possible to the nucleus. After all, the nucleus is positive and it attracts the negative electrons.

Now think of the situation whereby an electron is attracted to two nuclei at the same time. It seems reasonable that the two nuclei will have a greater attraction for an electron than will one nucleus. This is shown in Figure 15-1.

Thus the electrons would rather be in the space between the two nuclei of a molecule than surrounding only one nucleus of an atom. The greater attraction of two nuclei causes the electron to be at a lower energy than it would be were it surrounding only one nucleus. The energy that is lost by the electron appears as heat in a chemical reaction.

Each H atom has one electron, so the $H_2$ molecule has two electrons. What about the second electron? Where does it go? The second electron also goes into the space between the two nuclei, also getting its energy lowered. Figure 15-2 will give you some idea of what this looks like.

Notice that the electrons aren't found between the nuclei all of the time, just *most* of the time. This arrangement, in which two electrons are found between two nuclei most of the time, is called a **molecular orbital.**

The reaction $H + H \rightarrow H_2$ gives off heat because the electrons have lost energy in going from the individual atoms to the molecule.

You might wonder, "If two H atoms combine so readily, why not combine three or even more H atoms? Then the electrons would have even more nuclei to go around, and their energy would be lowered even further." To answer this, we have to go back to the discussion of *spin* in Chapter 13. We said there that "Two electrons can be in the same atomic orbital only if their spins are different." This rule also applies to molecular orbitals

H atom's 1s orbital    H atom's 1s orbital    $H_2$ molecule's electron cloud

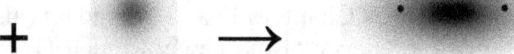

**Figure 15-2** Formation of an $H_2$ molecule. The shading represents the place where the electrons spend about 95% of their time. The electrons spend more time where the shading is the darkest. The H atom's electron cloud is called an *atomic orbital* because the electrons are associated with an atom. In the $H_2$ molecule, the shading is darkest between the nuclei (in this case protons, represented by black dots). The two electrons that spend most of their time between the nuclei make up the chemical bond. The electron cloud of the $H_2$ molecule is called a *molecular orbital* since the electron cloud is associated with a molecule.

See page xxiv for the attributions of these images.

because electrons are in the same region of space in a molecular orbital. **Two electrons can be in the same molecular orbital only if their spins are different.** Now you can see why more H atoms don't add onto the $H_2$ molecule—the additional electrons could *not* join the first two in their molecular orbital because there is already a spin-up and a spin-down electron in that molecular orbital. The additional electrons would have to go into other molecular orbitals that are farther from the nucleus and higher in energy. There is no room for more molecular orbitals at the same energy as the first one. These higher-energy molecular orbitals do *not* lower the energy of the electrons, and so there is no reason for more H atoms to add to $H_2$.[1]

The kind of chemical bond that we have been discussing, whereby the electrons spend most of their time between two nuclei, is called a covalent chemical bond, or simply a **covalent bond.** *In a covalent bond, two electrons are shared between two nuclei.* Before we can discuss the covalent bond in more detail, however, we must discuss the Lewis formulas of atoms and molecules.

## 15-2
## LEWIS FORMULAS OF ATOMS IN THE FIRST AND SECOND PERIODS SHOW THE VALENCE ELECTRONS

It is time-consuming to draw the fuzzy cloud picture of a covalent bond each time we want to discuss the bonding in a molecule. So chemists use a symbolic notation to show the electrons. This notation is called the Lewis electron-dot formula, or simply the **Lewis formula.** It was first developed in 1916 by Gilbert Newton Lewis (1875–1946), a professor of chemistry at the Berkeley campus of the University of California.

In Lewis formulas, each valence electron is represented by a dot. (Remember that valence electrons are electrons in the valence shell, which is the highest occupied energy level.) A covalent bond would be represented by two dots. For a hydrogen atom, we would write H·. The hydrogen atom has one valence electron. For a hydrogen molecule, we would write H:H. The two dots drawn between the atoms represent the covalent bond. They are drawn between the atoms because that is where the electrons spend most of their time.

H·  
↑  
This dot represents the $1s$ electron.

H:H  
↑  
These dots represent the covalent bond formed with two electrons.

---

[1] Actually, the $H_3$ molecule does exist. The additional electron does go into a higher-energy molecular orbital. But because this higher-energy molecular orbital doesn't lower the energy of the electron, the $H_3$ molecule is very unstable. It exists only for a fraction of a second in some chemical reactions.

# THE CHEMICAL BOND

> **Gilbert Newton Lewis** (1875 – 1946) was a professor of chemistry at the University of California, Berkeley. (He is usually referred to by his initials, G. N. Lewis.) He received his Ph.D. from Harvard in 1899 under the direction of Prof. T. W. Richards (see page 169). Lewis is known for the discovery, beginning in 1916, of the covalent bond and his concept of electron pairs; his Lewis dot formulas helped form the basis of modern theories of chemical bonding. Lewis contributed to thermodynamics, photochemistry, isotope separation, and is also known for his 1923 generalized concept of acids and bases, called Lewis acids and bases. This concept uses electron pairs to define acids and bases, not $H^+$ ions as discussed in Chapter 19 and page 187. In 1926, Lewis coined the term "photon" for the smallest unit of light energy, as discussed in Chapter 13. Though he was nominated a record 35 times, Lewis never won the Nobel Prize in Chemistry. See https://en.wikipedia.org/wiki/Gilbert_N._Lewis for more about his life. Read especially the section on "Valence Theory," which is relevant to the material in this chapter.

**WARNING:** In this chapter, don't confuse the period at the end of a sentence with a dot that represents an electron, or a colon in a sentence with an electron pair.

The reaction between two H atoms could be written with dots.

$$H\cdot + H\cdot \rightarrow H{:}H$$

**EXAMPLE 1** Write the Lewis formula for helium, He.

*Solution*: Helium has the electron configuration $1s^2$. So we draw two dots, that is, He: . The helium atom has two valence electrons. ∎

In the Lewis formula for helium in Example 1, we drew the dots on the same side of the symbol He. Why didn't we draw the dots separated like $\cdot$He$\cdot$? The reason is that the two electrons in He are in the same atomic orbital (the $1s$ orbital), and by drawing the Lewis formula as He:, we emphasize this fact. **In a Lewis formula, two dots drawn together always represent two electrons in either an atomic orbital or a molecular orbital. The spins of these electrons are always different.**

Let's draw Lewis formulas of some other atoms. For example, lithium, Li, has three electrons, $1s^2 2s^1$. What is the Lewis formula for Li? It is simply Li$\cdot$ because we have drawn in only the $2s^1$ electron. The reason for this is (please continue on page 263)

## 15-2 Lewis Formulas of Atoms Show the Valence Electrons

that the $1s^2$ electrons do not take part in chemical bonding for Li. They are already in a filled energy level (the first). There is no way that they could interact with the electrons of another atom unless one of them was raised to a higher-energy atomic orbital. This would take so much energy that it just doesn't happen. Only the $2s^1$ electron of Li is available for chemical bonding. So, electrons that are available for chemical bonding are the electrons in the highest-energy occupied orbitals or valence shell. These are called, as we have said, the valence electrons. Li has one valence electron in its valence shell.

We repeat our rule for drawing Lewis formulas. **Only valence electrons are drawn as dots. Electrons in lower, filled energy levels do not take part in chemical bonding and are not indicated by the use of dots.**

**EXAMPLE 2** Draw the Lewis formula for beryllium, Be.

**Solution:** The electron configuration for Be is $1s^2 2s^2$. Only the $2s^2$ electrons, the valence electrons, are available for bonding. The Lewis formula for Be is Be:. ∎

In Table 15-1, we have drawn the Lewis formulas for all the atoms in period 2 of the periodic table. We will also indicate the electron configurations in Table 15-1 so you can compare them with the Lewis formulas.

### TABLE 15-1

LEWIS FORMULAS OF SECOND-PERIOD ELEMENTS

| ELEMENT | ELECTRON CONFIGURATION | LEWIS FORMULA[a] | GROUP NUMBER[b] |
|---------|------------------------|------------------|-----------------|
| Li | $1s^2 2s^1$ | Li· | IA |
| Be | $1s^2 2s^2$ | Be: | IIA |
| B | $1s^2 2s^2 2p_x^1$ | :B· | IIIA |
| C | $1s^2 2s^2 2p_x^1 2p_y^1$ | :C̈· | IVA |
| N | $1s^2 2s^2 2p_x^1 2p_y^1 2p_z^1$ | :N̈· | VA |
| O | $1s^2 2s^2 2p_x^2 2p_y^1 2p_z^1$ | :Ö· | VIA |
| F | $1s^2 2s^2 2p_x^2 2p_y^2 2p_z^1$ | :F̈· | VIIA |
| Ne | $1s^2 2s^2 2p_x^2 2p_y^2 2p_z^2$ | :N̈e: | VIIIA |

[a] The number of dots is the number of electrons of all the orbitals in the second energy level.

[b] Note that the group number is the same as the number of dots (valence electrons) in the Lewis formula.

**EXAMPLE 3** Draw the Lewis formulas for Na and P.

*Solution:* The electron configurations are

$$\text{Na:} \quad 1s^2 2s^2 2p_x^2 2p_y^2 2p_z^2 3s^1$$

$$\text{P:} \quad 1s^2 2s^2 2p_x^2 2p_y^2 2p_z^2 3s^2 3p_x^1 3p_y^1 3p_z^1$$

Since the first and second energy levels are filled and do not take part in chemical bonding for Na and P, only the third-energy-level electrons make up the valence shell; only these electrons will be shown as dots.

$$\text{Na}\cdot \qquad :\dot{\text{P}}\cdot$$

Note the similarity between these Lewis formulas and those of Li and N. Compare Li with Na and compare N with P. Find these elements in the periodic table (Figure 14-10). Elements in the same *group* of the periodic table have the same Lewis formulas. ∎

## 15-3
## COVALENT BONDING FOR SOME MOLECULES IN THE FIRST AND SECOND PERIODS: THE ELEMENTS C, N, O, AND F OBEY THE OCTET RULE

In Section 15-2 we said that the Lewis formula for the molecule $H_2$ was H:H, where the two dots represent the covalent bond formed from two electrons. Helium, on the other hand, cannot form bonds because the atom already has two electrons in the $1s$ orbital and there is no room for any more electrons in that orbital. Helium could form bonds only if one of the electrons went into an orbital in the second energy level. This would take so much energy that it doesn't happen.

Most atoms in the periodic table do form compounds (He, Ne, and Ar are the exceptions), and we would like to be able to draw Lewis formulas of these compounds in order to understand bonding better. In this section we will develop the rules to do just that. Most of the ideas we will develop were first stated in the 1930s by Linus Pauling, John C. Slater, and others.

We have already seen that each hydrogen atom can only have two electrons around it. But atoms in the second period of the periodic table, from Li to Ne, have electrons in the second energy level. There is enough room for up to eight electrons around each of these atoms.

Since electrons lower their energy by being around more than one nucleus, it seems reasonable that an atom would form as many bonds as possible. Atoms in the second period can form *up to four* covalent bonds (each with two electrons) with other atoms. This gives each atom eight

## 15-3  Covalent Bonding for Some Molecules in the First and Second Periods

> **Linus Carl Pauling** (1901-1994), one of the most influential chemists of the 20th century, was a professor at the California Institute of Technology. His book, *The Nature of the Chemical Bond*, first published in 1939, had a profound influence on a generation of chemists. It was based on the Baker Lectures he gave during 1937-1938 at Cornell University. Pauling received the 1954 Nobel Prize in Chemistry "for his research into the nature of the chemical bond and its application to the elucidation of the structure of complex substances." He also received the Nobel Peace Prize in 1962 "for his campaign against nuclear weapons testing." His *alma mater*, Oregon State University, founded The Linus Pauling Institute in his honor. Its "mission is to promote optimal health through cutting-edge research on micronutrients and dietary supplements and trusted public outreach." His concept of electronegativity will be discussed in Chapter 17. A wonderful tribute to his life is at https://bit.ly/2WR2ENN. More about his life and times can be found at https://bit.ly/2UKtKmN.

electrons around it, the maximum number for which there is room. Some of these atoms form fewer than four bonds. We shall see why.

The first element in the second period of the periodic table is lithium. Li has one valence electron, and its Lewis formula is Li•. Lithium can react with a hydrogen atom to form lithium hydride, LiH, according to the equation

$$\text{Li}\cdot + \text{H}\cdot \rightarrow \text{Li:H}$$

Lithium doesn't form more than one bond. Apparently, the positive charge on the Li nucleus is strong enough to attract only one additional electron.

The second element in the second period is beryllium. Beryllium has two valence electrons, and its Lewis diagram is Be: . You might think that Be wouldn't form a compound with H atoms because both of the valence electrons in Be are paired in the 2s orbital. But Be actually does react with two H atoms to form beryllium hydride, BeH$_2$:

$$\text{Be:} + 2\text{H}\cdot \rightarrow \text{H:Be:H}$$

How do we explain this? Well, the valence electrons in the beryllium's 2s orbital can separate while Be is reacting, giving two unpaired electrons. The Lewis formula would then be •Be•. If you want to, you can think of one of these (please continue on page 266) electrons as going into one of the vacant 2p orbitals. You can see that there is room for the electron in unoccupied p orbitals of the second energy level. These vacant p orbitals are only slightly higher in energy than the 2s orbital. Notice that now the electrons in the Be atom are as far apart from one another as possible, on opposite sides of the atom. Since electrons repel one another, it is only natural for them to try to get as far apart from one another as possible. Thus the BeH$_2$ molecule is a **linear** molecule, which means that all the atoms are in a straight line.

Why do the electrons in Be separate in this way? The reason must have to do with energy. If the electrons separate, then Be can share two electrons from two H atoms. And since the 2p orbitals are only a little higher in energy than the 2s orbital, the overall process lowers the energy so much that it pays for the electrons to separate. (Electrons may be "lazy," but they "know" a good situation when they "see" one.)

We can now write the reaction between Be and two H atoms as

$$\cdot \text{Be} \cdot + 2\text{H} \cdot \rightarrow \text{H} : \text{Be} : \text{H}$$

Beryllium forms only two bonds. Apparently, the positive charge on the nucleus of the Be atom is strong enough to attract only two additional electrons.

The third element in the second period is boron. The Lewis formula for boron is :B· because boron has three valence electrons. You might think that boron would combine with one H atom to form BH according to the following equation:

$$:\text{B} \cdot + \text{H} \cdot \rightarrow :\text{B} : \text{H}$$

Alternatively, using the ideas just developed above for BeH$_2$, you might think that boron would combine with *three* H atoms to form BH$_3$:

$$\dot{\text{B}} + 3\text{H} \cdot \rightarrow \begin{matrix} & \text{H} & \\ & \cdot\cdot & \\ \cdot \dot{\text{B}} \cdot & & \\ \text{H} & & \text{H} \end{matrix}$$

This is actually what happens, although the BH$_3$ molecule, called borane, is unstable and exists only for short times during certain chemical reactions. The explanation for the formation of BH$_3$ is similar to the case of the BeH$_2$ molecule discussed above. The valence electrons in boron's 2s orbital (the paired dots) can separate while the boron atom is reacting, giving three unpaired electrons. The Lewis formula would then be $\dot{\text{B}}$. Again, you can think of a 2s electron as going into one of the vacant 2p orbitals. Notice that now the electrons in boron are as far apart from one another as possible, at the corners of an equilateral triangle.

## 15-3 Covalent Bonding for Some Molecules in the First and Second Periods

Since boron now has three unpaired electrons, it is easy to see why three H atoms can bond to the boron atom, forming three covalent bonds:

$$\dot{\underset{\cdot}{B}} + 3H\cdot \rightarrow \begin{array}{c} H \\ \cdot\ddot{B}\cdot \\ H \quad H \end{array}$$

The $BH_3$ molecule is a **triangular planar** molecule, which means that all the atoms are in the same plane and that the H atoms are at the corners of an equilateral triangle. A plane is a two-dimensional surface, such as a table top.

The boron atom in $BH_3$ has six electrons around it. It must be that the positive charge on the boron nucleus is strong enough to attract only three extra electrons. Boron *can* form compounds in which it has eight valence electrons. In these compounds the boron atom gets two electrons from another atom or molecule. We will not discuss these compounds in this book.

Carbon is the fourth element in the second period. Carbon has four electrons in its valence shell, and its Lewis formula is $:\dot{C}\cdot$. You might think that carbon would form a species like $CH_2$ according to the equation[2]

$$:\dot{C}\cdot + 2H\cdot \rightarrow :\ddot{C}:H \atop H$$

Carbon actually forms stable molecules like methane, $CH_4$. The explanation for this is similar to that mentioned for Be and B. The paired electrons in the $2s$ orbital of carbon split up to give

$$\cdot\dot{\underset{\cdot}{C}}\cdot$$

Now four H atoms can easily combine with a C atom to give four covalent bonds:

$$\cdot\dot{\underset{\cdot}{C}}\cdot + 4H\cdot \rightarrow \begin{array}{c} H \\ H:\ddot{C}:H \\ H \end{array}$$

Carbon now has eight electrons around it, the maximum number for which there is room in the second energy level.

Actually, the four separated electrons in carbon are not planar, that is, in the same plane. They are as far apart as they can be from one another in three dimensions. The geometric arrangement of $\cdot\dot{\underset{\cdot}{C}}\cdot$ and $CH_4$ is a **regular**

---

[2] The species $CH_2$, called methylene, does exist. It is *very* reactive and exists only for fractions of a second during certain chemical reactions. This type of reactive species is called a **radical.**

268   THE CHEMICAL BOND

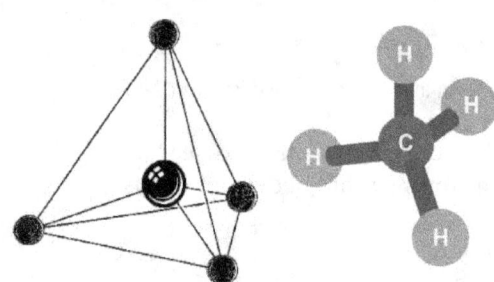

**Figure 15-3** Two diagrams of the methane ($CH_4$) molecule. The geometric structure shown at the far left is a **regular tetrahedron**. All the edges and faces of a regular tetrahedron are equal. The figure at the near left is a ball and stick model of methane. **NOTE:** The sticks represent chemical bonds and look somewhat like the one shown in Figure 15-2. The angle between the H atoms is 109.5°, which is called the **tetrahedral angle**, the farthest apart in space that four atoms can get from one another at a given distance from a central atom. This molecule is so symmetric that an H atom in one position is the same as an H atom in any other position. The molecule could be rotated so that if any H atom is at the top, the molecule would look the same. It you have access to a molecular model kit, try building it and see for yourself. See https://bit.ly/39VRcJA for a 3D model.*

**tetrahedron,** as shown in Figure 15-3. The corners of a regular tetrahedron are the farthest apart in space that four points can get from one another at a given distance from a central point. Since electrons repel one another, it is reasonable that they arrange themselves as far from one another as they can.

At this point, we should mention that equations like

$$\cdot \ddot{C} \cdot \ + \ 4H\cdot \ \rightarrow \ H \overset{H}{\underset{\ddot{H}}{\ddot{:}\ddot{C}:}} H$$

are *not* the chemical reactions that chemists use to make $CH_4$, $BH_3$, and so on. These equations are written here to help you see how the electrons act during the formation of a chemical bond. These equations are more like "thought" equations than real ones.

Nitrogen is the fifth element in the second period. Nitrogen has five electrons in its valence shell, and its Lewis formula is $:\dot{N}\cdot$. Nitrogen forms ammonia, $NH_3$, by combining with three H atoms:

$$:\dot{N}\cdot \ + \ 3H\cdot \ \rightarrow \ :\overset{H}{\underset{\ddot{H}}{\ddot{N}:}}H$$

There are eight electrons around the N atom in $NH_3$. Six electrons are in three covalent bonds, and two electrons are in a **nonbonded pair:**

$$\text{Nonbonded pair} \rightarrow \ :\overset{H}{\underset{\ddot{H}}{\ddot{N}:}}H$$

---

*The Wikipedia articles for all the molecules discussed in this chapter have 3D models.

15-3 Covalent Bonding for Molecules in the First and Second Periods    269

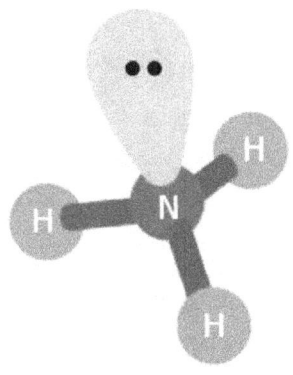

**Figure 15-4** The diagram at the left is a ball and stick model of the ammonia (NH₃) molecule. The nonbonded pair of electrons (the two dots – the blob they are in shows that they spead out) is at the top of the pyramid. **NOTE:** The blobs shown in this and later figures in this chapter really should be thought of as discussed at the bottom of page 230. The geometry of the figure formed by just the N and H atoms is called a **trigonal pyramid**. The angle between the H atoms is 107.3°. This angle is smaller than the tetrahedral angle because the nonbonded pair of electrons pushes away the H atoms more than the bonded pair in CH₄ does. See https://bit.ly/3OMAnzH for a 3D model.

This nonbonded pair does *not* split up, unlike the electron pairs in beryllium, boron, and carbon. There is room for only eight electrons in the second energy level of the nitrogen atom. The only place an electron could go if the pair split up is into a vacant orbital in the third energy level. This would require so much energy that it doesn't happen.

The structure of $NH_3$ is a **trigonal pyramid** (a pyramid whose faces are triangles), as shown in Figure 15-4. This structure is similar to a tetrahedron if we include the nonbonded pair of electrons at the top of the pyramid. As usual, the electrons are as far away from one another as possible. Actually, the nonbonded electron pair "wants" to be a little farther away from the rest, and it pushes the three H atoms down a bit. This is because the nonbonded electrons are only influenced by one positive nucleus, unlike bonded electrons, which are influenced by two positive nuclei. Thus, nonbonded electrons repel bonded electrons to a greater extent than bonded electrons repel other bonded electrons. That's why the angle between the H atoms in $NH_3$ is 107.3° instead of the tetrahedral angle of 109.5° in $CH_4$.

Oxygen has six valence electrons, and its Lewis formula is $:\ddot{O}\cdot$. Oxygen can combine with two H atoms to form water:

$$:\ddot{O}\cdot + 2H\cdot \rightarrow :\ddot{O}:H$$
$$\phantom{:\ddot{O}\cdot + 2H\cdot \rightarrow :\ddot{O}:}\ddot{H}$$

In water, oxygen has eight electrons around it, four in two nonbonded pairs and four in two covalent bonds with two H atoms. The structure of the water molecule is shown in Figure 15-5.

Fluorine, the next-to-last element in period 2, has seven valence electrons, and its Lewis formula is $:\ddot{F}\cdot$. Fluorine can combine with one H atom to form hydrogen fluoride, HF:

$$H\cdot + :\ddot{F}\cdot \rightarrow H:\ddot{F}:$$

## 270 THE CHEMICAL BOND

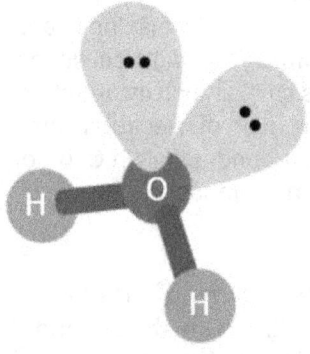

**Figure 15-5** The diagram at the left is a ball and stick model of the water (H₂O) molecule. The two pairs of nonbonded electrons (the two sets of dots – the blobs they are in show that they spread out) point to the corners of a tetrahedron. The angle between the H atoms is 105°. This is smaller than the angle between the H atoms in the NH₃ molecule because the two pairs of nonbonded electrons push away the H atoms more than the one pair of nonbonded electrons in NH₃. The geometry of the figure formed by just the H atoms and the O atom is called a **bent** or **angular** figure. See https://bit.ly/3OE0itd for a 3D model.

In HF the fluorine atom has eight electrons around it, six in three nonbonded pairs and two in one covalent bond with the H atom. HF is a **linear** molecule.

Neon, the last element in period 2, has eight valence electrons, and its Lewis formula is :Ṇe:. Since Ne already has eight electrons in the second energy level, there is no room for any more. Neon has never been observed to form molecules.

If we study the previous examples, we can find a rule for determining the number of covalent bonds that an atom in the second period of the periodic table can form. The rule is: **The number of covalent bonds an atom will form is equal to the maximum number of half-filled orbitals the atom can have.** Let's study Table 15-2 to see whether our rule is consistent with our previous discussion. As you can see from Table 15-2, this rule *is* consistent with everything that we said in our previous discussion.

Notice in Table 15-2 that the period-2 elements, C, N, O, and F have eight electrons around them when they form compounds. Their valence shell is now similar to that of neon, a noble gas that is unreactive chemically. The valence shell configuration of a noble gas is a low-energy configuration. This is one of the reasons why certain period-2 elements attain this configuration when they form compounds. It is so common for these atoms to have eight electrons around them in compounds that this tendency is referred to as the **octet rule** ("oct" means eight).

Students sometimes ask whether it matters where the dots go in a Lewis formula. Are :Ċ·, ·Ċ:, and Ċ· the same? Yes, all these are equivalent. So what should you draw? The best idea is to try to make your drawings like those of either the book or your instructor. But note that :Ċ· is not the same as ·Ċ·. You should keep the correct number of lone electrons and paired electrons in your Lewis formulas.

## TABLE 15-2
THE NUMBER OF COVALENT BONDS IN SECOND-PERIOD ELEMENTS

| ATOM | NUMBER OF VALENCE ELECTRONS | LEWIS FORMULA | REARRANGED LEWIS FORMULA | MAXIMUM NUMBER OF HALF-FILLED ORBITALS[a] | NUMBER OF COVALENT BONDS | TYPICAL COMPOUND WITH H ATOMS | GEOMETRY |
|---|---|---|---|---|---|---|---|
| Li | 1 | Li· | Li· | 1 | 1 | Li:H | linear |
| Be | 2 | Be: | ·Be· | 2 | 2 | H:Be:H | linear |
| B | 3 | :B· | ·Ḃ· | 3 | 3 | H<br>·Ḃ·<br>H   H | triangular planar |
| C | 4 | ·C̈· | ·C̈· | 4 | 4 | H<br>H:C̈:H<br>H | tetrahedral |
| N | 5 | ·N̈: | :N̈· | 3 | 3 | H<br>:N̈:H<br>H | trigonal pyramid |
| O | 6 | ·Ö: | :Ö· | 2 | 2 | :Ö:H<br>H | angular or bent |
| F | 7 | :F̈· | :F̈: | 1 | 1 | H:F̈: | linear |
| Ne | 8 | :N̈e: | :N̈e: | 0 | 0 | — | — |

[a] The maximum number of half-filled orbitals is the same as the number of lone dots in the column "rearranged Lewis formula."

## 15-4
## COVALENT BONDING IN OTHER MOLECULES OF THE SECOND-PERIOD ELEMENTS: THE ELEMENTS C, N, O, AND F OBEY THE OCTET RULE

Elements in the second period of the periodic table can form many different molecules. We will look at a few simple ones as examples of covalent bonding.

Two fluorine atoms can combine to form a fluorine molecule, $F_2$, as follows:

$$:\!\ddot{F}\!\cdot \;+\; :\!\ddot{F}\!\cdot \;\longrightarrow\; :\!\ddot{F}\!:\!\ddot{F}\!:$$

The unpaired electrons in each F atom join to form a covalent bond. This is shown in the following diagram, where we have drawn arrows to indicate the electrons that combine to form the covalent bond. One covalent bond connecting two atoms is called a **single bond**:

$$:\!\ddot{F}\!\cdot \;\rightarrow\; \leftarrow\; \cdot\ddot{F}\!:$$

Notice that one F atom has been turned around so that the two unpaired electrons face each other. This was done so you could easily see how they combine.

Two nitrogen atoms can combine to form a nitrogen molecule, $N_2$, as follows:

$$:\!\dot{N}\!\cdot \;+\; :\!\dot{N}\!\cdot \;\longrightarrow\; :\!N\!:\!:\!:\!N\!:$$

Here we have *three* covalent bonds connecting the two N atoms. Three covalent bonds connecting two atoms is called a **triple bond**:

$$:\!\dot{N}\!\cdot \;\rightrightarrows\; \leftleftarrows\; \cdot\dot{N}\!:$$

Again, one N atom has been turned around so that the unpaired electrons face each other.

Carbon can combine with two oxygen atoms to form carbon dioxide, $CO_2$, as follows:

$$\cdot\dot{C}\!\cdot \;+\; 2\;:\!\ddot{O}\!\cdot \;\longrightarrow\; :\!\ddot{O}\!:\!:\!C\!:\!:\!\ddot{O}\!:$$

There are two covalent bonds to each oxygen atom in $CO_2$. Two covalent bonds connecting two atoms is called a **double bond**:

$$:\!\ddot{O}\!\cdot \;\rightrightarrows\; \leftleftarrows\; \cdot\dot{C}\!\cdot \;\rightrightarrows\; \leftleftarrows\; \cdot\ddot{O}\!:$$

## 15-4 Covalent Bonding in Other Molecules of the Second-Period Elements

Two oxygen atoms can combine to form an oxygen molecule, $O_2$. You might think that you could draw its Lewis formula as $:\!\ddot{O}::\!\ddot{O}\!:$, but this is really wrong. Experiments show that the oxygen molecule has *two unpaired electrons*. So you might try drawing the Lewis formula as $:\!\ddot{O}\!:\!\ddot{O}\!:$. Now there are two unpaired electrons, seven electrons around each oxygen atom, and a single bond connecting the two atoms. But experiments and theory suggest that the bond between the oxygen atoms is a double bond. $O_2$ is one of the few molecules that cannot be represented by a Lewis formula. Lewis formulas work *most* of the time, but *not* all the time.

We have not discussed the molecules $Li_2$, $B_2$, and $C_2$. These molecules are unstable and exist only as gases at very high temperatures. In $Li_2$ there is a single bond joining the two Li atoms, as you might expect. The bonding in $B_2$ and $C_2$ cannot be described by Lewis formulas. But in case you are curious, in $B_2$ there is a single bond, and in $C_2$ there is a double bond. If you tried to use Lewis formulas and followed the rules, you would probably end up with a triple bond for $B_2$ and a quadruple bond for $C_2$. The predictions from the Lewis formulas are wrong.

The molecule $Be_2$ has never been found to exist. As you study more chemistry, you may learn a theory of bonding called **molecular orbital theory.** This theory is more involved than the one we have been discussing, but it does explain the bonding in molecules like $O_2$, $B_2$, and $C_2$ as well as explaining why $Be_2$ does not exist.

We can summarize the rules for forming covalent bonds for period-2 elements.

1. Each covalent bond contains two electrons whose spins are different.
2. The number of covalent bonds an atom will form is equal to the maximum number of half-filled orbitals that the atom can have.
3. The maximum number of electrons that a period-2 atom can have in its second energy level is eight. Period-2 atoms will form covalent bonds with other atoms in order to get eight electrons in their second energy level. Exceptions: Li and Be can have only two and four electrons, respectively, in their valence shell. Boron can form molecules with both six electrons and eight electrons in its valence shell.
4. The statement that many atoms tend to have eight valence electrons in their valence shell when they form compounds is called the **octet rule.**

**EXAMPLE 4**  Draw Lewis formulas for the following molecules: $NF_3$, $CF_4$, and $OF_2$.

*Solution:* We will write the "thought" equations to make things clearer:

274 THE CHEMICAL BOND

$$NF_3 \quad :\!\overset{..}{N}\!\cdot\; +\; 3\; :\!\overset{..}{F}\!\cdot\; \rightarrow\; :\!\overset{\overset{..}{:\!\ddot F\!:}}{\underset{:\!\ddot F\!:}{N\!:\!\ddot F\!:}}$$

$$CF_4 \quad \cdot\!\overset{.}{\underset{.}{C}}\!\cdot\; +\; 4\; :\!\overset{..}{F}\!\cdot\; \rightarrow\; :\!\overset{\overset{..}{:\!\ddot F\!:}}{\underset{:\!\ddot F\!:}{\ddot F\!:\!C\!:\!\ddot F\!:}}$$

$$OF_2 \quad :\!\overset{..}{\underset{.}{O}}\!\cdot\; +\; 2\; :\!\overset{..}{F}\!\cdot\; \rightarrow\; :\!\overset{}{\underset{:\!\ddot F\!:}{\ddot O\!:\!\ddot F\!:}} \quad \blacksquare$$

## 15-5
### LEWIS FORMULAS USING LINES INSTEAD OF DOTS IS COMMON

Chemists find that drawing all the dots in Lewis formulas is rather tedious. So sometimes they draw lines in for the dots according to the following rules.

1. A covalent bond consisting of two electrons is replaced by a line.
2. The dots representing nonbonded electrons are sometimes left out.

For example, let's draw all the molecules discussed in the previous section with lines instead of dots (see Table 15-3).

**EXAMPLE 5** Draw the Lewis formulas, as was done in Table 15-3, for the following molecules: $NF_3$, $CF_4$, and $OF_2$.

*Solution:*

| LEWIS FORMULA USING LINES | LEWIS FORMULA USING DOTS AND LINES |
|---|---|
| $\begin{array}{c} F \\ \vert \\ N\!-\!F \\ \vert \\ F \end{array}$ | $\begin{array}{c} :\!\ddot F\!: \\ \vert \\ :\!\ddot N\!-\!\ddot F\!: \\ \vert \\ :\!\ddot F\!: \end{array}$ |
| $\begin{array}{c} F \\ \vert \\ F\!-\!C\!-\!F \\ \vert \\ F \end{array}$ | $\begin{array}{c} :\!\ddot F\!: \\ \vert \\ :\!\ddot F\!-\!C\!-\!\ddot F\!: \\ \vert \\ :\!\ddot F\!: \end{array}$ |
| $\begin{array}{c} O\!-\!F \\ \vert \\ F \end{array}$ | $\begin{array}{c} :\!\ddot O\!-\!\ddot F\!: \\ \vert \\ :\!\ddot F\!: \end{array}$ |

■

## TABLE 15-3

**LEWIS FORMULAS OF SOME MOLECULES USING BOTH LINES AND DOTS**

| MOLECULE | LEWIS FORMULA USING LINES | LEWIS FORMULA USING DOTS AND LINES[a] |
|---|---|---|
| $H_2$ | H—H | H—H |
| $BH_3$ | H\B/H with H on top | H\B/H with H on top |
| $CH_4$ | H—C—H with H above and below | H—C—H with H above and below |
| $NH_3$ | H—N—H with H above | H—:N—H with H above |
| $H_2O$ | O—H with H below | :Ö—H with H below |
| HF | H—F | H—F̈: |
| $CO_2$ | O=C=O | :Ö=C=Ö: |
| $F_2$ | F—F | :F̈—F̈: |
| $N_2$ | N≡N | :N≡N: |

[a] Bonding electrons are drawn with lines. Nonbonding electrons are drawn as dots.

## 15-6
## COVALENT BONDING IN THE THIRD-PERIOD ELEMENTS: THE ELEMENTS P, S AND Cl MAY OR MAY NOT OBEY THE OCTET RULE

The third period of the periodic table consists of the elements from Na to Ar. Much, but not all, of their bonding is similar to the second-period elements.

The group IA and IIA elements Na and Mg, respectively, generally form ionic compounds, as will be discussed in Section 15-7.

The group IIIA element aluminum can form compounds, some with ionic bonds and some with covalent bonds.

## 276  THE CHEMICAL BOND

The group IVA through group VIIA elements Si to Cl form covalent bonds. Chlorine and sulfur can also form ionic compounds. These will be discussed in Section 15-7. Examples of the kinds of compounds with covalent bonds these elements can form are listed below. Notice the similarity with the corresponding period-2 elements. In many compounds, these period-3 elements obey the octet rule:

$$\cdot \ddot{S}i \cdot + 4\, H \cdot \rightarrow H\!:\!\overset{H}{\underset{H}{\ddot{S}i}}\!:\!H \quad \text{(silane)}$$

$$:\!\dot{P}\cdot + 3\, H \cdot \rightarrow :\!\overset{H}{\underset{H}{\ddot{P}}}\!:\!H \quad \text{(phosphine)}$$

$$:\!\ddot{S}\cdot + 2\, H \cdot \rightarrow :\!\overset{}{\underset{H}{\ddot{S}}}\!:\!H \quad \text{(hydrogen sulfide)}$$

$$:\!\ddot{C}l\cdot + H \cdot \rightarrow H\!:\!\ddot{C}l\!: \quad \text{(hydrogen chloride)}$$

However, there are some differences between the bonding of the period-2 elements and that of the period-3 elements. The period-3 elements in the A groups of the periodic table have up to eight electrons (two $s$ and six $p$ electrons) in their valence shell ($n = 3$). But some compounds formed by these elements cannot be explained using just eight electrons and four bonds. In other words, the octet rule does not apply to these compounds.

For example, phosphorus can form a compound with chlorine; this compound is called phosphorus pentachloride, $PCl_5$. In this molecule, the phosphorus atom has *five* covalent bonds and *ten* electrons in its valence shell. A diagram of this molecule is shown in Figure 15-6.

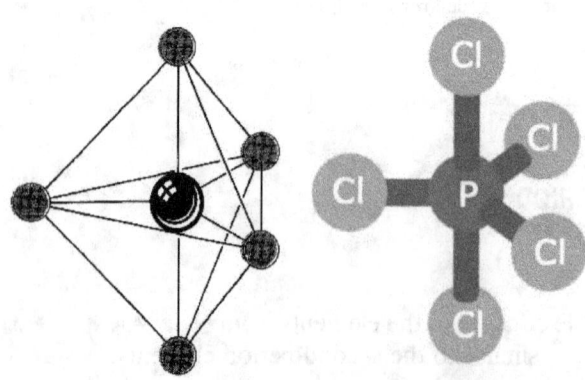

**Figure 15-6** Two diagrams of the phosphorus pentachloride (PCl₅) molecule. The geometric structure shown at the far left is a **trigonal bipyramid**. The structure can be thought of as two trigonal pyramids "back to back." The figure at the near left is a ball and stick model of PCl₅. The angle between the three Cl atoms in the middle of the figure is 120°. The angle between the top and bottom Cl atoms and three middle Cl atoms is 90°. See https://bit.ly/3NgHSxu for a 3D model.

## 15-6 Covalent Bonding in the Third-Period Elements

How can the phosphorus atom have 5 covalent bonds and ten electrons around it? Let's look at the Lewis formula of phosphorus: . If the :P̈· pair of valence electrons separated, five covalent bonds could easily form with five Cl atoms, and the results would be the molecule phosphorus pentachloride, $PCl_5$.

$$:P\cdot\cdot + 5 \; :\ddot{\underset{\cdot\cdot}{Cl}}\cdot \rightarrow \begin{array}{c} :\ddot{\underset{\cdot\cdot}{Cl}}:\; :\ddot{\underset{\cdot\cdot}{Cl}}: \\ :\ddot{\underset{\cdot\cdot}{Cl}}:P:\ddot{\underset{\cdot\cdot}{Cl}}: \\ :\ddot{\underset{\cdot\cdot}{Cl}}: \end{array}$$

Phosphorous can also form an ion with six chlorine atoms around it, called the phosphorus hexachloride ion, $PCl_6^-$. There is only room for six atoms around the phosphorous atom at the given covalent bond distance. The Lewis structure for this ion is at https://bit.ly/3dVo4kT.

In nitrogen :N̈· as we discussed before, this separation doesn't happen because only 8 electrons can be around second-period atoms. But third-period atoms can have up to 18 electrons around them.

Sulfur can form a compound with 6 fluorine atoms; the result is the molecule sulfur hexafluoride, $SF_6$. The Lewis formula of sulfur is: :S̈· . If *both* pairs of valence electrons separate, we have :Ṡ: and sulfur can form covalent bonds with six atoms. The structure of $SF_6$ is shown in Figure 15-7. At the covalent bond distances possible around S, there is only room for 6 atoms.

Chlorine can form a compound with 5 fluorine atoms; this compound is called chlorine pentafluoride, $ClF_5$. The Lewis formula of chlorine is: :C̈l· If two of the paired valence electrons separate, we get :Ċl· and the Cl atom could form 5 covalent bonds with five F atoms. The structure of $ClF_5$ is shown in Figure 15-8. As far as we know, Cl does not form a stable molecule with more than 5 atoms around it. At the covalent bond distances possible around Cl, there is not enough room for more than 5 atoms.

Because electrons are smaller than atoms, Cl can have 7 covalent bonds around it, some of which must be double bonds, as shown in the Lewis formula of the perchlorate ion, $ClO_4^-$, at https://bit.ly/3dJCF2B.

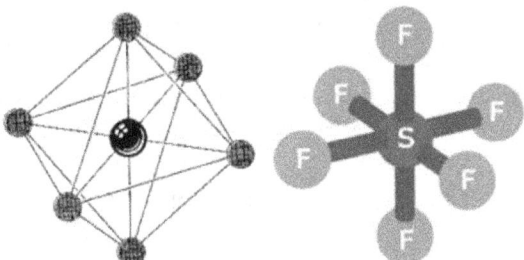

**Figure 15-7** Two diagrams of the sulfur hexafluoride ($SF_6$) molecule. The geometric structure on the far left is a **regular octahedron**. This structure can be thought of as two square pyramids "back-to-back" (see Figure 15-8). The figure at the near left is a ball and stick model of $SF_6$. The angle between the four F atoms in the center is 90°. The angle between the top and bottom F atoms and the four F atoms in the center is also 90°. This is the farthest 6 atoms can get from each other at a given distance from a central atom. This molecule is so symmetric that an F in one position is equivalent to an F in any other position. The molecule could be rotated so that if any F atom is at the top, the molecule would look the same. If you have access to a molecular model kit, try building it and see for yourself. See https://bit.ly/3NgHSxu for a 3D model.

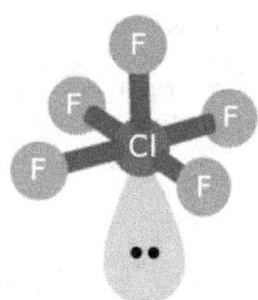

**Figure 15-8** A ball and stick model of the chlorine pentafluoride (ClF$_5$) molecule. This molecule would normally be a **square pyramid**, where the angle between the four F atoms at the bottom would be 90°, and the angle between the top F atom and any of the 4 bottom F atoms would also be 90°. But the nonbonded pair of electrons in the blob repel the four F atoms so the angle should be less than 90°. The angle is 86.5° as given by NIST at https://bit.ly/3yj0V4N. See https://bit.ly/3OEDGZw for a 3D model.

In elements higher than the 3$^{rd}$ period, we do find molecules with more than six atoms bonded to the central atom. In these higher-period atoms, there is much more room for such bonding. An example is iodine heptafluoride, IF$_7$, in which the iodine atom is surrounded by seven F atoms. All seven valence electrons of the iodine atom have separated to allow seven covalent bonds to be formed. The Lewis formula and 3D structure of IF$_7$ can be found at https://bit.ly/32W0DBK.

Since there are only 8 electrons in the 3$s$ and 3$p$ orbitals, you might think that the additional electrons in the above molecules occupy 3$d$ orbitals. Quantum mechanical computer calculations indicate that there is little participation of 3$d$ orbitals in the bonding of the 3$^{rd}$ period elements mentioned in this section (P, S, Cl). A discussion of what the bonding might look like is beyond the scope of this book, and a simple explanation does not seem possible. But if you are interested, see https://bit.ly/3LJKQe2. **NOTE:** Some general chemistry texts use $d$ orbitals to illustrate the bonding in the molecules in this section (https://bit.ly/3EFQdKg).

## 15-7
## THE STRONG ELECTRICAL ATTRACTION BETWEEN OPPOSITELY CHARGED IONS FORMS THE IONIC BOND

Elements in groups IA and IIA (the alkali metals and the alkaline earth metals) of the periodic table generally form **ionic bonds** with hydrogen and with elements in groups VIA and VIIA. Note that elements in groups VIA and VIIA can form *both* ionic and covalent compounds, depending on what element they are reacting with.

**Ionic bonds are bonds in which one or more electrons actually transfer from one atom to another. There is no sharing of electrons as there is in covalent bonds.**

For example, in sodium chloride, NaCl, which is common table salt, the bonding takes place this way:

$$\text{Na}\cdot \; + \; :\!\ddot{\text{Cl}}\!\cdot \; \rightarrow \; \text{Na}^+ \; + \; :\!\ddot{\text{Cl}}\!:^-$$

The sodium atom has lost its 3$s^1$ electron, and the chlorine atom has gained that electron; thus the chlorine now has a complete set of eight electrons in

## 15-7 The Ionic Bond

its valence shell. The $3s^1$ electron in Na comes off fairly easily, leaving sodium with a filled second energy level.

We have just written NaCl as

$$Na^+ \quad \text{and} \quad :\!\ddot{\underset{..}{Cl}}\!:^-$$

to emphasize the transfer of the electron from the Na atom to the Cl atom. Since the Na atom now has one less electron than it has protons, the Na atom will have a charge of $+1$ (it is common to leave off the "1" in writing $Na^+$). We can write an equation to represent the sodium atom giving up an electron:

$$\begin{bmatrix} \text{this Na} \\ \text{atom has} \\ 11\ e^- \end{bmatrix} \rightarrow Na\!\cdot\, \rightarrow Na^+ + e^- \leftarrow \begin{bmatrix} \text{this } e^- \\ \text{came off the} \\ \text{Na atom} \end{bmatrix}$$
$$\uparrow$$
$$\begin{bmatrix} \text{this } Na^+ \text{ has} \\ 10\ e^- \end{bmatrix}$$

The chlorine atom, on the other hand, has taken one electron from the sodium atom. Therefore, it has one more electron than it has protons and so has a charge of $-1$ (again, it is common to leave off the "1" in writing $:\!\ddot{\underset{..}{Cl}}\!:^-$). We can write an equation to represent this taking of an electron:

$$\begin{bmatrix} \text{this Cl atom} \\ \text{has 17 } e^- \end{bmatrix} \rightarrow :\!\ddot{\underset{..}{Cl}}\!\cdot\, + e^- \rightarrow :\!\ddot{\underset{..}{Cl}}\!:^- \leftarrow \begin{bmatrix} \text{this } Cl^- \text{ has} \\ 18\ e^- \end{bmatrix}$$
$$\uparrow$$
$$\begin{bmatrix} \text{this electron} \\ \text{adds to the} \\ \text{Cl atom} \end{bmatrix}$$

It is customary to leave off the dots when writing such reactions:

$$Na \rightarrow Na^+ + e^-$$
$$Cl + e^- \rightarrow Cl^-$$
$$Na + Cl \rightarrow Na^+ + Cl^-$$

Charged atoms like $Na^+$ and $Cl^-$ are called **ions.** $Na^+$ is called the "sodium ion," and $Cl^-$ is called the "chloride ion." Compounds that have ionic bonds are called **ionic compounds.** *It is the strong electrical attraction between positive and negative ions that causes the bonding force of an ionic compound.*

**EXAMPLE 6** Write the equation for removing an electron from an Li atom.

*Solution:* The Lewis formula for lithium is Li·. Li can easily lose the lone $2s$ electron:

$$\text{Li·} \rightarrow \text{Li}^+ + e^-$$ ∎

**EXAMPLE 7** Write the equation for removing two electrons from an Mg atom.

*Solution:* The Lewis formula for magnesium is Mg:. Mg can easily lose its two $3s$ electrons:

$$\text{Mg:} \rightarrow \text{Mg}^{2+} + 2e^-$$

The charge on the Mg ion is +2 because the Mg atom has lost two electrons. ∎

At this point we should say a word about writing ionic formulas. Notice that in writing $Mg^{2+}$ in Example 7, we wrote the "2" before the "+." This is the approved way of writing the ionic charge of an ion in an ionic formula. However, when we said "the charge on the Mg ion is +2 . . .," we wrote the "+" before the "2." We did so because we weren't writing an ionic formula—we were just referring to a "plus two" charge.

**EXAMPLE 8** Write the equation for the addition of an electron to an F atom.

*Solution:* The Lewis formula for fluorine is :F̈·. Fluorine can take one electron to fill up its valence shell:

$$:\!\ddot{\text{F}}\!\cdot + e^- \rightarrow :\!\ddot{\text{F}}\!:^-$$ ∎

**EXAMPLE 9** Write the equation for the addition of two electrons to an oxygen atom.

*Solution:* The Lewis formula for oxygen is :Ö·. Oxygen can acquire two electrons to fill up its valence shell:

$$:\!\ddot{\text{O}}\!\cdot + 2e^- \rightarrow :\!\ddot{\text{O}}\!:^{2-}$$

The charge on the oxygen ion is −2 because it has gained two electrons. ∎

## 15-7 The Ionic Bond

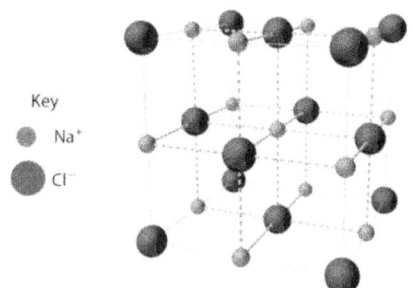

**FIGURE 15-9** A diagram of the crystal structure of NaCl. Each Na⁺ ion is surrounded by six Cl⁻ ions. Each Cl⁻ ion is surrounded by six Na⁺ ions. In an actual crystal that you could see there are about 10 quadrillion ions[3] arranged in the pattern shown. *Diagram courtesy of http://physicsopenlab.org/, Creative Commons Attribution 4.0 International.*

The electrical attraction between the positively charged Na⁺ ion and the negatively charged Cl⁻ ion is what holds the NaCl crystal together. The NaCl crystal is a giant array (see Chapter 4) of Na⁺ ions and Cl⁻ ions. A diagram of the crystal structure of NaCl is shown in Figure 15-9.

How do we know that there are separate Na⁺ ions and Cl⁻ ions in the NaCl crystal? As early as 1922, x-ray crystallography (see page 322) studies of the sodium chloride crystal had shown that this was the case. It was shown that the Na⁺ ion has 10 electrons around it and that the Cl⁻ ion has 18 electrons around it. This was very strong evidence that separate Na⁺ and Cl⁻ ions exist and that the bonding is ionic. See https://bit.ly/38fX8tz for the original paper,[4] especially page 443.

Why do ionic bonds form? After all, it must require some energy to pull an electron out of a Na atom and put it onto a Cl atom. It does indeed require some energy. But when sodium and chlorine react to form NaCl, a great deal of heat energy is given off. This lowering of the total energy as a result of the attraction of the Na⁺ ion and the Cl⁻ ion in the crystal is very large. It easily compensates for the energy needed to pull an electron out of an Na atom and add it to a Cl atom. The NaCl crystal is very stable.

Why don't atoms like sodium and chlorine form covalent bonds with one another? It must be that the total energy is lowest when an ionic bond is formed.

Elements in group IA of the periodic table have one *s* electron in their highest energy level. This lone electron can be easily removed, forming a +1 ion. Elements in group IIA have two *s* electrons in their highest energy level and can easily lose both of them, forming +2 ions. Elements in group IIIA have three electrons, namely, two *s* electrons and one *p* electron in their highest energy level. Boron forms three covalent bonds. Aluminum can form the +3 ion, $Al^{3+}$, and both ionic bonds and covalent bonds, depending on what it is bonding to. The rest of the group IIIA elements also form both covalent and ionic bonds.

As you go down a group in the periodic table, it gets easier to remove the valence electrons. This is because they are farther from the nucleus and aren't attracted to it as much as electrons that are closer to the nucleus. The valence electrons are also partially shielded from the positive charge of the nucleus by the inner electrons.

---

[3] This assumes that the smallest object you could see is about 60 μm. The period at the end of this sentence is about 200 μm.

[4] "The Distribution of Electrons around the Nucleus in the Sodium and Chlorine Atoms." By **W. Lawrence Bragg**, **R. W. James**, **C.H. Bosanquet**," *Phil. Mag.* S. 6 Vol. 44. No. 361. Sept. 1922. **NOTE:** In 1922, determining the structure of a simple substance like NaCl was a major achievement. Compare this to what we can now do because we have high-speed computers and much more advanced equipment (see pp. 285 and 322).

**EXAMPLE 10**  Write the equation for the formation of the ionic compound containing lithium and fluorine.

*Solution:*

$$\text{Li}\cdot + :\!\ddot{\text{F}}\!\cdot \rightarrow \text{Li}^+ + :\!\ddot{\text{F}}\!:^-$$

Without the dots we have

$$\text{Li} + \text{F} \rightarrow \text{Li}^+ + \text{F}^-$$

Under normal conditions, the fluorine gas occurs as $F_2$ molecules, and we should write the reaction as

$$2\text{Li} + \text{F}_2 \rightarrow 2\text{Li}^+ + 2\text{F}^-$$

This would be more realistic. ∎

**EXAMPLE 11**  Write the equation for the formation of the ionic compound containing magnesium and fluorine.

*Solution:* The Mg atom can give up two electrons to become the $Mg^{2+}$ ion. Two F atoms are needed to take these two electrons, since each F atom takes only one electron:

$$\text{Mg}: + 2\, :\!\ddot{\text{F}}\!\cdot \rightarrow \text{Mg}^{2+} + 2\, :\!\ddot{\text{F}}\!:^-$$

Without dots, we have

$$\text{Mg} + 2\text{F} \rightarrow \text{Mg}^{2+} + 2\text{F}^-$$

Again, since fluorine gas occurs as the $F_2$ molecule, we should write the reaction as

$$\text{Mg} + \text{F}_2 \rightarrow \text{Mg}^{2+} + 2\text{F}^-\ \blacksquare$$

    In closing, we should say that most chemical bonds are not purely ionic (transfer of electrons) or purely covalent (sharing of electrons). They are usually a mixture of the two kinds of chemical bonds.

    In this chapter we have only touched on the subject of chemical bonding. We have not discussed metallic bonding, bonding in the transition metals, and many other things. As you study more chemistry, you will learn more about the nature of the chemical bond and perhaps realize what a vast subject it is.

# PROBLEMS

## KEYED PROBLEMS

1. Write the Lewis formula for lithium, Li. Only the electrons in the second energy level should be shown since the electrons in the first energy level don't take part in the chemical reactions of lithium.

2. Draw the Lewis formula for magnesium, Mg.

3. Draw the Lewis formula for K and As.

4. Draw Lewis formulas for the following molecules: $NCl_3$, $BCl_3$, and $Cl_2O$.

   (**NOTE:** $Cl_2O$ has a structure similar to $OF_2$. The reason we write it $Cl_2O$ and not $OCl_2$ is indicated in Chapter 16.)

5. Draw the Lewis formulas with lines only, as well as with lines and dots, for the following molecules: $NCl_3$, $BCl_3$, and $Cl_2O$.

6. Write the equation for removing an electron from a K atom.

7. Write the equation for removing two electrons from a Ca atom.

8. Write the equation for the addition of an electron to a Br atom.

9. Write the equation for the addition of two electrons to an S atom.

10. Write the equation for the formation of the ionic compound containing potassium and bromine.

11. Write the equation for the formation of the ionic compound containing calcium and chlorine.

## SUPPLEMENTAL PROBLEMS

12. What do we mean when we say that electrons are "lazy"?

13. Define *molecular orbital*.

14. Why can't more than two electrons be in the same molecular orbital?

15. Define *covalent bond*.

16. Define *valence electrons*.

17. In the Lewis formula of an atom, which electrons are represented as dots and which electrons are not?

18. Draw Lewis formulas for all the third-period A-group elements.

19. Draw Lewis formulas with lines only, as well as with lines and dots, for the following molecules.

a. $CCl_4$    c. $H_2S$    e. $H_2CO$ (C in the center)
b. $NBr_3$    d. $HBr$    f. $I_2$

20. Draw Lewis formulas with lines only for the following molecules.

    a. $C_2H_6$ (carbon–carbon single bond)
    b. $C_2H_4$ (carbon–carbon double bond)
    c. $C_2H_2$ (carbon–carbon triple bond)

21. Draw the Lewis formula for $PCl_3$.

22. Draw the Lewis formula for $SF_4$.

    (**NOTE:** S has 10 electrons around it. Two electrons form a nonbonded pair.)

23. Define an *ionic bond*.

24. Write the equation for the loss of three electrons by Al.

25. Write the equation for the gain of two electrons by Se.

26. Write the equations for the formation of the ionic compounds containing the following pairs of elements.

    a. Al and $Cl_2$    c. Be and $Br_2$    e. Li and $S_8$ (elemental sulfur consists of molecules with eight atoms)
    b. K and $O_2$    d. Ba and $I_2$
    f. Al and $H_2$

27. Give an example of a molecule with each of the following geometric structures.

    a. linear
    b. angular or bent
    c. triangular planar
    d. trigonal pyramid
    e. regular tetrahedron

28. Give an example of a molecule with each of the following geometric structures.

    a. trigonal bipyramid    c. square pyramid
    b. octahedron

---

**PHOSPHINE (pronounced "faas-feen") ON VENUS?**

In 2017, signs of phosphine (see pp. 276, 297) *may* have been discovered in the atmosphere of the planet Venus (https://bit.ly/3nKVVjf). The team leader, Professor of Astronomy **Jane Greaves** of Cardiff University in Wales, said the concentration was very low, only 20 parts per billion. Phosphine is easily destroyed by sunlight and sulfuric acid, both of which are found in the atmosphere of Venus. It thus must be made continuously, possibly, as it is on earth in sewage sludge, by anaerobic bacteria — these are bacteria that can live without oxygen. **So there *may* be life on Venus!** Other scientists have questioned this discovery because it is so difficult to measure such a low concentration of phosphine from earth (https://nyti.ms/3LeM3tm).

# THE STRUCTURE OF SMALL COMPLEX MOLECULES (Part 1)

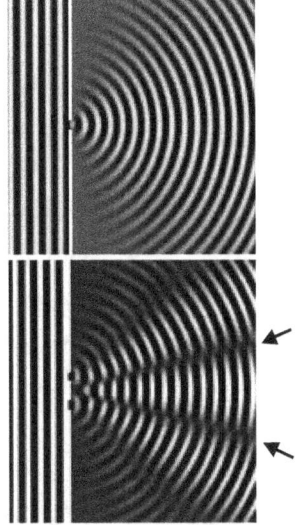

Waves (see page 257) have two properties that are of interest here, diffraction and interference. **Diffraction** is the ability of a wave to go around corners, as shown in the graphic to the right. In the top image, the vertical black and white lines are plane waves – a plane wave is like the water waves that hit a beach. They are all parallel to each other. When a plane wave goes through a slit (the open space in the vertical white line), it spreads out into semi-circles. **Interference**, discussed in the next paragraph, is the ability of waves to cancel each other out and to reinforce each other. See https://bit.ly/2JGN7WQ for some nice graphics that show this. See the attribution web sites below for these images in motion.

The bottom image at the right shows a plane wave going through two slits. Notice the areas where the wave is bright and where it is dark (at the arrows). The bright striped areas are where the two semi-circular waves (one set from each slit) reinforce each other, and the dark areas are where the waves cancel each other out. This reinforcing and cancelling of is called **interference**. You can see this effect for light waves and water waves in the following video: https://bit.ly/1re64Df. Water waves are discussed in the 2nd half of the video. To see laser light interference, see https://bit.ly/2UUEpNg. *Graphics above are CC BY-SA 3.0. Attribution is at https://bit.ly/2RFX92d and https://bit.ly/2FkGTOg.*

Prof. **Tamir Gonen** and his group at UCLA have used a Thermo Scientific™ Talos™ Arctica™ electron microscope (which uses 200,000 volts to accelerate the electrons as discussed on page 257) to rapidly determine the structure of small complex organic molecules using single nanocrystals, which are about 100 nm in size. This technique is called **electron crystallography**.[1] The nanocrystals form by themselves and don't need to be grown in the lab. It's a good thing that nanocrystals can be used in this technique, because growing larger crystals of some of these molecules can sometimes be **really** hard. Crystals (see pages 57 and 281) are needed because the regular arrangement of the atoms in the crystal allows the kind of diffraction and interference needed to get the spots (see the third image in the graphic on page 323) used to analyze the structure of the molecules in the crystal.[2]

In their first attempt, Prof. Gonen's group took some progesterone (a female sex hormone) powder out of a jar "as is" after removing any lumps. The little black speck in the second image on page 322 with a circle around it is a nanocrystal.[3] The electron microscope cools the crystals with liquid nitrogen (77 K) so the electron beam doesn't heat them causing damage. This technique is called **cryo-electron microscopy (cryo-EM)**. (Part 2 is on page 322.)

---

[1]**Crystallography** is the experimental science of determining the 3-dimensional arrangement of atoms in crystals.

[2]For information on using this technique for large molecules like proteins see https://bit.ly/2IUyK2V.

[3]The 7-small-grey circles are holes and serve no purpose for this experiment.

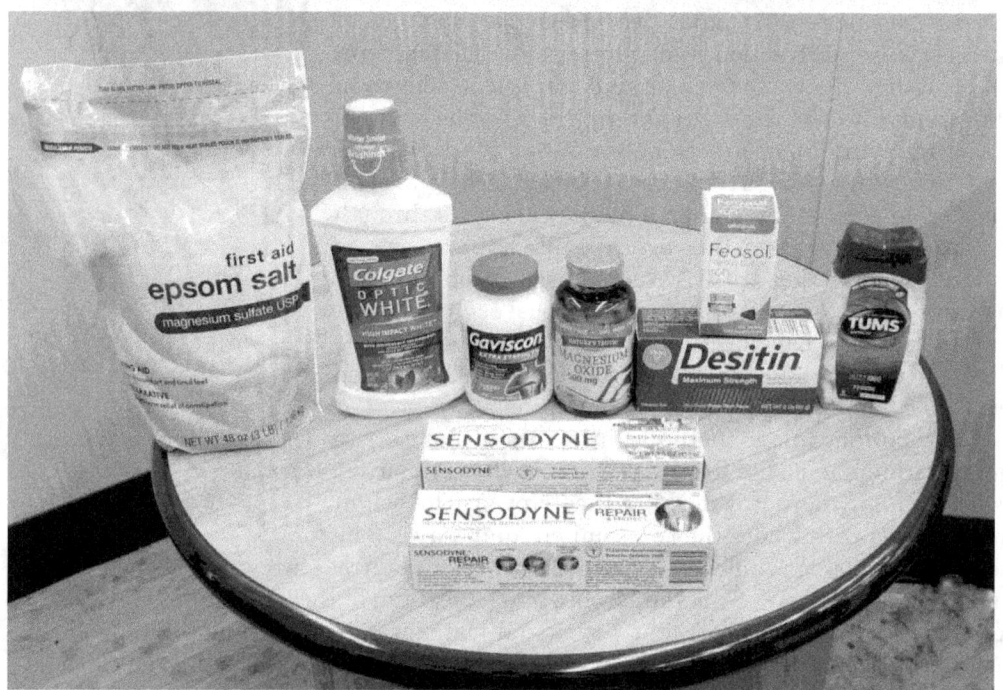

**Items from the Shelves of a Pharmacy:** After reading this chapter, if you are given the name of an ingredient as listed below, you should be able to write the chemical formula (and *vice versa*). Chemical formulas are written at the bottom of the page in the order listed for the ingredient names.

*Photo by the author.*

Back row from left to right (only the active ingredients are listed):

> first aid epsom salt: magnesium sulfate
> Colgate OPTIC WHITE®: hydrogen peroxide
> Gaviscon®: aluminum hydroxide, magnesium hydroxide
> MAGNESIUM OXIDE: magnesium oxide
> Desitin®: zinc oxide
> Feosol® (above the Desitin): ferrous sulfate*
> TUMS®: calcium carbonate

In front center from back to front (ingredients are the active ingredients):

> SENSODYNE Extra Whitening®: potassium nitrate, sodium fluoride
> SENSODYNE REPAIR & PROTECT®: stannous fluoride*

Chemical Formulas listed from top to bottom: $MgSO_4$, $H_2O_2$, $Al(OH)_3$, $Mg(OH)_2$, MgO, ZnO, $FeSO_4$, $CaCO_3$, $KNO_3$, NaF, $SnF_2$

*__NOTE:__ The older names are used on these labels. The use of the older names is common in over-the-counter drugs, supplements and most consumer products. This may be the only reason to learn the older names.

# NOMENCLATURE OF SIMPLE INORGANIC COMPOUNDS

In previous chapters we have mentioned many compounds. In this chapter we will explain the rules for naming various inorganic compounds and the rules for writing their formulas. By the end of the chapter, you should be able to write the name of a compound if you are given the formula and to write the formula if you are given the name.

This discussion will be limited to inorganic compounds, that is, compounds that do not involve carbon atoms (there are a few exceptions, such as $CO_2$, $H_2CO_3$, and HCN, which chemists normally regard as inorganic compounds). Most carbon compounds (methane, $CH_4$, is an example) are called organic compounds and are named according to a system that we will not discuss in this book.[1]

There are three classes of inorganic compounds that we will discuss here: (1) ionic compounds (formed between a metal and a nonmetal), (2) covalent compounds (formed between two nonmetals), and (3) covalent compounds called acids and oxyacids.

## 16-1
### NAMING IONIC COMPOUNDS FORMED BETWEEN A METAL AND A NONMETAL

In this section we will explain the rules for naming ionic compounds and writing the formulas of ionic compounds. The basic rules are:

1. **The formulas of ionic compounds are written without showing charges on the ions.** For example, sodium chloride is written NaCl, not $Na^+Cl^-$

2. **In writing formulas, we put the positive ion first and the negative ion second.** Thus sodium chloride is written NaCl, *never* ClNa.

[1] Naming organic compounds is sometimes briefly discussed in general chemistry courses. But a complete discussion must wait for a course in organic chemistry. The International Union of Pure and Applied Chemistry (IUPAC) is the group that sets nomenclature rules in both inorganic and organic chemistry.

288   NOMENCLATURE OF SIMPLE INORGANIC COMPOUNDS

3. **The same is true in writing the name of a compound: The name of the positive ion is written first, and the name of the negative ion is written second.** Thus NaCl is always called sodium chloride, *never* chlorine sodide.

4. **The formula of the compound must be electrically neutral; that is, the total number of positive charges on the positive ions must be equal to the total number of negative charges on the negative ions.** For example, in NaCl, the Na$^+$ has one positive charge and the Cl$^-$ has one negative charge. Thus the total number of positive charges equals the total number of negative charges in the formula NaCl.

5. **The smallest whole number of atoms possible to satisfy rule 4 is used to write the formula.** Thus we write NaCl, not Na$_2$Cl$_2$.

6. **To decide which elements are considered metals and which are considered nonmetals in this naming system, look at the metals and the nonmetals in a periodic table.** (Figures 14-10 and 14-12). Semimetals are considered nonmetals for naming purposes.

NOTE: In Table 16-1, the NH$_4^+$ ion is an example of a positive ion that consists of more than one element. It is formed when ammonia reacts with an acid (say HCl) according to the equation NH$_3$ + HCl → NH$_4$Cl. NH$_4$Cl completely ionizes in a water solution to give NH$_4^+$ and Cl$^-$.

## TABLE 16-1

CHARGES ON IONS OF GROUPS IA, IIA, AND IIIA ELEMENTS AND NH$_4^+$

| GROUP IA | GROUP IIA | GROUP IIIA | ANOTHER POSITIVE ION |
|---|---|---|---|
| H$^+$ | Be$^{2+}$ | B[a] | NH$_4^+$ |
| Li$^+$ | Mg$^{2+}$ | Al$^{3+}$ [b] | (ammonium) |
| Na$^+$ | Ca$^{2+}$ | Ga$^{3+}$ | |
| K$^+$ | Sr$^{2+}$ | In$^{3+}$ [c] | |
| Rb$^+$ | Ba$^{2+}$ | Tl$^{3+}$ [c] | |
| Cs$^+$ | Ra$^{2+}$ | | |

[a] Considered a semimetal—only forms covalent bonds. Boron compounds will not be considered in this section. Boron does not form the +3 ion.
[b] Definitely a metal, but forms both ionic and covalent bonds.
[c] Indium and thallium also form the +1 ions In$^+$ and Tl$^+$, respectively.

As you can see, it is important to know the ionic charge that an atom can have if you want to be able to write the correct formula. So we must now discuss how to determine the ionic charge that an atom can have in order to understand rules 4 and 5 above.

For group IA, IIA, and IIIA metals, the positive charge that the ion can have is the same as the group number. Table 16-1 lists the positive charge formed in these three groups and another important positive ion. For the nonmetals in groups VA, VIA, and VIIA, the negative ionic charge is found by the following formula:

$$\text{negative ionic charge} = \text{group number} - 8$$

For example, oxygen is in group VIA. Thus the ionic charge is $6 - 8 = -2$. So oxygen forms $O^{2-}$ ions. Notice that this formula simply tells us the number of electrons that must be added to the valence shell to fill it to eight electrons:

$$:\ddot{O}\cdot + 2e^- \rightarrow :\ddot{O}:^{2-}$$

In compounds, the names of the negative ions of groups IVA, VA, VIA, and VIIA end in *ide*. Table 16-2 lists the name and charge of these ions.

In Table 16-2, only carbon is listed in group IVA and only nitrogen and phosphorus are listed in group VA because only these elements in these groups sometimes form negative ions. As we mentioned in Chapter 15, carbon, nitrogen, and phosphorus usually form covalent bonds. All the other elements listed in Table 16-2 can also form covalent bonds under the appropriate conditions. Notice that hydrogen is mentioned in both Table 16-1 and Table 16-2. Hydrogen can either take an electron to form the hydride ion, $H:^-$, or lose an electron to form the $H^+$ ion.

We now give a number of examples that illustrate the use of the basic rules of writing formulas and names. All the elements used are from Tables 16-1 and 16-2.

## TABLE 16-2

NAME AND CHARGE ON IONS OF GROUPS IVA, VA, VIA, AND VIIA[a]

| GROUP IV | GROUP VA | GROUP VIA | GROUP VIIA |
|---|---|---|---|
| $C^{4-}$ (carbide)[b] | $N^{3-}$ (nitride) | $O^{2-}$ (oxide) | $F^-$ (fluoride) |
| | $P^{3-}$ (phosphide) | $S^{2-}$ (sulfide) | $Cl^-$ (chloride) |
| | | $Se^{2-}$ (selenide) | $Br^-$ (bromide) |
| | | $Te^{2-}$ (telluride) | $I^-$ (iodide) |

[a] Hydrogen can also form the negative ion $H^-$, which is called the hydride ion.
[b] Carbon also forms the negative ion $C_2^{2-}$, which is also called the carbide ion.

**EXAMPLE 1** Write the name and formula of the compound made from potassium and bromine.

*Solution:* The potassium ion is $K^+$, and the bromide ion is $Br^-$. One $K^+$ combines with one $Br^-$ as follows:

$$K^+ + Br^- \rightarrow KBr$$

KBr is called potassium bromide. ∎

**EXAMPLE 2** Write the name and formula of the compound made from calcium and iodine.

*Solution:* The calcium ion is $Ca^{2+}$, and the iodide ion is $I^-$. Two iodide ions combine with one calcium ion to give electrically neutral calcium iodide:

$$Ca^{2+} + 2I^- \rightarrow CaI_2$$ ∎

**EXAMPLE 3** Write the name and formula of the compound made from Al and F.

*Solution:* The ions are $Al^{3+}$ and $F^-$. Thus we need three $F^-$ ions for each $Al^{3+}$ ion to form electrically neutral aluminum fluoride:

$$Al^{3+} + 3F^- \rightarrow AlF_3$$ ∎

**EXAMPLE 4** Write the name and formula of the compound made from Mg and N.

*Solution:* The ions are $Mg^{2+}$ and $N^{3-}$. Thus we need three $Mg^{2+}$ ions to two $N^{3-}$ ions, for a total of six positive charges and six negative charges. The name of the compound is magnesium nitride:

$$3Mg^{2+} + 2N^{3-} \rightarrow Mg_3N_2$$ ∎

**EXAMPLE 5** Write the name and formula of the compound made from Al and O.

*Solution:* The ions are $Al^{3+}$ and $O^{2-}$. We need two $Al^{3+}$ ions to three $O^{2-}$ ions, for a total of six positive charges and six negative charges. The name of the compound is aluminum oxide:

$$2Al^{3+} + 3O^{2-} \rightarrow Al_2O_3$$ ∎

## 16-1 Naming Ionic Compounds Formed Between a Metal and a Nonmetal

**EXAMPLE 6** Write the correct names of the following compounds: $K_2S$, $CaBr_2$, $LiH$, $BaS$, $Al_2Se_3$, and $NH_4Cl$.

*Solution:*

| | | | |
|---|---|---|---|
| $K_2S$ | potassium sulfide | $BaS$ | barium sulfide |
| $CaBr_2$ | calcium bromide | $Al_2Se_3$ | aluminum selenide |
| $LiH$ | lithium hydride | $NH_4Cl$ | ammonium chloride ∎ |

**EXAMPLE 7** Write the correct formulas for the following names: cesium iodide, beryllium oxide, gallium sulfide, strontium bromide, potassium oxide, calcium hydride, radium iodide, and aluminum sulfide.

*Solution:*

| | | | |
|---|---|---|---|
| cesium iodide | $CsI$ | potassium oxide | $K_2O$ |
| beryllium oxide | $BeO$ | calcium hydride | $CaH_2$ |
| gallium sulfide | $Ga_2S_3$ | radium iodide | $RaI_2$ |
| strontium bromide | $SrBr_2$ | aluminum sulfide | $Al_2S_3$ ∎ |

In addition to the simple monatomic negative ions listed in Table 16-2, there are many common **polyatomic negative ions** ("poly" means many). These are listed in Table 16-3. The bonding within each polyatomic ion is covalent, but the whole negative ion can form either ionic or covalent bonds. Since there is no way to figure out their names or formulas, you should learn at least some of the ions listed in Table 16-3. The asterisked (*) ions are the most common.

In Table 16-3, we have grouped ions together in such a way that you can see some patterns. When nitrogen or sulfur form negative ions with oxygen, the names have a pattern. We list these ions as follows so that you can compare their names; we also list the nitride and sulfide ions for comparison:

| | | | |
|---|---|---|---|
| $N^{3-}$ | nitride | $S^{2-}$ | sulfide |
| $NO_2^-$ | nitrite | $SO_3^{2-}$ | sulfite |
| $NO_3^-$ | nitrate | $SO_4^{2-}$ | sulfate |

If an ion has no oxygen, its ending is *ide* ($N^{3-}$ and $S^{2-}$). With the larger number of oxygen atoms, the ending is *ate* ($NO_3^-$ and $SO_4^{2-}$). With the smaller number of oxygen atoms, the ending is *ite* ($NO_2^-$ and $SO_3^{2-}$). In the case of chlorine, which forms more than two negative ions with oxygen, there are slight complications, as the following list indicates:

| | | | |
|---|---|---|---|
| $Cl^-$ | chloride | $ClO_3^-$ | chlorate |
| $ClO^-$ | hypochlorite | $ClO_4^-$ | perchlorate |
| $ClO_2^-$ | chlorite | | |

## TABLE 16-3

| SOME POLYATOMIC NEGATIVE IONS | | | |
|---|---|---|---|
| $NO_2^-$ | nitrite* [a] | $PO_4^{3-}$ | phosphate* |
| $NO_3^-$ | nitrate* | $HPO_4^{2-}$ | hydrogen phosphate |
| $SO_3^{2-}$ | sulfite* | $H_2PO_4^-$ | dihydrogen phosphate |
| $SO_4^{2-}$ | sulfate* | $CrO_4^{2-}$ | chromate* |
| $HSO_3^-$ | hydrogen sulfite | $Cr_2O_7^{2-}$ | dichromate* |
| $HSO_4^-$ | hydrogen sulfate | $ClO^-$ | hypochlorite |
| $CO_3^{2-}$ | carbonate* | $ClO_2^-$ | chlorite |
| $HCO_3^-$ | hydrogen carbonate* | $ClO_3^-$ | chlorate* |
| | | $ClO_4^-$ | perchlorate* |
| $C_2H_3O_2^-$ | acetate* | $C_2O_4^{2-}$ | oxalate |
| $AsO_4^{3-}$ | arsenate | $O_2^{2-}$ | peroxide* |
| $BO_3^{3-}$ | borate | $MnO_4^-$ | permanganate* |
| $OCN^-$ | cyanate | $SiO_4^{4-}$ | silicate |
| $CN^-$ | cyanide* | $S_4O_6^{2-}$ | tetrathionate |
| $OH^-$ | hydroxide* | $SCN^-$ | thiocyanate |
| | | $S_2O_3^{2-}$ | thiosulfate |

[a] The asterisked (*) ions are the most common.

The prefix *hypo* (meaning "below") is used for the smallest number of oxygen atoms, and the prefix *per* is used for the largest number of oxygen atoms. So, in order of increasing number of oxygen atoms, we have *ide*, *hypo---ite*, *ite*, *ate*, and *per---ate*.

If hydrogen is included in the formula of the negative ion, this is usually stated in the name. For example,

| | | | |
|---|---|---|---|
| $CO_3^{2-}$ | carbonate | $PO_4^{3-}$ | phosphate |
| $HCO_3^-$ | hydrogen carbonate | $HPO_4^{2-}$ | hydrogen phosphate |
| | | $H_2PO_4^-$ | dihydrogen phosphate |

$HCO_3^-$ is sometimes called the *bicarbonate* ion, where *bi* is an old-fashioned term that stands for hydrogen. You may have heard of sodium bicarbonate ($NaHCO_3$, bicarbonate of soda, baking soda). It is more properly called sodium hydrogen carbonate.

The peroxide ion, $O_2^{2-}$, may be a little puzzling at first glance. To form the peroxide ion, two oxygen atoms are bonded together. A typical compound with the peroxide ion is hydrogen peroxide, $H_2O_2$. The structure is shown at the top of page 293. The H atoms are not in the plane of the page—one is in front of the page and one is behind the page:

structural formula
of hydrogen peroxide

Oxygen also forms a superoxide ion, $O_2^-$, which we will not discuss except to offer Problem 27 at the end of Chapter 17.

The following examples all use ions taken from Tables 16-1 and 16-3.

**EXAMPLE 8** Write the correct formula for the compounds that have the following names: sodium nitrate, calcium hypochlorite, potassium permanganate, aluminum acetate, magnesium hydroxide, rubidium peroxide, barium perchlorate, potassium dichromate, ammonium phosphate, and aluminum hydrogen sulfite.

*Solution:*

| | | | |
|---|---|---|---|
| sodium nitrate | $NaNO_3$ | rubidium peroxide | $Rb_2O_2$ |
| calcium hypochlorite | $Ca(ClO)_2$ | barium perchlorate | $Ba(ClO_4)_2$ |
| potassium permanganate | $KMnO_4$ | potassium dichromate | $K_2Cr_2O_7$ |
| aluminum acetate | $Al(C_2H_3O_2)_3$ | ammonium phosphate | $(NH_4)_3PO_4$ |
| magnesium hydroxide | $Mg(OH)_2$ | aluminum hydrogen sulfite | $Al(HSO_3)_3$ ∎ |

**EXAMPLE 9** Write the correct names of the compounds that have the following formulas: $NaNO_2$, $K_2O_2$, $Be(OH)_2$, $CaCO_3$, $Sr(ClO_2)_2$, $AlPO_4$, $Na_2S_2O_3$, and $MgC_2O_4$.

*Solution:*

| | | | |
|---|---|---|---|
| $NaNO_2$ | sodium nitrite | $Sr(ClO_2)_2$ | strontium chlorite |
| $K_2O_2$ | potassium peroxide | $AlPO_4$ | aluminum phosphate |
| $Be(OH)_2$ | beryllium hydroxide | $Na_2S_2O_3$ | sodium thiosulfate |
| $CaCO_3$ | calcium carbonate | $MgC_2O_4$ | magnesium oxalate ∎ |

Now we must consider the problem of naming compounds containing the transition metals and some other metals of the periodic table. These present a slight problem because *many of them can form more than one positive ion*. We will consider only the most common ions, and fortunately

## 294  NOMENCLATURE OF SIMPLE INORGANIC COMPOUNDS

there are no more than two of these for a metal. Consider iron, which can form the +2 and the +3 ions:

$$Fe^{2+} \text{ and } Fe^{3+}$$

There are two ways of naming these ions, the modern system and the old system. Both systems are in use today and you should learn both.

**In the modern system, the ionic charge is indicated after the name of the metal by a Roman numeral in parentheses.** For iron we have

$$Fe^{2+} \quad \text{iron(ll)}$$
$$Fe^{3+} \quad \text{iron(lll)}$$

Notice that there is no space between the name and the parentheses. $FeCl_2$ is called iron(ll) chloride, and $FeCl_3$ is called iron(lll) chloride. Read them as "iron two chloride" and "iron three chloride." As another example, FeS is called iron(ll) sulfide. See Table 16-2 for the charge on the negative ions.

**In the old system, the Latin name of the metal is sometimes (see the NOTE at the bottom of Table 16-4) used together with the suffixes "ous" and "ic."** "Ous" stands for the smaller positive charge and "ic" stands for the larger positive charge. For iron we have

$$Fe^{2+} \quad \text{ferrous}$$
$$Fe^{3+} \quad \text{ferric}$$

**NOTE:** The Latin name for iron is ferrum.

In the old system, $FeCl_2$ is called ferrous chloride and $FeCl_3$ is called ferric chloride. Our second example above, FeS, is called ferrous sulfide.

Many transition metals can form more than two positive ions. For these, the old system can be used to name only the most common positive ions. Table 16-4 lists only the two most common positive ions for some transition metals. Some of them can form not-so-common positive ions also. The not-so-common ions are named only by the modern system.

Table 16-5 lists the transition metals that have only one positive ion. These metals are named just like the ones in Table 16-1, that is, with no Roman numerals and no "ous" or "ic" forms.

Examples 10 and 11 use ions from Tables 16-2 through 16-5.

**EXAMPLE 10** Write the correct formulas for the following names: gold(l) chloride, nickel(lll) hydroxide, stannous fluoride, silver thiocyanate, lead(ll) sulfate, chromic carbonate, copper(l) phosphate, mercuric nitrate, zinc hydrogen carbonate, and mercurous chloride.

## TABLE 16-4

**METALS THAT CAN FORM TWO COMMON POSITIVE IONS**

| ION[a] | MODERN NAME | OLD NAME | LATIN NAME OF METAL |
|---|---|---|---|
| $Au^+$ | gold(I) | aurous | aurum |
| $Au^{3+}$ | gold(III) | auric | aurum |
| $Co^{2+}$ | cobalt(II) | cobaltous | |
| $Co^{3+}$ | cobalt(III) | cobaltic | |
| $Cr^{2+}$ | chromium(II) | chromous | |
| $Cr^{3+}$ | chromium(III) | chromic | |
| $Cu^+$ | copper(I) | cuprous | cuprum |
| $Cu^{2+}$ | copper(II) | cupric | cuprum |
| $Fe^{2+}$ | iron(II) | ferrous | ferrum |
| $Fe^{3+}$ | iron(III) | ferric | ferrum |
| $Hg_2^{2+}$ | mercury(I)[b] | mercurous | hydrargyrum |
| $Hg^{2+}$ | mercury(II) | mercuric | hydrargyrum |
| $Mn^{2+}$ | manganese(II) | manganous | magnes |
| $Mn^{3+}$ | manganese(III) | manganic | magnes |
| $Ni^{2+}$ | nickel(II) | nickelous | |
| $Ni^{3+}$ | nickel(III) | nickelic | |
| $Pb^{2+}$ | lead(II) | plumbous | plumbum |
| $Pb^{4+}$ | lead(IV) | plumbic | plumbum |
| $Sn^{2+}$ | tin(II) | stannous | stannum |
| $Sn^{4+}$ | tin(IV) | stannic | stannum |

[a] The most common ones, namely, Cr, Cu, Fe, Hg, Pb, and Sn, should be memorized.

[b] The $Hg_2^{2+}$ ion is unusual. Each Hg atom has a charge of +1 in the $Hg_2^{2+}$ ion, but two $Hg^+$ ions always join together. A typical compound is mercury(I) chloride, $Hg_2Cl_2$.

## TABLE 16-5

**TRANSITION METALS WITH ONE COMMON POSITIVE ION**

| ION | NAME | LATIN NAME |
|---|---|---|
| $Ag^+$ | silver | argentum |
| $Cd^{2+}$ | cadmium | cadmia |
| $Zn^{2+}$ | zinc | |

**NOTE:** In naming metal-nonmetal compounds, if the metal was not known in ancient times, no Latin name is used for the "ous" and "ic" forms. Only the English name is used. See Table 16-4. Mercury seems to be an exception. Since hydrargyrum is so awkward to pronounce, the old names for mercury compounds use mercurous and mercuric.

*Solution:*

| | | | |
|---|---|---|---|
| gold(I) chloride | AuCl | chromic carbonate | $Cr_2(CO_3)_3$ |
| nickel(III) hydroxide | $Ni(OH)_3$ | copper(I) phosphate | $Cu_3PO_4$ |
| | | mercuric nitrate | $Hg(NO_3)_2$ |
| stannous fluoride | $SnF_2$ | zinc hydrogen carbonate | $Zn(HCO_3)_2$ |
| silver thiocyanate | AgSCN | | |
| lead(II) sulfate | $PbSO_4$ | mercurous chloride | $Hg_2Cl_2$ |

In Example 11, use the known charge on the negative ion to figure out the charge on the positive ion. Once you know the charge on the positive ion, you can use Table 16.4 to look up its name. Thus, when given the formula $Co_2S_3$, you know that each sulfide ion has a $-2$ charge. You can then write an equation breaking up the chemical formula into its ions to help you figure out the charge on the cobalt ion:

$$Co_2S_3 \rightarrow 2Co^? + 3S^{2-}$$

$\uparrow$ $\quad\quad\quad\quad\quad$ $\uparrow$

+6 total charge $\quad$ −6 total charge

+6/2 = +3

As you can see, three $S^{2-}$ ions have a total $-6$ charge. Both cobalt ions must then have a total $+6$ charge, and each cobalt ion has a $+3$ charge. The name of $Co_2S_3$ is therefore cobalt(III) sulfide or cobaltic sulfide.

**EXAMPLE 11** Write the correct names for the following formulas: $Co_2S_3$, $FeSO_4$, $PbO_2$ (oxide, not peroxide), $Ni(ClO_4)_2$, $CdBr_2$, $CuO$, $Mn(NO_3)_3$, $Hg_2SO_4$, $CrAsO_4$, and $AuH_2PO_4$.

*Solution:* We shall list the modern name and the old name if commonly used.

| | | |
|---|---|---|
| $Co_2S_3$ | cobalt(III) sulfide | cobaltic sulfide |
| $FeSO_4$ | iron(II) sulfate | ferrous sulfate |
| $PbO_2$ | lead(IV) oxide | lead dioxide[2] |

---

[2] The name lead dioxide for $PbO_2$ is a common name and doesn't follow any of the rules we have mentioned. There are a number of compounds that have common names—we will mention some others in Section 16-2.

| | | |
|---|---|---|
| Ni(ClO$_4$)$_2$ | nickel(II) perchlorate | nickelous perchlorate |
| CdBr$_2$ | cadmium bromide | |
| CuO | copper(II) oxide | cupric oxide |
| Mn(NO$_3$)$_3$ | manganese(III) nitrate | manganic nitrate |
| Hg$_2$SO$_4$ | mercury(I) sulfate | mercurous sulfate |
| CrAsO$_4$ | chromium(III) arsenate | chromic arsenate |
| AuH$_2$PO$_4$ | gold(I) dihydrogen phosphate | aurous dihydrogen phosphate ∎ |

## 16-2
### NAMING COVALENT COMPOUNDS FORMED BETWEEN NONMETALS[3]

The second class of inorganic compounds whose names and formulas we will discuss are the covalent compounds formed from two nonmetals. To decide which elements are nonmetals, look at a periodic table (Figures 14-10 and 14-12). Semimetals and hydrogen are considered nonmetals for naming purposes. **Since very often more than one compound can form from two elements, prefixes are used to tell how many of each atom there are.** The prefixes used are shown in the box to the right.

| Prefixes for Naming | | | |
|---|---|---|---|
| mono | 1 | hexa | 6 |
| di | 2 | hepta | 7 |
| tri | 3 | octo | 8 |
| tetra | 4 | nona | 9 |
| penta | 5 | deca | 10 |

**If the two nonmetals form only one compound, no prefixes are used.** Thus H$_2$S is called hydrogen sulfide, not dihydrogen monosulfide.

The second element of a two-element compound always has the "ide" ending. The prefix "mono" is usually left out if it would normally precede the first element. Thus for CO, we say carbon monoxide, never monocarbon monoxide.

Another deviation from the rules is that many compounds have common names that are used almost all the time. Examples are:

| | | | |
|---|---|---|---|
| water | H$_2$O | phosphine (the IUPAC name is phosphane) | PH$_3$ |
| hydrogen peroxide | H$_2$O$_2$ | arsine (the IUPAC name is arsane) | AsH$_3$ |
| ammonia | NH$_3$ | nitric oxide | NO |
| hydrazine | N$_2$H$_4$ | nitrous oxide | N$_2$O |

Nobody seems to call water "dihydrogen monoxide." It's always called water.

---

[3] The prefixes "di", "tri", "tetra", etc. can also be used for naming elemental molecules. Examples are **dioxygen** for O$_2$, **dihydrogen** for H$_2$, **trioxygen** for O$_3$, also called ozone, and **tetraphosphorus** for P$_4$, also called white phosphorus. **NOTE: When referring to any elements or molecules, it is assumed that all the atoms have their naturally-occurring isotopic percentages.** See Table 2-1 for examples. If they don't, it will be made very clear by showing the isotope(s) involved. For example, consider dihydrogen. If all the H atoms are replaced with deuterium atoms, we would write D$_2$. If only one H atom is replaced with a deuterium atom, we would write HD. It is called hydrogen deuteride. A dioxygen molecule containing both $^{16}$O and $^{18}$O atoms would be written as $^{16}$O$^{18}$O or $^{34}$O$_2$, where 16 + 18 = 34. Heavy water has both H atoms replaced with deuterium atoms and would be written as D$_2$O (called deuterium oxide), where it is assumed that the O atoms have their naturally-occurring isotopic percentages. There is even D$_2$$^{18}$O, called "doubly labeled" water. It is very useful in measuring energy expenditure in humans (https://bit.ly/3IH9qdR). Isotopically substituted molecules have revolutionized chemistry and biochemistry, as they allow specific atoms to be followed in a chemical reaction **NOTE: The term "monooxygen" is occasionally used for oxygen atoms or atomic oxygen. The enzyme called "monooxygenase" is very common. (This footnote is copied from page 43.)**

| FORMULA | MODERN NAME | COMMON NAME |
|---|---|---|
| $N_2O$ | dinitrogen monoxide | nitrous oxide (laughing gas) |
| $NO$ | nitrogen monoxide | nitric oxide |
| $NO_2$ | nitrogen dioxide | |
| $N_2O_3$ | dinitrogen trioxide | |
| $N_2O_4$ | dinitrogen tetraoxide♦ | |
| $N_2O_5$ | dinitrogen pentaoxide♦ | |

♦You may see these names written as dinitrogen tetroxide and dinitrogen pentoxide. The IUPAC does not recommend removing the "a" in these names.

Above, the six oxides of nitrogen illustrate the rules for naming nonmetal-nonmetal compounds. Two of these are usually known by their common names.

Following the recommendations of the IUPAC, the order of the nonmetals used for naming nonmetal — nonmetal compounds that is used in this book is (semimetals are included and some radioactive elements are omitted):

**Rn, Xe, Kr, B, Ge, Si, C, Sb, As, P, N, H, Te, Se, S, I, Br, Cl, O, F**

That is why we wrote $BF_3$ for boron trifluoride, HF for hydrogen fluoride, $NH_3$ for ammonia, $OF_2$ for oxygen difluoride, and $H_2O$ for water in Chapter 15. We were following the order of this list.

The order of the elements in this list may seem sort of arbitrary. It has its origins, more or less, on electronegativity considerations (see Section 17-2) so that the names of compounds that were known before the list was made up wouldn't have to be changed. This list is arranged by the groups (or columns) of the nonmetals and semimetals in the periodic table.

The following discussion will make more sense if you look at Fig. 14-12. The list starts with radon at the bottom of group VIIIA and moves up to Kr. (Ar, Ne and He don't form stable compounds and are therefore not included in the list.) Boron in group IIIA comes next. We then move from the bottom up of groups IVA through VIIA. The exceptions are (1) H, which is put between N (group VA) and Te (group VIA), and (2) O, which is put between Cl and F. And there are also electronegativity considerations as mentioned in the previous paragraph. These exceptions are so that the names and formulas of compounds that have been known for a long time wouldn't have to be changed. (But see the *Update* after the Problems for a recent change in the above naming order that is recommended by the IUPAC.)

**EXAMPLE 12** Write the formulas of the following names: carbon dioxide, carbon monoxide, hydrogen sulfide\*, chlorine dioxide, silicon carbide\* (also known commercially as carborundum), boron nitride\*, phosphorus pentachloride, silicon tetrafluoride, diboron hexahydride (the common name is diborane). In the above names, compounds with an asterisk form only one compound from the two elements.

*Solution:*

| | | | |
|---|---|---|---|
| carbon dioxide | $CO_2$ | boron nitride | BN |
| carbon monoxide | CO | phosphorus pentachloride | $PCl_5$ |
| hydrogen sulfide | $H_2S$ | silicon tetrafluoride | $SiF_4$ |
| chlorine dioxide | $ClO_2$ | diboron hexahydride | $B_2H_6$ |
| silicon carbide | SiC | (or diborane) | |

■

**EXAMPLE 13** Write the names for the following formulas: $PCl_3$, $SO_2$, $SO_3$, $NO_2$, BP, $P_4O_{10}$, $BrO_3$, $I_2O_7$, $BrF_5$, and $As_4O_6$.

*Solution:*

| | | | |
|---|---|---|---|
| $PCl_3$ | phosphorus trichloride | $P_4O_{10}$ | tetraphosphorus decaoxide |
| $SO_2$ | sulfur dioxide | $BrO_3$ | bromine trioxide |
| $SO_3$ | sulfur trioxide | $I_2O_7$ | diiodine heptaoxide |
| $NO_2$ | nitrogen dioxide | $BrF_5$ | bromine pentafluoride |
| BP | boron monophosphide | $As_4O_6$ | tetraarsenic hexaoxide ■ |

As mentioned in the footnote of the table at the top of page 298, an "a" is *not* left out of the names for $N_2O_4$ and $N_2O_5$ which are dinitrogen tetraoxide and dinitrogen pentaoxide, respectively. In Example 13 above, both vowels next to each other are kept in the names for $P_4O_{10}$, $BrO_3$, $I_2O_7$, and $As_4O_6$. But in Example 12 and the table at the top of page 298, an "o" *is* left out in "monoxide." CO would *not* be named "carbon monooxide." In fact for any compound, monoxide is the proper form. **Because of long-time general use, "monoxide" is an exception to the IUPAC rule that vowels next to each other are not omitted.**

## 16-3
### NAMING ACIDS AND OXYACIDS

The third class of compounds we will discuss are the covalent compounds known as acids and oxyacids. Oxyacids are just acids that have oxygen in the molecule. These acids are compounds that contain hydrogen and give up a hydrogen ion ($H^+$) rather easily.

The name of an acid often depends on whether the acid is a solid, liquid, or gas that can be dissolved in water. For example, HCl is called hydrogen chloride or hydrochloric acid. Hydrogen chloride is pure gaseous HCl, and hydrochloric acid is a water solution of HCl gas. On the other hand, $H_2SO_4$, which is a liquid, is always called sulfuric acid, never hydrogen sulfate. And $H_3BO_3$, which is a solid, is always called boric acid, never hydrogen borate.

The following is a list of some gaseous acids and their two names. The names with an asterisk after them are less commonly used than the pure acid names.

| FORMULA | NAME OF PURE ACID | NAME OF ACID DISSOLVED IN WATER |
|---|---|---|
| HF | hydrogen fluoride | hydrofluoric acid |
| HCl | hydrogen chloride | hydrochloric acid |
| HBr | hydrogen bromide | hydrobromic acid* |
| HI | hydrogen iodide | hydroiodic acid* |
| $H_2S$ | hydrogen sulfide | hydrosulfuric acid* |
| HCN | hydrogen cyanide | hydrocyanic acid* |

## TABLE 16-6

### OXYACIDS AND THEIR CORRESPONDING NEGATIVE IONS

| FORMULA | NAME | NEGATIVE ION | NAME OF NEGATIVE ION |
|---|---|---|---|
| $HNO_2$ | nitrous acid | $NO_2^-$ | nitrite |
| $HNO_3$ | nitric acid | $NO_3^-$ | nitrate |
| $H_2SO_3$ | sulfurous acid | $SO_3^{2-}$ | sulfite |
| $H_2SO_4$ | sulfuric acid | $SO_4^{2-}$ | sulfate |
| $H_2CO_3$ | carbonic acid | $CO_3^{2-}$ | carbonate |
| $H_3PO_4$ | phosphoric acid | $PO_4^{3-}$ | phosphate |
| $HClO$ | hypochlorous acid | $ClO^-$ | hypochlorite |
| $HClO_2$ | chlorous acid | $ClO_2^-$ | chlorite |
| $HClO_3$ | chloric acid | $ClO_3^-$ | chlorate |
| $HClO_4$ | perchloric acid | $ClO_4^-$ | perchlorate |
| $H_3BO_3$ | boric acid | $BO_3^{3-}$ | borate |

The names with an asterisk after them are much less commonly used than the pure acid name.

The oxyacids listed in Table 16-6 have only one name. This name refers to both the pure acid and the acid dissolved in water. Notice the patterns in the names of some of the series, especially that the *ic* of the acid becomes the *ate* of the negative ion, and the *ous* becomes an *ite*.

## 16-4
### CHEMICAL NAMES SHOULD BE AS SIMPLE AS POSSIBLE BUT COMPLETELY CLEAR

Chemists name compounds with the least amount of information needed to be completely clear. For example, $Al_2O_3$ is called aluminum oxide, *never* aluminum(III) oxide. There is no need to specify anything more than aluminum oxide, because there is only one possible formula, namely, $Al_2O_3$. On the other hand, $Fe_2O_3$ is called iron(III) oxide or ferric oxide. If you only called it iron oxide, nobody would know whether you meant $Fe_2O_3$, FeO, or $Fe_3O_4$.

As for common names, they specify a definite compound. The names are very old but are still commonly used.

## PROBLEMS
KEYED PROBLEMS

1. Write the name and formula of the compound made from cesium and iodine.

2. Write the name and formula of the compound made from magnesium and bromine.

3. Write the name and formula of the compound made from Al and Cl.

4. Write the name and formula of the compound made from Ca and N.

5. Write the name and formula of the compound made from Al and S.

6. Write the names of the following compounds: $Na_2S$, $MgI_2$, $NaH$, $RaO$, $Al_2S_3$, $NH_4F$.

7. Write the correct formulas for the following names: rubidium nitrate, magnesium oxide, calcium chloride, sodium oxide, beryllium hydride, barium bromide, aluminum oxide.

8. Write the correct formulas for the following names: potassium nitrate, magnesium hypochlorite, lithium permanganate, barium acetate, calcium hydroxide, cesium peroxide, beryllium perchlorate, sodium dichromate, ammonium carbonate, potassium hydrogen sulfite.

9. Write the correct names for the following formulas: $KNO_2$, $Na_2O_2$, $Mg(OH)_2$, $Ca(ClO_2)_2$, $K_3PO_4$, $NaCN$, $Rb_2S_2O_3$, $BaC_2O_4$, $Li_2CrO_4$.

10. Write the correct formulas for the following names: gold(III) chloride, nickel(II) hydroxide, stannic fluoride, cadmium thiocyanate, lead(IV) sulfate, chromous carbonate, copper(II) phosphate, mercurous nitrate, silver hydrogen carbonate.

11. Write the correct names for the following formulas: $CoS$, $Fe_2(SO_4)_3$, $PbO$, $Ni(ClO_4)_3$, $CdBr_2$, $Cu_2O$, $Mn(NO_3)_2$, $HgSO_4$, $Cr_3(AsO_4)_2$, $Au(H_2PO_4)_3$.

12. Write the correct formulas for the following names: sulfur dioxide, sulfur trioxide, hydrogen selenide, dichlorine monoxide, phosphorus trichloride, carbon tetrafluoride, tetraboron decahydride (commonly called tetraborane).

13. Write the correct names for the following formulas: $PCl_3$, $CO$, $CO_2$, $N_2O_4$, $P_4O_6$ (the common name is phosphorus trioxide), $I_2O_5$, $BrF_3$, $Sb_2O_3$.

## SUPPLEMENTAL PROBLEMS

14. Name the following compounds:

   a. $CaF_2$   b. $AlCl_3$   c. $MgO$   d. $SrCl_2$   e. $CoCl_2$   f. $CoCl_3$

15. Name the following compounds.

    a. NO   b. $N_2O_3$   c. $N_2O_5$   d. $SF_4$   e. $SF_2$   f. $SF_6$

16. Name the following compounds.

    a. $SO_3$   b. CO   c. $SiF_4$   d. $N_2O_5$   e. $P_4S_{10}$   f. $XeF_4$

17. Name the following compounds.

    a. $PCl_5$   b. $IF_7$   c. $P_4O_6$   d. $As_4O_{10}$   e. $Cl_2O_3$   f. $Cl_2O_7$

18. Name the following oxyacids (assume they are dissolved in water).

    a. $H_3BO_3$   b. $HNO_2$   c. $H_2CO_3$   d. $HClO_3$   e. $HIO_3$   f. $HNO_3$
    g. $H_3PO_4$

19. Name the following acids (assume they are dissolved in water).

    a. HBr   b. HBrO   c. $HBrO_2$   d. $HBrO_3$   e. $HBrO_4$

20. Name the following compounds.

    a. $PbSO_3$        e. LiBrO          i. $CaCO_3$
    b. $Mg_2SiO_4$     f. $NH_4ClO_4$    j. $Ba(NO_2)_2$
    c. $Na_3PO_4$      g. $Fe_2(SO_4)_3$
    d. $Ca(NO_3)_2$    h. $FeSO_4$

21. Write the correct formulas for the following names.

    a. iron(III) sulfate              e. tellurium hexafluoride
    b. chromium(II) sulfite           f. mercury(I) acetate
    c. cupric sulfide                 g. disulfur dichloride
    d. silver dihydrogen phosphate    h. ammonium nitrate

22. Write the correct formulas for the following names.

    a. cobalt(III) sulfide         e. calcium bicarbonate
    b. bromine pentafluoride       f. tetraphosphorus trisulfide
    c. ferrous carbonate           g. cuprous cyanide
    d. gold(I) sulfate             h. lanthanum(III) phosphate

23. Write the correct formulas for the following names.

    a. iron(II) oxide      c. magnesium nitride
    b. potassium oxide     d. magnesium nitrite

e. pentasulfur dinitride  g. ferric sulfite  i. silver dichromate
f. calcium permanganate  h. lead(II) chromate

24. Sodium azide, along with KNO₃ and SiO₂, was used (until the late 1990s) in automobile airbags to generate nitrogen gas in a crash. The formula of the azide ion is $N^{3-}$. Write the formula of sodium azide. (In the late 1990s, Takata Corp. began using ammonium nitrate in their airbags. It caused trouble due to instability and led to many recalls, https://bit.ly/31Q9dA1. Currently, the organic molecule guanidinium nitrate is used to generate the nitrogen gas that inflates airbags.)

25. Many noble gas compounds have been prepared since 1962.
    a. Name the following: $XeF_6$, $XeF_4$, $XeF_2$.
    b. Write the correct formulas for the following names: krypton difluoride, xenon trioxide, xenon tetraoxide.

26. In 1987, the first interstellar (between the stars) compound containing phosphorus was discovered in a large gas cloud in the constellation Orion by **Lucy Ziurys**, who was then a postdoctoral student at the University of Massachusetts. It is a compound of nitrogen and phosphorus. Ziurys is now Regent's Professor of Chemistry & Biology and of Astronomy at the University of Arizona, https://bit.ly/2TnheJy. (Almost 200 different molecules have been observed in interstellar space. The building blocks for life as we know it are out there!) Write the name and formula for the simplest compound of nitrogen and phosphorus. (**HINT:** N and P form more than one compound. Others are $P_3N_3$ [a new form in the shape of a prism was discovered in 2021, https://go.nature.com/3uIDVus], $P_4N_6$ and $P_3N_5$.) **NOTE:** Prof. Ziurys' research group has found another simple phosphorus compound, PO, in evolved stars. Can you name it knowing that there are many phosphorus and oxygen compounds?

    From what you learned in Chapter 15, you should be able to draw the Lewis structure of PN. But you probably could not draw one for PO. See https://bit.ly/2VxFETa and https://bit.ly/2U0xMeF for more details.

---

## UPDATE ON NAMING NONMETAL — NONMETAL COMPOUNDS

In 2005, the IUPAC changed the recommendation about where oxygen should be in the order of naming nonmetal — nonmetal compounds. Before 2005, the order was that shown on page 298:

**Rn, Xe, Kr, B, Ge, Si, C, Sb, As, P, N, H, Te, Se, S, I, Br, Cl, O, F**

The new order as of 2005 is (following the periodic table more closely):

**Rn, Xe, Kr, B, Ge, Si, C, Sb, As, P, N, H, Te, Se, S, O, I, Br, Cl, F**

Oxygen would now be between S and I, following the periodic table more closely. Before 2005 it was between Cl and F. This new order would mean that, for example, a compound like chlorine dioxide, $ClO_2$, would be called dioxygen monochloride, $O_2Cl$. The change in order only affects compounds of O with I, Br and Cl. See Pages 42 and 260 in the IUPAC's *Nomenclature of Inorganic Chemistry: IUPAC recommendations 2005*, known affectionately as the *Red Book* because the cover of the printed version is red. The complete *Red Book* is available for a free download at https://bit.ly/3rLXgdU. **NOTE:** The sequence described on page 298 as the rational for the order of naming compounds is the reverse of that shown on page 260 in the *Red Book*. But the order of names comes out the same.

I have chosen to keep the older ordering list in this book because that is what many chemists are using.

**NantEnergy Rechargeable Zinc-Air Batteries**

The zinc-air battery shown above is rechargeable and these are now used in undeveloped areas of the world for electricity storage. The battery shown above is the size of a small suitcase. In many applications, 40 of these batteries are installed in a specially designed cabinet that connects them together. They are charged by solar panels. During the day, the solar panels supply electricity and charge all the batteries. During the night, the batteries supply electricity. As of 2018, over 3000 systems of 40 batteries per system are in use in 9 countries, mostly in Africa and Asia, and are the sole source of electricity for 200,000 people who never had electricity before. The batteries (with solar panels) are also in use in over 1000 cell towers worldwide. See https://nantenergy.com/ and https://bwnews.pr/2DV6fPU for more info.

The reaction taking place during discharge (making electricity) in this type of battery is $2Zn + O_2 \rightarrow 2ZnO$. The oxygen comes from the air, giving the battery a very high energy density. Also, these batteries use a mixture of water, potassium hydroxide, and ionic liquids (https://bit.ly/2r2TBKt) as the electrolyte.

They are not flammable, and do not need expensive temperature control systems as do the current lithium-ion batteries used for mass electricity storage and electric vehicles. Zinc is a non-toxic metal that is fairly inexpensive and in plentiful supply. Because of these properties, and because they are less expensive, rechargeable zinc-air batteries may become serious competition for lithium-ion batteries. See https://bit.ly/2GuV2sO. **NOTE:** Zinc-air batteries have been used in hearing aids for years; they are not rechargeable.

*Photo courtesy of NantEnergy, Inc*

# 17

## OXIDATION NUMBERS

Many chemical reactions involve the transfer of electrons from one atom to another. This is the way batteries supply electrons, which are electricity, to power a flashlight, your calculator, laptop computer, smartphone, and even a battery-powered car. It is also the way the cells in your body make high-energy ATP molecules from both glucose and the Krebs cycle, also called the citric acid cycle (https://bit.ly/3Su10v8). This chapter will show you how to calculate oxidation numbers, an important concept that will allow you to keep track of where electrons go in chemical reactions.

## 17-1
**Good Battery, Bad Battery: Oxidation and an Environmental Lesson**

In 1942, a new battery called the **mercury battery** came on the market. It was a great battery, much used in World War II by the United States Armed Forces in their portable electronic equipment such as the walkie talkie and handy talkie. After the war it found extensive use in consumer electronics. There were at least 23 different sizes of these batteries, from small button batteries that could be used in hearing aids to larger ones for portable and two-way radios. See https://bit.ly/3UC5hyh for more details.

A button-type mercury battery. Ruler in centimeters. Photo is CC by-SA 3.0. Attribution is "Andshel – Own work."

No matter how good they were, **mercury batteries were banned in Europe in 1991 and in the United States in 1996.** That is because *mercury is very toxic* (https://bit.ly/3r4SmY6) and dead batteries were being incinerated, thus polluting the air with mercury, or getting into landfills where mercury was leaching into groundwater. Even though mercury batteries were excellent in electronic components, there were other, and less toxic batteries, such as Li ion batteries, alkaline batteries, zinc-air batteries, and silver-oxide batteries, that could be used as substitutes. I discuss the mercury battery here because the chemistry is simple and it is a good way to introduce some serious environmental and health concerns.

The above discussion shows how a great product[1] can be an environmental disaster. Other examples of great products that can be environmental disasters are fossil fuels (oil and natural gas which give off the greenhouse gas carbon dioxide[2] when burned) and plastics, which get into the ocean and are degraded into microplastics that can adversely affect sea life (https://bit.ly/3DThfO8). It is not clear if microplastics are dangerous for humans. **NOTE:** Even though the data about the toxicity of mercury amalgam (It contains around 50% mercury, the rest being silver, tin, copper and zinc.)

---

[1]What I mean by a "great product" is one that is really good at doing the job it was designed for. It does not mean that it is nontoxic or good for the environment. **NOTE:** Mercuric oxide batteries are allowed in the United State for specialized medical and military uses if the user is licensed and agrees to recycle them properly.

[2]Carbon dioxide gas, in the proper concentration in the atmosphere, is vital for supporting the growth of living plants by allowing them to make glucose and recycle oxygen gas through photosynthesis (the reverse reaction of the one shown on p. 146, example 8). The energy to run this reverse reaction comes from sunlight. Only when $CO_2$'s concentration in the atmosphere becomes too high is it a problem.

# 306 OXIDATION NUMBERS

for filling cavities in teeth are not clear, as a precaution, I do not use mercury amalgam fillings in my teeth (https://bit.ly/3JCnTrO). Depending upon one's mouth chemistry and other conditions, mercury amalgam may leach out small amounts of mercury vapor that can be inhaled. Also, some people seem to be allergic to amalgam fillings. I prefer other restorative materials for tooth decay which do not contain mercury (https://bit.ly/3tRBarJ).

Now back to batteries. Let's write the chemical equations that describe what happens in a mercury battery. The reactions are as follows:

$Zn^0 \rightarrow Zn^{2+} + 2e^-$     Zinc metal ($Zn^0$) loses 2 electrons to give a zinc ion ($Zn^{2+}$).

$Hg^{2+} + 2e^- \rightarrow Hg^0$     The mercury ion ($Hg^{2+}$) takes 2 electrons to give mercury metal ($Hg^0$).

In the battery, the electrons do not go directly from zinc to mercury. They are forced to go through wires that connect to a resistance where they can do work or give off heat and/or light. The resistance can be a light bulb, a calculator, a digital watch, a motor, or any other electronic device. **NOTE:** In the actual battery, the zinc ions are in the compound zinc oxide and the mercury ions are in the compound mercury(II) oxide. Equations that show this are in Chapter 18.

In a chemical reaction, chemists need to know how many electrons each atom has lost or gained. To do this, chemists use the concept of oxidation numbers. **Oxidation** is defined as the losing of electrons. The **oxidation number** of an atom represents how many electrons an atom has lost or gained, assuming that there is no sharing of electrons. (This is, of course, not true for covalent compounds, but we can still make use of oxidation numbers as will be shown later in this chapter.) **NOTE:** The oxidation numbers for the atoms in the mercury battery above are: $Zn^0 = 0$, $Zn^{2+} = +2$, $Hg^{2+} = +2$, $Hg^0 = 0$. An oxidation number of zero means that the atom is neutral. It has not gained or lost any electrons.

Before we proceed with a more detailed discussion of oxidation numbers, we will introduce the concept of electronegativity.

## 17-2
### ELECTRONEGATIVITY IS A MEASURE OF THE ABILITY OF AN ATOM TO ATTRACT ELECTRONS TOWARD ITSELF IN A COVALENT BOND

The best way to decide which element has gained electrons and which has lost them is to use a table of electronegativities. **Electronegativity is a measure of the ability of an atom to attract electrons toward itself in a covalent bond.**[3] Figure 17-1 lists the electronegativities of the elements arranged according to the periodic table. Since the noble gases He, Ne and Ar don't form stable chemical bonds, electronegativity values for them can't be determined.

The concept of electronegativity and the method for calculating their values was introduced in 1932 by **Linus Pauling** (see page 265). The exact details of how he did this are beyond the scope of this book. The abbreviation for

---

[3]Even though the concept of electronegativity was originally formulated for covalent bonds, it can also be used for ionic bonds to help in determining oxidation numbers. It can also be used to help determine if a bond is ionic or covalent as will be discussed in a general chemistry course.

|   | IA | IIA | IIIB | IVB | VB | VIB | VIIB | VIII | | | IB | IIB | IIIA | IVA | VA | VIA | VIIA | VIIIA |
|---|---|---|---|---|---|---|---|---|---|---|---|---|---|---|---|---|---|---|
| 1 | H 2.1 | | | | | | | | | | | | | | | | | He — |
| 2 | Li 1.0 | Be 1.5 | | | | | | | | | | | B 2.0 | C 2.5 | N 3.0 | O 3.5 | F 4.0 | Ne — |
| 3 | Na 0.9 | Mg 1.2 | | | | | | | | | | | Al 1.5 | Si 1.8 | P 2.1 | S 2.5 | Cl 3.0 | Ar — |
| 4 | K 0.8 | Ca 1.0 | Sc 1.3 | Ti 1.5 | V 1.6 | Cr 1.6 | Mn 1.5 | Fe 1.8 | Co 1.8 | Ni 1.8 | Cu 1.9 | Zn 1.6 | Ga 1.6 | Ge 1.8 | As 2.0 | Se 2.4 | Br 2.8 | Kr — |
| 5 | Rb 0.8 | Sr 1.0 | Y 1.2 | Zr 1.4 | Nb 1.6 | Mo 1.8 | Tc 1.9 | Ru 2.2 | Rh 2.2 | Pd 2.2 | Ag 1.9 | Cd 1.7 | In 1.7 | Sn 1.8 | Sb 1.9 | Te 2.1 | I 2.5 | Xe — |
| 6 | Cs 0.7 | Ba 0.9 | La 1.1 | Hf 1.3 | Ta 1.5 | W 1.7 | Re 1.9 | Os 2.2 | Ir 2.2 | Pt 2.2 | Au 2.4 | Hg 1.9 | Tl 1.8 | Pb 1.8 | Bi 1.9 | Po 2.0 | At 2.2 | Rn — |

**FIGURE 17-1** A table of electronegativities arranged according to the periodic table. The number below the symbol of an element is its electronegativity. Notice that as you go up and to the right in the periodic table, the electronegativity values increase. Fluorine, in the upper right-hand part of the periodic table, has the largest electronegativity value of 4.0. Electronegativity values for the noble gases (group VIIIA) are not listed.

308  OXIDATION NUMBERS

electronegativity (or the plural, electronegativities) that we will use in this book is EN.

Referring to Figure 17-1, notice that the EN numbers range from 4.0 for fluorine (F) to 0.7 for cesium (Cs). The higher the EN number, the greater the tendency of an atom to attract electrons toward itself in a covalent bond. Thus fluorine has the greatest tendency to attract electrons since it has the largest EN.

**EXAMPLE 1**  Which element has a greater tendency to attract electrons toward itself, nitrogen (N) or hydrogen (H)?

*Solution:* From Figure 17-1, the EN of nitrogen is 3.0. The EN of hydrogen is 2.1. Nitrogen has the greater tendency to attract electrons because its EN is greater than hydrogen's. ∎

**EXAMPLE 2**  Of all the elements in the periodic table, which element has the greatest tendency to attract electrons to itself?

*Solution:* It is the element with the largest EN, namely, fluorine (F). ∎

Since EN is a measure of the ability of an atom to attract electrons toward itself in a covalent bond, and the oxidation number represents how many electrons an atom has lost, we can say the following: In a compound, the atom with the lower EN will have the more positive oxidation number, and the atom with the higher EN will have a less positive (or more negative) oxidation number.

Remember the following:

**lower electronegativity → more positive oxidation number**

**higher electronegativity → more negative oxidation number**

Let's look at the molecule ammonia, $NH_3$. Which element has the greater tendency to take electrons? From Example 1, the EN of nitrogen is 3.0 and the EN of hydrogen is 2.1. The nitrogen has the greater EN; thus it has the greater tendency to take electrons. The hydrogen has the smaller EN; thus it has a lesser tendency to take electrons. The nitrogen will have the more negative oxidation number, and the hydrogen will have the more positive oxidation number.

At this point, a word of clarification is in order. In a covalent molecule like $NH_3$, there really isn't a "loss" and a "gain" of electrons. There is an unequal sharing. The nitrogen, being more electronegative than the hydrogen, tends to pull the electrons toward itself in the molecule. This "pulling" is noticeable and affects the properties of the molecule. With this in mind, and to simplify things in this chapter, we will say that in $NH_3$ the N has "gained" electrons and the H has "lost" electrons. In ionic compounds like NaCl, the Na (EN = 0.9) really does lose an electron to the Cl (EN = 3.0).

**EXAMPLE 3** In methane, $CH_4$, which element has "lost" electrons and which has "gained" them? Which element has the more positive oxidation number?

*Solution:* The EN are C = 2.5 and H = 2.1. Therefore, since C has the greater EN, C has "gained" electrons and H has "lost" them. The hydrogen has the more positive oxidation number. ∎

You might have noticed that the EN difference of the elements in covalent molecules like $CH_4$ is small (2.5 − 2.1 = 0.4), whereas the EN difference in ionic compounds like NaCl is large (3.0 − 0.9 = 2.1). However, when we use EN to help us determine oxidation numbers, the distinction between ionic and covalent compounds is not important. In fact, you may want to think of all bonds as being ionic when you are determining oxidation numbers. Just remember in the back of your mind that many bonds are covalent.

In calculating oxidation numbers, we always assume a complete gain or loss of electrons. As we have said, this is not true for most bonds. Therefore, the oxidation numbers are somewhat artificial. Nevertheless, they are useful in keeping track of the *changes* in oxidation numbers that occur to atoms during a chemical reaction.

# 17-3
## CALCULATING OXIDATION NUMBERS IS BASED ON THE ELECTRONEGATIVITIES OF THE ELEMENTS

There are a few rules that will help you to calculate oxidation numbers. A listing and discussions of these rules follow.

### RULE 1
**Atoms of elements in their elemental form have an oxidation number of zero.**
Thus the atoms in $H_2$, $O_2$, $Cl_2$, $N_2$, $S_8$, Fe, Na, and Ne all have an oxidation number of zero. Their oxidation number is zero because there is complete and equal sharing of the valence electrons in $H_2$, $O_2$, $Cl_2$, $N_2$, and $S_8$. Nothing has lost electrons and nothing has gained them. In Fe and Na the

valence electrons are shared throughout the metal in a special kind of bond called a **metallic bond.** In Ne the atoms exist individually, and no sharing takes place.

## RULE 2
**The oxidation number of a monatomic (one atom) ion is equal to the charge of the ion.** In $Cl^-$, the oxidation number of chlorine is $-1$. In $Al^{3+}$, the oxidation number of aluminum is $+3$.

## RULE 3
**In a neutral molecule, the oxidation numbers of all the individual atoms add up to zero. In an ion, the oxidation numbers add up to the charge on that ion.** Thus in $CH_4$, the sum of the oxidation numbers of the one C atom and the four H atoms equals zero. In $SO_4^{2-}$, the sum of the numbers of each type of atom equals $-2$.

## RULE 4
**In a molecule or ion consisting of two atoms, the atom with the larger EN has the more negative oxidation number.**

## RULE 5
**Fluorine, F, which is the most electronegative element, always has an oxidation number of $-1$ in compounds.** Notice that F is in group VIIA and can attract one electron to complete its valence shell.

## RULE 6
**Hydrogen, H, usually has an oxidation number of $+1$ in compounds. However, when hydrogen combines with metals to form compounds called hydrides, it has an oxidation number of $-1$.** Most metals have an EN that is less than hydrogen's, so hydrogen would be negative with respect to these metals. Remember that hydrogen has one electron which it can lose. Hydrogen can also gain one electron to complete its $s$ orbital.

## RULE 7
**Oxygen usually has an oxidation number of $-2$ in compounds. When oxygen combines with fluorine, it has an oxidation number of $+2$. In peroxides (such as $H_2O_2$), oxygen has an oxidation number of $-1$.** Oxygen is in group VIA and needs two electrons to complete its valence shell. The EN of oxygen is 3.5, which is the second highest EN. Thus oxygen has a positive oxidation number *only* with respect to fluorine.

## RULE 8
**Metals in group IA have an oxidation number of +1 in compounds. Metals in group IIA have an oxidation number of +2 in compounds.**

A few examples will illustrate the application of these rules.

**EXAMPLE 4**   Calculate the oxidation number of each atom in the following: He, $P_4$, $F_2$, Au, K, Kr, $Br_2$.

*Solution:* Since these are all elements, the oxidation number of each atom is zero (Rule 1).   ■

**EXAMPLE 5**   Calculate the sum of the oxidation numbers in each of the following: $CO_2$, Ar, $PCl_5$, $H_2SO_4$, $MgCl_2$, $CO_3^{2-}$, $NH_4^+$, $Fe^{3+}$, $S^{2-}$.

*Solution:* Using Rules 2 and 3, we have

| FORMULA | SUM OF OXIDATION NUMBERS |
|---|---|
| $CO_2$ | 0 |
| Ar | 0 |
| $PCl_5$ | 0 |
| $H_2SO_4$ | 0 |
| $MgCl_2$ | 0 |
| $HCO_3^-$ | $-1$ |
| $CO_3^{2-}$ | $-2$ |
| $NH_4^+$ | $+1$ |
| $Fe^{3+}$ | $+3$ |
| $S^{2-}$ | $-2$ |

■

Now we must learn how to calculate the actual numerical value of the oxidation number for each atom in a molecule or polyatomic ion. To do this, we will refer to the periodic table and the table of electronegativities (Figure 17-1). We will also make use of the eight rules just discussed.

**EXAMPLE 6**   What is the oxidation number of each atom in HCl?

*Solution:* The EN are H = 2.1 and Cl = 3.0. Thus H has an oxidation number of +1 (Rule 6). Cl must have an oxidation number of −1 (Rule 3). Notice that

# 312 OXIDATION NUMBERS

Cl, like the other halogens, needs one electron to complete its valence shell. Thus in compounds without oxygen or fluorine, Cl will have an oxidation number of $-1$. Oxygen and fluorine have a larger EN than Cl; thus in compounds with oxygen and fluorine, Cl will have a positive oxidation number. ∎

**EXAMPLE 7** What is the oxidation number of each atom in sodium hydride, NaH?

*Solution:* The EN are Na = 0.9 and H = 2.1. Therefore, H has an oxidation number of $-1$ (Rule 6), and Na has an oxidation number of $+1$ (Rules 3 and 8). ∎

**EXAMPLE 8** What is the oxidation number of each atom in boron nitride, BN?

*Solution:* The EN are B = 2.0 and N = 3.0. Thus N is negative with respect to B. To determine the numerical values of the oxidation numbers, we see that N is in group VA of the periodic table and needs three electrons to complete its valence shell. Thus the oxidation numbers are N = $-3$ and B = $+3$. ∎

**EXAMPLE 9** What is the oxidation number of each element in water, $H_2O$?

*Solution:* The EN values are H = 2.1 and O = 3.5. Thus the oxidation number of O is $-2$ (Rule 7), and that of each H is $+1$ (Rule 6). To see whether Rule 3 is obeyed, add up all the oxidation numbers of each atom: $+1 + 1 - 2 = 0$. ∎

A good way to keep track of oxidation numbers is to write them in the following way.

1. Write the chemical formula.
2. Above each atom, write its oxidation number if you know it.
3. Then use Rule 3 to calculate the oxidation numbers of the remaining atoms. Usually there will be only one remaining atom.
4. On the first line write the oxidation number of each single atom.
5. Above these numbers write the total oxidation number for all the atoms of that element.

17-3 Calculating Oxidation Numbers    313

Let's see how this would look for water:

$$+2\ -2 \leftarrow \text{sum of the oxidation numbers of all the atoms of an element}$$

$$+1\ -2 \leftarrow \text{oxidation number of each single atom}$$

$$H_2O$$

The top line of numbers should add up to zero for a neutral molecule. If you have an ion, the top line will add up to the charge on the ion.

**EXAMPLE 10**    What is the oxidation number of sulfur in sulfur trioxide, $SO_3$?

*Solution:* The EN are S = 2.5 and O = 3.5. Therefore, S is positive with respect to oxygen. The oxidation number of oxygen is $-2$ (Rule 7). Therefore, we have

$$+6\ -6 \leftarrow \text{sum of oxidation numbers}$$

$$+6\ -2 \leftarrow \text{individual oxidation numbers}$$

$$S\ O_3$$

Thus the oxidation number of S is +6, the number directly above the S.   ■

**EXAMPLE 11**    What is the oxidation number of sulfur in sulfur dioxide, $SO_2$?

*Solution:* The EN are S = 2.5 and O = 3.5. Sulfur is positive with respect to oxygen; thus we have

$$+4\ -4$$

$$+4\ -2$$

$$S\ O_2$$

The oxidation number of S in $SO_2$ is +4, which is the number directly above the S.   ■

Examples 10 and 11 illustrate an important idea. The oxidation number of a given element may vary, depending on what compound it is in. In fact, the oxidation number of sulfur can range from +6 to −2 (see Problem 28).

314  OXIDATION NUMBERS

**EXAMPLE 12**   What is the oxidation number of sulfur in hydrogen sulfide, $H_2S$?

*Solution:* The EN are H = 2.1 and S = 2.5. Thus S is negative with respect to H. The oxidation number of H is +1 (Rule 6). Therefore, we have

$$\begin{array}{c} +2\ -2 \\ +1\ -2 \\ H_2\ S \end{array}$$

The oxidation number of S in $H_2S$ is −2, the number directly above the S.  ∎

**EXAMPLE 13**   What is the oxidation number of sulfur in the sulfate ion, $SO_4^{2-}$?

*Solution:* The EN are S = 2.5 and O = 3.5. The S is positive with respect to oxygen. $SO_4^{2-}$ is an ion with a charge of −2, so the sum of all the oxidation numbers must add up to −2 (Rule 3):

$$+6\ -8 = -2 \leftarrow \text{this is the charge on the ion}$$

$$\begin{array}{c} +6\ -2 \\ S\ O_4 \end{array}$$

The sum of the oxidation numbers is +6 − 8 = −2. The oxidation number of S = +6, which is the number directly above the S.  ∎

**EXAMPLE 14**   What is the oxidation number of each atom in the thiosulfate ion, $S_2O_3^{2-}$?

*Solution:*

$$+4\ -6 = -2 \leftarrow \text{this is the charge on the ion}$$

$$\begin{array}{c} +2\ -2 \\ S_2O_3 \end{array}$$

The sum of the oxidation numbers of the three oxygen atoms is −6. Thus the two S atoms must share a +4 total oxidation number. Each S atom has a +4/2 = +2 oxidation number, the number directly above the S.  ∎

17-3 Calculating Oxidation Numbers    315

**EXAMPLE 15**    What is the oxidation number of each element in $OF_2$?

*Solution:* The EN are O = 3.5 and F = 4.0. F is negative with respect to oxygen. The oxidation number of F is always −1 (Rule 5).

$$+2\ -2$$
$$+2\ -1$$
$$O\ F_2$$

The oxidation numbers are O = +2 and F = −1. You might check Rule 7 at this point. ■

**EXAMPLE 16**    What is the oxidation number of each element in $NaClO_4$?

*Solution:* The EN are Na = 0.9, Cl = 3.0, and O = 3.5. The Na and Cl are positive with respect to the oxygen. The oxidation number of Na is +1 (Rule 8), and that of O is −2 (Rule 7):

$$+1\ +7\ -8$$
$$+1\ +7\ -2$$
$$Na\ Cl\ O_4$$

The oxidation number of Cl is +7, since +1 + 7 − 8 = 0. ■

In Rule 7 we mentioned peroxides. In peroxides the oxygen atoms are bound to each other. The structure of hydrogen peroxide, $H_2O_2$, is

$$H-O-O-H$$

You can see that the oxygen atoms are bonded to each other. This increases the oxidation number of each oxygen atom (makes the oxidation number less negative) as the following example shows.

**EXAMPLE 17**    What is the oxidation number of each element in $H_2O_2$?

*Solution:* The EN are H = 2.1 and O = 3.5. Hydrogen is positive with respect to oxygen. Hydrogen has an oxidation number of +1 (Rule 6).

$$\begin{array}{c} +2\ -2 \\ +1\ -1 \\ H_2O_2 \end{array}$$

The oxidation number of each oxygen atom is $-1$, since $-2/2 = -1$.

## 17-4
### CALCULATING THE OXIDATION NUMBER OF ATOMS IN CARBON COMPOUNDS USES THE ELECTRONEGATIVITIES OF THE ATOMS

We will discuss one more interesting point about oxidation numbers. This is how to calculate the oxidation number of carbon in certain organic compounds. For many organic compounds, we have already learned the proper technique. For instance, in $CH_4$ the oxidation numbers are $C = -4$ and $H = +1$. In $CO_2$ the oxidation numbers are $C = +4$ and $O = -2$. Notice that carbon can have oxidation numbers ranging from $+4$ to $-4$. This is because carbon is in group IVA in the periodic table. It has four valence electrons. Thus a carbon atom can either "gain" four electrons to get eight electrons in its valence shell or "lose" four electrons to have none in its valence shell. Carbon can also "gain" or "lose" any number of electrons between $+4$ and $-4$.

The problems we will discuss here is calculating oxidation numbers in compounds like acetic acid ($C_2H_4O_2$) and acetone ($C_3H_6O$). For acetic acid, we can write (EN are $C = 2.5$, $H = 2.1$, and $O = 3.5$)

$$\begin{array}{c} 0\ +4\ -4 \\ 0\ +1\ -2 \\ C_2H_4O_2 \end{array}$$

Since the oxidation numbers of the H and O atoms add up to zero, the total oxidation number of both carbon atoms appears to be zero. But since there are two carbon atoms, it is possible that the apparent zero oxidation number is simply the average of equal positive and negative oxidation numbers. As we shall see later, this is actually the case. One carbon atom has an oxidation number of $+3$, and the other has an oxidation number of $-3$. The average of $+3$ and $-3$ is zero. We will discuss acetic acid in detail after we make a few more points.

Now let's look at acetone, $C_3H_6O$:

$$\begin{array}{c} -4\ +6\ -2 \\ -1.33\ +1\ -2 \\ C_3H_6O \end{array}$$

## 17-4 Calculating Oxidation Numbers of Atoms in Carbon Compounds

The oxidation number of O is $-2$, and that of H is $+1$. Thus the sum of the oxidation numbers of hydrogen and oxygen is $6(+1) - 2 = +4$. The total oxidation number of the three carbon atoms must equal $-4$. If we divide $-4$ equally among three carbon atoms, we get $-4/3 = -1.33$. It appears that each carbon atom has an oxidation number of $-1.33$. How can this be? Oxidation numbers are usually whole numbers. (However, see Problem 27 for an exception.)

The answer is that the oxidation numbers in acetone *are* whole numbers, but the oxidation numbers on different carbon atoms are not necessarily the same. To see this, we must resort to structural formulas.

Let's start with methane ($CH_4$) and carbon dioxide ($CO_2$). The structural formulas are

$$\begin{array}{c} H \\ | \\ H-C-H \\ | \\ H \end{array} \qquad O=C=O$$

**A good way to figure out oxidation numbers if you have structural formulas is to put a "+" sign on each atom's bond that has the lower EN and to put a "−" sign on each atom's bond that has the higher EN.**

Doing this for methane and carbon dioxide, we have (the EN are C = 2.5, H = 2.1, and O = 3.5)

$$\begin{array}{c} H \\ {\scriptstyle +}|{\scriptstyle -} \\ H \overset{+}{=} C \overset{-}{=} H \\ {\scriptstyle -}|{\scriptstyle +} \\ H \end{array} \qquad O \overset{-}{\underset{-}{\overset{+}{=}}} C \overset{-}{\underset{-}{\overset{+}{=}}} O$$

Add up the "+" signs and "−" signs around each atom. The result is the oxidation number of each atom. In methane the C atom has four "−" signs around it; therefore, the oxidation number of C is $-4$. The oxidation number of each H is $+1$. For $CO_2$ the oxidation number for C is $+4$, and that of each O is $-2$.

Now back to acetic acid. The structural formula with "+" signs and "−" signs is

$$\begin{array}{c} H \\ {\scriptstyle +}|{\scriptstyle -} \\ H \overset{+}{=} C \overset{0}{-} \overset{0}{C} \overset{+}{\diagup} O \\ {\scriptstyle -}|{\scriptstyle +} \qquad \diagdown \\ H \qquad\qquad O \overset{-}{=} H \end{array}$$

**NOTE:** If a bond connects two identical atoms, such as C—C, there is no "gaining" or "losing" of electrons in this bond. There is a complete sharing, and we put a zero at each end of the bond.

Now add up all the "+" signs and "−" signs around each atom. All the H atoms have +1. All the O atoms have −2. These are the oxidation numbers of H and O atoms, respectively. Now look at the carbon atoms. The C atom on the left has a total charge of −3. Its oxidation number is −3. The C atom on the right has a total charge of +3, so its oxidation number is +3. The average of +3 and −3 is zero, just what we found before as the average oxidation number of both carbon atoms. (If you work Problem 29, you will see a carbon atom that actually has a zero oxidation number.)

For acetone the structural formula with "+" signs and "−" signs is

$$\begin{array}{c} \text{H} \quad\; \text{O} \quad\; \text{H} \\ | \quad\;\; || \quad\;\; | \\ \text{H}-\text{C}-\text{C}-\text{C}-\text{H} \\ | \quad\;\;\;\;\;\;\; | \\ \text{H} \quad\;\;\;\;\; \text{H} \end{array}$$

The oxidation number of each H atom is +1, and that of the O atom is −2. The oxidation number of the leftmost carbon is $-3 + 0 = -3$. The oxidation number of the rightmost carbon atom is the same, −3. However, the oxidation number of the central carbon atom is $+2 + 0 = +2$. The average oxidation number of all three carbon atoms is

$$\frac{-3 + 2 - 3}{3} = \frac{-4}{3} = -1.33$$

which is the same as we found before.

If you study the structural formulas of acetic acid and acetone, you can see that the carbon atoms bonded to oxygen atoms (which are very electronegative) are more positive than carbon atoms bonded to hydrogen atoms (which are much less electronegative than oxygen). Carbon atoms with different oxidation numbers undergo very different kinds of chemical reactions, as you may learn in your further study of chemistry.

**EXAMPLE 18** What is the oxidation number of each atom in a molecule of chloroacetic acid, $CH_2ClCOOH$? The structural formula is

## 17-4 Calculating Oxidation Numbers of Atoms in Carbon Compounds

*Solution:* The difference between chloroacetic acid and acetic acid is the Cl atom replacing an H atom. The EN are Cl = 3.0, C = 2.5, and O = 3.5. Putting in "+" signs and "−" signs we have

$$\text{Cl}\overset{-}{-}\overset{+}{\underset{\underset{H}{|}}{\overset{\overset{H}{|}}{\underset{+}{\overset{+}{C}}}}}\overset{0}{-}\overset{0}{\underset{+}{C}}\overset{-}{\overset{\diagup\hspace{-0.3em}\diagdown}{\underset{\diagdown\hspace{-0.3em}\diagup}{}}}\begin{matrix}O\\ \\O\overset{-}{-}\overset{+}{H}\end{matrix}$$

The oxidation numbers are Cl = −1, H = +1, and O = −2. The oxidation number of the left-hand carbon atom is −1 − 1 + 1 + 0 = −1. The oxidation number of the right-hand carbon is +3. Notice that in acetic acid the oxidation number of the left-hand carbon atom was −3. Since Cl is more electronegative than H, it has the effect of making the oxidation number of the left-hand carbon atom in chloroacetic acid more positive (−1 is more positive than −3). The oxidation number of the right-hand carbon hasn't changed when compared to acetic acid, because the atoms bonded to it haven't changed. ∎

As we said earlier, carbon atoms can have oxidation numbers ranging from +4 to −4. If you look at the Lewis formula of a carbon atom, $\ddot{\text{C}}\cdot$, you see that carbon can "lose" up to four electrons and "gain" up to four electrons in compounds. This is why the oxidation number of carbon can range from +4 to −4.

## PROBLEMS

### KEYED PROBLEMS

1. Which element has a greater tendency to attract electrons toward itself in a covalent bond, oxygen (O) or hydrogen (H)?

2. Of all the elements in the periodic table, which element has the second greatest tendency to attract electrons toward itself in a covalent bond?

3. In the molecule HF, which atom has "lost" electrons and which has "gained" electrons? Which atom has the more positive oxidation number?

4. What are the oxidation numbers of each atom in the following: Ne, $O_2$, Ca, $Cl_2$, $S_8$, $O_3$?

5. What is sum of the oxidation numbers in each of the following: $H_2O$, CO, $HNO_3$, $CaCl_2$, $H_2SO_4$, $HSO_4^-$, $PO_4^{3-}$, $NO^+$, $NH_3$, $NH_4^+$, $Br^-$, $PCl_3$?

6. What is the oxidation number of each atom in HBr?

7. What is the oxidation number of each atom in lithium hydride, LiH?

8. What is the oxidation number of each atom in aluminum nitride, AlN?

9. What is the oxidation number of each atom in hydrogen sulfide, $H_2S$?

10. What is the oxidation number of each atom in chlorine trioxide, $ClO_3$?

11. What is the oxidation number of each atom in chlorine dioxide, $ClO_2$?

12. What is the oxidation number of each atom in $H_2Se$?

13. What is the oxidation number of each atom in the perchlorate ion, $ClO_4^-$?

14. What is the oxidation number of each atom in the tetrathionate ion, $S_4O_6^{2-}$?

15. What is the oxidation number of each atom in $ClF_3$?

16. What is the oxidation number of each atom in $NaClO_3$?

17. What is the oxidation number of each atom in sodium peroxide, $Na_2O_2$?

18. What is the oxidation number of each atom in fluoroacetic acid, whose structural formula is

$$\text{F}-\underset{\underset{\text{H}}{|}}{\overset{\overset{\text{H}}{|}}{\text{C}}}-\text{C}\underset{\text{O}-\text{H}}{\overset{\text{O}}{\diagup\!\!\!\diagdown}}$$

## SUPPLEMENTAL PROBLEMS

19. What is the oxidation number of each atom in the following?
    a. BrCl     c. $BrF_3$     e. ClF
    b. BrF      d. $BrF_5$     f. $ClF_5$

20. What is the oxidation number of each atom in the following?
    a. $HClO_4$    c. $HClO_2$    e. $Cl_2$
    b. $HClO_3$    d. HClO        f. HCl

21. What is the oxidation number of each atom in the following?
    a. $N_2O$     c. $N_2O_3$    e. $N_2O_5$    g. $NH_2OH$
    b. NO         d. $NO_2$      f. $NH_3$      h. $N_2H_4$

**NOTE:** If you do all of Problem 21, you will see that the N atom can have oxidation numbers ranging from +5 to −3. If you look at the Lewis formula of a nitrogen atom, $\cdot\ddot{\text{N}}\cdot$, you see that nitrogen can "lose" up to five electrons or "gain" up to three electrons in compounds. This is why the oxidation number of nitrogen can go from +5 to −3.

22. What is the oxidation number of each atom in the following?
   a. $H_3PO_4$    c. $HPO_4^{2-}$    e. $H_3PO_2$
   b. $H_2PO_4^-$  d. $PO_4^{3-}$    f. $H_2PO_2^-$

23. What is the oxidation number of each atom in the following?
   a. MnO         c. $Mn_2O_7$      e. $MnCl_3$
   b. $Mn_2O_3$   d. $MnCl_2$       f. $KMnO_4$

24. What is the oxidation number of each atom in the following?
   a. CuCl        c. $Hg_2Cl_2$     e. $AuCl_4^-$
   b. $CuCl_2$    d. $HgCl_2$       f. AuCl

25. What is the oxidation number of each atom in the following?

   a.

   H   H
   |   |
   H—C—C—O—H
   |   |
   H   H

   b.

   H        O
   |       //
   H—C—C
   |       \
   H        H

   c.

   H        O
   |       //
   H—C—C
   |       \
   H        O—H

   NOTE: As the number of oxygen bonds attached to a carbon atom increases, the oxidation number of this carbon atom increases. The more oxygen atoms attached to a carbon atom, the more the carbon atom is oxidized.

26. What is the oxidation number of each atom in the following?

   a.

   H       H
   |       |
   H—C—O—C—H
   |       |
   H       H

   b.

   F   H
   |   |
   F—C—C—H
   |   |
   F   H

   c.

   H
   |
   H—C—C≡N
   |
   H

   NOTE: In Problem 25, going from left to right, the molecules are called **ethanol**, **acetaldehyde**, and **acetic acid**. The rightmost carbon in each molecule has fewer H atoms attached to it as we go from a to c, thus increasing the oxidation number of that carbon. Ethanol is oxidized in the liver by an enzyme called **alcohol dehydrogenase** (means removes hydrogen) to acetaldehyde. The acetaldehyde is oxidized by the enzyme **aldehyde dehydrogenase** to acetic acid. Acetic acid is non-toxic. However, if a person drinks **methanol**, the final oxidized product is **formic acid**, a highly toxic substance. See https://theskepticalchemist.com/methanol-toxic-ethanol/.

   d.

   H  H      H
   |  |     /
   H—C—C—N
   |  |     \
   H  H      H

There is also an intermediate oxidation product from the oxidation of methanol. It is called formaldehyde. Its chemical structure is similar to acetaldehyde (problem 25b above) except that the middle carbon atom along with its 2 hydrogen atoms are removed. Formaldehyde is very toxic. One use is to preserve dead biological specimens. You don't want it building up in your body!

27. The superoxide ion, $O_2^-$, is important in some biological systems. What is the oxidation number of the oxygen atoms in the superoxide ion?

28. As Examples 10-14 show, sulfur can have an oxidation number that ranges from −2 to +6. Draw the Lewis formula of the sulfur atom and convince yourself that this range is reasonable. See problem 21 for some hints.

## 322 OXIDATION NUMBERS

29. Calculate the oxidation number of the carbon atom in formaldehyde, whose structural formula is at the right. Here is a compound where the carbon atom has a zero oxidation number.

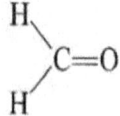

## THE STRUCTURE OF SMALL COMPLEX MOLECULES (Part 2, Part 1 is on page 285)

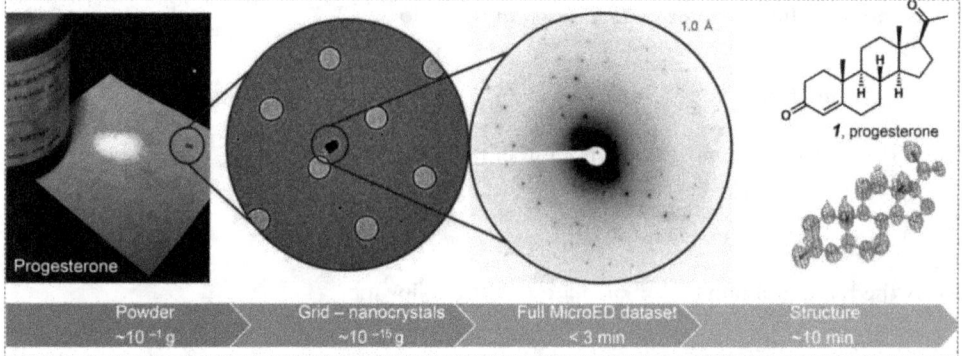

*Progesterone graphic above courtesy of the Gonen Laboratory.*

"Cryo" stands for cryogenic(s), which is the study of materials at very low temperatures. The resolution of this system is about 0.1 nm. The approximate distance between the carbon atoms in the progesterone molecule is 0.14 nm. Since the resolution of the system is smaller than the distance between carbon atoms, it is possible to resolve individual carbon atoms in the progesterone molecule. In the graphic above, 1 Å = 0.1 nm and is intended to show the resolution that the spots will give in determining the structure of progesterone. *

   **NOTE:** The Ångström unit (Å) is named after the Swedish physicist **Anders Jonas Ångström** (1814–1874). The Ångström unit is not an SI unit but is commonly used when measuring atomic and molecular sizes. It is also not in the metric system of units. The other commonly used unit to measure atomic sizes is the picometer (pm) where 1000 pm = 1 nm.

   The amazing thing is that the structure could be determined in less than 30 minutes—from taking the powder out of the bottle to seeing the structure on the computer monitor. With other common methods, determining such a complex structure could take hours or days. See https://bit.ly/2RKGKYy for more details. Also check out Prof. Gonen's home page (https://cryoem.ucla.edu/) and look at the link to "Our Equipment" to see not only his lab equipment, but also the powerful computer needed to perform the calculations for his research. His paper at https://bit.ly/2H7McCh shows some of the molecules whose structures his group has determined. See https://bit.ly/3qnllE4 for an interview with Prof. Gonen.

   As mentioned in the footnote on page 285, cryo-EM is also used to determine the structure of proteins. In early 2020, researchers used cryo-EM to determine how the novel coronavirus (SARS-CoV-2, the cause of the Covid-19 pandemic) uses its "spike" protein to latch onto the ACE2 receptor protein on the surface of lung cells. **NOTE:** The omicron variant is more contagious than other variants (so far) because it latches onto lung cells more efficiently (https://bit.ly/3rPisiK).

   When the spike protein is attached to a virus, it has a stubby shape. After it latches on to a cell, it springs open to an elongated "spear" form that fuses with the cell. If the spike protein is not attached to a virus but on the surface of a cell

---

*Norethindrone, a synthetic form of progesterone, used in one of the earliest birth control pills, was first synthesized by the 26-yr old Mexican chemistry student **Luis E. Miramontes** while doing his undergraduate bachelor's thesis research at the company Syntex in Mexico City.

exposing it to blood, as it would be after a person receives an mRNA vaccine (https://bit.ly/3k2zSUM), it spontaneously converts to the spear form, which would not create the proper antibodies to fight the virus. Antibodies must neutralize the stubby-shaped spike protein before it can latch onto a cell. Using information from cryo-EM, **Jason McLellan** at U Texas/Austin and **Barney Graham** at the NIH created a mutation of the spike protein that doesn't convert to the spear form when it is not attached to the virus but on the surface of a cell. They inserted two proline molecules, the most rigid amino acid, into the spike protein that stiffens it and keeps it from elongating. **Moderna and Pfizer-BioNTech use this mutation, called the 2P mutation, in their highly effective mRNA vaccines, https://bit.ly/2YD9HJt.**

The other **great innovation used in making the successful Covid-19 vaccines by Moderna and Pfizer-BioNTech** was discovered by **Katalin (Kati) Karikó**, who is a Senior Vice President at BioNTech, and her collaborator **Drew Weissman** of the University of Pennsylvania. Natural mRNA that is free in the blood is attacked by the immune system and destroyed. See https://nyti.ms/3jc2nia. To use an mRNA vaccine, the mRNA must be shielded from attack by the immune system. Karikó and Weissman discovered a modification they could make in the structure of mRNA so that it is not attacked by the immune system, thus allowing it to get into cells where the cells make copies of the spike protein. The spike protein then migrates to the surface of the cell where the immune system can see it and make antibodies against it. See https://bit.ly/3r5jlTZ.

Much research on finding mRNA vaccines against cancer is being done. Preliminary trials are encouraging but much more work is needed. See https://bit.ly/3jV4szw.

Cryo-EM has been used to determine the structure of the virus's RNA polymerase, a protein which helps the coronavirus make copies of its RNA. (Coronaviruses don't have DNA.) If the virus can't make copies of its RNA, it can't make copies of itself, and can't spread the infection. Knowing the structure of the virus's RNA polymerase could lead to a treatment for Covid-19 if researchers could find a small molecule that inhibits the virus's RNA polymerase, https://bit.ly/3c3NYzd. In 2021, **Marti Head**, Director, Joint Institute of Biological Sciences, Oak Ridge National Laboratory and her large team, using ORNL's Summit supercomputer (from 6/2018 to 6/2020 it was the fastest supercomputer in the world) discovered a small molecule that blocks the coronavirus's RNA polymerase from working. See https://bit.ly/3x4tMKU. **NOTE:** In June 2022, ORNL's Frontier supercomputer became the fastest in the world.

See https://nyti.ms/3lzgLQI for a wonderful article showing high-resolution cryo-EM studies of the coronavirus, complete with animations.

**Another very important method of determining the structure of molecules is to use x-rays with a technique called *x-ray crystallography*.** The crystals must be about a billion times larger (in volume) than for electron crystallography.[2] These crystals are usually grown in the laboratory until they are at least 0.1 mm in size.

Perhaps the most famous x-ray crystallography photo, Photograph 51 of DNA (https://bit.ly/2QHxHVi), was taken in May 1952 by **Rosalind Franklin** (1920-1958) and her doctoral student **Raymond Gosling** (1926-2015). The X pattern of the "dots" is typical of a stretched spring's diffraction pattern (https://bit.ly/2UIf5Kh), giving **James Watson** (b. 1928) and **Francis Crick** (1916-2004) a critical clue that they needed to figure out the spiral (double helix) structure of DNA. They, along with **Maurice Wilkins** (1916-2004), won the 1962 Nobel Prize in Physiology or Medicine for this discovery. **Rosalind Franklin**, whom many believe should also have received a Nobel Prize, died before it was awarded for this discovery (only living persons can receive Nobel Prizes). All the above scientists worked at the University of Cambridge in England and King's College in London. See https://bit.ly/3oyYkzP for the story of this research.

---

[2]This is because the interaction of x-rays with atoms is much weaker than the interaction of electrons with atoms.

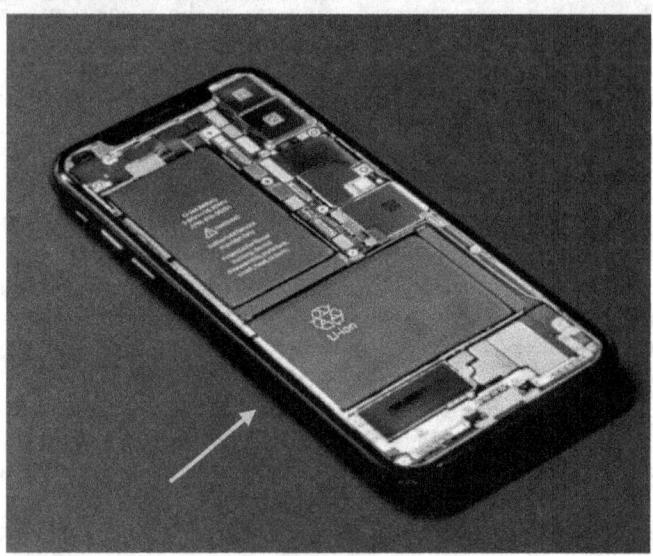

**A Lithium-Ion Battery in a Smartphone (at arrow):** They are one of the most popular types of rechargeable batteries for consumer electronics, with a high energy density, tiny memory effect and low self-discharge. Li-ion batteries are also very popular in military and aerospace applications, and in battery-powered electric vehicles,. The first commercial Li-ion battery was released in 1991. Lithium is the lightest of all metals and Li-ion batteries have the greatest voltage of any battery.

Lithium-ion batteries can pose unique safety hazards since they contain a flammable electrolyte (see page 404). A battery charged too quickly or that is defective could lead to a short circuit, causing a fire and/or an explosion. Even electric vehicles have burned due to defective batteries and/or software, leading to massive recalls. These events are very rare and when they happen, they make the news.

In an attempt to correct these problems and increase the energy density, the major auto companies and many smaller companies, including startups, are funding massive research projects to come up with safer, less environmentally harmful, and higher energy batteries that have a short recharge time and are less expensive (https://nyti.ms/36CZpkk).

Three promising approaches are: (1) Replace the carbon anode (see page 405) with silicon (see Sila Nanotechnologies at www.silanano.com. (2) Replace the flammable liquid electrolyte with a solid electrolyte (see page 358 and https://bit.ly/3kJrgUd). (3) Use a pure lithium anode and a special separator membrane (see Sepion Technologies at https://bit.ly/3EIxDB8). Some companies are trying to replace lithium with other metals since lithium is very expensive and mining it is difficult. See https://nyti.ms/3CuwH29 for more information on mining lithium.

The chemistry of Li-ion batteries is very complex and will not be discussed in this book.

**NOTE:** The lithium-iodide battery mentioned in Problem 26 is not a lithium-ion battery and is not rechargeable, but it is very reliable and long lived. Thus it has been used in pacemakers to maintain a normal heart beat. *Photo is in the Public Domain.*

# BALANCING OXIDATION–REDUCTION EQUATIONS

In our discussion of oxidation numbers in Section 17-1, we discussed a chemical reaction that involves the transfer of electrons from one substance to another. The example chosen was the chemistry of a mercury battery. In Section 17-1 we only wrote the parts of the equations that were transferring electrons. Here we write the full equations which gives a better description of the reactions. (Compare these equations with those on page 306. Also, why we used the OH⁻ ions and the water molecules will be explained in the next 15 pages in this chapter. After you have mastered these pages, the reactions below will be very clear.)

The reaction in which electrons are lost is

$$\text{Zn} + 2\text{OH}^- \rightarrow \text{ZnO} + \text{H}_2\text{O} + 2e^- \qquad (1)$$

The reaction in which electrons are gained is

$$\text{HgO} + \text{H}_2\text{O} + 2e^- \rightarrow \text{Hg} + 2\text{OH}^- \qquad (2)$$

Together these reactions make up the overall reaction involved in a mercury battery. Adding reactions (1) and (2), we get:

$$\text{Zn} + \cancel{2\text{OH}^-} + \text{HgO} + \cancel{\text{H}_2\text{O}} + \cancel{2e^-} \rightarrow \text{ZnO} + \cancel{\text{H}_2\text{O}} + \cancel{2e^-} + \text{Hg} + \cancel{2\text{OH}^-}$$

Subtracting H₂O, OH⁻, and the electrons from both sides of the equation we get the overall reaction

$$\text{Zn} + \text{HgO} \rightarrow \text{ZnO} + \text{Hg} \qquad (3)$$

This chapter will teach you how to write and balance reactions that involve the transfer of electrons from one substance to another.

## 18-1
### OXIDATION AND REDUCTION IN CHEMICAL REACTIONS INVOLVE THE LOSS AND GAIN OF ELECTRONS

Before we start to balance equations, we must define the terms oxidation and reduction. **Oxidation** is the loss of electrons, whereas **reduction** is the gain of electrons. In the mercury battery reactions, reaction 1 represents an oxidation:

$$Zn + 2OH^- \rightarrow ZnO + H_2O + 2e^- \qquad (1)$$

Notice that the electrons appear on the right side of the arrow. The electrons are products of the reaction. The Zn metal has lost two electrons to form ZnO.

Reaction 2 represents a reduction:

$$HgO + H_2O + 2e^- \rightarrow Hg + 2OH^- \qquad (2)$$

The electrons appear on the left side of the arrow. The electrons are reactants. The mercury in the HgO has gained two electrons to become mercury metal, Hg. **NOTE:** As discussed on page 305, mercury is very toxic and mercury batteries have not been legal for sale in the US since 1996 except in special cases.

In every battery, the oxidation and reduction reactions take place at **electrodes**. Every battery has two electrodes. In reaction 1 above, which is an oxidation, the electrons are flowing **out** of one of the electrodes into an external electrical circuit. This electrode, where oxidation is taking place, is defined as the **anode**. In reaction 2, which is a reduction, the electrons are flowing **into** the other electrode from the external electrical circuit. This electrode, where reduction is taking place, is defined as the **cathode**. For a full discussion of how a simple rechargeable battery works, see page 405.

To remember which electrode is which, think of the following: **Oxidation at the anode**. Then you will also know that reduction takes place at the cathode. **In a battery, the electrons flow from the anode to the cathode.** See pages 345 and 405 for more on anodes and cathodes.

Adding the oxidation reaction 1 and the reduction reaction 2 and subtracting $2OH^-$ and $2e^-$ from each side gives the overall reaction, reaction 3:

$$Zn + HgO \rightarrow ZnO + Hg \qquad (3)$$

Even though the electrons are not explicitly written, reaction 3 involves both an oxidation and a reduction. Therefore, it is called an **oxidation—reduction** reaction. The abbreviated name is **redox**, from **red**uction and **ox**idation. More about how batteries work is at https://bit.ly/3eWQdGR.

In every redox reaction, the oxidation numbers of some elements are changed. Let's assign oxidation numbers to each element in reaction 3:

$$\overset{0}{Zn} + \overset{+2\ -2}{HgO} \rightarrow \overset{+2\ -2}{ZnO} + \overset{0}{Hg} \qquad (3)$$

(Rules for assigning oxidation numbers were discussed in Section 17-3.)

The zinc goes from zero to +2. The mercury goes from +2 to zero. The zinc is oxidized (its oxidation number increases from zero to +2) and the mercury is reduced (its oxidation number decreases from +2 to zero). The way to tell whether a chemical reaction is a redox reaction is to calculate the oxidation number of each atom in the reaction. *If there is a change in oxidation numbers in a reaction, the reaction is a redox reaction.*[1]

The functioning of batteries and fuel cells, burning fuel for heat, electroplating of metals, production of metals such as aluminum (see page 345) and magnesium using electrolysis, production of hydrogen gas by electrolysis or

---

[1] Sometimes only one substance will be both oxidized and reduced. An example is in the reaction $2Cu^+ \rightarrow Cu^{2+} + Cu^o$. This is called a **disproportionation** or a **dismutation**. We will not discuss such reactions further except to offer Problem 17.

from sunlight for energy storage, some analytical chemistry techniques, and many reactions in biological systems involve redox reactions.

**EXAMPLE 1** Is the following a redox reaction?

$$Fe + Cu^{2+} \rightarrow Fe^{2+} + Cu$$

*Solution:* Assign oxidation numbers to each atom:

$$\overset{0}{Fe} + \overset{+2}{Cu^{2+}} \rightarrow \overset{+2}{Fe^{2+}} + \overset{0}{Cu}$$

Since oxidation numbers change, the reaction *is* a redox reaction. ∎

**EXAMPLE 2** Is the following a redox reaction?

$$HCl + NaOH \rightarrow NaCl + HOH$$

*Solution:* Assign oxidation numbers to each atom:

$$\overset{+1-1}{HCl} + \overset{+1-2+1}{Na\,O\,H} \rightarrow \overset{+1-1}{NaCl} + \overset{+1-2+1}{H\,O\,H}$$

Since there is no change in oxidation numbers, the reaction is *not* a redox reaction. ∎

## 18-2
### WHEN BALANCING REDOX EQUATIONS USING HALF-REACTIONS, LEAVE OUT THE SPECTATOR IONS

Now we will give the details of how to balance redox reactions. To start, we will use the reaction in Example 1, since it illustrates many of the features of redox reactions. The overall reaction, which takes place in a water solution, is

$$Fe + Cu^{2+} \rightarrow Fe^{2+} + Cu$$

We can write this reaction in two parts—the oxidation reaction and the reduction reaction:

$$Fe \rightarrow Fe^{2+} + 2e^-  \quad \text{oxidation (electrons lost)}$$
$$Cu^{2+} + 2e^- \rightarrow Cu \quad \text{reduction (electrons gained)}$$

## 328 BALANCING OXIDATION–REDUCTION EQUATIONS

Each of these reactions is called a **half-reaction,** since each is one-half of the overall reaction.

Let's look at the oxidation half-reaction in some detail. The Fe goes to $Fe^{2+}$ by giving up two electrons. Since each electron has a charge of $-1$, two of them are lost when Fe goes from an oxidation number of zero to $+2$. The total charge on each side of the arrow is the same. The total charge on the left-hand side is zero because Fe is neutral. The total charge on the right-hand side is zero:

$$+2 + (-2) = +2 - 2 = 0$$
$$\uparrow \qquad \uparrow$$
$$\text{from } Fe^{2+} \quad \text{from } 2e^-$$

Now look at the reduction half-reaction. The two electrons lost by the iron are gained by the copper in going from $+2$ to zero. The total charge on the left-hand side of the arrow is zero:

$$+2 + (-2) = +2 - 2 = 0$$
$$\uparrow \qquad \uparrow$$
$$\text{from } Cu^{2+} \quad \text{from } 2e^-$$

The total charge on the right-hand side is also zero because Cu is neutral.

Let's see how to add the two half-reactions to get the overall reaction. Line up the arrows and then add everything on each side of the arrow. The order of the terms on each side of the arrow doesn't matter.

$$\begin{array}{ll} Fe \rightarrow Fe^{2+} + 2e^- & \text{oxidation} \\ \underline{Cu^{2+} + 2e^- \rightarrow Cu} & \underline{\text{reduction}} \\ Fe + Cu^{2+} + 2e^- \rightarrow Fe^{2+} + Cu + 2e^- & \text{sum of half-reactions} \end{array}$$

Since two electrons are on each side of the arrow and we can always subtract the same thing from both sides of an equation, we can subtract two electrons from each side (we will cross them out),

$$Fe + Cu^{2+} + \cancel{2e^-} \rightarrow Fe^{2+} + Cu + \cancel{2e^-}$$

and obtain the overall reaction

$$Fe + Cu^{2+} \rightarrow Fe^{2+} + Cu \qquad \text{overall balanced equation}$$

**EXAMPLE 3** Write the two half-reactions of the following unbalanced overall reaction. This reaction doesn't take place in water; instead, it occurs in the molten (melted) state:

## 18-2 Balance Redox Equations Using Half Reactions

$$Na + Mg^{2+} \rightarrow Na^+ + Mg$$

*Solution:* The two half-reactions are

$$Na \rightarrow Na^+ + e^- \quad \text{oxidation}$$
$$Mg^{2+} + 2e^- \rightarrow Mg \quad \text{reduction}$$

Before we can add these up, we must equalize the number of electrons that are lost and gained. Thus we need to multiply the oxidation half-reaction by 2:

$$2Na \rightarrow 2Na^+ + 2e^- \quad \text{oxidation}$$
$$\underline{Mg^{2+} + 2e^- \rightarrow Mg \quad \text{reduction}}$$
$$2Na + Mg^{2+} + 2e^- \rightarrow 2Na^+ + Mg + 2e^- \quad \text{sum of half-reactions}$$
$$2Na + Mg^{2+} \rightarrow 2Na^+ + Mg \quad \text{balanced overall equation} \quad \blacksquare$$

You may wonder about the equations in Example 3. Where are the negative ions, which keep the solution electrically neutral? Well, they are there, but we didn't write them since they don't take an active part in the chemical reaction. Ions that don't take an active part in a chemical reaction are called **spectator ions.**

**EXAMPLE 4** Write the equation from Example 3 as a molecular equation and as a complete ionic equation. For simplicity, use $Cl^-$ as the spectator ion.

*Solution:* The molecular equation is

$$2Na + MgCl_2 \rightarrow 2NaCl + Mg$$

The complete ionic equation is

$$Na + Mg^{2+} + 2Cl^- \rightarrow 2Na^+ + 2Cl^- + Mg$$

In the complete ionic equation, we can see that $Cl^-$ doesn't undergo a change in oxidation number and can be subtracted from both sides. When we do this, we get the overall balanced equation without spectator ions:

$$Na + Mg^{2+} \rightarrow 2Na^+ + Mg \quad \blacksquare$$

It is common to leave out spectator ions when writing redox reactions because doing so makes it easier for us. Our attention is focused entirely on the species that undergo change.

In many reactions in water solution, oxygen and hydrogen are present in the reactants and products. It is thus necessary to have a convenient way to balance oxygen and hydrogen. The most realistic way is to use molecules and ions commonly present in water. They are $H_2O$, $H^+$, and $OH^-$. The $H^+$ and $OH^-$ come from water or from acid ($H^+$) or base ($OH^-$) put into the water. If the solution is acidic, there are many more $H^+$ ions than $OH^-$ ions. If the solution is basic, there are many more $OH^-$ ions than $H^+$ ions. So in acidic solution we will use $H_2O$ and $H^+$ to balance oxygen and hydrogen. Similarly, in basic solution we will use $H_2O$ and $OH^-$. Electrons ($e^-$) are used to balance charge.

Many reactions that occur in water solution are written as ionic reactions because the substances involved actually exist in water in an ionized form.

At this point, it might be useful to summarize the ideas we have presented for balancing redox reactions.

## 18-3
### THE RULES FOR BALANCING REDOX EQUATIONS ARE BASED ON MASS AND ELECTRICAL BALANCE

The rules for balancing redox equations are as follows.

1. Calculate the oxidation number of each element to decide which ones undergo oxidation and reduction.
2. If it isn't already written in this way, write the reaction such that the spectator ions are eliminated.
3. Write the two half-reactions—one for the oxidation and the other for the reduction.
4. Balance each half-reaction **materially;** that is, the number of atoms of each element must be equal on both sides of the equation. This procedure is also referred to as **mass balance.** (You learned how to balance equations materially in Chapter 3.)
5. Balance each half-reaction **electrically;** that is, the charge must be the same on each side of the equation. To balance the charge, add electrons ($e^-$) to the more positive side.
6. Equalize the number of electrons in each half-reaction by taking multiples of one or both half-reactions.
7. Add both balanced half-reactions. Subtract the electrons (which must be the same on each side of the equation) and chemical species that are common to both sides.
8. Check the final balanced overall equation to see whether it is materially and electrically balanced.

## 18-4
### BALANCE REDOX EQUATIONS IN ACIDIC SOLUTION BY USING $H^+$ AND $H_2O$ TO BALANCE HYDROGEN AND OXYGEN ATOMS

Let's look at the following reaction that takes place in acid solution:

*unbalanced molecular equation:*

$$KMnO_4 + FeCl_2 \rightarrow MnCl_2 + FeCl_3 + KCl$$

*unbalanced ionic equation:*

$$K^+ + MnO_4^- + Fe^{2+} + 2Cl^- \rightarrow Mn^{2+} + Fe^{3+} + 6Cl^- + K^+$$

The spectator ions are $K^+$ and $Cl^-$. In a redox reaction, the spectator ions don't change oxidation number. Notice that the oxidation numbers of both $K^+$ and $Cl^-$ don't change. Since the spectator ions are $K^+$ and $Cl^-$, we remove them to get the redox reaction without spectator ions (Rule 2):

$$MnO_4^- + Fe^{2+} \rightarrow Mn^{2+} + Fe^{3+}$$

The oxidation numbers of each atom are Mn (in $MnO_4^-$) = +7, $Fe^{2+}$ = +2, $Mn^{2+}$ = +2, and $Fe^{3+}$ = +3 (Rule 1). The unbalanced half-reactions are (Rule 3)

$$MnO_4^+ \rightarrow Mn^{2+} \quad \text{reduction (Mn goes from +7 to +2)}$$
$$Fe^{2+} \rightarrow Fe^{3+} \quad \text{oxidation (Fe goes from +2 to +3)}$$

Now we must balance each half-reaction materially (Rule 4). In the reduction half-reaction, there are four oxygen atoms in the $MnO_4^-$ ion on the left-hand side of the equation. This oxygen must be accounted for on the right-hand side of the equation. Since the reaction takes place in a water solution, it seems reasonable that $H_2O$ can be used as a source of oxygen. So we will use four $H_2O$ molecules to put four oxygen atoms on the right-hand side of the equation:

$$MnO_4^- \rightarrow Mn^{2+} + 4H_2O$$

But now we have added eight H atoms to the right side. Since there are no H atoms on the left-hand side, we must add eight. The best way to do this, since the reaction takes place in acidic solution, is to use eight $H^+$ ions. We now have

$$MnO_4^- + 8H^+ \rightarrow Mn^{2+} + 4H_2O$$

The next thing to do is to balance the half-reaction electrically (Rule 5). A good way to do this is to

1. Add up the charges on the left-hand side of the arrow.
2. Add up the charges on the right-hand side of the arrow.
3. Add enough electrons ($e^-$) to the more *positive* side so that both sides have the same charge.

On the left-hand side of the arrow, the $MnO_4^-$ has one negative charge. The eight $H^+$ contribute eight positive charges. We have $-1 + 8 = +7$ charges on the left-hand side. On the right-hand side the $Mn^{2+}$ has two positive charges. The four $H_2O$ molecules are neutral and contribute no charge. Thus we have a $+2$ charge on the right-hand side:

|  | **Left-hand side** | **Right-hand side** |
|---|---|---|
| **total charge:** | $+7$ | $+2$ |

The left-hand side is more positive. If we add five electrons to the left-hand side, then the charge on the left side will equal $+2$, the same as the charge on the right-hand side:

|  | **Left-hand side** | **Right-hand side** |
|---|---|---|
| **total charge after adding five electrons** | $+7 + (-5) =$ $+7 - 5 = +2$ | $+2$ |

The balanced reduction half-reaction now reads

$$MnO_4^- + 8H^+ + 5e^- \rightarrow Mn^{2+} + 4H_2O$$

The oxidation half-reaction is easier to balance. It is already balanced materially. All we have to do is add one electron to the right-hand side to balance the charge:

$$Fe^{2+} \rightarrow Fe^{3+} + e^-$$

The total charge on each side of the arrow is now $+2$.

Both balanced half-reactions now read

$$MnO_4^- + 8H^+ + 5e^- \rightarrow Mn^{2+} + 4H_2O \quad \text{reduction}$$
$$Fe^{2+} \rightarrow Fe^{3+} + e^- \quad \text{oxidation}$$

Before we can add the two half-reactions, we must equalize the number of electrons (Rule 6). To do this, we notice that the reduction gains *five* electrons and that the oxidation loses only *one* electron. So we must multiply the oxidation half-reaction by *five*. Writing both half-reactions and adding them (Rule 7), we get

## 18-5 Balancing Redox Equations in Basic Solution

$$MnO_4^- + 8H^+ + 5e^- \rightarrow Mn^{2+} + 4H_2O \qquad \text{reduction}$$

$$\underline{5Fe^{2+} \rightarrow 5Fe^{3+} + 5e^- \qquad \text{oxidation}}$$

$$MnO_4^- + 8H^+ + 5e^- + 5Fe^{2+} \rightarrow Mn^{2+} + 4H_2O + 5Fe^{3+} + 5e^- \qquad \text{sum of half-reactions}$$

We can now subtract $5e^-$ from each side:

$$MnO_4^- + 8H^+ + 5Fe^{2+} \rightarrow Mn^{2+} + 4H_2O + 5Fe^{3+} \qquad \text{overall balanced equation}$$

Since there are no chemical species common to both sides that can be subtracted, this is the final overall balanced equation.

You might be interested in seeing how the equation looks as a molecular equation after putting back all the spectator ions ($K^+$ and $Cl^-$) and writing molecular formulas. $K^+$ ions are used to combine with the negative ions, and $Cl^-$ ions are used to combine with the positive ions. You are not expected to be able to do the following from what you have learned so far:

$$KMnO_4 + 8HCl + 5FeCl_2 \rightarrow MnCl_2 + 4H_2O + 5FeCl_3 + KCl$$

## 18-5
### BALANCE REDOX EQUATIONS IN BASIC SOLUTION BY USING OH⁻ AND H₂O TO BALANCE HYDROGEN AND OXYGEN ATOMS

Now we will illustrate how to balance a redox reaction in basic solution. The unbalanced overall equation without spectator ions (the reactants and products are in ionic form) is

$$CrO_4^{2-} + SO_3^{2-} \rightarrow CrO_2^- + SO_4^{2-}$$

The oxidation number of each element is

$$\overset{+6-2}{CrO_4^{2-}} + \overset{+4-2}{SO_3^{2-}} \rightarrow \overset{+3-2}{CrO_2^-} + \overset{+6-2}{SO_4^{2-}}$$

The Cr goes from +6 to +3; thus it is reduced. The S goes from +4 to +6; thus it is oxidized. The two unbalanced half-reactions are

$$CrO_4^{2-} \rightarrow CrO_2^- \qquad \text{reduction}$$
$$SO_3^{2-} \rightarrow SO_4^{2-} \qquad \text{oxidation}$$

Let's balance the reduction half-reaction first. There are four oxygen atoms on the left-hand side and two on the right-hand side. Two oxygen atoms must be added to the right-hand side. This is done by adding $2H_2O$ to the right-hand side:

$$CrO_4^{2-} \rightarrow CrO_2^- + 2H_2O$$

Now we must add four H atoms to the left-hand side to compensate for the four H atoms in the $2H_2O$. Since the reaction takes place in basic solution, we can add four H atoms to the left-hand side by adding $4H_2O$ to the left-hand side and $4OH^-$ to the right-hand side. Notice what this does by looking at the individual atoms in $H_2O$ and $OH^-$ and subtracting common atoms (we have crossed them out):

$$H\cancel{O} \rightleftarrows \cancel{O}\cancel{H}^- \quad \text{or} \quad H-\cancel{O}-\cancel{H} \rightleftarrows \cancel{O}-\cancel{H}^-$$

Adding an $H_2O$ to the left-hand side and an $OH^-$ to the right-hand side of the equation is like adding one H atom to the left-hand side. Thus the reduction half-reaction becomes

$$CrO_4^{2-} + 4H_2O \rightarrow CrO_2^- + 2H_2O + 4OH^-$$

We can subtract $2H_2O$ from each side:

$$CrO_4^{2-} + 2H_2O \rightarrow CrO_2^- + 4OH^-$$

To balance the charges, note that the right-hand side has a $-5$ charge and that the left-hand side has a $-2$ charge. The left-hand side is more positive, so we add $3e^-$ to it. This completes the balancing of the reduction half-reaction:

$$CrO_4^{2-} + 2H_2O + 3e^- \rightarrow CrO_2^- + 4OH^-$$

The oxidation half-reaction is balanced as follows:

$$SO_3^{2-} \rightarrow SO_4^{2-} \qquad \text{unbalanced}$$

$$\left.\begin{array}{l} SO_3^{2-} + H_2O \rightarrow SO_4^{2-} \\ SO_3^{2-} + H_2O + 2OH^- \rightarrow SO_4^{2-} + 2H_2O \\ SO_3^{2-} + 2OH^- \rightarrow SO_4^{2-} + H_2O \end{array}\right\} \text{mass balance}$$

$$SO_3^{2-} + 2OH^- \rightarrow SO_4^{2-} + H_2O + 2e^- \qquad \text{electrical balance}$$

Writing both half-reactions together, we see that the oxidation gives up $2e^-$ and that the reduction takes $3e^-$:

$$CrO_4^{2-} + 2H_2O + 3e^- \rightarrow CrO_2^- + 4OH^- \quad \text{reduction}$$

$$SO_3^{2-} + 2OH^- \rightarrow SO_4^{2-} + H_2O + 2e^- \quad \text{oxidation}$$

Before we can add both half-reactions, we must ensure that the number of electrons lost equals the number of electrons gained. We can do this by multiplying the reduction half-reaction by 2 and by multiplying the oxidation half-reaction by 3:

$$2 \times [CrO_4^{2-} + 2H_2O + 3e^- \rightarrow CrO_2^- + 4OH^-]$$
$$3 \times [SO_3^{2-} + 2OH^- \rightarrow SO_4^{2-} + H_2O + 2e^-]$$

equalizing the number of electrons in each half-reaction

By multiplying out we get

$$2CrO_4^{2-} + 4H_2O + 6e^- \rightarrow 2CrO_2^- + 8OH^-$$

$$3SO_3^{2-} + 6OH^- \rightarrow 3SO_4^{2-} + 3H_2O + 6e^-$$

Adding both half-reactions gives

$$2CrO_4^{2-} + 4H_2O + 6e^- + 3SO_3^{2-} + 6OH^- \rightarrow$$
$$2CrO_2^- + 8OH^- + 3SO_4^{2-} + 3H_2O + 6e^-$$

sum of half-reactions

Subtracting $6e^-$, $3H_2O$, and $6OH^-$ from each side, we arrive at the overall balanced equation:

$$2CrO_4^{2-} + H_2O + 3SO_3^{2-} \rightarrow 2CrO_2^- + 3SO_4^{2-} + 2OH^-$$

overall balanced equation

## 18-6
### CHECK THE BALANCED REDOX EQUATION FOR MASS AND ELECTRICAL BALANCE IN THE FOLLOWING FOUR EXAMPLES

In the four examples that follow, what we have done at each step will be noted after that step. While you are following these examples and working the problems at the end of the chapter, it would be an excellent idea if you *checked* the final balanced equation to be sure it is correct. You can do this in two steps. (1) Count the number of each kind of atom on each side of the arrow and make sure the respective numbers are equal; (2) count the total charge on each side of the arrow and make sure the charges are the same.

## 336   BALANCING OXIDATION–REDUCTION EQUATIONS

**EXAMPLE 5**   Balance the following redox reaction in acidic solution. No spectator ions are written.

$$Cu + NO_3^- \rightarrow Cu^{2+} + NO$$

*Solution:*

$$\overset{0}{Cu} + \overset{+5-2}{NO_3^-} \rightarrow \overset{+2}{Cu^{2+}} + \overset{+2-2}{NO} \quad \text{oxidation number determination}$$

The two unbalanced half-reactions are

$$Cu \rightarrow Cu^{2+} \quad \text{oxidation}$$
$$NO_3^- \rightarrow NO \quad \text{reduction}$$

Balancing the oxidation half-reaction gives

$$Cu \rightarrow Cu^{2+} + 2e^- \quad \text{final mass and electrical balance}$$

Balancing the reduction half-reaction gives

$$NO_3^- \rightarrow NO + 2H_2O \quad \text{O balance}$$
$$NO_3^- + 4H^+ \rightarrow NO + 2H_2O \quad \text{H balance, final mass balance}$$
$$NO_3^- + 4H^+ + 3e^- \rightarrow NO + 2H_2O \quad \text{final mass and electrical balance}$$

$$\left. \begin{array}{l} 3 \times [Cu \rightarrow Cu^{2+} + 2e^-] \\ 2 \times [NO_3^- + 4H^+ + 3e^- \rightarrow NO + 2H_2O] \end{array} \right\} \quad \text{equalizing the number of electrons in each half-reaction}$$

---

$$3Cu + 2NO_3^- + 8H^+ + 6e^- \rightarrow 3Cu^{2+} + 6e^- + 2NO + 4H_2O \quad \text{sum of half-reactions}$$

$$3Cu + 2NO_3^- + 8H^+ \rightarrow 3Cu^{2+} + 2NO + 4H_2O \quad \text{overall balanced equation} \quad \blacksquare$$

**EXAMPLE 6**   Balance the following in acidic solution:

$$HCl + KMnO_4 \rightarrow MnCl_2 + Cl_2 + KCl$$

### 18-6 Check the Balanced Redox Equation for Mass and Electrical Balance

*Solution:* The ionic equation, without the spectator ions ($K^+$ and $Cl^-$—notice that $Cl^-$ is both a reactant and a spectator ion), is

$$Cl^- + MnO_4^- \rightarrow Mn^{2+} + Cl_2$$

Assigning oxidation numbers, we have

$$\overset{-1}{Cl^-} + \overset{+7\ -2}{MnO_4^-} \rightarrow \overset{+2}{Mn^{2+}} + \overset{0}{Cl_2}$$

The two half-reactions are

$$Cl^- \rightarrow Cl_2 \qquad \text{oxidation}$$
$$MnO_4^- \rightarrow Mn^{2+} \qquad \text{reduction}$$

Balancing the oxidation half-reaction, we get

$$2Cl^- \rightarrow Cl_2 \qquad \text{Cl balance, final mass balance}$$
$$2Cl^- \rightarrow Cl_2 + 2e^- \qquad \text{final mass and electrical balance}$$

The reduction reaction was balanced in Section 18-4, where we obtained

$$MnO_4^- + 8H^+ + 5e^- \rightarrow Mn^{2+} + 4H_2O \qquad \text{final mass and electrical balance}$$

Equalizing electrons and adding both half-reactions gives

$$\left. \begin{array}{l} 5 \times [2Cl^- \rightarrow Cl_2 + 2e^-] \\ 2 \times [MnO_4^- + 8H^+ + 5e^- \rightarrow Mn^{2+} + 4H_2O] \end{array} \right\} \text{equalizing the number of electrons in each half-reaction}$$

$$10Cl^- + 2MnO_4^- + 16H^+ + 10e^- \rightarrow 5Cl_2 + 10e^- + 2Mn^{2+} + 8H_2O \qquad \text{sum of half-reactions}$$

$$10Cl^- + 2MnO_4^- + 16H^+ \rightarrow 5Cl_2 + 2Mn^{2+} + 8H_2O \qquad \text{overall balanced equation}$$

To get the balanced molecular equation, use the spectator ions to neutralize charge. $K^+$ is combined with the negative ions, and $Cl^-$ is combined with the positive ions. (Again notice that in this reaction, $Cl^-$ is both a reactant and a spectator ion.) In this reaction, since both $Cl^-$ and $H^+$ are on the left-hand side of the arrow, they also combine to form HCl. You are not expected to know how to do this from what you have learned so far.

$$16HCl + 2KMnO_4 \rightarrow 5Cl_2 + 2MnCl_2 + 8H_2O + 2KCl \qquad \blacksquare$$

**EXAMPLE 7** Balance the following in basic solution. No spectator ions are written.

$$CN^- + MnO_4^- \rightarrow OCN^- + MnO_2$$

***Solution:*** The oxidation numbers of N and C in $OCN^-$ are not obvious. If you could draw the Lewis formula for $OCN^-$, you could figure out the oxidation numbers of C and N using the methods of Chapter 17. The Lewis formula is $-\ddot{\text{O}}-\text{C}\equiv\text{N}:$, which is slightly different than the ones you have seen because one of the lone pairs of electrons on the oxygen has been written as a line rather than as two dots. This was done because it really isn't a lone pair—it is a bond to a metal atom such as Na in a compound like NaOCN. Putting in the "+" and "−" signs, we get

$$\overset{+-}{\underset{+-}{-}}\ddot{\text{O}}\overset{+}{-}\overset{+}{\text{C}}\overset{+-}{\underset{+-}{\equiv}}\text{N}:$$

Counting the "+" and "−" signs gives $N = -3$ and $C = +4$. The oxygen has two "−" signs; the one on the left comes from its bond with a metal atom such as Na in a compound like NaOCN. We leave assigning oxidation numbers in $CN^-$ for Problem 33.

You could not be expected to be able to figure out the oxidation numbers in $CN^-$ and $OCN^-$ unless you were given the Lewis formulas and had studied Section 17-4. However, you could still figure out what the half-reactions are and balance the equation without knowing the oxidation numbers, since $CN^-$ must go to $OCN^-$ and $MnO_4^-$ must go to $MnO_2$:

$$\overset{+2-3}{\text{C N}^-} + \overset{+7-2}{\text{MnO}_4^-} \rightarrow \overset{-2+4-3}{\text{O C N}^-} + \overset{+4-2}{\text{MnO}_2} \quad \text{oxidation number determination}$$

The half-reactions are

$$CN^- \rightarrow OCN^- \quad \text{oxidation}$$

$$MnO_4^- \rightarrow MnO_2 \quad \text{reduction}$$

Balancing the oxidation half-reaction gives

| | |
|---|---|
| $CN^- + H_2O \rightarrow OCN^-$ | O balance |
| $CN^- + H_2O + 2OH^- \rightarrow OCN^- + 2H_2O$ | H balance |
| $CN^- + 2OH^- \rightarrow OCN^- + H_2O$ | final mass balance |
| $CN^- + 2OH^- \rightarrow OCN^- + H_2O + 2e^-$ | final mass and electrical balance |

## 18-6 Check the Balanced Redox Equation for Mass and Electrical Balance

Balancing the reduction half-reaction, we have

$$MnO_4^- \rightarrow MnO_2 + 2H_2O \qquad \text{O balance}$$

$$MnO_4^- + 4H_2O \rightarrow MnO_2 + 2H_2O + 4OH^- \qquad \text{H balance}$$

$$MnO_4^- + 2H_2O \rightarrow MnO_2 + 4OH^- \qquad \text{final mass balance}$$

$$MnO_4^- + 2H_2O + 3e^- \rightarrow MnO_2 + 4OH^- \qquad \text{final mass and electrical balance}$$

Equalizing electrons and adding the balanced half-reactions gives

$$\left.\begin{array}{l} 3 \times [CN^- + 2OH^- \rightarrow OCN^- + H_2O + 2e^-] \\ 2 \times [MnO_4^- + 2H_2O + 3e^- \rightarrow MnO_2 + 4OH^-] \end{array}\right\} \text{equalizing the number of electrons in each half-reaction}$$

$$3CN^- + 6OH^- + 2MnO_4^- + 4H_2O + 6e^- \rightarrow$$
$$3OCN^- + 3H_2O + 6e^- + 2MnO_2 + 8OH^- \qquad \text{sum of half-reactions}$$

$$3CN^- + 2MnO_4^- + H_2O \rightarrow 3OCN^- + 2MnO_2 + 2OH^- \qquad \text{overall balanced equation} \quad \blacksquare$$

**EXAMPLE 8** Balance the following in acidic solution. No spectator ions are written.

$$S_2O_3^{2-} + I_2 \rightarrow I^- + S_4O_6^{2-}$$

***Solution:*** The oxidation numbers are

$$\overset{+2\ -2}{S_2O_3^{2-}} + \overset{0}{I_2} \rightarrow \overset{-1}{I^-} + \overset{+2.5\ -2}{S_4O_6^{2-}} \qquad \text{oxidation number determination}$$

Don't worry about the fractional oxidation number of S in the tetrathionate ion, $S_4O_6^{2-}$. The S atoms are bonded differently, just as were some of the carbon atoms you may have studied in Section 17-4. Here we are only interested in the change in oxidation numbers, so the details of the bonding don't concern us. The half-reactions are

$$S_2O_3^{2-} \rightarrow S_4O_6^{2-} \qquad \text{oxidation}$$
$$I_2 \rightarrow I^- \qquad \text{reduction}$$

Balancing the oxidation half-reaction, we have

$$2S_2O_3^{2-} \rightarrow S_4O_6^{2-} \qquad \text{S and O balance, final mass balance}$$
$$2S_2O_3^{2-} \rightarrow S_4O_6^{2-} + 2e^- \qquad \text{final mass and electrical balance}$$

Balancing the reduction half-reaction gives

$$I_2 \rightarrow 2I^- \qquad \text{I balance, final mass balance}$$
$$I_2 + 2e^- \rightarrow 2I^- \qquad \text{final mass and electrical balance}$$

Equalizing electrons and adding the balanced half-reactions, we have

$$\left.\begin{array}{l}2S_2O_3^{2-} \rightarrow S_4O_6^{2-} + 2e^- \\ I_2 + 2e^- \rightarrow 2I^-\end{array}\right\} \text{equalizing the number of electrons (the electrons are already equal)}$$

$$2S_2O_3^{2-} + I_2 + 2e^- \rightarrow S_4O_6^{2-} + 2e^- + 2I^- \qquad \text{sum of half-reactions}$$

$$2S_2O_3^{2-} + I_2 \rightarrow S_4O_6^{2-} + 2I^- \qquad \text{overall balanced equation} \blacksquare$$

## 18-7
## OXIDIZING AND REDUCING AGENTS CAUSE OXIDATION AND REDUCTION

At this point we will introduce and define two important terms: **oxidizing agent** and **reducing agent**. Earlier in the chapter we defined **oxidation** as the loss of electrons, and we defined **reduction** as the gain of electrons. The name oxidation comes from the effect that oxygen has when it combines with most substances. The rusting of iron to give iron(III) oxide is an example. Oxygen increases the metal's oxidation number. Reduction is an old name for removing oxygen from metallic ores (many are oxides) to give the pure metals. The ores are "reduced" to the metal. Removing oxygen reduces the metal's oxidation number.

Using these definitions we will now define an **oxidizing agent** as something that causes oxidation. A **reducing agent** is something that causes reduction.

For example, in the reaction from Example 1, $Fe + Cu^{2+} \rightarrow Fe^{2+} + Cu$, the half-reactions are

$$Fe \rightarrow Fe^{2+} + 2e^- \qquad \text{oxidation}$$
$$Cu^{2+} + 2e^- \rightarrow Cu \qquad \text{reduction}$$

## 18-7 Oxidizing and Reducing Agents Cause Oxidation and Reduction

The $Cu^{2+}$ causes the Fe to be oxidized. So in this reaction, the $Cu^{2+}$ is the oxidizing agent. At the same time, the Fe causes the $Cu^{2+}$ to be reduced, so in this reaction the Fe is a reducing agent. **Notice that the oxidizing agent is reduced and the reducing agent is oxidized.** The following table will illustrate these definitions.

(**NOTE:** An "agent" is always a reactant.)

| SUBSTANCE | TYPE OF REACTION | SUBSTANCE IS | TYPE OF AGENT |
|---|---|---|---|
| Fe | oxidation | oxidized | reducing agent |
| $Cu^{2+}$ | reduction | reduced | oxidizing agent |

**EXAMPLE 9** For the reaction from Example 3, $Na + Mg^{2+} \rightarrow Na^+ + Mg$, fill in the following table.

| SUBSTANCE | TYPE OF REACTION | SUBSTANCE IS | TYPE OF AGENT |
|---|---|---|---|
| Na | | | |
| $Mg^{2+}$ | | | |

*Solution:* First write down the half-reactions to see what the oxidation and reduction steps are. Then fill in the table.

$$Na \rightarrow Na^+ + e^- \quad \text{oxidation}$$
$$Mg^{2+} + 2e^- \rightarrow Mg \quad \text{reduction}$$

| SUBSTANCE | TYPE OF REACTION | SUBSTANCE IS | TYPE OF AGENT |
|---|---|---|---|
| Na | oxidation | oxidized | reducing agent |
| $Mg^{2+}$ | reduction | reduced | oxidizing agent |

■

**EXAMPLE 10** For the reaction from Example 6, $Cl^- + MnO_4^- \rightarrow Cl_2 + Mn^{2+}$, fill in a table as was done in Example 9.

*Solution:* The half-reactions are

$$2Cl^- \rightarrow Cl_2 + 2e^- \quad \text{oxidation}$$
$$2MnO_4^- + 8H^+ + 5e^- \rightarrow Mn^{2+} + 4H_2O \quad \text{reduction}$$

| SUBSTANCE | TYPE OF REACTION | SUBSTANCE IS | TYPE OF AGENT |
|---|---|---|---|
| $Cl^-$ | oxidation | oxidized | reducing agent |
| $MnO_4^-$ | reduction | reduced | oxidizing agent |

Don't get the idea that a certain substance is always an oxidizing or reducing agent in every reaction. For instance, oxygen, $O_2$, is usually an oxidizing agent. It oxidizes things very well, as shown in the following reactions:

$$2H_2 + O_2 \rightarrow 2H_2O$$

$$4Fe + 3O_2 \rightarrow 2Fe_2O_3$$

$$C + O_2 \rightarrow CO_2$$

But if oxygen reacts with fluorine, $F_2$, we get

$$2F_2 + O_2 \rightarrow 2OF_2$$

In this reaction, oxygen is the *reducing agent* and fluorine is the *oxidizing agent*. Look at the oxidation numbers. The F goes from zero to $-1$. The O goes from zero to $+2$. The oxygen is thus oxidized. Since fluorine is the most electronegative element, it *always* acts as an oxidizing agent. Oxygen, the second most electronegative element, always acts as an oxidizing agent *except* when reacting with fluorine.

## PROBLEMS

## KEYED PROBLEMS

1. Is the following a redox reaction: $3K + Al^{3+} \rightarrow 3K^+ + Al$?

2. Is the following a redox reaction: $H_2SO_4 + Ca(OH)_2 \rightarrow CaSO_4 + 2H_2O$?

3. Write the two half-reactions of the following unbalanced overall reaction that takes place in a molten salt solution: $Li + Ca^{2+} \rightarrow Li^+ + Ca$.

4. Write the equation from Problem 3 as a molecular equation and as a complete ionic equation. For simplicity, use $Cl^-$ as the spectator ion.

5. Balance the following redox reaction in acidic solution: $Cu + NO_3^- \rightarrow Cu^{2+} + NO_2$.

6. Balance the following redox reaction in acidic solution: $C_2O_4^{2-} + MnO_4^- \rightarrow CO_2 + Mn^{2+}$.

7. Balance the following redox reaction in basic solution: $MnO_4^- + NH_3 \rightarrow MnO_2 + NO_3^-$.

8. Balance the following redox reaction in acidic solution: $I^- + NO_2^- \rightarrow I_2 + NO$.

9. For the reaction in Problem 3, fill in the following table.

| SUBSTANCE | TYPE OF REACTION | SUBSTANCE IS | TYPE OF AGENT |
|---|---|---|---|
| Li | | | |
| $Ca^{2+}$ | | | |

10. For the reaction in Problem 6, fill in the following table.

| SUBSTANCE | TYPE OF REACTION | SUBSTANCE IS | TYPE OF AGENT |
|---|---|---|---|
| $C_2O_4^{2-}$ | | | |
| $MnO_4^-$ | | | |

## SUPPLEMENTAL PROBLEMS

11. Balance the following redox reactions.
    a. $Ni + F_2 \rightarrow Ni^{2+} + F^-$    c. $Co^{2+} + Cl_2 \rightarrow Co^{3+} + Cl^-$
    b. $Fe + O_2 \rightarrow Fe^{2+} + O^{2-}$

12. Balance the following redox reactions in acidic solution.
    a. $ClO_3^- + SO_3^{2-} \rightarrow Cl^- + SO_4^{2-}$    c. $Cr_2O_7^{2-} + I^- \rightarrow Cr^{3+} + I_2$
    b. $MnO_2 + I^- \rightarrow Mn^{2+} + I_2$

13. Balance the following redox reaction in acidic solution: $MnO_4^- + H_2S \rightarrow Mn^{2+} + S$. Which element is being oxidized? Which element is being reduced? What substance is the oxidizing agent? What substance is the reducing agent?

14. Balance the following redox reactions in basic solution.
    a. $MnO_4^- + H_2O \rightarrow MnO_4^{2-} + O_2$    c. $Cl_2 + IO_3^- \rightarrow IO_4^- + Cl^-$
    b. $ClO_2 + H_2O_2 \rightarrow ClO_2^- + O_2$

15. Balance the following redox reactions in basic solution.
    a. $Mn^{2+} + H_2O_2 \rightarrow MnO_2 + H_2O$    c. $MnO_2 + SO_3^{2-} \rightarrow Mn(OH)_2 + SO_4^{2-}$
    b. $Bi(OH)_3 + SnO_2^- \rightarrow SnO_3^{2-} + Bi$

Problems 16 and 17 are harder to balance. Balance them under the conditions stated.

16. $CH_3Cl + MnO_4^- \rightarrow Cl_2 + CO_2 + Mn^{2+}$ (acidic solution). (Hint: The balanced equation will contain large coefficients.)

17. $P_4 \rightarrow PH_3 + H_2PO_2^-$ (basic solution). (Hint: P is both oxidized and reduced.)

18. Balance the following redox reactions in basic solution.

    a. $MnO_4^- + V^{2+} \rightarrow VO_2^+ + Mn^{2+}$   c. $NO_3^- + Zn \rightarrow NH_4^+ + Zn^{2+}$
    b. $Fe^{2+} + ClO_2^- \rightarrow Fe^{3+} + Cl^-$

19. Balance the following redox reactions in acidic solution.

    a. $Cr_2O_7^{2-} + C_2O_4^{2-} \rightarrow Cr^{3+} + CO_2$   c. $Sb + NO_3^- \rightarrow Sb_2O_3 + NO$
    b. $H_2O_2 + Fe^{2+} \rightarrow Fe^{3+} + H_2O$

Problems 20 through 26 describe the chemistry of various batteries. The equations vary in difficulty.

20. The usual dry cell (flashlight battery): $Zn + NH_4^+ + MnO_2 \rightarrow Zn^{2+} + NH_3 + Mn_2O_3$. Balance in basic solution. (Hint: The $NH_4^+$ and the $NH_3$ don't take part in the redox.)

21. Mercury battery: $Zn + HgO \rightarrow ZnO + Hg$. Balance in basic solution.

22. Lead storage battery (automobile battery): $Pb + PbO_2 + SO_4^{2-} \rightarrow PbSO_4$. Notice that both Pb and $PbO_2$ react to form $PbSO_4$. **HINT: Sol'n is acidic.**

23. Hydrogen–oxygen fuel cell: $H_2 + O_2 \rightarrow H_2O$. Balance in basic solution. This is the reaction that supplies electricity and drinking water to most space vehicles, including the Apollo moon ship and the space shuttle. The overall equation itself is easy to balance. But try doing it using the following half-reactions:

$$H_2 + OH^- \rightarrow H_2O \quad \text{and} \quad O_2 + H_2O \rightarrow OH^-$$

24. Propane–oxygen fuel cell: $C_3H_8 + O_2 \rightarrow CO_2 + H_2O$. Balance in basic solution. The overall equation is easy to balance. But try doing it using the following half-reactions:

$$C_3H_8 + H_2O \rightarrow CO_2 + OH^- \quad \text{and} \quad O_2 + OH^- \rightarrow H_2O$$

25. Nickel–cadmium (Nicad) battery: $Cd + NiO_2 \rightarrow Cd(OH)_2 + Ni(OH)_2$. Balance in basic solution.

26. Lithium–iodide ("Lithium") battery: $Li + I_2 \rightarrow Li^+ + I^-$.

27. Calculate the oxidation number of each atom in $CN^-$ given that the Lewis formula is $—C≡N:$ and a compound containing the $CN^-$ ion is NaCN.

## HOW TWO 23-YR OLD MEN CHANGED THE WORLD (continued on p. 358)

How did they change the world? They discovered an inexpensive way to make **aluminum** on an industrial scale. Think of all the things that are made out of aluminum or its alloys (an **alloy** is a mixture of two or more metals). Today we take for granted aluminum beer and soda cans, aluminum foil, aluminum cooking utensils and furniture, building materials, car parts, boats, airplane bodies, tubing, auto engine blocks, bicycle frames, antacids, antiperspirants, and many other products. Aluminum is non-toxic, corrosion resistant, light weight, and strong for its weight. Here's the story of how they did it.

In 1886, two 23-yr old engineers, the American **Charles Martin Hall** (1863-1914) and the Frenchman **Paul Héroult** (1863-1914), independently discovered a way to make aluminum from aluminum oxide ($Al_2O_3$). **NOTE:** By sheer coincidence, both men were born and died in the same years and made their discovery the same year, without knowing of each other's work. Their invention is now called the **Hall–Héroult process**. See https://bit.ly/3rvrdxT for the full story.

Why was this discovery so important? Previously, aluminum was made by reacting aluminum chloride with sodium metal: $AlCl_3 + 3Na \rightarrow 3NaCl + Al$. This process was very expensive. For example, in 1852, Al was the most expensive metal in the world, *costing 1.8 times more than gold*, and by 1884 total world production was only about 4 tons per year. Now, aluminum is *39,000 times cheaper* than gold, and is the 2nd most produced metal after iron, with about 60 million tons made each year at a cost of about $1400 *per ton* (1 ton = 2000 lb). In the summer of 2022, gold cost about $1800 *per ounce*.

What Hall and discovered was that $Al_2O_3$ could be dissolved in the melted mineral cryolite and that they could electrolyze the solution.[1] The $Al_2O_3$ ionizes in the melted (liquid) cryolite. Here is a simplified set of redox reactions for the reduction of $Al_2O_3$ to Al. All the electrons that reduce $Al^{3+}$ come from the outside by way of an electric power line. In principle, this process is similar to charging a battery as shown on pages 409.

> The reaction takes place in a large vat with carbon anodes and cathodes. The anodes are used up and periodically replaced because of the third reaction below the line. See https://bit.ly/3CeiOUd for a full description of the Bayer and Hall-Héroult processes.

$Al_2O_3 \rightarrow 2Al^{3+} + 3O^{2-}$   The $Al_2O_3$ ionizing.

$Al^{3+} + 3e^- \rightarrow Al$   Reduction (at the cathode).

$2O^{2-} \rightarrow O_2 + 4e^-$   Oxidation (at the anode).

$4 \times [Al^{3+} + 3e^- \rightarrow Al]$   Equalizing electrons.

$3 \times [2O^{2-} \rightarrow O_2 + 4e^-]$   Equalizing electrons.

---

$4Al^{3+} + 6O^{2-} + 12e^- \rightarrow 4Al + 3O_2 + 12e^-$   Sum with electrons.

$4Al^{3+} + 6O^{2-} \rightarrow 4Al + 3O_2$   Overall balanced redox reaction.

$3C + 3O_2 \rightarrow 3CO_2$   The oxygen reacts with the hot carbon anode.

$2Al_2O_3 + 3C \rightarrow 4Al + 3CO_2$   The overall reaction in the reaction vessel.

See https://bit.ly/3bC4Jl9 for a video about the Hall–Héroult process that shows where the anode and cathode are in the reaction vessel. **NOTE:** By definition, *oxidation takes place at the **anode**. Reduction takes place at the **cathode**.* Just remember: ***Oxidation at the anode.***                    (continued on page 358)

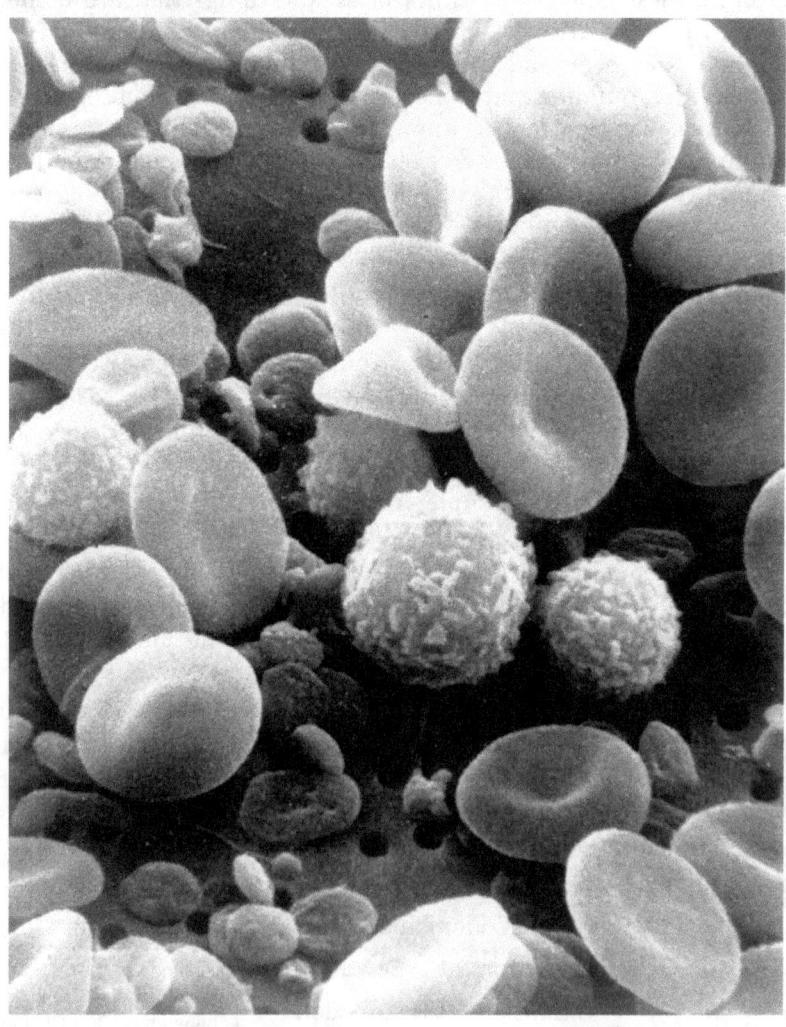

**A Scanning Electron Microscope (SEM) Photo of Blood Cells:** The doughnut shapes with a dimple in the middle are red blood cells. The white prickly balls are white blood cells, and the little discs are platelets. Red blood cells contain hemoglobin, a large red-colored molecule containing iron, that allows the red cells to shuttle oxygen from your lungs to your tissues and help transport carbon dioxide back. The pH of the blood in your tissues is around 7.2, and the pH of the blood in your lungs is around 7.6. This pH difference causes a change in hemoglobin's structure that allows it to perform properly. The higher pH in the lungs allows hemoglobin to pick up oxygen. The lower pH in tissues forces hemoglobin to release oxygen, which your cells can use for respiration. See https://bit.ly/2SAMOkQ for details. *Photo courtesy of the National Cancer Institute, Bruce Wetzel and Harry Schaefer photographers.*

# 19
# LOGARITHMS AND pH

As you know from our discussion of Avogadro's number, chemists work with very large and very small numbers. One area in which the numbers can get very small is in the calculation of hydrogen ions ($H^+$) in aqueous (water) solutions. The higher the concentration of $H^+$ ions, the more acidic the solution.

For example, the $H^+$ concentration in moles per liter in pure water is $1 \times 10^{-7}$ mol/L. Instead of always writing "$H^+$ concentration in mol/L," chemists use the notation $[H^+]$. Square brackets around any substance denote concentration in mol/L. So we can write that, in pure water, $[H^+] = 1 \times 10^{-7}$ mol/L or $[H^+] = 1 \times 10^{-7}$ M, where M stands for molarity, another name for mol/L.

**EXAMPLE 1** The $H^+$ concentration in normal human blood is $3.98 \times 10^{-8}$ M. Write this as an equation with square brackets.

***Solution:*** $[H^+] = 3.98 \times 10^{-8}$ M. ■

Scientists found that using such small numbers was awkward, so they decided to make things easier. Thus they turned to logarithms. You will understand why after reading this chapter.

## 19-1
### THE LOGARITHM OF A NUMBER IS THE POWER TO WHICH 10 MUST BE RAISED TO GIVE THAT NUMBER

Logarithms were invented by John Napier in 1614 and were improved by Henry Briggs a few years later. The word comes from the Greek "logos," meaning calculation, and "arithmos," meaning number. In this chapter, we will mainly discuss a special case of logarithms. These are called logarithms (abbreviated log) to the base 10. With this in mind, we can define the **logarithm** of a number as *the power to which 10 must be raised to give that number*. For example, what is the log of 100? This is the same as asking, "To what power must 10 be raised to give 100?" The answer is 2. Writing this statement as an equation, we have

$$10^2 = 100$$

From the definition of logarithms, it follows that

$$\log 100 = 2$$

Read this equation as "The logarithm of one hundred equals two."

**EXAMPLE 2** What are the logs of the following numbers: 10; 100; 1000; 10,000; 100,000; 1,000,000?

*Solution:*

| | |
|---|---|
| $10 = 10^1$ | $\log 10 = 1$ |
| $100 = 10^2$ | $\log 100 = 2$ |
| $1000 = 10^3$ | $\log 1000 = 3$ |
| $10,000 = 10^4$ | $\log 10,000 = 4$ |
| $100,000 = 10^5$ | $\log 100,000 = 5$ |
| $1,000,000 = 10^6$ | $\log 1,000,000 = 6$ ∎ |

**EXAMPLE 3** What are the logs of the following numbers: 1, 0.1, 0.01, 0.001, 0.0001?

*Solution:* Convert each number to scientific notation:

| | |
|---|---|
| $1 = 10^0$ | $\log 1 = 0$ |
| $0.1 = 10^{-1}$ | $\log 0.1 = -1$ |
| $0.01 = 10^{-2}$ | $\log 0.01 = -2$ |
| $0.001 = 10^{-3}$ | $\log 0.001 = -3$ |
| $0.0001 = 10^{-4}$ | $\log 0.0001 = -4$ ∎ |

Let's plot a graph of numbers versus their logarithms. The equation we'll plot is $y = \log x$. The horizontal axis will be $x$. The vertical axis will be $y$ or $\log x$. See Figure 19-1.

Notice some interesting characteristics of logs in Figure 19-1. They seem to compress numbers. For example, the range of numbers from 0.001 to 100 spans a factor of 100,000. This means that 100 is 100,000 times larger than 0.001. The logs of 0.001 and 100 are "compressed" into a span between $-3$ and $+2$. There are also no points plotted to the left of the $y$-axis. This means that there are no logs of negative numbers.

You might wonder why scientists went through all the trouble of devising logs if the only advantage they had was to make expressing very large and very small numbers easier. Well, in the days of Napier and Briggs, and even as recently as 1970, there were no hand-held calculators. The computers that were available after 1945 were large and expensive. Doing multiplication and division could be a tedious job, even with the help of mechanical calculators, which had been around for a century or so. Logarithms were

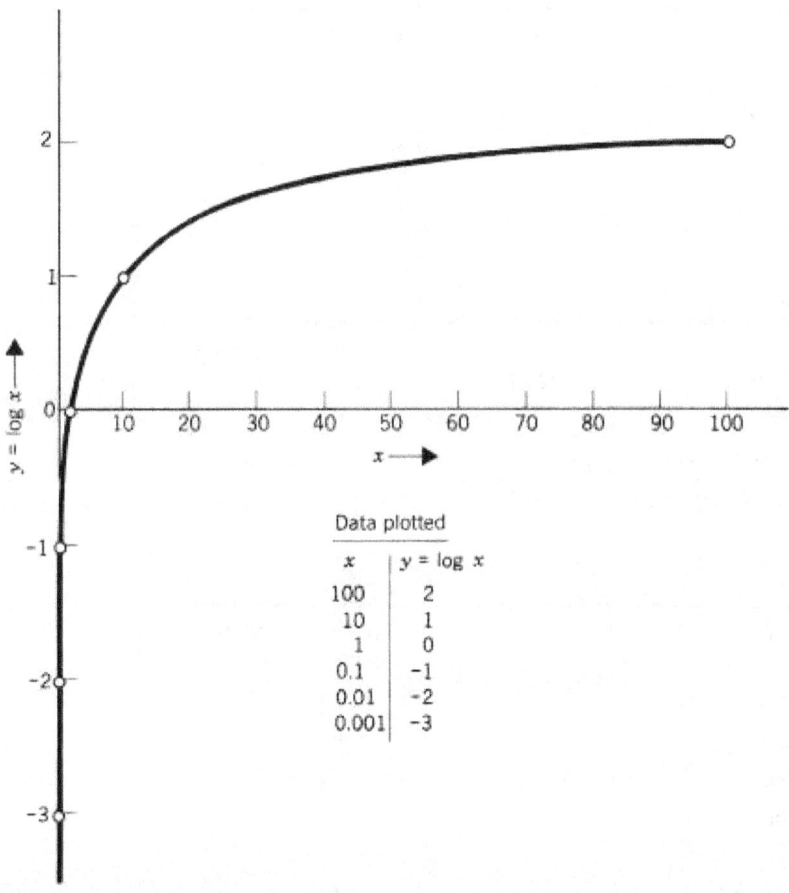

**FIGURE 19-1** A plot of $y = \log x$.

very useful in helping scientists do these calculations because when they added the logs of numbers, the numbers were multiplied. And when they subtracted the logs of numbers, the numbers were divided. It is much easier and faster to add and subtract numbers than to multiply and divide. Obtaining roots and powers is also easier with logarithms.

But now, scientific calculators and computers are available to all. And multiplications, divisions, powers, and roots are easy to do. Thus we will use logs only to make things easier and because they come up naturally in certain equations. We don't need them as calculation aids.

To obtain a log on your calculator, just enter the number and push the **log** button.

**EXAMPLE 4** Obtain the logs of the following numbers on your calculator: 100, 2.4, 58.5, $3.7 \times 10^6$, 1.33, $8.86 \times 10^4$.

*Solution:*

| NUMBER | BUTTON SEQUENCE ON CALCULATOR | LOG OF NUMBER (ROUNDED TO TWO DECIMAL PLACES) |
|---|---|---|
| 100 | **100, log** | 2.00 |
| 2.4 | **2.4, log** | 0.38 |
| 58.5 | **58.5, log** | 1.77 |
| $3.7 \times 10^6$ | **3.7, EE, 6, log** | 6.57 |
| 1.33 | **1.33, log** | 0.12 |
| $8.86 \times 10^4$ | **8.86, EE, 4, log** | 4.95 |

NOTE: Some calculators use **exp** instead of **EE**. ■

**EXAMPLE 5** Obtain the logs of the following numbers on your calculator: 1, 0.1, 0.46, 0.0037, $4.2 \times 10^{-5}$, $8.36 \times 10^{-8}$.

*Solution:*

| NUMBER | BUTTON SEQUENCE ON CALCULATOR | LOG OF NUMBER (ROUNDED TO TWO DECIMAL PLACES) |
|---|---|---|
| 1 | **1, log** | 0.00 |
| 0.1 | **0.1, log** | $-1.00$ |
| 0.46 | **0.46, log** | $-0.34$ |
| 0.0037 | **0.0037, log** | $-2.43$ |
| $4.2 \times 10^{-5}$ | **4.2, EE, +/−, 5, log** | $-4.38$ |
| $8.36 \times 10^{-8}$ | **8.36, EE, +/−, 8, log** | $-7.08$ |

■

**EXAMPLE 6** Obtain the log of $-2.4$ on your calculator.

*Solution:* The calculator button sequence is **2.4, +/−, log**. The answer is E or ERROR. As we said before, you cannot obtain the log of a negative number—it does not exist. ∎

## 19-2
### THE ANTILOGARITHM IS THE NUMBER CORRESPONDING TO A GIVEN LOGARITHM

Suppose you have the log of a number and want to find the number. On a calculator all you have to do is enter the log of the number, push the following button(s), and read the number. The buttons are either **INV, log,** or **10$^x$**, depending upon which brand of calculator you have. These operations convert the log of a number into the number. The mathematical operation is called **taking the antilogarithm.** An **antilogarithm** is the number corresponding to a given logarithm.

Before we work out a few examples, it would be instructive to see why taking $10^{(\log \text{ of a number})}$ gives the number. The reason is that

$$10^{\log x} = x$$

for any positive number $x$. Let's see how this works. Take the number $x = 100$. Then $\log x = \log 100 = 2$. Now

$$10^{\log x} = 10^{\log 100} = 10^2 = 100$$

You see, the log operation and raising to a power of 10 are "opposites" of each other and "cancel" each other out. They are called **inverses** of each other. This is the reason for the **INV** button on some calculators.

**EXAMPLE 7** What is the antilogarithm (abbreviated antilog) of 3?

*Solution:* To find the antilog, raise 10 to the power 3:

$$10^3 = 1000$$

The antilog of 3 is 1000. And conversely, the log of 1000 is 3. ∎

In the following examples, we take antilogs using the **INV** and **log** buttons on a calculator. If your calculator has a **10$^x$** button, use it instead.

## 352 LOGARITHMS AND pH

**EXAMPLE 8**  Find the antilogs of the following logarithms: 0, 0.45, 1, 1.5, 2, 2.7, 3.

*Solution:*

| LOG | BUTTON SEQUENCE ON CALCULATOR | ANTILOG OF NUMBER |
|---|---|---|
| 0 | 0, INV, log | 1 |
| 0.45 | 0.45, INV, log | 2.82 |
| 1 | 1, INV, log | 10 |
| 1.5 | 1.5, INV, log | 31.62 |
| 2 | 2, INV, log | 100 |
| 2.7 | 2.7, INV, log | 501.19 |
| 3 | 3, INV, log | 1000 |

**EXAMPLE 9**  Find the antilogs of the following logarithms: 0, −0.5, −1, −1.9, −2, −2.2, −3.

*Solution:*

| LOG | BUTTON SEQUENCE ON CALCULATOR | ANTILOG OF NUMBER |
|---|---|---|
| 0 | 0, INV, log | 1 |
| −0.5 | 0.5, +/−, INV, log | 0.316 |
| −1 | 1, +/−, INV, log | 0.1 |
| −1.9 | 1.9, +/−, INV, log | 0.0126 |
| −2 | 2, +/−, INV, log | 0.01 |
| −2.2 | 2.2, +/−, INV, log | 0.00631 |
| −3 | 3, +/−, INV, log | 0.001 |

**EXAMPLE 10**  Find the antilogs of the following logarithms: 5.8, 8, 10, 12.7, −7, −9.43, −13.11.

*Solution:*

| LOG | BUTTON SEQUENCE ON CALCULATOR | ANTILOG OF NUMBER |
|---|---|---|
| 5.8 | 5.8, INV, log | 630957.34 |
| 8 | 8, INV, log | $1 \times 10^8$ |
| 10 | 10, INV, log | $1 \times 10^{10}$ |
| 12.7 | 12.7, INV, log | $5.01 \times 10^{12}$ |
| −7 | 7, +/−, INV, log | $0.0000001 = 10^{-7}$ |
| −9.43 | 9.43, +/−, INV, log | $3.72 \times 10^{-10}$ |
| −13.11 | 13.11, +/−, INV, log | $7.76 \times 10^{-14}$ |

## 19-3
### THE NATURAL LOGARITHM IS USEFUL IN ADVANCED SCIENTIFIC WORK

In looking at your calculator, you may have noticed another button labeled **ln x.** This is the natural logarithm button and takes natural logs of numbers. The base of the natural log system is the number 2.7182818. . . . (Remember the base of the log system we have been using is 10.) This number is so important in science and mathematics that it has been given a special symbol, e. Thus, e = 2.7182818. . . . The inverse of ln $x$ is $e^x$. If you want to display the value of e on your calculator, push **1, $e^x$** or push **1, INV, ln x**, depending on the model you have. The value for e that you will get will be 2.7182818. When we wrote this number earlier, we put three dots after it. That's because it doesn't end with eight digits but goes on forever. We won't discuss the natural log system further in this book, but if you take more chemistry and mathematics (especially calculus), you will surely learn its importance.

## 19-4
### pH IS AN APPLICATION OF LOGARITHMS TO ACIDIC AND BASIC SOLUTIONS

In Example 1, we said that the normal pH of human blood, [H+] = 3.09 x $10^{-8}$ M. We promised to show you how scientists decided to make it easier to discuss such numbers. As you can see from Example 5, the logs of very small numbers become rather simple negative numbers as, for example, log (8.36 x $10^{-8}$) = –7.08. So scientists decided to take logs of [H+]. But they didn't like the negative numbers they got. So they put a "–" sign in front of the log: –log [H+].

The two minuses became a plus and all was well. Let's try it for blood. For normal human blood, [H+] = 3.98 x $10^{-8}$ M. Substituting 3.98 x $10^{-8}$ into –log [H+] gives

$$-\log [H+] = -\log (3.98 \times 10^{-8}) = -(-7.4) = 7.4.$$

The expression **–log [H+]** is defined as the **pH**. As an equation we have

$$\boxed{pH = -\log [H^+]}$$

The equation in the box for pH is read: **"pH equals the negative logarithm of the hydrogen ion concentration in mol/L (or molarity)."**

Don't worry about what to do with the unit M (molarity, or mol/L). You cannot take the log of a unit. (How would you enter it on a calculator?) Note that pH and all other logs don't have units; they are dimensionless. The calculation we performed above shows that the pH of normal human blood is 7.4.

## 354 LOGARITHMS AND pH

The symbol pH was first used by the Danish biochemist **Søren Sørensen** (1868–1939) in 1909. The "p" stands for "power." What Sørensen meant by "power" is described toward the bottom of the next page.

**EXAMPLE 11** What is the calculator button sequence for calculating the pH of normal human blood, whose [H$^+$] is $3.98 \times 10^{-8}$ M?

*Solution:* pH = $-\log$ [H$^+$] = $-\log (3.98 \times 10^{-8})$ = $-(-7.4)$ = 7.4. The calculator button sequence is **3.98, EE, +/–, 8, log, +/–**. ∎

**EXAMPLE 12** Calculate the pH of a milk of magnesia[1] solution, Mg(OH)$_2$, where [H$^+$] = $2.51 \times 10^{-11}$ M.

*Solution:* pH = $-\log$ [H$^+$] = $-\log (2.51 \times 10^{-11})$ = 10.6. The calculator button sequence is **2.51, EE, +/–, 11, log, +/–**. ∎

**EXAMPLE 13** Calculate the pH of orange juice. The [H$^+$] = $3.16 \times 10^{-4}$ M.

*Solution:* pH = $-\log$ [H$^+$] = $-\log (3.16 \times 10^{-4})$ = 3.5. The calculator sequence is **3.16, EE, +/–, 4, log, +/–**. ∎

### [1]Magnesium compounds as Antacids, Laxatives and Supplements

In medicine, a water suspension of Mg(OH)$_2$ (magnesium hydroxide) is sometimes called milk of magnesia. Magnesium hydroxide is only very slightly soluble in water, thus, to get a therapeutic dose the bottle of milk of magnesia will have a white solid at the bottom. When it is shaken the suspension will be cloudy, because most of the white magnesium hydroxide is suspended in the water. Thus, it is called "*milk* of magnesia," a very old name that comes from a region of Greece or Turkey. (The word *magnetism* is also derived from a similar ancient Greek word, because magnetic stones called lodestones [the mineral magnetite, Fe$_3$O$_4$] were also found in that region.) The small amount of Mg(OH)$_2$ that dissolves is completely ionized: Mg(OH)$_2$ → Mg$^{2+}$ + 2OH$^-$.

When used as an antacid, Mg(OH)$_2$ is sometimes combined with Al(OH)$_3$. Aluminum hydroxide is also a white solid that has a very low solubility in water. Since the Mg$^{2+}$ ion has laxative properties, and the Al$^{3+}$ ion is constipating, the combination works well for most people. It is the hydroxide ion that neutralizes stomach acid, HCl, in the reaction H$^+$ + OH$^-$ → H$_2$O (see problem 9 on page 154, and page 187). As the hydroxide is neutralized, more Mg(OH)$_2$ and Al(OH)$_3$ dissolve. Since the solubility of both hydroxides is very low, the pH can't get high enough to do any damage to the digestive tract.

The dose of Mg$^{2+}$ ion one gets when using the Al/Mg antacid combination is between 170 mg and 340 mg. When using Mg(OH)$_2$ by itself as a laxative, doses of Mg$^{2+}$ ion are between 800 mg and 2000 mg. The Mg$^{2+}$ ion is the laxative; it works by drawing water into the bowel by osmosis (https://bit.ly/336ElO4). Other compounds of Mg can also be used as a laxative, such as MgO, (very insoluble in water – sold in tablets or caplets) and magnesium citrate (very soluble in water – sold as a clear solution). Both magnesium oxide tablets and magnesium citrate are also sold as supplements, the latter in soft gel capsules (called gelcaps). The Daily Value (p. 121) of magnesium is 420 mg.

### 19-4 pH Is an Application of Logarithms to Acidic and Basic Solutions

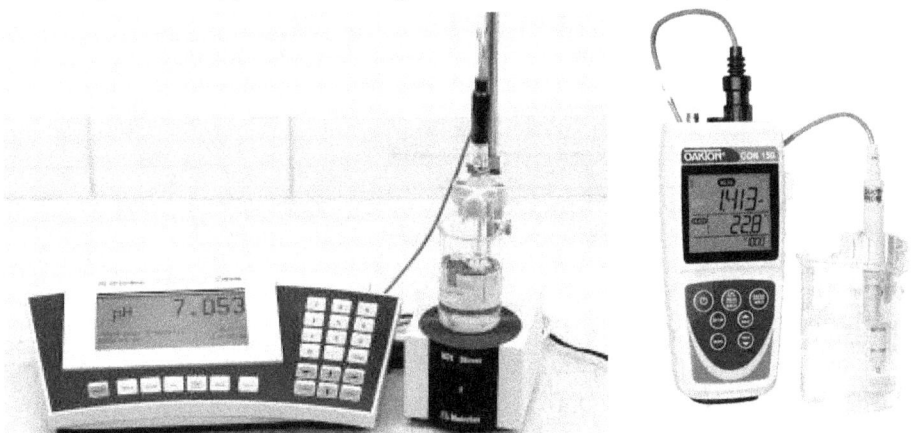

**Digital pH Meters:** Notice that the reading on the desktop pH meter (left photo) is pH = 7.053. The glass beaker on top of the hot plate equipped with a magnetic stirrer has a glass electrode in it. This picks up a signal from the H⁺ ions in the solution and transmits it to the meter. pH meters range in size from the desktop model (left photo) to rugged hand-held models for use in the field (right photo), to the 26 mm x 13 mm (a bit larger than a large vitamin capsule) SmartPill™ (https://bit.ly/3SsIS6d) that is swallowed. It passes completely through the gastrointestinal tract, broadcasting pH and other useful information to the physician. **Arnold O. Beckman** (1900–2004), a professor of chemistry at Cal Tech, built the first commercial pH meter in 1934 to measure the pH of lemon juice, https://bit.ly/3xNC86G. *Left photo is in the Public Domain. Right photo courtesy of Oakton Instruments.*

Now that you know how to calculate pH, you might like to look at Table 19-1, which lists the pH values for some common substances. The pH of a solution can be measured directly with an instrument called a pH meter such as the ones in the above photos.

If you know the pH of a solution, it is sometimes desirable to calculate the [H⁺]. This can be done as follows. Since **pH = –log [H⁺]**, we first put the minus sign on the other side by multiplying both sides of the equation by –1 and then we raise both sides to the power of 10:

$$\text{pH} = -\log [\text{H}^+] \rightarrow -\text{pH} = \log [\text{H}^+] \rightarrow 10^{-\text{pH}} = 10^{\log [\text{H}^+]}$$

Since $10^{\log [\text{H}^+]} = [\text{H}^+]$ (see section 19-2), we have $10^{-\text{pH}} = [\text{H}^+]$. Rearranging, we get

$$\boxed{[\text{H}^+] = 10^{-\text{pH}}}$$

In words, this equation reads: **"The negative of the pH is the *power* to which 10 must be raised to get the [H⁺]."**

This is the sense that Sørensen meant for the word "power." Given that the pH of human blood is 7.4 (see Example 11), calculate the [H⁺]. [H⁺] = $10^{-\text{pH}}$ = $10^{-7.4}$ = 3.98 x $10^{-8}$ M. The calculator button sequence is **7.4, +/-, INV, log** or on some calculators, **7.4, +/-, 10ˣ**. See https://bit.ly/3czUQ77.

## TABLE 19-1

[H⁺] AND THE pH OF SOME COMMON SUBSTANCES[a]

| $[H^+]$ | pH | SUBSTANCE |
|---|---|---|
| 10 | −1 | |
| 1 | 0 | |
| $10^{-1}$ | 1 | stomach acid, pH = 1 |
| $10^{-2}$ | 2 | lemon juice, pH = 2.2 |
| | | vinegar, pH = 2.4 |
| $10^{-3}$ | 3 | carbonated soft drinks, pH = 3 |
| | | grapefruit juice, pH = 3.2 |
| | | orange juice, pH = 3.5 |
| $10^{-4}$ | 4 | club soda, pH = 4 |
| | | tomato juice, pH = 4.2 |
| $10^{-5}$ | 5 | urine, pH = 4.8–7.5, |
| $10^{-6}$ | 6 | depending on what you eat and your metabolism |
| $10^{-7}$ | 7 | pure water, pH = 7 |
| $10^{-8}$ | 8 | human blood, pH = 7.4 |
| $10^{-9}$ | 9 | |
| $10^{-10}$ | 10 | milk of magnesia, pH = 10.6 |
| $10^{-11}$ | 11 | household ammonia, pH = 11.8 |
| $10^{-12}$ | 12 | |

[a] A solution with a pH of 7 is called a **neutral solution**. As the pH gets smaller than 7, the acidity increases. As the pH gets larger than 7, the solution becomes more basic.

**EXAMPLE 14** Calculate the [H⁺] in human erythrocytes (red blood cells), whose pH is 7.25.

*Solution:* $[H^+] = 10^{-pH} = 10^{-7.25} = 5.62 \times 10^{-8}$ M. The calculator sequence is **7.25, +/−, INV, log** or **7.25, +/−, 10ˣ**. ∎

**EXAMPLE 15** Calculate the [H⁺] of the urine of a person who takes large amounts of ascorbic acid (vitamin C). The pH of the urine is 4.5.

*Solution:* $[H^+] = 10^{-pH} = 10^{-4.5} = 3.16 \times 10^{-5}$ M. The calculator sequence is **4.5, +/−, INV, log** or **4.5, +/−, 10ˣ**. ∎

## PROBLEMS

### KEYED PROBLEMS

1. The $H^+$ concentration in human saliva is about $2.51 \times 10^{-7}$ M. Write this as an equation with square brackets around the $H^+$.

2. What are the logs of 10,000,000 and 100,000,000?

3. What are the logs of 0.00001 and 0.000001?

4. Take the logs of 962 and $7.76 \times 10^5$ on your calculator.

5. Take the logs of 0.087 and $3.24 \times 10^{-6}$ on your calculator.

6. Take the log of any negative number.

7. What is the antilog of 4?

8. Find the antilogs of 0.86 and 4.66.

9. Find the antilogs of $-0.67$ and $-2.33$.

10. Find the antilogs of 7.3 and $-7.3$.

11. What is your calculator button sequence for calculating the pH of blood in a person who has serious acidosis? The $[H^+]$ is $8.91 \times 10^{-8}$ M.

    (**NOTE:** Your button sequence will depend on which calculator you have.)

12. Calculate the pH of household ammonia. The $[H^+]$ is $1.58 \times 10^{-12}$ M.

13. Calculate the pH of lemon juice. The $[H^+]$ is $3.98 \times 10^{-3}$ M.

14. Calculate the $[H^+]$ of milk whose pH is 6.5.

15. Calculate the $[H^+]$ of human tears whose pH is 7.4.

### SUPPLEMENTAL PROBLEMS

16. What is the logarithm of the following numbers?

    a. $1 \times 10^8$  b. 1  c. $1 \times 10^{-8}$

17. Find the logarithms of each of the following.

    a. 327  c. $4.76 \times 10^3$  e. $6.5 \times 10^{13}$
    b. 8262  d. $9.8 \times 10^{23}$  f. $4.2 \times 10^9$

18. Find the logarithm of each of the following.

    a. 0.134  c. $6.22 \times 10^{-5}$  e. $9.21 \times 10^{-18}$
    b. 0.00467  d. $5.43 \times 10^{-10}$  f. $1.01 \times 10^{-23}$

## 358 LOGARITHMS AND pH

19. Find the antilogarithm of each of the following logarithms.

   a. 5   b. 0   c. −5

20. Find the antilog of each of the following.

   a. 3.22   c. 10.99   e. 0.432
   b. 7.58   d. 13.76   f. 0.30103

21. Find the antilog of each of the following.

   a. −0.3010   c. −0.0058   e. −22.37
   b. −4.53     d. −11.45    f. −1.11

22. What is the pH of tomato juice whose [$H^+$] is $7.94 \times 10^{-5}$ M?

23. What is the pH of wine whose [$H^+$] is $3.16 \times 10^{-4}$ M?

24. What is the [$H^+$] of black coffee whose pH is 5.1?

25. What is the [$H^+$] of milk whose pH is 6.9?

---

**HOW TWO 23-YR OLD MEN CHANGED THE WORLD (continued from p. 345)**

The reason that pure $Al_2O_3$ isn't just melted and electrolyzed is that its melting point is 2072 °C, much too high to be practical. The melting point of cryolite ($Na_3AlF_6$) is 1009 °C. When both are mixed, and some aluminum fluoride ($AlF_3$) is added to reduce the melting point even further,[2] the melting point of the mixture becomes around 960 °C, which is a practical temperature for commercial purposes, and is above the melting point of aluminum, which is 660 °C. The liquid aluminum is heavier than the electrolyte mixture, so the liquid Al sinks to the bottom of the reaction vessel and can be drawn off, making for a continuous process. Additional $Al_2O_3$ is added as it is used up. **NOTE:** An **electrolyte** is the liquid in which electrolysis takes place.

[1]Aluminum oxide is obtained from the ore **bauxite**. The Austrian chemist **Carl Josef Bayer** (1847-1904) invented a way of extracting pure aluminum oxide from bauxite, which made the Hall–Héroult process commercially feasible. **Electrolysis** is the process of adding to and taking away electrons from a solution using an outside electrical current. It is the opposite to the way a battery works, which generates its own electrical current.

[2]When a solid is dissolved in a liquid, the freezing point (f.p.) of the liquid is lowered. That's why, in the winter, salt is sprinkled on ice to melt it. Also, when two liquids are mixed, the freezing point of the mixture is lower than the freezing point of the individual liquids. That's why alcohol is mixed with water for wintertime use of windshield washer fluid, and why a (usually) 50%-50% mixture of water (f.p. = 32 °F or 0 °C) and ethylene glycol (f.p. = 8.8 °F or −13 °C) is used as an antifreeze in cars. The freezing point of the mixture is about −35 °F or −37 °C.

# GLOSSARY

**Absolute temperature scale** A temperature scale that uses absolute zero (zero kelvins) as the lowest temperature.

**Absolute zero** The lowest temperature possible. It cannot be reached by any apparatus.

**Abundance** The amount of a substance present in a sample. See also *percent abundance* and *fractional abundance*.

**Accuracy** The closeness of a measurement to the true value.

**Acid** A substance that yields one or more hydrogen ions (H⁺) when dissolved in water.

**Acidic solution** A water solution in which the hydrogen ion concentration is greater than the hydroxide ion concentration. Its pH is less than 7.

**Actinoid series** Elements 89 (Ac) through 103 (Lr). Also called the actinide series. See page 247 for more details.

**Alkali metals** Metals in group IA of the periodic table. These are Li, Na, K, Rb, Cs, and Fr. All have one $s$ electron in their valence shell.

**Alkaline Basic.** Pertaining to a water solution having more hydroxide ions than hydrogen ions. Its pH is greater than 7.

**Alkaline earth metals** Metals in group IIA of the periodic table. These are Be, Mg, Ca, Sr, Ba, and Ra. All have two $s$ electrons in their valence shell.

**Alpha particle** A helium-4 nucleus, $^4He^{2+}$.

**Anhydride (anhydrous)** Without water. For example, $CuSO_4$ is sometimes called anhydrous copper(II) sulfate, whereas $CuSO_4 \cdot 5H_2O$ is called copper(II) sulfate pentahydrate. The water after the "·" is called the "water of hydration."

**Antilogarithm** For logarithms to the base 10, it is the number obtained by raising 10 to a power. If $y = 10^x$, then y is the antilogarithm of $x$. Another way of saying this is the following: If $\log y = x$, then y is the antilogarithm of $x$.

**Aqueous** Relating to water; or "dissolved in water."

**Atmosphere (atm)** A unit of pressure based on standard atmospheric pressure at sea level; 1 atm = 14.7 lb/in.$^2$ = 760 torr = 101 kPa. Also, the atmosphere is the air surrounding the earth.

**Atmospheric pressure** The pressure on objects resulting from the layer of air surrounding our earth. At the earth's surface, this pressure is about 760 torr.

**Atom** The basic unit of an element that can enter into chemical combination (unless it is one of the three noble gases that do not form compounds: He, Ne, and Ar).

**Atomic mass unit (amu)** An older unit for the mass of atoms. See "unified atomic mass unit" (u) for the definition of the modern term.

**Atomic number** The number of protons in an atom.

**Atomic theory** The theory stating that substances are composed of atoms.

**Atomic weight** The weighted average mass of all the naturally occurring isotopes of an element.

**Avogadro's Law** Equal volumes of all (ideal) gases, at the same temperature and pressure, have the same number of atoms or molecules, or moles of particles.

**Avogadro's number or constant** The number of elementary entities in one mole. The value of Avogadro's constant is defined as: $6.022\ 140\ 76 \times 10^{23}$ mol$^{-1}$.

**Balanced chemical equation** A chemical
  equation having the same number of each
  kind of atom and the same total electrical
  charge on each side of the equation.

**Balancing numbers (or coefficients)** Numbers placed in front of atoms or molecules in a chemical equation to balance the equation. The term "coefficients" is preferred.

**Barometer** An instrument for measuring atmospheric pressure.

**Base** A substance that produces one or more hydroxide ions (OH$^-$) when
  dissolved in water. Another definition, which is generally more useful, is this: A base is a substance that can take hydrogen ions (H$^+$) from an acid.

**Basic solution** A water solution containing more hydroxide ions than hydrogen ions. Its pH is greater than 7.

**Battery** An electrochemical cell that can be used as a source of current and voltage (in other words, a battery is a source of electricity).

**Beta particle** An electron.

**Binding energy**
Energy equivalent to the mass difference between the sum of the masses of the individual free protons and neutrons in a nucleus and the actual mass of the
  nucleus.

**Bond** See chemical bond.

**Bonding electrons**
Electrons transferred or shared in forming a chemical bond. The valence electrons.

**Boyle's law** $P_1/V_1 = P_2/V_2$ at constant temperature and amount of gas.

**Buret** A device designed to allow the delivery of an accurate quantity of liquid. Any amount up to the buret's capacity can be accurately delivered.

**Calibrate** To compare your instrument with a standard instrument so that you know it is working correctly. You can also run a known sample through your instrument to see whether you get the correct value.

**Celsius scale (°C)** The temperature scale in which water freezes at 0 °C and boils at 100 °C at 1 atm pressure.

**Centrifugal force** A force directed outward from an object moving in a curved path. This is the force that throws you to the side when your car goes around a turn.

**Charge** An electrical property. There are positive charges and negative charges.

**Charles's law** $V_1/T_1 = V_2/T_2$ at constant pressure and amount of gas.

**Chemical bond** The attractive force that holds atoms together.

**Chemical change** A change that produces substances different from those originally present.

**Chemical equation** An expression showing the reactants and products in a chemical reaction.

**Chemical family** Elements appearing in the same column in the periodic table. They have similar valence electron configurations and therefore have similar chemical properties.

**Chemical formula** An expression showing the number of atoms in a molecule. It can also show the simplest ratio of atoms in a compound where no individual molecules are found (such as a "giant array"). Chemical formulas use the symbols of the elements.

**Chemical property** Any property of a substance that can be studied only by changing the substance to another substance by a chemical reaction.

**Chemical reaction** The process by which one or more substances are converted to different substances by making and breaking chemical bonds.

**Chemistry** The study of the composition, structure, properties, and reactions of atoms and molecules.

**Coefficients** Whole numbers or fractions placed before the chemical formulas in a chemical equation. When this is done correctly, the equation is said to be balanced. An older name is "balancing numbers."

**Combined gas law** $P_1V_1/T_1 = P_2V_2/T_2$ at constant amount of gas.

**Combustion** Usually the process of burning a substance with oxygen. The term can also apply to burning substances with other oxidizing agents such as $F_2$, $Cl_2$, $N_2O_4$, and $H_2O_2$.

**Common name (of a compound)** A very old name that is commonly used. It usually doesn't follow any of the nomenclature rules.

**Compound** Two or more atoms of different elements held together with one or more chemical bonds.

**Concentrated solution** A solution containing a relatively large amount of solute.

**Concentration of a solution** An expression describing the amount of dissolved solute in a certain quantity of solvent or solution.

**Constant** A number in a mathematical expression that doesn't change its value under the specified conditions.

**Constant of proportionality** A number that relates two different quantities. It generally has a number part and a unit part.

**Cosmic rays** Nuclei that come from outer space at extremely high speed. They collide with the air atoms in the atmosphere and produce, among other things, tritium and carbon-14.

**Covalent bond** A chemical bond in which two electrons are shared by two atoms.

**Crystalline solid** A solid in which the component atoms or molecules are arranged in an orderly, three-dimensional, repetitive structural pattern.

**Dalton's law of partial pressures** The total pressure exerted by a mixture of gases is equal to the sum of the partial pressure of each gas in the mixture.

**Decay (radioactive)** The spontaneous disintegration of an atomic nucleus by the emission of particles and/or radiation.

**Decompose** Break up into one or more parts.

**Density** The mass of a substance divided by its volume. $D = m/V$, where $D$ = density, $m$ = mass, and $V$ = volume.

**Deuterium** An isotope of hydrogen, written as $^2_1H$ or $^2_1D$.

**Diatomic molecule** A molecule consisting of two atoms.

**Diffusion** The process by which atoms or molecules of one substance gradually mix with another substance. The kinetic energy of the atoms or molecules gives rise to the motion that causes the diffusion.

**Dilute solution** A solution containing a relatively small amount of solute.

**Directly proportional** Pertaining to two quantities interacting so that one gets larger as the other gets larger. These quantities are said to be directly proportional to each. It is also implied that as one quantity gets smaller, the other gets smaller.

**Dismutation** See *disproportionation*.

**Disproportionation** The process in which a substance spontaneously reacts to produce one atom in a higher oxidation state and another atom of the same element in a lower oxidation state. Also called dismutation.

**Dissociation** The process by which a compound separates into individual ions when dissolved in water. Example: NaCl dissolved in water dissociates to give Na$^+$ and Cl$^-$ ions. Dissociation also refers to molecules breaking apart such as H$_2$ → H + H.

**Dissolve** To go into solution.

**Distributive law of algebra** $a(b + c) = ab + ac$ for any numbers $a$, $b$, and $c$.

**Double bond** Two covalent bonds joining two nuclei. Each bond consists of two electrons, giving a total of four electrons joining the two nuclei together.

**Dry cell** A battery in which the electrolyte solution is a paste rather than a liquid. A regular flashlight battery (a "D" cell) is a dry cell. An automobile battery is called a wet cell because it contains liquid sulfuric acid as the electrolyte.

**Einstein's mass–energy equation** An equation, published by Einstein in 1906, showing the relationship between mass and energy: $E = mc^2$, where $E$ = energy, $m$ = mass, $c$ = speed of light. Since the speed of light squared is so large, a small amount of mass can be converted into a large amount of energy.

**Electrolyte** A substance whose water solution will conduct electricity.

**Electron** A subatomic particle with low mass and unit negative charge.

**Electron cloud** A region of space around or between atoms that is occupied by electrons.

**Electron configuration** A listing of the occupied orbitals of an atom, going from the first energy level to the highest occupied energy level.

**Electronegativity** The ability of an atom to attract electrons toward itself in a covalent bond.

**Electron orbit** The circular or elliptical path followed by an electron around a nucleus. This term is not used anymore and has been replaced by electron orbital.

**Electron orbital** A description of the shape of the space in which an electron exists.

**Electron shell** See *energy levels of electrons*.

**Electron pair repulsion** The principle stating that molecular geometry is determined by the repulsion between electron pairs around a bonded atom.

**Electrostatic force** The force of attraction or repulsion between two electrically charged objects. Like charges repel and unlike charges attract each other.

**Element** A collection of atoms all having the same atomic number. Also, a pure substance that cannot be separated into simpler substances by chemical means.

**Elemental molecules** Molecules consisting of atoms that all have the same atomic number.

**Empirical formula** A chemical formula that gives the smallest whole number ratio of atoms in a substance.

**Empirical weight** The molecular (or formula) weight of an empirical formula.

**Energy** The ability to do work.

**Energy levels of electrons** Areas in which electrons are located at various distances from the nucleus.

**Energy sublevels** The orbitals that are in an energy level. Examples are the $s$, $p$, $d$, and $f$ orbitals.

**Exact number** A number that has an infinite number of significant figures.

**Excited state** An energy state of a system that is higher than the normal or ground state. An example of this is when an electron in an atom or molecule absorbs energy and is raised to a higher-energy orbital.

**Exponent** A superscript written after a quantity to indicate the power to which it is to be raised. Example: In $10^n$, $n$ is an exponent.

**Exponential notation** A notation in which exponents are used. See *exponent*.

**Family** See *chemical family*.

**Fahrenheit scale (°F)** The temperature scale in which water freezes at 32 °F and boils at 212 °F at 1 atm pressure.

**Fission, nuclear** The splitting of a heavy atomic nucleus into two or more pieces.

**Formula, chemical** See *chemical formula*.

**Formula unit** The smallest number of ions that represent the composition of an ionic compound. Example: NaCl. For other "giant arrays," the formula unit is the simplest formula that represents the composition.

**Formula weight** The mass of one mole of a substance. For a substance that consists of individual molecules, the formula weight and the molecular weight are the same. For a substance that is an ionic crystal or other "giant array," the term "formula weight" is used for the mass of one mole of the simplest formula.

**Fractional abundance** The decimal equivalent of the percent abundance. See *percent abundance*.

**Fusion, nuclear** The combination of small nuclei at a high temperature to form a larger, more stable nucleus.

**Gas** A state of matter that has neither definite shape nor definite volume.

**Giant array** A crystalline substance in which individual atoms or molecules don't exist in a separate state. The whole crystal can be considered a molecule or an array of molecules. Metals can also be considered as giant arrays of atoms.

**GIGO rule** Garbage In, Garbage Out. If you enter incorrect data into a calculator or computer, the results you get from calculation will be wrong.

**Graduated cylinder** A piece of apparatus that can contain a given amount of liquid. It is not used for the most accurate measurements.

**Gram (g)** A unit of mass in the metric system that is 0.001 of a kilogram.

**Graph** A pictorial representation of two or more variables.

**Ground state** The lowest energy state of an electron, an atom, or a molecule.

**Group (periodic table)** The elements in a vertical column in the periodic table. See *chemical family*.

**Half-reaction** Either the oxidation or the reduction step in a redox reaction.

**Halide ion** A negative ion of the halogen family, which consists of the elements F, Cl, Br, I, and At. The corresponding negative ions are $F^-$, $Cl^-$, $Br^-$, $I^-$, and $At^-$.

**Halogen** An element in group VIIA of the periodic table. These elements are F, Cl, Br, I, and At.

**Homogeneous** Having uniform properties (such as the same composition) throughout a sample.

**Horizontal orbital diagram** An orbital diagram drawn horizontally rather than vertically.

**Hormone** A chemical substance that acts as a control or regulatory agent in the body.

**Hydrate** A substance that contains water as part of the formula. Example: $CuSO_4 \cdot 5H_2O$ is called copper(II) sulfate pentahydrate.

**Hydration** A process in which water molecules are added to a substance.

**Hydrocarbon** Molecules consisting only of carbon and hydrogen atoms.

**Hypo** As a prefix it means "less than." It is also the common name for the "fixer" in photography.

**Ideal gas** A gas that can be described by the ideal gas equation. In an ideal gas, the molecules are assumed to be points with no force of attraction or repulsion between them, except when they collide. No real gas is ideal.

**Ideal gas equation** $PV = nRT$.

**Ideal gas law** A law described by the ideal gas equation.

**Imaginary number** The square root of a negative number.

**Indicator** A dye whose color changes over a specific pH range. Used as an aid in doing titrations and for measuring the pH of solutions. There are also indicators for other reactions, such as redox reactions.

**Inorganic chemistry** The branch of chemistry that, in general, studies all the elements and their compounds except for most carbon compounds.

**Integer** A whole number such as 1, 2, 3, 4, 5, and so on.

**Inversely proportional** Pertaining to two quantities interacting so that one becomes larger as the other becomes smaller.

**Ion** A charged particle formed by the loss or gain of electrons by a neutral atom or group of atoms.

**Ionic bond** A bond resulting from the electrostatic attraction between the positive and negative ions in an ionic compound.

**Ionic compound** A substance that contains positive and negative ions.

**Ionic equation** A chemical equation in which, as far as possible, ions, rather than molecules, are written.

**Ionization** The formation of an ion or ions from an atom or molecule.

**Isotonic solution** A solution that can be injected into the blood without damaging blood cells or blood vessels.

**Isotopes** Atoms of the same element having a different number of neutrons.

**Isotopic symbol** A notation for writing isotopes. The atomic number is put at the lower left of the element's symbol. The mass number is put at the upper left of the symbol.

**Kelvin temperature scale** A temperature scale starting at absolute zero, the lowest possible temperature. Zero kelvins equals $-273.15\ °C$.

**Kinetic energy** Energy of motion equal to ½$mv^2$, where $m$ = mass and $v$ = velocity.

**Lanthanoid series** Elements 57 (La) through 71 (Lu). Also called the rare earth elements or the lanthanide series. See page 247 for more information.

**Law of conservation of energy** Energy can neither be created nor destroyed, but it can be converted from one form to another. See Einstein's mass — energy equation for an extension of this law.

**Lewis formula** Representation of atoms, molecules, and ions in which the valence electrons are represented as dots and/or lines.

**Limiting reagent** The reactant that is used up completely in a chemical reaction. Also called "limiting reactant."

**Liquid** One of the three states of matter. A liquid can move about freely while it still retains a definite volume. Liquids flow and take the shape of their containers.

**Liter (L)** A unit of volume. The volume of one kilogram of water at $4^0$ C.

**Log** See logarithm.

**Logarithm** In base 10 logarithms, the power to which 10 must be raised to give the number of interest. Example: log 100 = 2. Thus 2 is the logarithm of 100. Note that $10^2 = 100$.

**Mantissa** The number written before the " sign in scientific notation. In 3.5 x $10^6$, 3.5 is the mantissa. In logarithms, the mantissa is the number written after the decimal point. The number before the decimal point is called the "characteristic."

**Mass** The quantity of matter in an object which causes the object to have weight in a gravitational field.

**Mass defect** The difference between the actual mass of an atom and the calculated mass of the protons, neutrons, and electrons that make up that atom.

**Mass number** The sum of the number of protons and neutrons in the nucleus of an atom.

**Matter** Anything that has mass and occupies space.

**Measurement** The amount of something; this is indicated by a number followed by a unit.

**Mechanism of a reaction** The route or steps by which a reaction takes place. The mechanism describes the manner in which atoms or molecules are transformed from reactants into products.

**Metal** An element that conducts electricity and heat. Metals are usually shiny and tend to lose valence electrons and become positive ions. Combinations of metallic elements are also called metals or alloys.

**Metalloid** The older name for semimetal.

**Meter (m)** The standard of length in the metric system of units.

**Metric system** A system of measurement based on factors of 10 between different units. See Sec 7-6.

**Mixture** Two or more substances combined without a chemical bond. They can be mixed in any proportion.

**Molarity** The number of moles of solute per liter of solution.

**Molar solution** A solution containing one mole of solute per liter of solution.

**Mole** An amount of substance containing 6.022 140 76 × $10^{23}$ particles or elementary entities.

**Molecular compound** A compound made up of individual discrete molecules.

**Molecular formula** An expression showing the number of atoms of each element in a molecule.

**Molecular geometry** The bond lengths and bond angles of the atoms in a molecule. The three-dimensional structure of a molecule with definite bond lengths and bond angles.

**Molecular orbital** A description of a chemical bond in which the atomic orbitals combine in a special way.

**Molecular weight** The sum of the atomic weights of all the atoms in a molecule.

**Molecule** Two or more atoms held together with one or more chemical bond(s).

**Molten state** The liquid state of a substance. Water is the molten state of ice. Liquid sodium chloride is the molten state of solid table salt.

**Monatomic** One atom.

**Net ionic equation** A chemical equation in which only the species (many of which are ions) that actually take part in the chemical reaction are shown.

**Neutralization** The reaction of an acid and base to form a salt and water.

**Neutron** A subatomic particle with a mass of about 1 u and no charge.

**Noble gas** An element in group VIIIA of the periodic table. These elements are He, Ne, Ar, Kr, Xe, and Ra. He, Ne, and Ar are unreactive chemically.

**Nomenclature** A systematic method of naming things.

**Nonbonded electron pair** An electron pair that doesn't take part in a chemical bond. These pairs of electrons play a role in determining molecular geometry.

**Nonmetal** Elements that are generally poor conductors of heat and electricity. They are not shiny and tend to form negative ions. Many nonmetals are gases.

**Nuclear fission** See *fission, nuclear*.

**Nuclear fusion** See *fusion, nuclear*.

**Nuclear power** Controllable energy produced by the use of nuclear fission or, probably in the future, nuclear fusion.

**Nuclear reactor** A device that uses fissionable elements (such as uranium-235) to produce large amounts of neutrons and heat.

**Nucleon** A nuclear particle such as a proton or a neutron.

**Nucleus** The central part of an atom that contains the protons and neutrons. The nucleus has most of the mass and all the positive charge of the atom.

**Octahedron (regular)** An eight-sided solid geometric figure with all edges and faces equal.

**Octet rule** Many atoms other than hydrogen tend to form bonds until they are surrounded by eight electrons in their valence shell.

**Orbit** See *electron orbit*.

**Orbital** See *electron orbital*.

**Orbital diagram** A diagram showing the kinds of orbitals and their relative energies for the electrons of an atom.

**Organic chemistry** The chemistry of carbon compounds.

**Orthomolecular nutrition** A nutritional theory stating that, in general, each person's enzyme systems have different properties and that supplying the optimum levels of nutrients for a person can help lead to optimal health for that person. Literally, orthomolecular means "the right molecule."

**Oxidation** The process in which a substance loses electrons.

**Oxidation number** A number assigned to each atom in a substance (or ion), using somewhat arbitrary rules; this number helps to keep track of what happens in redox reactions.

**Oxidation–reduction reaction** A reaction in which electrons are both lost and gained. Also called "redox" reaction.

**Oxidizing agent** The species that takes electrons in a redox reaction. The oxidizing agent is reduced.

**Oxyacid** An acid that contains oxygen.

**Partial pressure** The pressure of one component in a mixture of gases.

**Pascal (Pa)** A unit of pressure; 101,325 Pa = 1 atm.

**Pauli exclusion principle** A principle of quantum mechanics which states that no more than two electrons (each having opposite spin) can occupy the same atomic or molecular orbital.

**Per** As part of a chemical name, "the most" or the "greatest."

**Percent** Parts per hundred.

**Percent abundance** The percent by number of atoms of an isotope that makes up an element.

**Percent composition of a compound** The percent by weight of each element in a compound.

**Percent uncertainty** The error of a measurement divided by the value of the measurement and then multiplied by 100.

**Period (periodic table)** A horizontal row of the periodic table.

**Periodic law** The properties of the elements are periodic functions of their atomic number. This means that the properties repeat themselves regularly.

**Periodic table** A table arranging the elements in order of increasing atomic number. Elements with similar valence electron configurations and similar chemical and physical properties are arranged in columns.

**pH** A way of expressing acidity or alkalinity; $pH = -\log [H^+]$.

**Photon** The particle associated with a light wave.

**Photosynthesis** The process that green plants use to make sugars and elemental oxygen. In nature, the energy for this process comes from sunlight.

**Physical change** An alteration in form, such as size, shape, or physical state, without a change in composition. A change without chemical reaction.

**Physical property** A property of a substance that can be studied without causing a chemical reaction. Examples of physical properties are color, density, melting point, and boiling point.

**Pipet** A device used to deliver a known volume of a liquid.

**Planar molecule** A molecule whose atoms are all in the same plane.

**Polyatomic ion** An ion consisting of more than one atom.

**Potential energy** Stored energy or the energy an object has because of its position, chemical state, or electrical condition.

**Precipitate** An insoluble solid that can be separated from a solution.

**Precision** The closeness with which individual measurements agree with one another. This quality is related to how good an experimenter you are. In general, the greater the number of significant figures in a measurement, the greater the precision.

**Pressure** Force per unit area. Pressure can be expressed in the following units: pounds per square inch, atmospheres, torr, millimeters of mercury, and pascal.

**Principal energy levels** The main energy levels of electrons in an atom. These energy levels are quantized and are referred to by a set of integers, the principal quantum numbers $n$, which can take the value 1, 2, 3, 4, 5, 6, 7, and so on.

**Product (of a chemical reaction)** A substance produced from the reactants in a chemical reaction.

**Properties** The characteristics, or traits, of substances which can be observed or measured. Properties are classified as physical or chemical.

**Proportion** A mathematical relationship between two quantities. See *directly proportional* and *inversely proportional*.

**Proportionality constant** A constant that turns a proportional relationship into an equality.

**Protium** A somewhat uncommon name for an isotope of hydrogen, written as $_1^1H$.

**Proton** A subatomic particle having a mass of about 1 u and a charge of +1.

**Pure substance** A substance that consists of only one kind of compound, molecule, or atom.

**Pyramid (trigonal)** A geometric figure that has four triangular sides. In general, the triangles are not equilateral triangles.

**Pyramidal molecule** A molecule that has the structure of a pyramid.

**Quantization of energy** The existence of certain discrete (separate) energy levels in an atom or molecule.

**Quantum mechanics** The current theory of atomic structure which has the uncertainty principle and quantization of energy as basic principles.

**Quotient** The number produced by dividing a given number by another number.

**Radical** A reactive species that usually exists for only a short time during a chemical reaction. Also called a "free radical."

**Radioactivity** The spontaneous disintegration or decay of a nucleus by the emission of a particle and/or a wave.

**Radioactive decay** See *radioactivity*.

**Radioisotope** An isotope that undergoes radioactive decay.

**Ratio identity** A ratio made from an identity.

**Reactant** A starting substance in a chemical reaction.

**Reaction equivalent (REQ)** The number of moles of a reactant divided by the coefficient of that substance in a balanced chemical equation. The REQ is used to determine the limiting reagent.

**Redox reaction** A reaction in which there is a transfer of one or more electrons from one substance to another.

**Reducing agent** A substance that can donate electrons to another substance.

**Reduction** The process in which a substance gains electrons.

**Regular notation** The usual way of writing numbers, as contrasted to the scientific notation way of writing numbers.

**Relative atomic weight** The weighted average mass of all the naturally occurring isotopes of an element relative to an atom of $_6^{12}C$, which is given a mass of exactly 12 u.

**Rounding error** A small error introduced into a calculation caused by rounding some of the numbers before the calculation is finished.

**Salt** A compound formed by the reaction between an acid and a base. In the reaction KOH + HCl → KCl + H$_2$O, the KCl is the salt.

**Saturated solution** A solution that has as much solute in it as it can hold at a given temperature.

**Scientific notation** A means of expressing very small and very large numbers in two parts: a number between 1 and 10 (or 1) multiplied by 10 raised to some exponent or power.

**Semimetal** An element having properties between that of a metal and a nonmetal. Also called a "metalloid."

**Shell** Another name for an energy level of an atom.

**Shielding** A decrease in the nuclear charge felt by outer electrons owing to the presence of inner electrons.

**SI system of units** A system of units based on the metric system, which uses seven basic units. The seven basic units are the *meter, kilogram, second, kelvin, mole, ampere,* and *candela.*

**Significant figures** The number of digits in a number that are known for sure, plus one more that is uncertain.

**Single bond** A covalent bond consisting of two electrons.

**Solid** A state of matter that has definite shape and volume.

**Solubility** The maximum amount of solute that will dissolve in a given quantity of solvent.

**Soluble** Capable of being dissolved.

**Solute** The substance that is dissolved in a solvent to form a solution.

**Solution** A homogeneous mixture of two substances.

**Solvent** The component of a solution that is present in the largest amount. In a solution consisting of a solid dissolved in a liquid, the liquid is the solvent.

**Spectator ion** An ion that does not undergo chemical change during a chemical reaction.

**Spin (of an electron)** A quantity that is one of the properties of an electron. Two electrons with different spins can be in the same orbital.

**Spin state** A description of the spin of an electron. The two spin states are referred to as "up" and "down."

**Square pyramid** A geometric figure with a square base and four triangular sides. The Egyptian pyramids are square pyramids.

**Stoichiometry** The mass relationships of reactants and products in a chemical reaction.

**Structural formula** An expression showing the number of atoms in a molecule and giving information about how the atoms are bonded to one another.

**Substance** A form of matter that has a definite composition and distinct properties.

**Superconductor** A material that conducts electricity with no apparent resistance.

**Supersaturated solution** A solution that contains more solute than a saturated solution at the same temperature. These solutions tend to be unstable, and the excess solute can easily crystallize out.

**Suspension** A mixture that gradually separates on standing.

**Symbol (for an element)** An abbreviation for the name of an element.

**Temperature** A measure of how hot or cold a system is.

**Tetrahedral angle** The angle that the corners of a regular tetrahedron make with one another. The value of the angle is 109.5°.

**Tetrahedron (regular)** A regular four-sided solid. Each side consists of congruent (identical) equilateral triangles.

**Theory** A unifying principle that explains a large number of facts or observations.

**Titration** The process of measuring the volume of one reagent required to react with a measured weight or volume of another reagent.

**Torr** A unit of pressure equal to 1 mmHg or 1/760 atm.

**Total ionic equation** An equation that shows substances in the form in which they actually exist.

**Transition elements** The elements in groups IB through VIIB and in group VI in the periodic table.

**Transition metals** See transition elements.

**Transmutation (nuclear)** The conversion of one chemical element or an isotope into another chemical element or isotope. A transmutation can be achieved either by nuclear reactions (in which an outside particle reacts with a nucleus) or by radioactive decay, where no outside cause is needed.

**Transuranium elements** Elements with atomic numbers greater than that of uranium.

**Trigonal bipyramid** A geometric figure consisting of two trigonal pyramids connected by a common face.

**Trigonal pyramid** A four-sided geometric figure whose sides are triangles. In general, the sides are not equilateral triangles.

**Triple bond** A covalent bond that consists of three pairs of electrons that are shared between two atoms.

**Tritium** An isotope of hydrogen, whose chemical symbol is usually written as $^3H$ or T.

**Uncertainty principle** A basic principle of quantum mechanics which states that it is impossible to determine both the position and the velocity of an electron or other atomic-sized particles accurately at the same time.

**Unified atomic mass unit (u)** Exactly one-twelfth the mass of an atom of one of the isotopes of carbon, namely carbon-12.

**Unionized equation (molecular equation)** A chemical equation that shows the substances as molecules, even if this isn't the form in which they actually react.

**Unit** The scale of a measurement. For example, a unit of length is the meter.

**Unsaturated solution** A solution that contains less solute than its corresponding saturated solution at the same temperature.

**Valence** The combining power of an atom, that is the ability of an atom to form chemical bonds.

**Valence electrons** Electrons in the outermost occupied energy level that are available to form chemical bonds.

**Valence shell** The energy level containing the valence electrons.

**Variable** A mathematical quantity that can have different values in a formula or equation.

**Vitamins** Organic compounds that are needed by the body in small amounts and that perform vital biological functions. Many of the vitamins act as coenzymes; that is, they allow the protein part of an enzyme, called an apoenzyme, to work.

**Volume** The amount of space an object occupies.

**Volumetric flask** A flask that is designed to contain accurately a specific quantity of liquid.

**Volumetric pipet** A pipet that is designed to deliver accurately a specific quantity of liquid.

**Water of crystallization** Water molecules that are included as part of the structural parts of crystals formed from water solutions.

**Water of hydration** See "water of crystallization."

**Wavelength** The distance between identical points on successive waves.

**Weight** The force that gravity exerts on an object.

**Weight percent solution** The grams of solute in 100 g of solution.

**Weight–volume percent solution** The grams of solute in 100 mL of solution.

**Weighted average (to calculate atomic weight)** The sum of all the following terms: the fractional abundance multiplied by the mass of an isotope for each naturally occurring isotope of an element. (Note: Weighted average can be used for calculating things other than atomic weights.)

---

**Quotes by Famous People about Science and Life**
See pp. vi, viii, xv for more quotes. The lives of these people are *really* interesting. Check out the Wikipedia articles about them.

If we can't think for ourselves, if we're unwilling to question authority, then we're just putty in the hands of those in power. But if the citizens are educated and form their own opinions, then those in power work for us. In every country, we should be teaching our children the scientific method and the reasons for a Bill of Rights. With it comes a certain decency, humility and community spirit. In the demon-haunted world that we inhabit by virtue of being human, this may be all that stands between us and the enveloping darkness. **–Carl Sagan** (see page 401 in this book.)

The truth may be puzzling. It may take some work to grapple with. It may be counterintuitive. It may contradict deeply held prejudices. It may not be consonant with what we desperately want to be true. But our preferences do not determine what's true. **-Carl Sagan**

Extraordinary claims demand extraordinary evidence **-Carl Sagan**

The absence of evidence is not the evidence of absence.
**-Martin Rees** (b. 1942) **and Carl Sagan**

It takes a fearless, unflinching love and deep humility to accept the universe as it is. The most effective way he [her late husband Carl Sagan] knew to accomplish that, the most powerful tool at his disposal, was the scientific method (https://bit.ly/3sK8kb6), which over time winnows out deception. It can't give you absolute truth because science is a permanent revolution, always subject to revision, but it can give you successive approximations of reality.

**Ann Druyan** (b. 1949)

And what greater might do we possess as human beings than our capacity to question and to learn? **-Ann Druyan**

# SOLUTIONS TO PROBLEMS

**CHAPTER 1**

1. $^{85}_{37}Rb$

2. Atomic number = 79, mass number = 197.

3. Neutrons = 41 − 19 = 22, electrons = 19.

4. (9p⁺, 10n) 9e⁻   $^{19}_{9}F$

5. (7p⁺, 7n) 7e⁻  $^{14}_{7}N$,   (7p⁺, 8n) 7e⁻  $^{15}_{7}N$

6. (86p⁺, 136n) 86e⁻  $^{222}_{86}Rn$,   (86p⁺, 134n) 86e⁻  $^{220}_{86}Rn$,

   (86p⁺, 133n) 86e⁻  $^{219}_{86}Rn$

7. See glossary.

8. a. (10p⁺, 10n) 10e⁻  $^{20}_{10}Ne$,   (10p⁺, 11n) 10e⁻  $^{21}_{10}Ne$,

   (10p⁺, 12n) 10e⁻  $^{22}_{10}Ne$

   b. (16p⁺, 16n) 16e⁻  $^{32}_{16}S$,   (16p⁺, 17n) 16e⁻  $^{33}_{16}S$,

   (16p⁺, 18n) 16e⁻  $^{34}_{16}S$,   (16p⁺, 20n) 16e⁻  $^{36}_{16}S$

c. (22p⁺, 26n) 22e⁻  $^{48}_{22}Ti$,   (22p⁺, 24n) 22e⁻  $^{46}_{22}Ti$,

   (22p⁺, 25n) 22e⁻  $^{47}_{22}Ti$,   (22p⁺, 27n) 22e⁻  $^{49}_{22}Ti$,

   (22p⁺, 28n) 22e⁻  $^{50}_{22}Ti$

d. (29p⁺, 34n) 29e⁻  $^{63}_{29}Cu$,   (29p⁺, 36n) 29e⁻  $^{65}_{29}Cu$

e. (34p⁺, 46n) 34e⁻  $^{80}_{34}Se$,   (34p⁺, 44n) 34e⁻  $^{78}_{34}Se$,

   (34p⁺, 42n) 34e⁻  $^{76}_{34}Se$,   (34p⁺, 48n) 34e⁻  $^{82}_{34}Se$,

   (34p⁺, 43n) 34e⁻  $^{77}_{34}Se$,   (34p⁺, 40n) 34e⁻  $^{74}_{34}Se$

f. (39p⁺, 50n) 39e⁻  $^{89}_{39}Y$    g. (53p⁺, 74n) 53e⁻  $^{127}_{53}I$

h. (63p⁺, 90n) 63e⁻  $^{153}_{63}Eu$,  (63p⁺, 88n) 63e⁻  $^{151}_{63}Eu$

i. (77p⁺, 116n) 77e⁻  $^{193}_{77}Ir$,  (77p⁺, 114n) 77e⁻  $^{191}_{77}Ir$

j. (83p⁺, 126n) 83e⁻  $^{209}_{83}Bi$,  (83p⁺, 127n) 83e⁻  $^{210}_{83}Bi$,

   (83p⁺, 128n) 83e⁻  $^{211}_{83}Bi$,  (83p⁺, 129n) 83e⁻  $^{212}_{83}Bi$,

   (83p⁺, 131n) 83e⁻  $^{214}_{83}Bi$

9. (106p⁺, 157n) 106e⁻  $^{263}_{106}Sg$   (97p⁺, 152n) 97e⁻  $^{249}_{97}Bk$,

   (8p⁺, 10n) 8e⁻  $^{18}_{8}O$

10. (103p⁺, 158n) 103e⁻  $^{261}_{103}$Lr,   (103p⁺, 159n) 103e⁻  $^{262}_{103}$Lr,

(99p⁺, 155n) 99e⁻  $^{254}_{99}$Es,   (10p⁺, 12n) 10e⁻  $^{22}_{10}$Ne

11.

| ISOTOPIC SYMBOL | NUMBER OF PROTONS | NUMBER OF NEUTRONS | NUMBER OF ELECTRONS | ATOMIC NUMBER | MASS NUMBER |
|---|---|---|---|---|---|
| $^{24}_{12}$Mg | 12 | 12 | 12 | 12 | 24 |
| $^{31}_{15}$P | 15 | 16 | 15 | 15 | 31 |
| $^{40}_{20}$Ca | 20 | 20 | 20 | 20 | 40 |
| $^{37}_{17}$Cl | 17 | 20 | 17 | 17 | 37 |
| $^{40}_{18}$Ar | 18 | 22 | 18 | 18 | 40 |
| $^{72}_{32}$Ge | 32 | 40 | 32 | 32 | 72 |

**CHAPTER 2**

1. $(41,900)(0.26) + (49,900)(0.74) = \$47,820$.

2. $(34.96885)(0.7577) + (36.96590)(0.2423) = 35.453$ u. Atomic weight from Table 2.2 is 35.453 u.

3. $(19.99244)(0.9051) + (20.99385)(0.0027) + (21.99138)(0.0922) = 20.179$ u. Atomic weight from Table 2.2 is 20.179 u.

4. $(35)(0.7577) + (37)(0.2423) = 35.48$ u. Atomic weight from Table 2.2 is 35.453 u.

5. Br = 79.904 u, He = 4.00260 u, Co = 58.9332 u, Ag = 107.868 u, Sb = 121.75 u.

6. See glossary.

7. a. $(1.007825)(0.99985) + (2.01410)(0.00015) = 1.00798$ u. Atomic weight from Table 2.2 is 1.0079 u.
   b. $(10.01294)(0.20) + (11.00931)(0.80) = 10.81$ u. Atomic weight from Table 2.2 is 10.81 u.
   c. 12.011 u. Atomic weight from Table 2.2 is 12.011 u.
   d. 55.846 u. Atomic weight from Table 2.2 is 55.847 u.
   e. 118.69 u. Atomic weight from Table 2.2 is 118.69 u.
   f. 238.029 u. Atomic weight from Table 2.2 is 138.029 u.

8. a. $(1)(0.99985) + (2)(0.00015) = 1.00015$ u. Atomic weight from Table 2.2 is 1.0079 u.
   b. $(10)(0.20) + (11)(0.80) = 10.80$ u. Atomic weight from Table 2.2 is 10.81 u.
   c. 12.011 u. Atomic weight from Table 2.2 is 12.011 u.
   d. 55.911 u. Atomic weight from Table 2.2 is 55.847 u.
   e. 118.783 u. Atomic weight from Table 2.2 is 118.69 u.
   f. 237.978 u. Atomic weight from Table 2.2 is 238.029 u.

9. The mass of one proton is 1.0072765 u, the mass of one neutron is 1.0086650 u, and the mass of one electron is 0.00054858 u. Thus the mass of eight protons is 8.058212 u, the mass of eight neutrons is 8.06932 u, and the mass of eight electrons is 0.00438864 u. The total mass of eight protons, eight neutrons, and eight electrons is 16.13192 u. The

isotopic mass of oxygen-16 = 15.99491 u. The difference is 16.13192 u − 15.99491 u = 0.137 u.

10. Move the decimal point two places to the right. In other words, multiply by 100:
a. 0.75 = 75%   d. 0.031 = 3.1%
b. 0.42 = 42%   e. 0.101 = 10.1%
c. 1.61 = 161%  f. 0.002 = 0.2%

11. Move the decimal point two places to the left. In other words, divide by 100:
a. 38% = 0.38   d. 0.24% = 0.0024
b. 7% = 0.07    e. 1.46% = 0.0146
c. 135% = 1.35  f. 11.77% = 0.1177

12. (70)(0.25) = 17.5.

13. (250)(0.072) = 18.

14. (150)(0.0002) = 0.03.

15. (30)(1.50) = 45.

16. $7/12 \times 100 = 58.3\%$.

17. $30/180 \times 100 = 16.7\%$.

18. $0.03/7 \times 100 = 0.43\%$.

## CHAPTER 3

1. $N_2$ is the reactant and 2N are the products.

2. $2Cl \rightarrow Cl_2$

3. In one $NH_3$ molecule, there is one N atom and three H atoms. In two $NH_3$ molecules, there are two N atoms and six H atoms.

4. $NO + \frac{1}{2}O_2 \rightarrow NO_2$ or $2NO + O_2 \rightarrow 2NO_2$

5. $C_2H_6 + \frac{7}{2}O_2 \rightarrow 2CO_2 + 3H_2O$ or $2C_2H_6 + 7O_2 \rightarrow 4CO_2 + 6H_2O$

6. $2Fe + \frac{3}{2}O_2 \rightarrow Fe_2O_3$ or $4Fe + 3O_2 \rightarrow 2Fe_2O_3$

7. $C_{12}H_{26} + \frac{37}{2}O_2 \rightarrow 12CO_2 + 13H_2O$ or $2C_{12}H_{26} + 37O_2 \rightarrow 24CO_2 + 26H_2O$

8. In the formula $Fe_2(SO_3)_3$, there are 2 Fe atoms, 3 S atoms, and 9 O atoms.

9. $2Al(OH)_3 + 3H_2SO_4 \rightarrow Al_2(SO_4)_3 + 6HOH$

10. See glossary.

11. a. 1 K atom, 1 Cl atom
    b. 1 K atom, 1 Cl atom, 4 O atoms
    c. 1 Ag atom, 1 N atom, 3 O atoms
    d. 6 C atoms, 14 H atoms
    e. 4 C atoms, 8 H atoms, 2 O atoms
    f. 2 H atoms, 1 S atom
    g. 2 N atoms, 8 H atoms, 1 C atom, 3 O atoms
    h. 1 Zn atom, 2 N atoms, 6 O atoms
    i. 1 Mg atom, 2 C atoms, 2 N atoms
    j. 1 Ca atom, 2 H atoms, 2 C atoms, 6 O atoms

12. a. $SO_2 + \frac{1}{2}O_2 \rightarrow SO_3$
    b. $PCl_5 \rightarrow PCl_3 + Cl_2$
    c. $CaH_2 + 2H_2O \rightarrow Ca(OH)_2 + 2H_2$
    d. $(NH_4)_2Cr_2O_7 \rightarrow Cr_2O_3 + N_2 + 4H_2O$
    e. $4Na + O_2 \rightarrow 2Na_2O$
    f. $H_2 + Cl_2 \rightarrow 2HCl$
    g. $4P + 3O_2 \rightarrow 2P_2O_3$
    h. $2NH_3 + H_2SO_4 \rightarrow (NH_4)_2SO_4$
    i. $Zn + Pb(NO_3)_2 \rightarrow Zn(NO_3)_2 + Pb$
    j. $2Cu + S \rightarrow Cu_2S$
    k. $2Al + 2H_3PO_4 \rightarrow 3H_2 + 2AlPO_4$
    l. $NaNO_3 \rightarrow NaNO_2 + \frac{1}{2}O_2$
    m. $H_2O_2 \rightarrow H_2O + \frac{1}{2}O_2$
    n. $BaO_2 \rightarrow BaO + \frac{1}{2}O_2$
    o. $2Al + 3Cl_2 \rightarrow 2AlCl_3$
    p. $P_4 + 5O_2 \rightarrow P_4O_{10}$
    q. $3H_2 + N_2 \rightarrow 2NH_3$
    r. $BaCl_2 + (NH_4)_2CO_3 \rightarrow BaCO_3 + 2NH_4Cl$
    s. $PbO_2 \rightarrow PbO + \frac{1}{2}O_2$
    t. $2Al + 6HCl \rightarrow 2AlCl_3 + 3H_2$
    u. $Fe_2(SO_4)_3 + 3Ba(OH)_2 \rightarrow 3BaSO_4 + 2Fe(OH)_3$
    v. $2KClO_3 \rightarrow 2KCl + 3O_2$
    w. $3Mg + N_2 \rightarrow Mg_3N_2$
    x. $C_3H_7CHO + \frac{11}{2}O_2 \rightarrow 4CO_2 + 4H_2O$
    y. $NaHCO_3 + HCl \rightarrow NaCl + H_2O + CO_2$
    z. $Zn(OH)_2 + H_2SO_4 \rightarrow ZnSO_4 + 2HOH$
    aa. $C_4H_9OH + 6O_2 \rightarrow 4CO_2 + 5H_2O$
    bb. $CaC_2 + 2H_2O \rightarrow C_2H_2 + Ca(OH)_2$
    cc. $3CaCO_3 + 2H_3PO_4 \rightarrow Ca_3(PO_4)_2 + 3CO_2 + 3H_2O$

dd. $C_3H_7COOH + 5O_2 \rightarrow 4CO_2 + 4H_2O$

13. $N_2H_4 + 2H_2O_2 \rightarrow N_2 + 4H_2O$
14. $C_6H_{12}O_6 \rightarrow 2C_2H_5OH + 2CO_2$
15. $CO_2 + 2LiOH \rightarrow Li_2CO_3 + H_2O$
16. $CaO + H_2O \rightarrow Ca(OH)_2$
17. $Fe_2O_3 + 3CO \rightarrow 2Fe + 3CO_2$

## CHAPTER 4
1. Elements: $F_2$, $P_4$, Na, Cr, $Br_2$. Compounds: $N_2O_5$, $C_3H_8$, $PCl_5$, $SO_2$, $CO_2$.
2. Mixtures: salt and pepper; iron powder and charcoal; polluted air; cherry soda. Compounds: sugar; $SF_6$; steam (gaseous $H_2O$). Elements: liquid nitrogen (liquid $N_2$); rubidium metal; $P_4$.
3. See glossary.
4. a. mercury and bromine
   b. He and Ne
   c. $N_2$ and $O_2$
   d. $H_2O$ and ethanol
   e. $SO_2$ and $NO_2$
   f. NaCl and KCl
   g. carbon (diamond); metals such as Fe or Na

## CHAPTER 5
1. $5 \times 10^7$
2. $2 \times 5 \times 10 = 2 \times 10 \times 5 = 5 \times 2 \times 10 = 5 \times 10 \times 2 = 10 \times 2 \times 5 = 10 \times 5 \times 2 = 100$
3. $6 \times 7 \times 10 = (6 \times 7) \times 10 = 42 \times 10 = 420$; $6 \times 7 \times 10 = 6 \times (7 \times 10) = 6 \times 70 = 420$
4. $\frac{1}{4} \times 16 \times 5 = (\frac{1}{4} \times 16) \times 5 = 4 \times 5 = 20$; $\frac{1}{4} \times 16 \times 5 = \frac{1}{4} \times (16 \times 5) = \frac{1}{4} \times 80 = 20$
5. $20 \times \frac{1}{4} = \frac{20}{1} \times \frac{1}{4} = \frac{20 \times 1}{1 \times 4} = \frac{20 \times 1}{4 \times 1} = \frac{20}{4} \times \frac{1}{1} = 5$
6. $3.8423 \times 10^4$
7. $2.56 \times 10^1$
8. $7.360000 \times 10^6$
9. 820,000
10. 572,000,000
11. 588,500,000
12. $10^6 \times 10^{13} = 10^{6+13} = 10^{19}$
13. $8 \times 10^5 \times 3 \times 10^{15} = 8 \times 3 \times 10^5 \times 10^{15} = 24 \times 10^{20} = 2.4 \times 10^{21}$
14. $5 \times 7 \times 10^{12} \times 10^6 = 35 \times 10^{18} = 3.5 \times 10^{19}$
15. $10^7/10^5 = 10^{7-5} = 10^2$
16. $\frac{4 \times 10^8}{3 \times 10^5} = \frac{4}{3} \times \frac{10^8}{10^5} = 1.33 \times 10^3$
17. $\frac{5 \times 8 \times 10^5 \times 10^8}{9 \times 3 \times 10^2 \times 10^4} = \frac{40 \times 10^{13}}{27 \times 10^6} = 1.48 \times 10^7$
18. $10^7/10^8 = 10^{7-8} = 10^{-1}$
19. $10^9/10^9 = 10^{9-9} = 10^0 = 1$
20. $\frac{8 \times 10^{-6}}{6 \times 10^5} = \frac{8}{6} \times \frac{10^{-6}}{10^5} = 1.33 \times 10^{-11}$
21. $\frac{4}{3} \times \frac{10^8}{10^{-7}} = 1.33 \times 10^{15}$
22. $\frac{7}{3} \times \frac{10^{-4}}{10^{-9}} = 2.33 \times 10^5$
23. $\frac{9.3}{3} \times 10^{-15} = 3.1 \times 10^{-15}$
24. $8.34 \times \frac{10^{-12}}{10^{-9}} = 8.34 \times 10^{-3}$
25. $\frac{9}{4} \times \frac{1}{10^{15}} = 2.25 \times 10^{-15}$
26. $0.72 = 7.2 \times 10^{-1}$
27. $0.00048 = 4.8 \times 10^{-4}$
28. $4.25 \times 10^{-3} = 0.00425$
29. $(3 \times 27)^{1/2} = 81^{1/2} = 9$
30. $(8 \times 2)^2 = 16^2 = 256$
31. **5, √** gives 2.24
32. **27, $y^x$, 0.666 . . .,** = gives 9. Note: $27^{1/3} = 3$ and $3^2 = 9$.
33. **7, $y^x$, 0.666 . . .,** = gives 3.66
34. **5, $y^x$, 6.71,** = gives 48987.9
35. **17, $y^x$, 0.055, +/−,** = gives 0.856
36. $(10^4)^5 = 10^{20}$
37. $(10^{16})^{1/4} = 10^{16 \times 1/4} = 10^4$

Solutions to Problems   377

38. $(25 \times 10^{12})^{1/2} = 25^{1/2} \times (10^{12})^{1/2}$
    $= 5 \times 10^6$

39. $(64 \times 10^{21})^{1/3} = 64^{1/3} \times 10^{21 \times 1/3}$
    $= 4 \times 10^7$

40. **7.2, EE, 9,** √ gives $8.48 \times 10^4$

41. **3.6, EE, +/−,** √ gives $6 \times 10^{-4}$. By hand: $(3.6 \times 10^{-7})^{1/2} = (36 \times 10^{-8})^{1/2} = 36^{1/2} \times (10^{-8})^{1/2} = 6 \times 10^{-4}$

42. **1, EE, 15, y$^x$, 0.25, =** gives $5.62 \times 10^3$

43. **5.73, EE, 15, y$^x$, 0.2, =** gives $1.42 \times 10^3$

44. **7.47, EE, +/−, 16, y$^x$, 5.41, =** gives $1.46 \times 10^{-82}$

45. a. $3.264 \times 10^3$   e. $4.67 \times 10^6$
    b. $5.82 \times 10^2$   f. $9 \times 10^{-6}$
    c. $4.3 \times 10^{-2}$   g. $6 \times 10^9$
    d. $5.72 \times 10^{-4}$   h. $7.001 \times 10^3$

46. a. 370           e. 511,700
    b. 48,900,000    f. 32,400
    c. 0.051         g. 0.0001
    d. 0.00000000892 h. 3200

47. a. $4.2 \times 3.6 \times 10^2 \times 10^8 = 15.12 \times 10^{10} = 1.512 \times 10^{11}$
    b. $8 \times 6 \times 10^{15} \times 10^{23} = 48 \times 10^{38} = 4.8 \times 10^{39}$
    c. $5.3 \times 6 \times 10^{-2} \times 10^5 = 31.8 \times 10^3 = 3.18 \times 10^4$
    d. $3.1 \times 2 \times 10^{-5} \times 10^{-10} = 6.2 \times 10^{-15}$
    e. $4.9 \times 8 \times 10^6 \times 10^{-12} = 39.2 \times 10^{-6} = 3.92 \times 10^{-5}$
    f. $3 \times 4 \times 10^{-10} = 12 \times 10^{-10} = 1.2 \times 10^{-9}$

48. a. $8/4 \times 10^5/10^2 = 2 \times 10^3$
    b. $6/4 \times 10^{-2}/10^7 = 1.5 \times 10^{-9}$
    c. $4.7/8.2 \times 10^{-12}/10^{-15} = 0.573 \times 10^3 = 5.73 \times 10^2$
    d. $7.43/2 \times 10^{10}/10^{-4} = 3.715 \times 10^{14}$
    e. $3.2/4 \times 1/10^3 = 0.8 \times 10^{-3} = 8 \times 10^{-4}$
    f. $2.7/4 \times 10^{-2} = 0.675 \times 10^{-2} = 6.75 \times 10^{-3}$

49. **4.07, EE, +/−, 8, x, 3.26, EE, +/−, 5, ÷, 8.99, EE, +/−, 7, =** gives $1.48 \times 10^{-6}$

50. **5.88, EE, 5, ÷, 3.16, EE, 7, ÷, 7.02, EE, +/−, 6 =** gives $2.65 \times 10^3$.

51. **3.27, EE, 4, ×, 8.53, EE, 7, ÷, 5.55, EE, 8, ÷, 7.76, EE, +/−, 5, =** gives $6.48 \times 10^7$

52. $7.61 \times 10^4 + 92.3 \times 10^4 = 99.9 \times 10^4$

53. $2.21 \times 10^{-5} − 0.890 \times 10^{-5} = 1.32 \times 10^{-5} = 99.9 \times 10^5$

54. $5.66 \times 10^8$

55. a. **4.26, y$^x$, 0.8, =** gives 3.19
    b. **8.99, y$^x$, 1.26, =** gives 15.9
    c. **6.25, y$^x$, 0.011, +/−, =** gives 0.980
    d. **5.42, EE, +/−, 11, y$^x$, 4.45, +/−, =** gives $4.83 \times 10^{45}$
    e. **4.77, EE, 9, y$^x$, 0.76, =** gives $2.27 \times 10^7$
    f. **3.2, EE, +/−, 4, y$^x$, 0.015, +/−, =** gives 1.13

**CHAPTER 6**

1. 2.3
2. $59.9 \pm 0.2$
3. 59.3
4. 2.001 has four significant figures; 97.300 has five significant figures; 0.001161 has four significant figures.
5. 1.26 (Yes, to three significant figures, not two.)

6. 1.0342 g     9. 75.9      12. 34.0
7. 1.134 g     10. 28.9      13. 96.8
8. 437.2       11. 4.89      14. 440

15. a. 4.27     d. $3.99 \times 10^{12}$
    b. 25.1     e. $9.23 \times 10^{-4}$
    c. 0.0306   f. $3.26 \times 10^4$

16. a. 27.1            d. $2.63 \times 10^{-1}$
    b. $2.21 \times 10^3$   or 0.263
    c. $1.21 \times 10^3$   e. $4.02 \times 10^{11}$
                       f. $1.91 \times 10^{-11}$

17. a. 1.79     d. 1.91
    b. 1.27     e. $1.18 \times 10^{-11}$
    c. 173      f. $3.16 \times 10^{-2}$

18. a. 15       d. 0.078
    b. 78.3     e. $1.36 \times 10^9$
    c. 31.6     f. $3.8 \times 10^{-6}$

378  SOLUTIONS TO PROBLEMS

19. a. 0.45
 b. 0.0633
 c. 1.18
 d. $1.2 \times 10^2$
 e. $2.0 \times 10^{-1}$, or 0.20
 f. $3.65 \times 10^3$

20. a. 7.3
 b. 13.77
 c. 252.7
 d. 12.3
 e. 129.6
 f. 197.00

21. The coefficients are exact numbers because they represent individual molecules.

## CHAPTER 7

1. 20 mi/5 mi = 4 times farther

2. 200 g × 1 lb/454 g = 0.441 lb

3. 5.0 lb × 454 g/1 lb = $2.3 \times 10^3$ g (to two significant figures)

4. 40 min × 60 s/1 min = $2.4 \times 10^3$ s

5. 8.0 in. × 2.54 cm/1 in. = $2.0 \times 10^1$ cm (to two significant figures)

6. 50 cm × 1 in./2.54 cm = 20 in.

7. 200 m × 1000 mm/1 m = $2.00 \times 10^5$ mm

8. 200 g × 1 kg/1000 g = 0.200 kg

9. 725 mL × 1 L/1000 mL = 0.725 L

10. 24 hr × 60 min/1 hr × 60 s/1 min = $8.6 \times 10^4$ s

11. 0.200 gal × 4 qt/1 gal × 1 L/1.057 qt × 1000 mL/1 L = $7.57 \times 10^2$ mL

12. 60 mg × 1 g/1000 mg = 0.060 g

13. 3 g × 1000 mg/1 g = $3 \times 10^3$ mg

14. 3 μg × 1 g/$10^6$ μg = $3 \times 10^{-6}$ g

15. $3.0 \times 10^3$ mi × 5280 ft/1 mi × 12 in./1 ft × 1 m/39.37 in. × 1 km/1000 m = $4.8 \times 10^3$ km

16. 150 lb × 1 kg/2.2 lb = 68.2 kg

17. 38 in. × 2.54 cm/1 in. = 97 cm; 26 in. = 66 cm; 34 in. = 86 cm

18. 26 mi × 5280 ft/1 mi × 12 in./1 ft = $1.65 \times 10^6$ in. 385 yd × 3 ft/1 yd × 12 in./1 ft = $1.39 \times 10^4$ in. Total = $1.66 \times 10^6$ in.

19. 1000 s × 1 min/60 s × 1 hr/60 min = 0.278 hr

20. 300 mL × 1 L/1000 mL × 1.057 qt/1 L = 0.317 qt

21. 400 IU × 1 mg/1.49 IU = 268 mg

22. 0.1 lb × 454 g/1 lb = $5 \times 10^1$ g

23. ½ gal × 4 qt/1 gal × 1 L/1.057 qt = 1.89 L (The number of significant figures in ½ is unknown.)

24. 750 mL × 1 L/1000 mL × 1.057 qt/1 L = 0.793 qt; ½ gal × 4 qt/1 gal = 0.800 qt. Thus you're getting 0.007 qt less now than before.

25. 48 L × 1.057 qt/1 L × 1 gal/4 qt = 13 gal

26. 2500 kcal × 4.184 kJ/1 kcal = $1.046 \times 10^4$ kJ; $1.046 \times 10^4$ kJ × 1000 J/1 kJ = $1.046 \times 10^7$ J

27. 9200 kJ × 1 kcal/4.184 kJ = 2199 kcal; 2199 kcal × 0.18 = 396 kcal/day from refined sugar.

28. 589 nm × 1 m/$10^9$ nm × 100 cm/1 m = $5.89 \times 10^{-5}$ cm

29. 240 mg/dL × [(0.02586 mmol/L)/(1 mg/dL)] = 6.21 mmol/L

## CHAPTER 8

1. 40/12 = 3.3

2. The atomic weights are Na = 22.98977 u; Xe = 131.30 u; W = 183.85 u; Pb = 207.2 u.

3. 55.847 g Fe = 1 atomic weight of Fe = $6.02 \times 10^{23}$ Fe atoms

4. $6.02 \times 10^{23}$ Pd atoms = 106.4 g Pd

5. 1 mol Ni weighs 58.70 g.

6. 1 mol Ni contains $6.02 \times 10^{23}$ Ni atoms.

7. Since the author doesn't know the date you are reading this, here is the calculation for the number of seconds in one year. All you need to do is multiply by the number of years. If you count leap years, notice that 1900 was *not* a leap year. 1 year × 365 days/1 year × 24 h/1 day × 60 min/1 h × 60 s/1 min = 31,536,000 s

8. 2.0 mol Si × 28.1 g Si/1 mol Si = 56 g Si

9. 14 g Si × 1 mol Si/28.1 g Si = 0.50 mol Si

Solutions to Problems   379

10. $14 \text{ g Si} \times 6.02 \times 10^{23} \text{ Si atoms}/28.1 \text{ g Si} = 3.0 \times 10^{23}$ Si atoms

11. $2.0 \text{ mol Si} \times 6.02 \times 10^{23} \text{ Si atoms}/1 \text{ mol Si} = 1.2 \times 10^{24}$ Si atoms

12. $3.0 \times 10^{23} \text{ Si atoms} \times 1 \text{ mol Si}/6.02 \times 10^{23} \text{ Si atoms} = 0.50$ mol Si

13. $1.2 \times 10^{24} \text{ Si atoms} \times 28.1 \text{ g Si}/6.02 \times 10^{23} \text{ Si atoms} = 56$ g Si

14. $19.0 \text{ g} \times 2 = 38.0$ g

15. $14.0 \text{ g} \times 3(1.01 \text{ g}) = 17.0$ g

16. $2(12.0 \text{ g}) + 6(1.01 \text{ g}) + 16.0 \text{ g} = 46.1$ g

17. $2(27.0 \text{ g}) + 3(32.1 \text{ g}) + 12(16.0 \text{ g}) = 342$ g

18. $34 \text{ g NH}_3 \times 1 \text{ mol NH}_3/17.0 \text{ g NH}_3 = 2.0 \text{ mol NH}_3$; $34 \text{ g NH}_3 \times 6.02 \times 10^{23} \text{ NH}_3 \text{ molecules}/17.0 \text{ g NH}_3 = 1.2 \times 10^{24}$ $NH_3$ molecules

19. $23 \text{ g ethanol} \times 1 \text{ mol ethanol}/46.1 \text{ g ethanol} = 0.50$ mol ethanol

20. $7 \text{ mol CH}_4 \times 1 \text{ mol C}/1 \text{ mol CH}_4 = 7$ mol C; $7 \text{ mol CH}_4 \times 4 \text{ mol H}/1 \text{ mol CH}_4 = 28$ mol H

21. $5 \text{ C}_2\text{H}_6 \text{ molecules} \times 2 \text{ C atoms}/1 \text{ molecule C}_2\text{H}_6 = 10$ C atoms; $5 \text{ C}_2\text{H}_6 \text{ molecules} \times 6 \text{ H atoms}/1 \text{ C}_2\text{H}_6 \text{ molecule} = 30$ H atoms

22. $8 \text{ mol C}_6\text{H}_{14} \times 14 \text{ mol H}/1 \text{ mol C}_6\text{H}_{14} \times 6.02 \times 10^{23} \text{ H atoms}/1 \text{ mol H} = 6.7 \times 10^{25}$ H atoms

23. $49 \text{ g H}_3\text{PO}_4 \times 6.02 \times 10^{23} \text{ H}_3\text{PO}_4 \text{ molecules}/98.0 \text{ g H}_3\text{PO}_4 \times 4 \text{ O atoms}/1 \text{ H}_3\text{PO}_4 \text{ molecule} = 1.2 \times 10^{24}$ O atoms

24. $2.0 \text{ mol e}^- \times 6.02 \times 10^{23} \text{ e}^-/1 \text{ mol e}^- = 1.2 \times 10^{24} \text{ e}^-$

25. $5.0 \text{ mol photons} \times 6.02 \times 10^{23} \text{ photons}/1 \text{ mol photons} = 3.0 \times 10^{24}$ photons

26. $2 \text{ mol NO} + 1 \text{ mol O}_2 \rightarrow 2 \text{ mol NO}_2$

27. $1 \text{ mol Mg(OH)}_2 + 2 \text{ mol HCl} \rightarrow 1 \text{ mol MgCl}_2 + 2 \text{ mol H}_2\text{O}$

28. $19 \text{ g F} \times 6.02 \times 10^{23} \text{ F atoms}/19.0 \text{ g F} = 6.0 \times 10^{23}$ F atoms

29. $40.1 \text{ g} + 32.1 \text{ g} + 4(16.0 \text{ g}) = 136.2$ g

30. $25 \text{ g MnO}_2 \times 1 \text{ mol MnO}_2/86.9 \text{ g MnO}_2 = 0.29$ mol $MnO_2$

31. $4.0 \text{ mol Na}_2\text{S}_2\text{O}_3 \times 158 \text{ g Na}_2\text{S}_2\text{O}_3/1 \text{ mol Na}_2\text{S}_2\text{O}_3 = 632$ g $Na_2S_2O_3$

32. $3.0 \text{ mol SO}_3 \times 6.02 \times 10^{23} \text{ SO}_3 \text{ molecules}/1 \text{ mol SO}_3 \times 1 \text{ S atom}/1 \text{ SO}_3 \text{ molecule} = 1.8 \times 10^{24}$ S atoms; $3.0 \text{ mol SO}_3 \times 6.02 \times 10^{23} \text{ SO}_3 \text{ molecules}/1 \text{ mol SO}_3 \times 3 \text{ O atoms}/1 \text{ SO}_3 \text{ molecule} = 5.4 \times 10^{24}$ O atoms

33. $50 \text{ g ClF}_3 \times 6.02 \times 10^{23} \text{ ClF}_3/92.5 \text{ g ClF}_3 \times 1 \text{ atom Cl}/1 \text{ molecule ClF}_3 = 3.3 \times 10^{23}$ Cl atoms; $50 \text{ g ClF}_3 \times 6.02 \times 10^{23} \text{ ClF}_3 \text{ molecules}/92.5 \text{ g ClF}_3 \times 3 \text{ F atoms}/1 \text{ ClF}_3 \text{ molecule} = 9.8 \times 10^{23}$ F atoms

34. $86 \text{ g HN}_3 \times 1 \text{ mol HN}_3/43.0 \text{ g HN}_3 = 2.0$ mol $HN_3$

35. $9.50 \times 10^{24} \text{ CO}_2 \text{ molecules} \times 44.0 \text{ g CO}_2/6.02 \times 10^{23} \text{ CO}_2 \text{ molecules} = 694 \text{ g CO}_2$

36. $3.2 \times 10^{21} \text{ C}_4\text{H}_{10}\text{O molecules} \times 74.1 \text{ g C}_4\text{H}_{10}\text{O}/6.02 \times 10^{23} \text{ C}_4\text{H}_{10}\text{O molecules} = 0.394$ g $C_4H_{10}O$

37. $4.0 \text{ g AA} \times 1 \text{ mol AA}/176 \text{ g AA} = 0.023$ mol AA

38. $1.0 \times 10^{-10} \text{ mol B}_{12} \times 63 \text{ mol C}/1 \text{ mol B}_{12} \times 6.02 \times 10^{23} \text{ C atoms}/1 \text{ mol C} = 3.8 \times 10^{15}$ C atoms

39. $5 \text{ μg T}_4 \times 1 \text{ g T}_4/10^6 \text{ μg T}_4 \times 1 \text{ mol T}_4/777 \text{ g T}_4 = 6 \times 10^{-9}$ mol $T_4$

40. Abbreviate progesterone as P: $1.0 \text{ mg P} \times 1 \text{ g P}/1000 \text{ mg P} \times 1 \text{ mol P}/314 \text{ g P} = 3.2 \times 10^{-6}$ mol P

41. $4 \times 10^{-3} \text{ mol K} \times 39.1 \text{ g K}/1 \text{ mol K} = 0.2$ g K

42. $4 \text{ mol NH}_3 + 5 \text{ mol O}_2 \rightarrow 4 \text{ mol NO} + 6 \text{ mol H}_2\text{O}$

43. $1 \text{ mol PBr}_3 + 3 \text{ mol HOH} \rightarrow 1 \text{ mol P(OH)}_3 + 3 \text{ mol HBr}$

44. $1 \text{ Au atom} \times 1 \text{ mol Au}/6.02 \times 10^{23} \text{ Au atoms} \times 197 \text{ g Au}/1 \text{ mol Au} = 3.27 \times 10^{-22}$ g Au

**CHAPTER 9**

1. $1 \text{ mol S} = 1 \text{ mol O}_2$

2. $2 \text{ mol SO}_2 = 1 \text{ mol O}_2$

3. $4 \text{ mol Al} = 3 \text{ mol O}_2 = 2 \text{ mol Al}_2\text{O}_3$

4. 5.0 mol Al × 3 mol $O_2$/4 mol Al = 3.8 mol $O_2$

5. 10 mol $Cl_2$ × 2 mol HCl/1 mol $Cl_2$ = 20 mol HCl

6. 128 g $SO_2$ × 1 mol $SO_2$/64.0 g $SO_2$ × 1 mol $O_2$/2 mol $SO_2$ × 32.0 g $O_2$/1 mol $O_2$ = 32.0 g $O_2$

7. 0.50 mol octane × 9 mol $H_2O$/1 mol octane × 18.0 g $H_2O$/1 mol $H_2O$ = 81 g $H_2O$

8. 23 g ethanol × 1 mol ethanol/46.0 g ethanol × 2 mol $CO_2$/1 mol ethanol = 1.0 mol $CO_2$

9. 40.0 g HCl × 1 mol HCl/36.5 g HCl × 1 mol $Mg(OH)_2$/2 mol HCl × 58.3 g $Mg(OH)_2$/1 mol $Mg(OH)_2$ = 31.9 g $Mg(OH)_2$

10. 4 mol S/1 = 4 and 3 mol $O_2$/1 = 3. Thus $O_2$ is the limiting reagent and is completely used up. There is 1 mol of S left over; 3 mol of $SO_2$ are produced.

11. 10 mol A/3 = 3.3, and 8 mol B/7 = 1.1. B will be completely used up—it is the limiting reagent.

12. 40 mol Al/2 = 20, and 15 mol $Fe_2O_3$/1 = 15. $Fe_2O_3$ is completely used up—it is the limiting reagent.

13. 8.0 g $CH_4$ × 1 mol $CH_4$/16.0 g $CH_4$ = 0.50 mol $CH_4$; 64 g $O_2$ × 1 mol $O_2$/32.0 g $O_2$ = 2.0 mol $O_2$; 0.50 mol $CH_4$/1 = 0.50; and 2.0 mol $O_2$/2 = 1.0. Thus $CH_4$ is the limiting reagent.

14. 8.0 g $CH_4$ × 1 mol $CH_4$/16.0 g $CH_4$ × 2 mol $O_2$/1 mol $CH_4$ × 32.0 g $O_2$/1 mol $O_2$ = 32 g $O_2$ used up. There were 64 g $O_2$ to start with, so 32 g $O_2$ are left over after the reaction. 8.0 g $CH_4$ × 1 mol $CH_4$/16.0 g $CH_4$ × 1 mol $CO_2$/1 mol $CH_4$ × 44.0 g $CO_2$/1 mol $CO_2$ = 22 g $CO_2$ formed. 8.0 g $CH_4$ × 1 mol $CH_4$/16.0 g $CH_4$ × 2 mol $H_2O$/1 mol $CH_4$ × 18.0 g $H_2O$/1 mol $H_2O$ = 18 g $H_2O$ formed.

15. 20 g acid × 1 mol acid/60.0 g acid × 1 mol bicarb/1 mol acid × 84.0 g bicarb/1 mol bicarb = 28 g bicarb

16. Abbreviate salicylic acid as SA: 100 g SA × 1 mol SA/138 g SA × 1 mol aspirin/1 mol SA × 180 g aspirin/1 mol aspirin = 130 g aspirin

17. 7.0 g $N_2$ × 1 mol $N_2$/28.0 g $N_2$ × 3 mol $H_2$/1 mol $N_2$ × 2.02 g $H_2$/1 mol $H_2$ = 1.5 g $H_2$

18. 1.0 g $CaCO_3$ × 1 mol $CaCO_3$/100 g $CaCO_3$ × 2 mol HCl/1 mol $CaCO_3$ × 36.5 g HCl/1 mol HCl = 0.73 g HCl

19. First determine the limiting reagent: 10 g Na × 1 mol Na/23 g Na = 0.44 mol Na; 10 g $AlCl_3$ × 1 mol $AlCl_3$/134 g $AlCl_3$ = 0.075 mol $AlCl_3$; 0.44 mol Na/3 = 0.15; and 0.075 mol $AlCl_3$/1 = 0.075. Thus $AlCl_3$ is the limiting reagent. 10 g $AlCl_3$ × 1 mol $AlCl_3$/134 g $AlCl_3$ × 1 mol Al/1 mol $AlCl_3$ × 27.0 g Al/1 mol Al = 2.0 g Al metal produced.

20. 0.50 g AgBr × 1 mol AgBr/188 g AgBr × 2 mol $Na_2S_2O_3$/1 mol AgBr × 158 g $Na_2S_2O_3$/1 mol $Na_2S_2O_3$ = 0.84 g $Na_2S_2O_3$

21. 5.0 g CaO × 1 mol CaO/56.1 g CaO = 0.089 mol CaO; 10 g $H_2O$ × 1 mol $H_2O$/18.0 g $H_2O$ = 0.56 mol $H_2O$. Since the coefficients of both reactants in the balanced chemical equation are 1, CaO is the limiting reagent. No CaO will be left after the reaction. 5.0 g CaO × 1 mol CaO/56.1 g CaO × 1 mol $H_2O$/1 mol CaO × 18.0 g $H_2O$/1 mol $H_2O$ = 1.6 g $H_2O$ used up. Therefore, 10 g − 1.6 g = 8.4 g $H_2O$ left after the reaction.

22. First determine the limiting reagent: 25 g $C_3H_4$ × 1 mol $C_3H_4$/40.1 g $C_3H_4$ = 0.62 mol $C_3H_4$; 20 g $H_2$ × 1 mol $H_2$/2.02 g $H_2$ = 9.9 mol $H_2$; 0.62 mol $C_3H_4$/1 = 0.62; and 9.9 mol $H_2$/2 = 5.0. Thus propyne is the limiting reagent. 25 g $C_3H_4$ × 1 mol $C_3H_4$/40.0 g $C_3H_4$ × 1 mol $C_3H_8$/1 mol $C_3H_4$ × 44.0 g $C_3H_8$/1 mol $C_3H_8$ = 28 g $C_3H_8$

23. 50 g $CaC_2$ × 1 mol $CaC_2$/64.1 g $CaC_2$ × 1 mol $C_2H_2$/1 mol $CaC_2$ × 26.0 g $C_2H_2$/1 mol $C_2H_2$ = 20 g $C_2H_2$

24. 100 g alcohol × 1 mol alcohol/46.1 g alcohol × 1 mol glucose/2 mol alcohol ×

180 g glucose/1 mol glucose = 196 g glucose

25. 250 g Al × 1 mol Al/27.0 g Al × 3 mol $Br_2$/2 mol Al × 160 g $Br_2$/1 mol $Br_2$ = 2.22 × $10^3$ g $Br_2$

**CHAPTER 10**

1. The fraction correct is 88/100 = 0.88. The percent correct is 88/100 × 100 = 88%.

2. 72/89 × 100 = 81%

3. You got 40 − 12 = 28 correct. Your grade is 28/40 × 100 = 70%.

4. 32.1/64.1 × 100 = 50.1%

5. 32.0/64. × 100 = 49.9%

6. The percent composition of $SO_2$ is 50.1% sulfur and 49.9% oxygen. See Problems 4 and 5.

7. %H = 2.02/34.0 × 100 = 5.94%; %O = 32.0/34.0 × 100 = 94.1%

8. %C = 48.0/74.1 × 100 = 64.8%; %H = 10.1/74.1 × 100 = 13.6%; %O = 16.0/74.1 × 100 = 21.6%

9. Using the results of Problem 8, we have 50.0 g × 0.648 = 32.4 g C; 50.0 g × 0.136 = 6.80 g H; 50.0 g × 0.216 = 10.8 g O

10. The empirical weight of CH is 12.0 g + 1.01 g = 13.0 g, and 78.1 g/13.0 g = 6. Thus the molecular formula of benzene is $C_6H_6$.

11. The empirical weight of $NO_2$ is 14.0 g + 2(16.0) = 46.0 g, and 46.0 g/92.0 g = 2. Thus the molecular formula of the oxide of nitrogen is $N_2O_4$, which is called dinitrogen tetraoxide.

12. For carbon: 15.78 g C × 1 mol C/12.01 g C = 1.314 mol C. For sulfur: 84.22 g S × 1 mol S/32.06 g S = 2.627 mol S. Divide each mole value by the smaller of the two: 1.314/1.314 = 1; and 2.627/1.314 = 1.999, which can be rounded to 2. The empirical formula is $CS_2$.

13. For carbon: 8.56 g C × 1 mol C/12.01 g C = 7.13 mol C. For hydrogen: 14.4 g H × 1 mol H/1.01 g H = 14.3 mol H. Divide each mole value by the smaller of the two: 7.13/7.13 = 1; and 14.3/7.13 = 2.01, or 2. The empirical formula is $CH_2$. The empirical weight is 14.0 g. Dividing the molecular weight by the empirical weight, we get 65.1 g/14.0 g = 4.01, or 4. The molecular formula is 4 × $CH_2$ = $C_4H_8$.

14. For carbon: 54.5 g C × 1 mol C/12.0 g C = 4.54 mol C. For oxygen: 36.3 g O × 1 mol O/16.0 g O = 2.27 mol O. For hydrogen: 9.5 g H × 1 mol H/1.01 g H = 9.06 mol H. Divide by the smallest mole value: 4.54/2.27 = 2; 2.27/2.27 = 1; and 9.06/2.27 = 3.99, or 4. The empirical formula is $C_2H_4O$. The empirical weight is 44.1 g. Dividing the molecular weight by the empirical weight, we get 88.1 g/44.1 g = 2. The molecular formula is 2 × $C_2H_4O$ = $C_4H_8O_2$.

15. For sulfur: 25.2 g S × 1 mol S/32.1 g S = 0.785 mol S. For fluorine: 74.8 g F × 1 mol F/19.0 g F = 3.94 mol F. Divide by the smallest mole value: 0.785/0.785 = 1; and 3.94/0.785 = 5.02, or 5. The empirical formula is $SF_5$. The empirical weight is 127 g. Dividing the molecular weight by the empirical weight, we get 254 g/127 g = 2. The molecular formula is 2 × $SF_5$ = $S_2F_{10}$.

16. For $ClF_3$: %Cl = 35.5 g/92.5 g × 100 = 38.4%; %F = 57.0 g/92.5 g × 100 = 61.6%. For $BrF_5$: %Br = 79.9 g/174.9 g × 100 = 45.7%; %F = 95.0 g/174.9 g × 100 = 54.3%. For $IF_7$: %I = 127 g/260 g × 100 = 48.8%; %F = 133 g/260 g × 100 = 51.2%.

17. The formula weight of $(CaSO_4)_2 \cdot H_2O$ is 290 g. %Ca = 80.2 g/290 g × 100 = 27.7%; %S = 64.2 g/290 g × 100 = 22.1%; %O = 144 g/290 g × 100 = 49.7%; %H = 2.02 g/290 g × 100 = 0.70%. The sum of the percents is 100.2%. This is due to rounding error. The percent of water is 18.0 g/290 g × 100 = 6.21%.

18. The formula weight of $CuSO_4 \cdot 5H_2O$ is 250 g. %Cu = 63.5 g/250 g × 100 =

25.4%; %S = 32.1 g/250 g × 100 = 12.8%; %O = 144 g/250 g = 57.6%; %H = 10.1 g/250 g × 100 = 4.04%. The sum of the percents is 99.8%. This is due to rounding error.

19. The formula weight of $Ca_3N_2$ is 148 g. %Ca = 120 g/148 g × 100 = 81.1%; %N = 28.0 g/148 g × 100 = 18.9%.

20. For carbon: 40.9 g C × 1 mol C/12.0 g C = 3.41 mol C. For oxygen: 54.5 g O × 1 mol O/16.0 g O = 3.41 mol O. For hydrogen: 4.55 g H × 1 mol H/1.01 g H = 4.50 mol H. Divide by the smallest mole value: 3.41/3.41 = 1, 3.41/3.41 = 1, and 4.50/3.41 = 1.32. Multiplying by 3 gives the empirical formula $C_3H_4O_3$. The empirical weight is 88.0 g. Dividing the molecular weight by the empirical weight gives 176 g/88.0 g = 2. Thus the molecular formula is $2 \times C_3H_4O_3 = C_6H_8O_6$.

21. For hydrogen: 2.04 g H × 1 mol H/1.01 g H = 2.02 mol H. For sulfur: 32.65 g S × 1 mol S/32.1 g S = 1.02 mol S. For oxygen: 65.31 g O × 1 mol O/16.0 g O = 4.08 mol O. Divide by the smallest mole value: 2.02/1.02 = 1.98, 1.02/1.02 = 1, and 4.08/1.02 = 4. The empirical formula is $H_2SO_4$. The empirical weight is 98.1 g, which is the same as the molecular weight. Thus the molecular formula is $H_2SO_4$.

22. For phosphorus: 22.54 g P × 1 mol P/30.97 g P = 0.7278 mol P. For chlorine: 77.46 g Cl × 1 mol Cl/35.45 g Cl = 2.185 mol Cl. Divide by the smallest mole value: 0.7278/0.7278 = 1; and 2.185/0.7278 = 3.002, or 3. The empirical formula is $PCl_3$; the empirical weight is 137.3 g, which is the same as the molecular weight. Thus the molecular formula is $PCl_3$.

23. For silver: 87.1 g Ag × 1 mol Ag/108 g Ag = 0.806 mol Ag. For sulfur: 12.9 g S × 1 mol S/32.1 g S = 0.402 mol S. Divide by the smallest mole value: 0.806/0.403 = 2 and 0.403/0.403 = 1. The empirical formula is $Ag_2S$. The empirical weight is 248 g, which is the same as the molecular weight. Thus the molecular formula is $Ag_2S$.

24. %C in $CO_2$: 12.01 g/44.01 g × 100 = 27.29%. %H in $H_2O$: 2.016 g/18.02 g × 100 = 11.19%. Grams of C: 13.72 g × 0.2729 = 3.744 g C. Grams of H: 11.23 g × 0.1119 = 1.257 g H. %C in hydrocarbon (HC): 3.744 g/5.000 g × 100 = 74.88%. %H in HC: 1.257 g/5.000 g × 100 = 25.14%. Moles of C in HC: 74.88 g C × 1 mol C/12.01 g C = 6.235 mol C. Moles of H in HC: 25.14 g H × 1 mol H/1.008 g H = 24.94 mol H. Divide by the smallest mole value: 6.235/6.235 = 1 and 24.94/6.235 = 4. The empirical formula of the HC is $CH_4$.

25. The formula weights are KBr = 119 g, SiC = 40.1 g.

26. All molecular weights are formula weights, but not all formula weights are molecular weights.

## CHAPTER 11

1. The salt is the solute, and the water is the solvent.

2. M = mol/L = 1.5 mol sucrose/5.0 L solution = 0.30 M sucrose

3. M = mol/L = 0.250 mol $KMnO_4$/0.350 L solution = 0.714 M $KMnO_4$

4. M = mol/L = 0.400 mol fructose/0.400 L solution = 1.00 M fructose

5. 1 mol $CaSO_4$ = 136 g $CaSO_4$; 8.00 g $CaSO_4$ × 1 mol $CaSO_4$/136 g $CaSO_4$ = 0.0588 mol $CaSO_4$; M = mol/L = 0.0588 mol $CaSO_4$/0.800 L solution = 0.0735 M $CaSO_4$

6. M = mol/L and mol = mol/L × L = 0.12 mol KBr/L × 0.075 L = 0.0090 mol KBr. Also, 1 mol KBr = 119 g KBr = 119 g KBr and 0.0090 mol KBr × 119 g KBr/1 mol KBr = 1.1 g KBr.

7. 1 mol $MgCl_2$ = 95.2 g $MgCl_2$; 5.2 g $MgCl_2$ × 1 mol $MgCl_2$/95.2 g $MgCl_2$ = 0.055 mol $MgCl_2$; L = 0.055 mol $MgCl_2$/0.50 M $MgCl_2$ = 1.1 L

8. mol KOH = 0.0950 mol KOH/L × 0.235 L = 0.00223 mol KOH; 1 mol ox-

alic acid = 2 mol KOH; 0.00223 mol KOH × 1 mol oxalic acid/2 mol KOH = 0.00112 mol oxalic acid; M = 0.00112 mol oxalic acid/0.025 L = 0.0446 M oxalic acid

9. The balanced equation is $3NaOH + H_3PO_4 \rightarrow Na_3PO_4 + 3HOH$. Then we have the following: mol $H_3PO_4$ = 0.110 mol $H_3PO_4$/L × 0.175 L = 0.0192 mol $H_3PO_4$; 3 mol NaOH = 1 mol $H_3PO_4$; 0.0192 mol $H_3PO_4$ × 3 mol NaOH/1 mol $H_3PO_4$ = 0.0578 mol NaOH; and L = 0.0578 mol NaOH/0.250 M NaOH = 0.231 L = 231 mL.

10. The number of moles of $KMnO_4$ used is 0.125 M $KMnO_4$ × 0.0273 L = 0.00341 mol $KMnO_4$. From the balanced chemical equation in Example 10, we have 2 mol $KMnO_4$ = 5 mol $H_2C_2O_4$. The number of moles of $H_2C_2O_4$ is 0.00341 mol $KMnO_4$ × 5 mol $H_2C_2O_4$/2 mol $KMnO_4$ = 0.00853 mol $H_2C_2O_4$. The molarity of the oxalic acid solution is M = 0.00853 mol $H_2C_2O_4$/0.0250 L = 0.341 M $H_2C_2O_4$.

11. $M_iL_i = M_fL_f$ and (0.150 M)(100 mL) = $(M_f)$(500 mL). $M_f$ = (0.150 M)(100 mL)/500 mL = 0.0300 M.

12. $M_iL_i = M_fL_f$ and (16 M)$(L_i)$ = (0.23 M)(600 mL). $L_i$ = (0.23 M)(600 mL)/16 M = 8.6 mL. You need 8.6 mL of concentrated nitric acid and *about* 600 mL − 8.6 mL = 591.4 mL of water. You would prepare the solution as described at the end of Example 12.

13. 0.075 M $CuSO_4 \cdot 5H_2O$ × 0.250 L = 0.0188 mol $CuSO_4 \cdot 5H_2O$; 0.0188 mol $CuSO_4 \cdot 5H_2O$ × 250 g $CuSO_4 \cdot 5H_2O$/1 mol $CuSO_4 \cdot 5H_2O$ = 4.7 g $CuSO_4 \cdot 5H_2O$

14. 0.07362 M $FeSO_4$ × 0.05000 L = 0.003681 mol $FeSO_4$; 0.003681 mol $FeSO_4$ × 2 mol $Ce(SO_4)_2$/2 mol $FeSO_4$ = 0.003681 mol $Ce(SO_4)_2$; 0.003681 mol $Ce(SO_4)_2$/0.1422 M $Ce(SO_4)_2$ = 0.02589 L $Ce(SO_4)_2$ = 25.89 mL $Ce(SO_4)_2$

15. This will be solved in one step: 0.1018 M $KMnO_4$ × 0.03500 L × 16 mol HCl/2 mol $KMnO_4$ × 1/0.5125 M HCl × 1000 mL/1 L = 55.62 mL HCl.

16. This will be solved in one step: 0.987 M $H_2SO_4$ × 0.00934 L × 2 mol NaOH/1 mol $H_2SO_4$ × 1/0.102 M NaOH × 1000 mL/1 L = 18.1 mL NaOH.

17. 1 mol $ZnCl_2$ = 136 g $ZnCl_2$; 15.2 g $ZnCl_2$ × 1 mol $ZnCl_2$/136 g $ZnCl_2$ = 0.112 mol $ZnCl_2$; 0.112 mol $ZnCl_2$/0.850 M $ZnCl_2$ = 1.31 L = 1.31 × $10^3$ mL. The final volume of the solution is 1.31 × $10^3$ mL, so you would probably add slightly less than 1.31 × $10^3$ mL of water.

18. $M_iL_i = M_fL_f$ and (12 M)$(L_i)$ = (0.50 M)(1.0 L); $L_i$ = (0.50 M)(1.0 L)/12 M = 0.042 L = 42 mL. Add 42 mL of 12 M HCl to a 1-L graduated cylinder. Then add enough water to bring the volume up to 1.0 L. There is no need to use a volumetric flask, since we need the concentration to only two significant figures.

19. $M_iL_i = M_fL_f$ and (1.50 M)(300 mL) = (0.350 M)$(L_f)$; $L_f$ = (1.50 M)(300 mL)/0.350 M = 1286 mL = 1.29 × $10^3$ mL. The approximate amount of water needed is 1286 mL − 300 mL = 986 mL.

20. 0.85 g NaCl × 1 mol NaCl/58.5 g NaCl = 0.0145 mol NaCl; M = 0.0145 mol NaCl/0.10 L = 0.15 M NaCl

21. 0.5103 M NaOH × 0.01702 L = 0.008685 mol NaOH; 0.008685 mol NaOH × 1 mol acetic acid/1 mol NaOH = 0.008685 mol acetic acid; 0.008685 mol acetic acid/0.01000 L = 0.8685 M acetic acid

22. 0.600 M NaOH × 0.253 L = 0.152 mol NaOH; 0.152 mol NaOH × 1 mol $Cl_2$/2 mol NaOH = 0.0759 mol $Cl_2$; 0.0759 mol $Cl_2$ × 70.9 g $Cl_2$/1 mol $Cl_2$ = 5.38 g $Cl_2$

23. 1.32 g Zn × 1 mol Zn/65.4 g Zn × 2 mol HCl/1 mol Zn × 1/0.500 M HCl × 1000 mL/1 L = 80.7 mL HCl

24. 0.4216 M $AgNO_3$ × 0.02632 L × 1 mol $Cl^-$/1 mol $AgNO_3$ × 1/0.02500 L = 0.4439 M $Cl^-$

25. 0.296 M $H_3PO_4$ × 0.0428 L = 0.0127 mol $H_3PO_4$; 0.0127 mol $H_3PO_4$ × 3 mol KOH/1 mol $H_3PO_4$ = 0.0381 mol KOH;

M = 0.0381 mol KOH/0.0249 L = 1.53 M KOH

**CHAPTER 12**

1. 1500 Torr × 1 atm/760 Torr = 1.97 atm
2. 12 lb/in.$^2$ × 760 Torr/14.7 lb = 620 or 6.2 × 10$^2$
3. 3 atm × 101 kPa/atm = 303 kPa or 3 × 10$^2$ kPa
4. °F = 1.8 °C + 32 = (1.8)(25) + 32 = 77 °F
5. °C = (°F − 32)/1.8 = (86 − 32)/1.8 = 30 °C
6. K = °C + 273 = 100 °C + 273 = 373 K
7. K = °C + 273; 77 K = °C + 273; °C = 77 K − 273 = −196 °C
8. $a = qb$
9. $a = t/b$
10. $P \propto n$. If $n$ doubles (10 mol → 20 mol), then $P$ doubles (5 atm → 10 atm).
11. $V \propto n$. If $n$ doubles (3 mol → 6 mol), then $V$ doubles (15 L → 30 L).
12. $P = \dfrac{nRT}{V} = \dfrac{(2.0 \text{ mol})(0.0821 \text{ atm·L/mol·K})(400 \text{ K})}{7.0 \text{ L}} = 9.4$ atm
13. 80 °C + 273 = 353 K, and 500 Torr × 1 atm/760 Torr = 0.66 atm; thus

$V = \dfrac{nRT}{P} = \dfrac{(0.25 \text{ mol})(0.0821 \text{ atm·L/mol·K})(353 \text{ K})}{0.66 \text{ atm}} = 11$ L

14. 30 °C + 273 = 303 K, and 80.0 kPa × 1 atm/101 kPa = 0.792 atm; thus

$n = \dfrac{PV}{RT} = \dfrac{(0.792 \text{ atm})(0.200 \text{ L})}{(0.0821 \text{ atm·L/mol·K})(303 \text{ K})} = 6.37 \times 10^{-3}$ mol

15. 25 °C + 273 = 298 K; thus $n = \dfrac{PV}{RT} = \dfrac{(1.30 \text{ atm})(1.50 \text{ L})}{(0.0821 \text{ atm·L/mol·K})(298 \text{ K})} = 0.0797$ mol

MW = g/mol = 4.40 g/0.0797 mol = 55.2 g/mol

16. $P_1V_1/T_1 = P_2V_2/T_2$; thus

$V_2 = \dfrac{P_1V_1T_2}{T_1P_2} = \dfrac{(1.5 \text{ atm})(5.0 \text{ L})(200 \text{ K})}{(273 \text{ K})(4.0 \text{ atm})} = 1.4$ L

17. $P_1V_1/T_1 = P_2V_2/T_2$; thus $P_2 = \dfrac{P_1V_1T_2}{T_1V_2} = \dfrac{(0.20 \text{ atm})(5.0 \text{ L})(310 \text{ K})}{(210 \text{ K})(0.70 \text{ L})} = 2.1$ atm

18. 150 Torr + 150 Torr + 500 Torr = 800 Torr
19. (7 atm)(0.208) = 1.46 atm; thus 1.46 atm × 760 Torr/1 atm = 1110 Torr, or 1 × 10$^3$
20. 500 kPa × 1 atm/101 kPa = 4.95 atm
21. 0.50 atm × 760 Torr/1 atm = 3.8 × 10$^2$ Torr
22. °C = (°F − 32)/1.8 = (104 °F − 32)/1.8 = 40 °C
23. °F = 1.8 °C + 32 = (1.8)(41) + 32 = 105.8 °F
24. °C = (°F − 32)/1.8 = (68 °F − 32)/1.8 = 20 °C. Then 20 °C + 273 = 293 K.
25. 892 °C + 273 = 1165 K

Solutions to Problems   385

26. $-195.8\ °C + 273 = 77.2$ K

27. $2800\ K - 273 = 2527\ °C$; thus °F = $1.8\ °C + 32 = (1.8)(2527\ °C) + 32 = 4581$ °F

28. $T = 1.8T + 32$, $T - 1.8T = 32$, $-0.8T = 32$, $T = -32/0.8 = -40$. Therefore, $-40\ °C = -40\ °F$.

29. $°C = (°F - 32)/1.8 = (100\ °F - 32)/1.8 = 37.8\ °C$; $37.8\ °C + 273 = 311$ K; and 400 Torr × 1 atm/760 Torr = 0.526 atm. Also, 8.0 g $O_2$ × 1 mol $O_2$/32.0 g $O_2$ = 0.25 mol $O_2$; thus

$$V = \frac{nRT}{P} = \frac{(0.25\ \text{mol})(0.0821\ \text{atm·L/mol·K})(311\ \text{K})}{0.526\ \text{atm}} = 12\ \text{L}$$

30. $25\ °C + 273 = 298$ K, and 15 g $F_2$ × 1 mol $F_2$/38.0 g $F_2$ = 0.40 mol $F_2$; thus

$$P = \frac{nRT}{V} = \frac{(0.40\ \text{mol})(0.0821\ \text{atm·L/mol·K})(298\ \text{K})}{2.0\ \text{L}} = 4.9\ \text{atm}$$

31. 200 kPa × 1 atm/101 kPa = 1.98 atm, and $40.0\ °C + 273 = 313.0$ K; thus

$$n = \frac{PV}{RT} = \frac{(1.98\ \text{atm})(0.700\ \text{L})}{(0.0821\ \text{atm·L/mol·K})(313.0\ \text{K})} = 0.0539\ \text{mol}$$

0.0539 mol $N_2O$ × 44.0 g $N_2O$/1 mol $N_2O$ = 2.37 g $N_2O$

32. 300 Torr × 1 atm/760 Torr = 0.395 atm, and $22\ °C + 273 = 295$ K; thus

$$n = \frac{PV}{RT} = \frac{(0.395\ \text{atm})(0.400\ \text{L})}{(0.0821\ \text{atm·L/mol·K})(295\ \text{K})} = 0.00652\ \text{mol}$$

MW = g/mol = 3.00 g/0.00652 mol = 460 g/mol

33. 1 atm = 760 Torr, and 50 Torr + 150 Torr + $P_{H_2}$ = 760 Torr; thus $P_{H_2}$ = 560 Torr

34. $50\ °C + 273 = 323\ K = T_1$; $100\ °C + 273 = 373\ K = T_2$; and $P_1/T_1 = P_2/T_2$; thus $P_2 = P_1T_2/T_1 = (150\ \text{kPa})(373\ \text{K})/323\ \text{K} = 173$ kPa

35. $P_1V_1 = P_2V_2$; thus $V_2 = P_1V_1/P_2 = (800\ \text{Torr})(2.0\ \text{L})/760\ \text{Torr} = 2.1$ L

36. $23\ °C + 273 = 296\ K = T_1$; $30\ °C + 273 = 303\ K = T_2$; and $P_1V_1/T_1 = P_2V_2/T_2$; thus

$$P_2 = \frac{P_1V_1T_2}{T_1V_2} = \frac{(20.0\ \text{kPa})(800\ \text{mL})(303\ \text{K})}{(296\ \text{K})(400\ \text{mL})} = 40.9\ \text{kPa}$$

37. $3.0 \times 10^{10}$ molecules × 1 mol/$6.02 \times 10^{23}$ molecules = $5.0 \times 10^{-14}$ mol; thus

$$P = \frac{nRT}{V} = \frac{(5.0 \times 10^{-14}\ \text{mol})(0.0821\ \text{atm·L/mol·K})(298\ \text{K})}{0.0030\ \text{L}} = 4.1 \times 10^{-10}\ \text{atm}$$

38. $PV = (m/MW)RT$, and 95 kPa × 1 atm/101 kPa = 0.94 atm; thus

$$MW = \frac{mRT}{PV} = \frac{(7.2 \text{ g})(0.0821 \text{ atm·L/mol·K})(373 \text{ K})}{(0.94 \text{ atm})(2.5 \text{ L})} = 94 \text{ g/mol}$$

39. 1 mol $C_2H_4$ = 28.0 g $C_2H_4$, and 50 g $C_2H_4$ × 1 mol $C_2H_4$/28.0 g $C_2H_4$ = 1.8 mol $C_2H_4$; thus

$$P = \frac{nRT}{V} = \frac{(1.8 \text{ mol})(0.0821 \text{ atm·L/mol·L})(307 \text{ K})}{0.100 \text{ L}} = 4.5 \times 10^3 \text{ atm}$$

40. 50 Torr × 1 atm/760 Torr = 0.066 atm; thus

$$n = \frac{PV}{RT} = \frac{(0.066 \text{ atm})(5.0 \text{ L})}{(0.0821 \text{ atm·L/mol·K})(348 \text{ K})} = 0.012 \text{ mol}$$

Since 1 mol $C_6H_6$ = 78.1 g $C_6H_6$, we have 0.012 mol $C_6H_6$ × 78.1 g $C_6H_6$/1 mol $C_6H_6$ = 0.94 g $C_6H_6$.

41. $PV = nRT$ and $n = m/MW$. (Here $m$ stands for mass.) Substituting the value of $n$ into $PV = nRT$, we get $PV = (m/MW)RT$. Solving for MW we get $MW = mRT/PV$.

42. $P_1V_1/T_1 = P_2V_2/T_2$. Let $T_1 = T_2 = T$, since $T$ is constant. Thus $P_1V_1/\cancel{T} = P_2V_2/\cancel{T}$, or $P_1V_1 = P_2V_2$.

43. $P_1V_1/T_1 = P_2V_2/T_2$. Let $P_1 = P_2 = P$, since $P$ is constant. Thus $\cancel{P}V_1/T_1 = \cancel{P}V_2/T_2$, or $V_1/T_1 = V_2/T_2$.

44. The total pressure is about 7 atm. Air is 78.1% $N_2$. Thus 7 atm × 0.781 = 5.5 atm of $N_2$.

## CHAPTER 13

1. Uncertainty principle: There is a limit to the accuracy of any measurement, and this limit becomes important when we try to measure (or "see") the electron.

2. Wavelength: The distance between nearest identical parts of a wave.

3. As the photon energy increases, the wavelength of light decreases.

4. Optical effect: To see an electron, the wavelength of the light must be very small. In fact, it must be a few times smaller than the size of the electron. Energy effect: As the wavelength of light gets shorter, the photon energy increases. Combining these effects, the energetic photon needed to see the electron will clearly change the electron's position.

5. The effects of the uncertainty principle are not noticeable in everyday living because the photon energy needed to see large objects is very low compared to the mass of the object. For example, the uncertainty in measuring a one-kilogram object with photons is a very small fraction of the size of an atom's nucleus. This is not noticeable.

6. An orbital is a description of the shape of the space in which the electrons exist. Each orbital can contain up to two electrons.

7. Rule 1: Electrons surrounding a nucleus are arranged in orbitals. Rule 2: An orbital can contain a maximum of two electrons. Rule 3: Electrons like to be as far apart from one another as possible because they have negative charges and like charges repel. Rule 4: Electrons like to be as close to the nucleus as possible because being close to the nucleus means that they need less energy. Rule 5: The farther away from the nucleus, the more room there is for electrons.

8. Energy level, $n = 8$; number of orbitals, $n^2 = 64$; number of electrons, $2n^2 = 128$.

9. The quantum numbers of the two spin states for an electron are $\frac{1}{2}$ and $-\frac{1}{2}$.

10. Two electrons can be in the same orbital *only* if they have opposite spins.

11.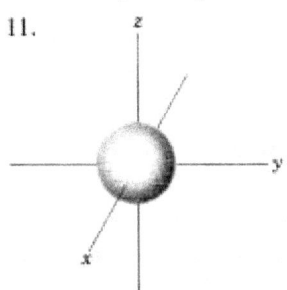

12.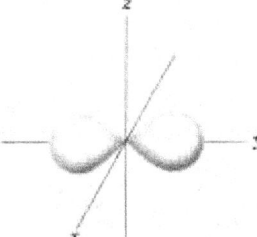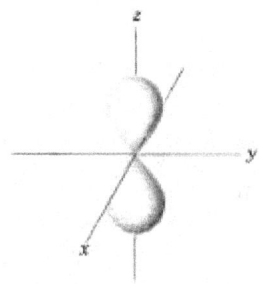

13. a. There are nine g orbitals in the fifth and higher energy levels.
    b. The nine g orbitals can contain up to 18 electrons.

## CHAPTER 14

1. A 5p orbital.
2. A 6d orbital.
3. 4s: K, Ca. 3d: Sc, Ti, V, Cr, Mn, Fe, Co, Ni, Cu, Zn. 5f: Th, Pa, U, Np, Pu, Am, Cm, Bk, Cf, Es, Fm, Md, No, Lr.

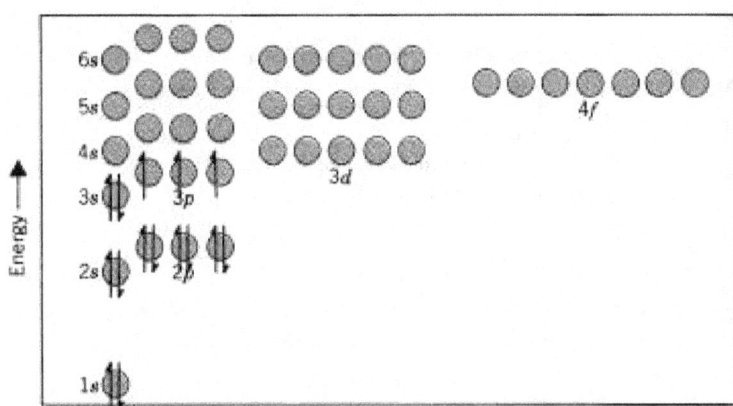

4. P: $1s^2 2s^2 2p_x^2 2p_y^2 2p_z^2 3s^2 3p_x^1 3p_y^1 3p_z^1$

5. ↑↓  ↑↓  ↑↓  ↑↓  ↑↓  ↑↓  ↑   ↑
   $1s^2$ $2s^2$ $2p_x^2$ $2p_y^2$ $2p_z^2$ $3s^2$ $3p_x^1$ $3p_y^1$

6. a. This diagram is incorrect because the p electrons have different spins.
   b. This diagram is correct.

388 SOLUTIONS TO PROBLEMS

c. This diagram is incorrect because one of the $2p_y$ electrons should be in the $2p_z$ orbital.
d. This diagram is correct. However, we usually draw this with the electrons in the $2p_x$ and $2p_y$ orbitals.

7. Group IIIA: B, Al, Ga, In, Tl, Nh.
Group VB: V, Nb, Ta, Db.
Group VIA: O, S, Se, Te, Po, Lv.

8. Period 4: K, Ca, Sc, Ti, V, Cr, Mn, Fe, Co, Ni, Cu, Zn, Ga, Ge, As, Se, Br, Kr. Period 5: Rb, Sr, Y, Zr, Nb, Mo, Tc, Ru, Rh, Pd, Ag, Cd, In, Sn, Sb, Te, I, Xe.

9. $5d$: La, Hf, Ta, W, Re, Os, Ir, Pt, Au, Hg.

10. $5f$: Th, Pa, U, Np, Pu, Am, Cm, Bk, Cf, Es, Fm, Md, No, Lr.

11. Group IIA: 2 electrons. Group VA: 5 electrons. Group VIA: 6 electrons.

12. a. $3p_x^2$: Two electrons in a third energy level $p$ orbital.
b. $4f^7$: Seven electrons in fourth energy level $f$ orbitals.
c. $2s^1$: One electron in a second energy level $s$ orbital.
d. $4d^{10}$: Ten electrons in fourth energy level $d$ orbitals.

13. a. Li; b. C; c. Na; d. S; e. K; f. Co

14. a. C: $\frac{\uparrow\downarrow}{1s^2}\frac{\uparrow\downarrow}{2s^2}\frac{\uparrow}{2p_x^1}\frac{\uparrow}{2p_y^1}$
b. O: $\frac{\uparrow\downarrow}{1s^2}\frac{\uparrow\downarrow}{2s^2}\frac{\uparrow\downarrow}{2p_x^2}\frac{\uparrow}{2p_y^1}\frac{\uparrow}{2p_z^1}$
c. Mg: $\frac{\uparrow\downarrow}{1s^2}\frac{\uparrow\downarrow}{2s^2}\frac{\uparrow\downarrow}{2p_x^2}\frac{\uparrow\downarrow}{2p_y^2}\frac{\uparrow\downarrow}{2p_z^2}\frac{\uparrow\downarrow}{3s^2}$
d. P: $\frac{\uparrow\downarrow}{1s^2}\frac{\uparrow\downarrow}{2s^2}\frac{\uparrow\downarrow}{2p_x^2}\frac{\uparrow\downarrow}{2p_y^2}\frac{\uparrow\downarrow}{2p_z^2}\frac{\uparrow\downarrow}{3s^2}\frac{\uparrow}{3p_x^1}\frac{\uparrow}{3p_y^1}\frac{\uparrow}{3p_z^1}$

e. Ar: $\frac{\uparrow\downarrow}{1s^2}\frac{\uparrow\downarrow}{2s^2}\frac{\uparrow\downarrow}{2p_x^2}\frac{\uparrow\downarrow}{2p_y^2}\frac{\uparrow\downarrow}{2p_z^2}\frac{\uparrow\downarrow}{3s^2}\frac{\uparrow\downarrow}{3p_x^2}\frac{\uparrow\downarrow}{3p_y^2}\frac{\uparrow\downarrow}{3p_z^2}$
f. V: $\frac{\uparrow\downarrow}{1s^2}\frac{\uparrow\downarrow}{2s^2}\frac{\uparrow\downarrow}{2p_x^2}\frac{\uparrow\downarrow}{2p_y^2}\frac{\uparrow\downarrow}{2p_z^2}\frac{\uparrow\downarrow}{3s^2}\frac{\uparrow\downarrow}{3p_x^2}\frac{\uparrow\downarrow}{3p_y^2}\frac{\uparrow\downarrow}{3p_z^2}$
$\frac{\uparrow\downarrow}{4s^2}\frac{\uparrow}{3d^1}\frac{\uparrow}{3d^1}$

15. a. Li, Na, K, Rb, Cs, Fr.
b. Be, Mg, Ca, Sr, Ba, Ra.
c. F, Cl, Br, I, At.
d. He, Ne, Ar, Kr, Xe, Rn.
e. Ce, Pr, Nd, Pm, Sm, Eu, Gd, Tb, Dy, Ho, Er, Tm, Yb, Lu.
f. Th, Pa, U, Np, Pu, Am, Cm, Bk, Cf, Es, Fm, Md, No, Lr.

16. B, Si, Ge, As, Sb, Te, Po, At.

17. scandium, Sc; titanium, Ti; vanadium, V; chromium, Cr; manganese, Mn; iron, Fe; cobalt, Co; nickel, Ni; copper, Cu; zinc, Zn.

18. carbon, C; silicon, Si; germanium, Ge; tin, Sn; lead, Pb.

19. a. This diagram is incorrect. The $2s$ orbital should have two electrons.
b. This diagram is incorrect. The $2p_x$ orbital should have only one electron.
c. This diagram is incorrect. The $2p_x$ orbital should have only two electrons.
d. This diagram is correct. The usual way of writing it would be to have the $2p$ electrons with spin up.
e. This diagram is correct. The usual way of writing it would be to have the $2p$ electron placed in the $2p_x$ orbital.

**CHAPTER 15**
1. Li·
2. Mg:

3. K·   :Äs·

4. :C̈l:N̈:C̈l:   :C̈l:B̈:C̈l:   :C̈l:Ö:C̈l:
   :C̈l:        :C̈l:

5. Cl—N—Cl    Cl—B—Cl    Cl—O—Cl
       |              |
       Cl            Cl

   :Cl—N̈—Cl:   :Cl—B—Cl:   :Cl—Ö—Cl:
       |              |
      :Cl:          :Cl:

6. Na → Na⁺ + e⁻
7. Ca → Ca²⁺ + 2e⁻
8. Br + e⁻ → Br⁻
9. S + 2e⁻ → S²⁻
10. K + Br → KBr. Since bromine occurs in nature as Br$_2$, we should write 2K + Br$_2$ → 2KBr.
11. Ca + Cl$_2$ → CaCl$_2$
12. We mean that electrons tend to exist in the lowest energy level available to them.
13. A molecular orbital is a chemical bond in which the atomic orbitals combine in a special way.
14. Only electrons with opposite spins can be in the same molecular orbital. Since there are only two spin states, up and down, there can be only two electrons in a molecular orbital.
15. A covalent bond is a chemical bond in which two electrons are shared by two atoms.
16. Valence electrons are electrons in the outermost energy level of an atom that are available for forming chemical bonds.
17. Valence electrons are represented as dots. The other electrons, those below the outermost energy level, are not shown.
18. Na·   Mg:   :Al·   :Si·
    :P·   :S:   :Cl·   :Ar:
19. a.       Cl                :Cl:
             |                   |
         Cl—C—Cl           :Cl—C—Cl:
             |                   |
             Cl                :Cl:

    b.   Br—N—Br        :Br—N̈—Br:
             |                  |
             Br                :Br:

c. H—S—H         H—S̈—H
d. H—Br          H—B̈r:

e.      O                   :O:
        ‖                    ‖
    H—C—H           H—C—H

f. I—I           :Ï—Ï:

20. a.    H  H
          |  |
      H—C—C—H
          |  |
          H  H

    b.  H           H
         \         /
          C=C
         /         \
        H           H

    c. H—C≡C—H

21. :Cl:P:Cl:
        :Cl:

22.     :F:
    :F:S:F:
        :F:

23. An ionic bond is a bond resulting from the electrostatic attraction between positive and negative ions.
24. Al → Al³⁺ + 3e⁻
25. Se + 2e⁻ → Se²⁻
26. a. 2Al + 3Cl$_2$ → 2AlCl$_3$
    b. 4K + O$_2$ → 2K$_2$O
    c. Be + Br$_2$ → BeBr$_2$
    d. Ba + I$_2$ → BaI$_2$
    e. 16Li + S$_8$ → 8Li$_2$S (Notice that elemental sulfur forms molecules with eight atoms.)
    f. 2Al + 3H$_2$ → 2AlH$_3$
27. a. BeF$_2$ or BeH$_2$; b. OF$_2$ or H$_2$O; c. BF$_3$; d. NH$_3$; e. CH$_4$.
28. a. PCl$_5$; b. SF$_6$; c. ClF$_5$.

## CHAPTER 16

1. CsI, cesium iodide
2. MgBr$_2$, magnesium bromide
3. AlCl$_3$, aluminum chloride
4. Ca$_3$N$_2$, calcium nitride
5. Al$_2$S$_3$, aluminum sulfide
6. Na$_2$S, sodium sulfide; MgI$_2$, magnesium iodide; NaH, sodium hydride; RaO, radium oxide; Al$_2$S$_3$, aluminum sulfide; NH$_4$F, ammonium fluoride
7. rubidium nitrate, RbNO$_3$; magnesium oxide, MgO; calcium chloride, CaCl$_2$; sodium oxide, Na$_2$O; beryllium hydride, BeH$_2$; barium bromide, BaBr$_2$; aluminum oxide, Al$_2$O$_3$
8. potassium nitrate, KNO$_3$; magnesium hypochlorite, Mg(ClO)$_2$; lithium permanganate, LiMnO$_4$; barium acetate, Ba(C$_2$H$_3$O$_2$)$_2$; calcium hydroxide, Ca(OH)$_2$; cesium peroxide, Cs$_2$O$_2$; beryllium perchlorate, Be(ClO$_4$)$_2$; sodium dichromate, Na$_2$Cr$_2$O$_7$; ammonium carbonate, (NH$_4$)$_2$CO$_3$; potassium hydrogen sulfite, KHSO$_3$
9. KNO$_2$; potassium nitrite; Na$_2$O$_2$, sodium peroxide; Mg(OH)$_2$, magnesium hydroxide; Ca(ClO$_2$)$_2$, calcium chlorite; K$_3$PO$_4$, potassium phosphate; NaCN, sodium cyanide; Rb$_2$S$_2$O$_3$, rubidium thiosulfate; BaC$_2$O$_4$, barium oxalate; Li$_2$CrO$_4$, lithium chromate
10. gold(III) chloride, AuCl$_3$; nickel(II) hydroxide, Ni(OH)$_2$; stannic fluoride, SnF$_4$; cadmium thiocyanate, Cd(SCN)$_2$; lead(IV) sulfate, Pb(SO$_4$)$_2$; chromous carbonate, CrCO$_3$; copper(II) phosphate, Cu$_3$(PO$_4$)$_2$; mercurous nitrate, Hg$_2$(NO$_3$)$_2$; silver hydrogen carbonate, AgHCO$_3$
11. CoS, cobalt(II) sulfide; Fe$_2$(SO$_4$)$_3$, iron(III) sulfate; PbO, lead(II) oxide; Ni(ClO$_4$)$_3$, nickel(III) perchlorate; CdBr$_2$, cadmium bromide; Cu$_2$O, copper(I) oxide; Mn(NO$_3$)$_2$, manganese(II) nitrate; HgSO$_4$, mercury(II) sulfate; Cr$_3$(AsO$_4$)$_2$, chromium(II) arsenate; Au(H$_2$PO$_4$)$_3$, gold(III) dihydrogen phosphate
12. sulfur dioxide, SO$_2$; sulfur trioxide, SO$_3$; hydrogen selenide, H$_2$Se; dichlorine monoxide, Cl$_2$O; phosphorus trichloride, PCl$_3$; carbon tetrafluoride, CF$_4$; tetraboron decahydride, B$_4$H$_{10}$
13. PCl$_5$, phosphorus pentachloride; CO, carbon monoxide; CO$_2$, carbon dioxide; N$_2$O$_4$, dinitrogen tetraoxide; P$_4$O$_6$, tetraphosphorus hexaoxide; I$_2$O$_5$, diiodine pentaoxide; BrF$_3$, bromine trifluoride; Sb$_2$O$_5$, diantimony pentaoxide
14. a. CaF$_2$, calcium fluoride
    b. AlCl$_3$, aluminum chloride
    c. MgO, magnesium oxide
    d. SrCl$_2$, strontium chloride
    e. CoCl$_2$, cobalt(II) chloride
    f. CoCl$_3$, cobalt(III) chloride
15. a. NO, nitric oxide (a common name)
    b. N$_2$O$_3$, dinitrogen trioxide
    c. N$_2$O$_5$, dinitrogen pentaoxide
    d. SF$_4$, sulfur tetrafluoride
    e. SF$_2$, sulfur difluoride
    f. SF$_6$, sulfur hexafluoride
16. a. SO$_3$, sulfur trioxide
    b. CO, carbon monoxide
    c. SiF$_4$, silicon tetrafluoride
    d. N$_2$O$_5$, dinitrogen pentaoxide
    e. P$_4$S$_{10}$, tetraphosphorus decasulfide
    f. XeF$_4$, xenon tetrafluoride
17. a. PCl$_5$, phosphorus pentachloride
    b. IF$_7$, iodine heptafluoride
    c. P$_4$O$_6$, tetraphosphorus hexaoxide
    d. As$_4$O$_{10}$, tetraarsenic decaoxide
    e. Cl$_2$O$_3$, dichlorine trioxide
    f. Cl$_2$O$_7$, dichlorine heptaoxide
18. a. H$_3$PO$_4$, phosphoric acid
    b. HNO$_2$, nitrous acid
    c. H$_2$CO$_3$, carbonic acid
    d. HClO$_3$, chloric acid
    e. HIO$_3$, iodic acid
    f. HNO$_3$, nitric acid
    g. H$_3$PO$_4$, phosphoric acid
19. a. HBr, hydrogen bromide
    b. HBrO, hypobromous acid
    c. HBrO$_2$, bromous acid
    d. HBrO$_3$, bromic acid
    e. HBrO$_4$, perbromic acid

Solutions to Problems    391

20. a. $PbSO_3$, lead (II) sulfite
    b. $Mg_2SiO_4$, magnesium silicate
    c. $Na_3PO_4$, sodium phosphate
    d. $Ca(NO_3)_2$, calcium nitrate
    e. LiBrO, lithium hypobromite
    f. $NH_4ClO_4$, ammonium perchlorate
    g. $Fe_2(SO_4)_3$, iron(III) sulfate
    h. $FeSO_4$, iron(II) sulfate
    i. $CaCO_3$, calcium carbonate
    j. $Ba(NO_2)_2$, barium nitrite

21. a. iron(III) sulfate, $Fe_2(SO_4)_3$
    b. chromium(II) sulfite, $CrSO_3$
    c. cupric sulfide, CuS
    d. silver dihydrogen phosphate, $AgH_2PO_4$
    e. tellurium hexafluoride, $TeF_6$
    f. mercury(I) acetate, $Hg_2(C_2H_3O_2)_2$
    g. disulfur dichloride, $S_2Cl_2$
    h. ammonium nitrate, $NH_4NO_3$

22. a. cobalt(III) sulfide, $Co_2S_3$
    b. bromine pentafluoride, $BrF_5$
    c. ferrous carbonate, $FeCO_3$
    d. gold(I) sulfate, $Au_2SO_4$
    e. calcium bicarbonate, $Ca(HCO_3)_2$
    f. tetraphosphorus trisulfide, $P_4S_3$
    g. cuprous cyanide, CuCN
    h. lanthanum(III) phosphate, $LaPO_4$

23. a. iron(II) oxide, FeO
    b. potassium oxide, $K_2O$
    c. magnesium nitride, $Mg_3N_2$
    d. magnesium nitrite, $Mg(NO_2)_2$
    e. pentasulfur dinitride, $S_5N_2$
    f. calcium permanganate, $Ca(MnO_4)_2$
    g. ferric sulfite, $Fe_2(SO_3)_3$
    h. lead(II) chromate, $PbCrO_4$
    i. silver dichromate, $Ag_2Cr_2O_7$

24. $Na_3N$

25. a. $XeF_6$, xenon hexafluoride; $XeF_4$, xenon tetrafluoride; $XeF_2$, xenon difluoride
    b. krypton difluoride, $KrF_2$; xenon trioxide, $XeO_3$; xenon tetraoxide, $XeO_4$

26. Both N and P have three unpaired electrons in their valence shell. These two elements can combine to form a compound with bonding similar to that of $N_2$; this compound is PN, which is called phosphorus mononitride. PN is the first phosphorus compound identified in interstellar space. PO is called phosphorus monoxide.

CHAPTER 17

1. oxygen

2. oxygen

3. H has given an electron. F has taken an electron. H has the more positive oxidation number.

4. The oxidation numbers are all zero, since these are all uncombined elements.

5. For $H_2O$, CO, $HNO_3$, $CaCl_2$, $H_2SO_4$, $NH_3$, and $PCl_3$: The sum of the oxidation numbers is zero. For $HSO_2^-$ and $Br^-$: The sum of the oxidation numbers is $-1$. For $PO_4^{3-}$: The sum of the oxidation numbers is $-3$. For $NO^+$ and $NH_4^+$: The sum of the oxidation numbers is $+1$.

6. $\overset{+1}{H}\overset{-1}{Br}$

7. $\overset{+1}{Li}\overset{-1}{H}$

8. $\overset{+3}{Al}\overset{-3}{N}$

9. $\overset{+2\ -2}{H_2}\overset{}{S}$ ; $\overset{+1\ -2}{H_2}S$

10. $\overset{+6\ -6}{Cl}\overset{}{O_3}$ ; $\overset{+6\ -2}{Cl O_3}$

11. $\overset{+4\ -4}{Cl O_2}$ ; $\overset{+4\ -2}{Cl O_2}$

12. $\overset{+2\ -2}{H_2}\overset{}{Se}$ ; $\overset{+1\ -2}{H_2 Se}$

13. $\overset{+7\ -8=-1}{Cl O_4^-}$ ; $\overset{+7\ -2}{Cl O_4^-}$

14. $\overset{+10\ -12=-2}{S_4 O_6^{2-}}$ ; $\overset{+2.5\ -2}{S_4 O_6^{2-}}$

15. $\overset{+3\ -3}{Cl}\overset{}{F_3}$ ; $\overset{+3\ -1}{Cl F_3}$

16. $\overset{+1\ +5\ -6}{Na\ Cl\ O_3}$ ; $\overset{+1\ +5\ -2}{Na Cl O_3}$

17. $\overset{+2\ -2}{Na_2}\overset{}{O_2}$ ; $\overset{+1\ -1}{Na_2 O_2}$

18. 
$$\overset{+}{F}-\overset{H}{\underset{H}{\overset{|}{C}}}-\overset{O}{\underset{O-H}{\overset{\parallel}{C}}}$$

Left-hand C = $-1$; right-hand C = $+3$; F = $-1$; each O = $-2$; each H = $+1$. The sum of the oxidation numbers of all the atoms is zero.

## 392 SOLUTIONS TO PROBLEMS

19. a. $\overset{+1}{Br}\overset{-1}{Cl}$  c. $\overset{+3}{Br}\overset{-3}{F_3}$ $\overset{+3}{\phantom{Br}}\overset{-1}{\phantom{F_3}}$  e. $\overset{+1}{Cl}\overset{-1}{F}$  e. $\overset{+3}{Au}\overset{-4=-1}{Cl_4^-}$ $\overset{+3}{\phantom{Au}}\overset{-1}{\phantom{Cl_4^-}}$  f. $\overset{+1}{Au}\overset{-1}{Cl}$

b. $\overset{+1}{Br}\overset{-1}{F}$  d. $\overset{+5}{Br}\overset{-5}{F_5}$ $\overset{+5}{\phantom{Br}}\overset{-1}{\phantom{F_5}}$  f. $\overset{+5}{Cl}\overset{-1}{F_5}$ $\overset{+5}{\phantom{Cl}}\overset{-1}{\phantom{F_5}}$

25. a.

$H \overset{+-}{=} \overset{H}{\underset{H}{\overset{|+}{C}}} \overset{00}{=} \overset{H}{\underset{H}{\overset{|+}{C}}} \overset{}{=} O \overset{-+}{=} H$

Left-hand C = −3; right C = −1; all H = +1; O = −2.

20. a. $\overset{+1\ +7\ -8}{H\ Cl\ O_4}$ $\overset{+1\ +7\ -2}{\phantom{H\ Cl\ O_4}}$  d. $\overset{+1\ +1\ -2}{H\ Cl\ O}$

b. $\overset{+1\ +5\ -6}{H\ Cl\ O_3}$ $\overset{+1\ +5\ -2}{\phantom{H\ Cl\ O_3}}$  e. $\overset{0}{Cl_2}$

c. $\overset{+1\ +3\ -4}{H\ Cl\ O_2}$ $\overset{+1\ +3\ -2}{\phantom{H\ Cl\ O_2}}$  f. $\overset{+1\ -1}{H\ Cl}$

b.

Left-hand C = −3; right-hand C = +1; all H = +1; O = −2.

21. a. $\overset{+2\ -2}{N_2\ O}$ $\overset{+1\ -2}{\phantom{N_2\ O}}$  e. $\overset{+10\ -10}{N_2\ O_5}$ $\overset{+5\ -2}{\phantom{N_2\ O_5}}$

b. $\overset{+2\ -2}{N\ O}$  f. $\overset{-3\ +3}{N\ H_3}$ $\overset{-3\ +1}{\phantom{N\ H_3}}$

c. $\overset{+6\ -6}{N_2\ O_3}$ $\overset{+3\ -2}{\phantom{N_2\ O_3}}$  g. $\overset{-1\ +2\ -2\ +1}{N\ H_2\ O\ H}$ $\overset{-1\ +1\ -2\ +1}{\phantom{N\ H_2\ O\ H}}$

d. $\overset{+4\ -4}{N\ O_2}$ $\overset{+4\ -2}{\phantom{N\ O_2}}$  h. $\overset{-4\ +1}{N_2\ H_4}$ $\overset{-2\ +1}{\phantom{N_2\ H_4}}$

c.

Left-hand C = −3; right-hand C = +3; all H = +1; each O = −2.

22. a. $\overset{+3\ +5\ -8}{H_3\ P\ O_4}$ $\overset{+1\ +5\ -2}{\phantom{H_3\ P\ O_4}}$  d. $\overset{+5\ -8\ =\ -3}{P\ O_4^{3-}}$ $\overset{+5\ -2}{\phantom{P\ O_4^{3-}}}$

b. $\overset{+2\ +5\ -8\ =\ -1}{H_2\ P\ O_4^-}$ $\overset{+1\ +5\ -2}{\phantom{H_2\ P\ O_4^-}}$  e. $\overset{+3\ +1\ -4}{H_3\ P\ O_2}$ $\overset{+1\ +1\ -2}{\phantom{H_3\ P\ O_2}}$

c. $\overset{+1\ +5\ -8\ =\ -2}{H\ P\ O_4^{2-}}$ $\overset{+1\ +5\ -2}{\phantom{H\ P\ O_4^{2-}}}$  f. $\overset{+2\ +1\ -4\ =\ -1}{H_2\ P\ O_2^-}$ $\overset{+1\ +1\ -2}{\phantom{H_2\ P\ O_2^-}}$

26. a.

$H \overset{+-}{=} \overset{H}{\underset{H}{\overset{|+}{C}}} \overset{+-}{=} O \overset{-+}{=} \overset{H}{\underset{H}{\overset{|+}{C}}} \overset{-+}{=} H$

All C = −2; oxygen = −2; all H = +1.

23. a. $\overset{+2\ -2}{Mn\ O}$  d. $\overset{+2\ -2}{Mn\ Cl_2}$ $\overset{+2\ -1}{\phantom{Mn\ Cl_2}}$

b. $\overset{+6\ -6}{Mn_2\ O_3}$ $\overset{+3\ -2}{\phantom{Mn_2\ O_3}}$  e. $\overset{+3\ -3}{Mn\ Cl_3}$ $\overset{+3\ -1}{\phantom{Mn\ Cl_3}}$

c. $\overset{+14\ -14}{Mn_2\ O_7}$ $\overset{+7\ -2}{\phantom{Mn_2\ O_7}}$  f. $\overset{+1\ +7\ -8}{K\ Mn\ O_4}$ $\overset{+1\ +7\ -2}{\phantom{K\ Mn\ O_4}}$

b.

All F = −1; all H = +1; left-hand C = +3; right-hand C = −3.

24. a. $\overset{+1\ -1}{Cu\ Cl}$  c. $\overset{+2\ -2}{Hg_2\ Cl_2}$ $\overset{+1\ -1}{\phantom{Hg_2\ Cl_2}}$

b. $\overset{+2\ -2}{Cu\ Cl_2}$ $\overset{+2\ -1}{\phantom{Cu\ Cl_2}}$  d. $\overset{+2\ -2}{Hg\ Cl_2}$ $\overset{+2\ -1}{\phantom{Hg\ Cl_2}}$

c.
$$H \stackrel{+}{-} \underset{H}{\overset{H}{\underset{|}{C}}} \stackrel{00}{-} C \equiv N$$

All H = +1; N = −3; left-hand C = −3; right-hand C = +3.

d.
$$H \stackrel{+}{-} \underset{H}{\overset{H}{\underset{|}{C}}} \stackrel{00}{-} \underset{H}{\overset{H}{\underset{|}{C}}} - N \underset{H}{\overset{H}{\diagdown}}$$

Left-hand C = −3; right-hand C = −1; all H = +1; N = −3.

27. −1
    −1/2
    $O_2^-$

28. :S̈· Sulfur can "attract" as many as two electrons or "give" as many as six electrons.

29.
$$\underset{H}{\overset{H}{\diagup}} C = O$$

C = 0; all H = +1; O = −2.

**CHAPTER 18**

1. Yes, the reaction *is* a redox reaction, because there is a change in oxidation number.

2. No, the reaction is *not* a redox reaction, because there is no change in oxidation number.

3. $Li \rightarrow Li^+ + e^-$ and $Ca^{2+} + 2e^- \rightarrow Ca$

4. The balanced molecular equation is $2Li + CaCl_2 \rightarrow 2LiCl + Ca$. The balanced ionic equation is $2Li + Ca^{2+} + 2Cl^- \rightarrow 2Li^+ + 2Cl^- + Ca$.

5. 
$$1 \times [Cu \rightarrow Cu^+ + 2e^-]$$
$$\underline{2 \times [e^- + NO_3^- + 2H^+ \rightarrow NO_2 + H_2O]}$$
$$Cu + 2NO_3^- + 4H^+ \rightarrow Cu^{2+} + 2NO_2 + 2H_2O$$

6.
$$5 \times [C_2O_4^{2-} \rightarrow 2CO_2 + 2e^-]$$
$$\underline{2 \times [5e^- + MnO_4^- + 8H^+ \rightarrow Mn^{2+} + 4H_2O]}$$
$$5C_2O_4^{2-} + 2MnO_4^- + 16H^+ \rightarrow 10CO_2 + 2Mn^{2+} + 8H_2O$$

7.
$$8 \times [3e^- + MnO_4^- + 2H_2O \rightarrow MnO_2 + 4OH^-]$$
$$\underline{3 \times [NH_3 + 9OH^- \rightarrow NO_3^- + 6H_2O + 8e^-]}$$
$$8MnO_4^- + 16H_2O + 3NH_3 + 27OH^- \rightarrow 8MnO_2 + 32OH^- + 3NO_3^- + 18H_2O$$
$$8MnO_4^- + 3NH_3 \rightarrow 8MnO_2 + 5OH^- + 3NO_3^- + 2H_2O$$

8. 
$$2 \times [e^- + NO_2^- + H_2O \rightarrow NO + 2OH^-]$$
$$\underline{1 \times [2I^- \rightarrow I_2 + 2e^-]}$$
$$2NO_2^- + 2H_2O + 2I^- \rightarrow 2NO + 4OH^- + I_2$$

9.

| SUBSTANCE | TYPE OF REACTION | SUBSTANCE IS | TYPE OF AGENT |
|---|---|---|---|
| Li | oxidation | oxidized | reducing agent |
| $Ca^{2+}$ | reduction | reduced | oxidizing agent |

10.

| SUBSTANCE | TYPE OF REACTION | SUBSTANCE IS | TYPE OF AGENT |
|---|---|---|---|
| $C_2O_4^{2-}$ | oxidation | oxidized | reducing agent |
| $MnO_4^-$ | reduction | reduced | oxidizing agent |

11. a. $\quad Ni \rightarrow Ni^{2+} + 2e^-$
$\quad\quad\quad \underline{F_2 + 2e^- \rightarrow 2F^-}$
$\quad\quad\quad Ni + F_2 \rightarrow Ni^{2+} + 2F^-$

b. $\quad 2 \times [Fe \rightarrow Fe^{2+} + 2e^-]$
$\quad\quad \underline{1 \times [O_2 + 4e^- \rightarrow 2O^{-2}]}$
$\quad\quad\quad 2Fe + O_2 \rightarrow 2Fe^{2+} + 2O^{2-}$

c. $\quad 2 \times [Co^{2+} \rightarrow Co^{3+} + e^-]$
$\quad\quad \underline{1 \times [Cl_2 + 2e^- \rightarrow 2Cl^-]}$
$\quad\quad\quad 2Co^{2+} + Cl_2 \rightarrow 2Co^{3+} + 2Cl^-$

12. a. $\quad 1 \times [6e^- + ClO_3^- + 6H^+ \rightarrow Cl^- + 3H_2O]$
$\quad\quad\quad\quad \underline{3 \times [SO_3^{2-} + H_2O \rightarrow SO_4^{2-} + 2H^+ + 2e^-]}$
$\quad\quad ClO_3^- + 6H^+ + 3SO_3^{2-} + 3H_2O \rightarrow Cl^- + 3H_2O + 3SO_4^{2-} + 6H^+$
$\quad\quad\quad\quad ClO_3^- + 3SO_3^{2-} \rightarrow Cl^- + 3SO_4^{2-}$

b. $\quad 2e^- + MnO_2 + 4H^+ \rightarrow Mn^{2+} + 2H_2O$
$\quad\quad\quad\quad \underline{2I^- \rightarrow I_2 + 2e^-}$
$\quad\quad MnO_2 + 4H^+ + 2I^- \rightarrow Mn^{2+} + 2H_2O + I_2$

c. $1 \times [6e^- + Cr_2O_7^{2-} + 14H^+ \rightarrow 2Cr^{3+} + 7H_2O]$
$\quad\quad\quad\quad \underline{3 \times [2I^- \rightarrow I_2 + 2e^-]}$
$\quad\quad Cr_2O_7^{2-} + 14H^+ + 6I^- \rightarrow 2Cr^{3+} + 7H_2O + 3I_2$

13. $2 \times [5e^- + MnO_4^- + 8H^+ \rightarrow Mn^{2+} + 4H_2O]$
    $\underline{5 \times [H_2S \rightarrow S + 2H^+ + 2e^-]}$
    $2MnO_4^- + 16H^+ + 5H_2S \rightarrow 2Mn^{2+} + 8H_2O + 5S + 10H^+$

S is oxidized. Mn is reduced. $MnO_4^-$ is the oxidizing agent. $H_2S$ is the reducing agent.

14. a. $4 \times [MnO_4^- + e^- \rightarrow MnO_4^{2-}]$
    $\underline{1 \times [4OH^- \rightarrow O_2 + 2H_2O + 4e^-]}$
    $4MnO_4^- + 4OH^- \rightarrow 4MnO_4^{2-} + O_2 + 2H_2O$

   b. $2 \times [ClO_2 + e^- \rightarrow ClO_2^-]$
    $\underline{1 \times [H_2O_2 + 2OH^- \rightarrow O_2 + 2H_2O + 2e^-]}$
    $2ClO_2 + H_2O_2 + 2OH^- \rightarrow 2ClO_2^- + O_2 + 2H_2O$

   c. $2e^- + Cl_2 \rightarrow 2Cl^-$
    $\underline{IO_3^- + 2OH^- \rightarrow IO_4^- + H_2O + 2e^-}$
    $Cl_2 + IO_3^- + 2OH^- \rightarrow 2Cl^- + IO_4^- + H_2O$

15. a. $Mn^{2+} + 4OH^- \rightarrow MnO_2 + 2H_2O + 2e^-$
    $\underline{H_2O_2 + 2e^- \rightarrow 2OH^-}$
    $Mn^{2+} + 2OH^- + H_2O_2 \rightarrow MnO_2 + 2H_2O$

   b. $1 \times [Bi(OH)_3 + 3e^- \rightarrow Bi + 3OH^-]$
    $\underline{3 \times [SnO_2^- + 2OH^- \rightarrow SnO_3^{2-} + H_2O + e^-]}$
    $Bi(OH)_3 + 3SnO_2^- + 3OH^- \rightarrow Bi + 3SnO_3^{2-} + 3H_2O$

   c. $2e^- + MnO_2 + 2H_2O \rightarrow Mn(OH)_2 + 2OH^-$
    $\underline{SO_3^{2-} + 2OH^- \rightarrow SO_4^{2-} + H_2O + 2e^-}$
    $MnO_2 + H_2O + SO_3^{2-} \rightarrow Mn(OH)_2 + SO_4^{2-}$

16. $14 \times [5e^- + MnO_4^- + 8H^+ \rightarrow Mn^{2+} + 4H_2O]$
    $\underline{5 \times [2CH_3Cl + 4H_2O \rightarrow Cl_2 + 2CO_2 + 14H^+ + 14e^-]}$
    $14MnO_4^- + 42H^+ + 10CH_3Cl \rightarrow 14Mn^{2+} + 36H_2O + 5Cl_2 + 10CO_2$

17. $1 \times [12e^- + P_4 + 12H_2O \rightarrow 4PH_3 + 12OH^-]$
    $\underline{3 \times [P_4 + 8OH^- \rightarrow 4H_2PO_2^- + 4e^-]}$
    $4P_4 + 12H_2O + 12OH^- \rightarrow 4PH_3 + 12H_2PO_2^-$

18. a. $3 \times [5e^- + MnO_4^- + 4H_2O \rightarrow Mn^{2+} + 8OH^-]$
    $\underline{5 \times [V^{2+} + 4OH^- \rightarrow VO_2^+ + 2H_2O + 3e^-]}$
    $3MnO_4^- + 2H_2O + 5V^{2+} \rightarrow 3Mn^{2+} + 4OH^- + 5VO_2^+]$

## 396 SOLUTIONS TO PROBLEMS

b.  $\quad 4 \times [Fe^{2+} \rightarrow Fe^{3+} + e^-]$

$\underline{1 \times [4e^- + ClO_2^- + 2H_2O \rightarrow Cl^- + 4OH^-]}$

$4Fe^{2+} + ClO_2^- + 2H_2O \rightarrow 4Fe^{3+} + Cl^- + 4OH^-$

c.  $1 \times [8e^- + NO_3^- + 7H_2O \rightarrow NH_4^+ + 10OH^-]$

$\underline{\quad\quad\quad 4 \times [Zn \rightarrow Zn^{2+} + 2e^-]}$

$NO_3^- + 7H_2O + 4Zn \rightarrow NH_4^+ + 10OH^- + 4Zn^{2+}$

19. a. $1 \times [6e^- + Cr_2O_7^{2-} + 14H^+ \rightarrow 2Cr^{3+} + 7H_2O]$

$\underline{\quad\quad\quad 3 \times [C_2O_4^{2-} \rightarrow 2CO_2 + 2e^-]}$

$Cr_2O_7^{2-} + 14H^+ + 3C_2O_4^{2-} \rightarrow 2Cr^{3+} + 7H_2O + 6CO_2$

b. $1 \times [2e^- + H_2O_2 + 2H^+ \rightarrow H_2O + H_2O]$

$\underline{\quad\quad\quad 2 \times [Fe^{2+} \rightarrow Fe^{3+} + e^-]}$

$H_2O_2 + 2H^+ + 2Fe^{2+} \rightarrow 2H_2O + 2Fe^{3+}$

c.  $\quad 3 \times [2Sb + 5H_2O \rightarrow Sb_2O_5 + 10H^+ + 10e^-]$

$\underline{10 \times [3e^- + NO_3^- + 4H^+ \rightarrow NO + 2H_2O]}$

$6Sb + 10NO_3^- + 10H^+ \rightarrow 3Sb_2O_5 + 10NO + 5H_2O$

20. $\quad\quad\quad Zn \rightarrow Zn^{2+} + 2e^-$

$\underline{2MnO_2 + 2H_2O + 2e^- \rightarrow Mn_2O_3 + H_2O + 2OH^-}$

$Zn + 2MnO_2 + H_2O \rightarrow Zn^{2+} + Mn_2O_3 + 2OH^-$

Now insert the $NH_4^+$ and the $NH_3$. Use $H^+$ to balance H, since this is easy to do and the $H^+$ will disappear in the end anyway:

$Zn + 2MnO_2 + H_2O + NH_4^+ \rightarrow Zn^{2+} + Mn_2O_3 + 2OH^- + NH_3 + H^+$

Combine $H^+$ and $OH^-$ to give $H_2O$:

$Zn + 2MnO_2 + H_2O + NH_4^+ \rightarrow Zn^{2+} + Mn_2O_3 + OH^- + NH_3 + H_2O$

Subtract $H_2O$ from each side:

$Zn + 2MnO_2 + NH_4^+ \rightarrow Zn^{2+} + Mn_2O_3 + OH^- + NH_3$

21. $Zn + H_2O + 2OH^- \rightarrow ZnO + 2H_2O + 2e^-$

$\underline{HgO + 2H_2O + 2e^- \rightarrow Hg + H_2O + 2OH^-}$

$\quad\quad\quad Zn + HgO \rightarrow ZnO + Hg$

Notice that the equation was balanced as given.

22. $\quad\quad\quad Pb + SO_4^{2-} \rightarrow PbSO_4 + 2e^-$

$\underline{PbO_2 + 4H^+ + SO_4^{2-} + 2e^- \rightarrow PbSO_4 + 2H_2O}$

$Pb + PbO_2 + 4H^+ + 2SO_4^{2-} \rightarrow 2PbSO_4 + 2H_2O$

To get the molecular equation, combine the $H^+$ and the $SO_4^{2-}$ to give $H_2SO_4$ (sulfuric acid). The liquid in the lead–acid battery in an automobile is about 10 M sulfuric acid:
$Pb + PbO_2 + 2H_2SO_4 \rightarrow 2PbSO_4 + 2H_2O$

23. $2 \times [H_2 + 2OH^- \rightarrow 2H_2O + 2e^-]$
$\underline{1 \times [4e^- + O_2 + 2H_2O \rightarrow 4OH^-]}$
$2H_2 + O_2 \rightarrow 2H_2O$

24. $5 \times [O_2 + 2H_2O + 4e^- \rightarrow 4OH^-]$
$\underline{1 \times [C_3H_8 + 20OH^- \rightarrow 3CO_2 + 14H_2O + 20e^-]}$
$C_3H_8 + 5O_2 \rightarrow 3CO_2 + 4H_2O$

25. $Cd + 2OH^- \rightarrow Cd(OH)_2 + 2e^-$
$\underline{2e^- + NiO_2 + 2H_2O \rightarrow Ni(OH)_2 + 2OH^-}$
$Cd + NiO_2 + 2H_2O \rightarrow Cd(OH)_2 + Ni(OH)_2$

26. $2 \times [Li \rightarrow Li^+ + e^-]$
$\underline{1 \times [I_2 + 2e^- \rightarrow 2I^-]}$
$2Li + I_2 \rightarrow 2Li^+ + 2I^-$

27. $Na^+ {-} C{\equiv}N{:}$

There are $+3 - 1 = +2$ charges around the C atom. There are $-3$ charges around the N atom. The oxidation number of C is $+2$, and the oxidation number of N is $-3$.

## CHAPTER 19

1. $[H^+] = 2.51 \times 10^{-7}$ M
2. $\log 10{,}000{,}000 = 7$ and $\log 100{,}000{,}000 = 8$
3. $\log 0.00001 = -5$ and $\log 0.000001 = -6$
4. $\log 962 = 2.9832$ and $\log 7.76 \times 10^5 = 5.8899$
5. $\log 0.087 = -1.0605$ and $\log 3.24 \times 10^{-6} = -5.4896$
6. You get an ERROR message. Logs of negative numbers don't exist.
7. antilog $4 = 10^4 = 10{,}000$
8. antilog $0.86 = 10^{0.86} = 7.24$ and antilog $4.66 = 10^{4.66} = 45{,}709$
9. antilog $(-0.67) = 10^{-0.67} = 0.214$ and antilog $(-2.33) = 10^{-2.33} = 0.00468$
10. antilog $7.3 = 10^{7.3} = 1.995 \times 10^7$ and antilog $(-7.3) = 10^{-7.3} = 5.01 \times 10^{-8}$
11. **8.91, EE, +/−, 8, log, +/−**   pH = 7.05
12. pH = 11.80
13. pH = 2.40
14. $[H^+] = 3.16 \times 10^{-7}$
15. $[H^+] = 3.98 \times 10^{-8}$
16. a. 8   c. −8
    b. 0
17. a. 2.515   d. 23.991
    b. 3.917   e. 13.813
    c. 3.678   f. 9,623
18. a. −0.873   d. −9.265
    b. −2.331   e. −17.036
    c. −4.206   f. −22.996
19. a. $1 \times 10^5$   c. $1 \times 10^{-5}$
    b. 1

# 398  SOLUTIONS TO PROBLEMS AND TWO RIDDLES

20. a. 1660
    b. $3.80 \times 10^7$
    c. $9.77 \times 10^{10}$
    d. $5.75 \times 10^{13}$
    e. 2.70
    f. 2.00

21. a. 0.500
    b. $2.95 \times 10^{-5}$
    c. 0.987
    d. $3.55 \times 10^{-12}$
    e. $4.27 \times 10^{-23}$
    f. 0.0776

22. pH = 4.10
23. pH = 3.50
24. $[H^+] = 7.94 \times 10^{-6}$ M
25. $[H^+] = 1.26 \times 10^{-7}$ M

---

## TWO RIDDLES THAT REQUIRE LOGICAL THINKING

Logical thinking is an important part of the skill set needed to be a scientist. I don't know if puzzling over riddles will make you a better chemist, but I hope so. At least you should have fun trying to solve these.

### 1. The lightbulb and three switches.*

There is an incandescent lightbulb inside a closet. The door is closed, and you cannot see if the light is on or off through the door. The light is off to start. Outside the closet, there are three light switches, all in the off position to start. One of the switches controls the lightbulb in the closet. You can flip the switches however you want and as many times as you want, but once you open the door, you can no longer touch the switches.

How do you decide for sure which switch controls the light?

**HINT:** Incandescent lightbulbs give off a lot of heat in addition to light. When left on for a few minutes, they get very hot. The solution can be found at https://bit.ly/3vBf0dJ.

### 2. The farmer, the wolf, the goat, and the cabbage.

Once upon a time a farmer purchased a wolf, a goat, and a cabbage. On her way home, she came to the bank of a river and rented a boat. When crossing the river by boat, she could carry herself and only one of her purchases — either the wolf, the goat, or the cabbage.

If left unattended together, *the wolf would eat the goat*, or *the goat would eat the cabbage*.

The farmer's challenge was to carry herself and her purchases to the far bank of the river, leaving all her purchases uneaten. How did she do it?

**HINT:** The farmer can make as many trips back and forth across the river as necessary. Strangely, animals left alone do not run away. The solution can be found at https://bit.ly/3MKsDgj.

---

*This riddle was inspired by and modified from a similar one in *Analytical Chemistry* by Gary D. Christian, et al, 7th edition, 2014, page 19, https://amzn.to/3Os0blH. The version of this riddle given in the book is a favorite of Stanford University chemistry professor Richard N. Zare.

# APPENDIX

| | |
|---|---|
| The Front Cover Photograph Explained | 400 |
| Scientific Calculators | 402 |
| Fundamental Physical Constants, SI Base Units, and SI Prefixes | 403 |
| How a Rechargeable Battery Works | 405 |
| The Double Slit Experiment with Electrons | 412 |

## About the Author

Harvey F. Carroll received a Ph.D. in physical chemistry from Cornell University. He was a professor of physical sciences at Kingsborough Community College of The City University of New York where he served as department chair and founding director of the Engineering Science Program. For 35 years he taught courses at the level of this book, general chemistry, environmental chemistry, chemistry and technology, chemistry and physics for allied health professionals, physics (with and without calculus), history and philosophy of physical sciences, mathematics, and nutrition. After retiring to the Seattle area, he taught courses in nutrition and alternative energy sources to retirees, and a course in *Chemistry for Pre-Nursing Students* at a local community college. He is the author of the previous version of this textbook, coauthor of both *General Chemistry I and II Laboratory Manuals* (which incorporated computer acquisition of data along with graphical analysis of the data), and the lab manual for a *Chemistry and Physics for Allied Health Professionals* course. His research interests were in studying high temperature chemical reactions behind shock waves, and he is the coauthor of numerous publications in peer-reviewed professional journals. He has done research in the chemistry departments at the University of Washington and the Hebrew University of Jerusalem (https://bit.ly/3US8DNg), and presented his research to three international conferences, one of which was in Beijing. (Yes, he and his wife are heroes of the Great Wall.)

# THE FRONT COVER PHOTOGRAPH EXPLAINED[1]

The cover photo, courtesy of NASA/CXC/SAO, was taken by NASA's Chandra X-ray Observatory that was launched into orbit around the earth on July 23, 1999. It is a picture of the **Cassiopeia A supernova remnant**. It gives off a lot of X-rays that can be used to determine some of the elements in the remnant. Since X-rays don't have color, this means that the colors are added by computers—see https://s.si.edu/3NwvsSl for what this photo looks like with the colors separated by element—silicon is red, sulfur is yellow, calcium is green, and iron is purple. The white dot in the center is a colorized x-ray image of the remnant's neutron star (https://s.si.edu/3nuiMRa). The blue lines in the photo are shock waves that travel at about 1/10th the speed of light[2] and heat the gas behind them to an incredibly high temperature. This heat creates new elements and cosmic rays (mostly protons). The main shock wave is at the edge of the remnant, but shock waves also bounce back and forth in the remnant. Some supernovas can be as bright as all the stars in a galaxy put together before fading after a few weeks.

A supernova remnant is the expanding cloud that remains after a supernova explosion. What else remains is a neutron star (see p.4). Stars that become this type of supernova when they die are about 8 to about 15 times the mass of our sun. (Heavier stars form black holes.) Light from the explosion reached earth around 1667 CE but was not reported. This supernova remnant, which is seen in the constellation Cassiopeia, is about 11,000 light years away from the earth. A light year (1 light year = 5.88 trillion miles) is the distance that light can travel in a vacuum in one year. The supernova remnant is expanding at the rate of about 11 million miles per hour. It is in our galaxy and is about 10 light years in diameter. In comparison, our solar system is only about 0.0013 light years in diameter, and the distance to our nearest star, Proxima Centauri, is about 4.3 light years. You can see how immense this supernova remnant is! See https://go.nasa.gov/2LuOKNe for more information.

You might wonder why I used an image of a supernova remnant on the cover of this book. As stated above, some of the elements in the universe, from carbon to zirconium, are made in these types of explosions. These elements got blown into space in the explosion and a very small fraction of them became part of our solar system and our earth (https://go.nasa.gov/35frmYY).

As **Carl Sagan** (1934-1996), the David Duncan Professor of Astronomy and Space Sciences at Cornell University, said (for more about Carl Sagan, see https://bit.ly/3rHnrCl):

> The nitrogen in our DNA, the calcium in our teeth, the iron in our blood, the carbon in our apple pies were made in the interiors of collapsing stars.[3] The cosmos is within us. **We are made of starstuff.**[4] (https://bit.ly/2QAqf26)

[1]This discussion will be more understandable if you read Chapter 1 and the links given in this discussion.

[2] Shock waves are also found on earth as the thunder after a lightning bolt and are given off when an airplane flies faster than the speed of sound (the "sonic boom"). The speed of light is 186,000 mi/s or 300,000 km/s.

## THE FRONT COVER PHOTOGRAPH EXPLAINED

**Margaret Burbidge** (1919-2020), Professor of Astrophysics at the University of California, San Diego, was the lead author of a most influential 1957 paper (over 100 pages long) in the journal *Reviews of Modern Physics*, "Synthesis of the Elements in Stars" (referred to as B²FH) along with her husband **G. R. Burbidge, W. A. Fowler, and F. Hoyle.** See https://bit.ly/3xal2Tx and https://bit.ly/3xeenaE for the story of her life when she turned 100 years old. She was a trailblazer in getting equal rights for women in astronomy and was the first woman in the US to be allowed to use the large telescope in a major observatory at night. Now women are treated as equals in astronomy.

To continue our discussion of where the elements are made, we look at some recent articles. In the article at https://bit.ly/2XspJV4 there is a periodic table that shows where all the elements are made. A more technical article by **Jennifer A. Johnson**, Prof. of Astronomy at Ohio State University, called "Populating the periodic table: Nucleosynthesis of the elements," is in the February 1, 2019 issue of *Science* magazine. The 4th reference in this paper is to B²FH. Unfortunately, this article is available only to registered users. **NOTE:** This issue of *Science* commemorates the 150th anniversary of the periodic table (https://bit.ly/2O1VtNJ). Prof. Johnson has a less technical article at https://bit.ly/2KCdvEc.

A summary of the processes that make the elements, taken from Prof. Johnson's paper, is below. (A free astronomy text is at https://bit.ly/34pQV9S.)

- Big Bang fusion (p 34 for fusion): H, He, Li (The Big Bang is when the universe appeared as a rapidly expanding "point." This seems to have happened about 13.8 billion years ago. Where it came from or how it started is a great mystery. Some physicists have come up with a few theories, none of which seem to satisfy other physicists. See https://bit.ly/37s6Ue9 for details.)
- Comic ray fission (p 34): Li, Be, B
- Exploding massive stars (the most energetic supernova): C to Zr
- Exploding white dwarfs (another, less energetic, type of supernova – this occurs when a white dwarf accumulates mass from a nearby dying star – they are circling each other): Si to Zn
- Merging neutron stars (see pp. 4, 15): Nb to Pu
- Dying low mass stars (lone white dwarfs, which is what our sun will become when it dies): Li, C, N, Sr to Bi

**NOTE:** All the isotopes of Tc, Pm, Po to Ra, Ac, Pa, Np, and Am to Og are so radioactive (they have short half-lives) that there is nothing left on earth from the stars (see pp 12, 15). Helium and some other elements are also made in stars by nuclear fusion (see pp 33, 34).

**NOTE:** Our bodies are 10% H by mass but 62 % of all the atoms in our bodies are H atoms (https://bit.ly/2OfdqZf). For a discussion and table of the abundance of the elements in the universe, the earth, and our bodies, see https://bit.ly/3oF4vlR.

[3] Before a star explodes to become a supernova, it collapses.

[4] H and some He and Li appeared after the Big Bang when there were no stars. The first stars "switched on" (the technical term for starting fusion and giving off heat and light) around 250-350 million years after the Big Bang (https://bbc.in/3mxLv7i). This is when the giant expanding clouds of hydrogen produced in the Big Bang had cooled enough so gravity could condense atoms to form the first stars. Also, the cooling of the gas clouds stopped fusion reactions from taking place and no additional elements were made. From here on elements would be made in stars. The universe is now made up of about 73% hydrogen and 25% helium; all the other elements in the universe have been made in stars (remember that a bit of lithium was also made in the hot gas clouds) and amount to about 2% of the total. See https://bit.ly/3zcOwl5 for more information.

# SCIENTIFIC CALCULATORS

A scientific calculator is necessary to get the most benefit from this book, and I highly recommend a solar model. These are available for under $20 and should serve you throughout your studies. Unless you love to play around with complicated things, I'd suggest that you buy a simple solar-powered scientific calculator. An example is the TI-30XIIS at around $12 on Amazon during the late Fall of 2022. It has all the functions you will need for a general chemistry course and more. I am not endorsing Amazon or this or any other calculator but just mentioning it as an example.

Good scientific calculators are also available as apps for smartphones and online, but you may not be allowed to use smartphones on exams because you could look up information on the internet or text an expert for help. You may not be allowed to use some graphing calculators on exams because they might have enough memory to enter notes and solved problems. Check before you buy one!

As of 2020, calculators for Android are reviewed at https://bit.ly/3b2CH1r. My Android phone has a built-in basic scientific calculator that can be accessed by opening the simple calculator app that comes with Android and turning the phone sideways. Note that what I call EE in this book is an ^ in this calculator.

I use the free RealCalc scientific calculator on my Android phone which is available at https://bit.ly/33ahRJM. Note that the FSE button converts a number in the display from regular notation (called Fixed in calculators) to scientific notation to engineering notation.* The EXP button is what I call EE in this book. The DRG button allows entering angles in degrees, radians or grads.* Also, the SHIFT button acts as an INV, 2nd, or 2ndF button that are on various free-standing calculators.

Calculators for the iPhone are reviewed at https://bit.ly/3d5aetJ. PCalc is full featured; it costs $10. iPhones also have a built-in scientific calculator. Open the simple calculator app that comes with iOS and turn the phone sideways.

Windows 10 has a built-in scientific calculator that is very good. It is at https://bit.ly/2ysftjT. Note that the up arrow in this calculator converts some keys to the inverse function, and the exp button is what I call EE in this book. Also, the F-E button converts a number from regular notation to scientific notation, and *vice versa*.

A simple online scientific calculator can be found at https://bit.ly/2MZV7Fg. A guide to scientific calculators for Mac computers is at https://apple.co/2tzMyf9.

There are two free equation graphing calculators online that are easy to use: https://bit.ly/1i8XxM4 & https://bit.ly/2m9PSIh.

Unfortunately, there is no agreed-upon format for scientific calculators, and some of the instructions in this book may not match the buttons and/or the display on your calculator.

* Engineering notation, radians and grads are not used in this book.

# FUNDAMENTAL PHYSICAL CONSTANTS, SI BASE UNITS, AND SI PREFIXES

## Fundamental Physical Constants

| Quantity | Value | Unit |
|---|---|---|
| speed of light* | 299 792 458 | $m \cdot s^{-1}$ |
| electron mass | $9.109\ 383\ 701\ 5 \times 10^{-31}$ | kg |
| proton mass | $1.672\ 621\ 923\ 69 \times 10^{-27}$ | kg |
| neutron mass | $1.674\ 927\ 498\ 04 \times 10^{-27}$ | kg |
| electron charge | $1.602\ 176\ 634 \times 10^{-19}$ | C |
| proton charge – same as electron charge but positive | | |
| Avogadro's constant* | $6.022\ 140\ 76 \times 10^{23}$ | $mol^{-1}$ |
| Ideal gas law constant | 0.082 057 366 081 | $L \cdot atm \cdot K^{-1} \cdot mol^{-1}$ |

**NOTE:** In the numbers above, spaces are used for clarity instead of commas. The unit abbreviation "C" stands for coulomb, a unit of electrical charge. It is not discussed in this book. Quantities with an * are defined and thus are considered exact numbers. (See Section 6-6.) The following link, https://bit.ly/2KOmXWR, lists more constants with an indication of those that are defined and thus are exact numbers.

## The Seven SI Base Units*

| Symbol | Name | Quantity |
|---|---|---|
| s | second | time |
| m | meter | length |
| kg | kilogram | mass |
| A | ampere | electric current |
| K | kelvin | temperature |
| mol | mole | amount of substance |
| cd | candela | luminous intensity |

*The seven base units are a basic set of units from which all other SI units can be derived. For example, speed is meters per second. The ampere and the candela are not used in this book.

## SI Prefixes

| Prefix Name | Prefix Symbol* | | Value | Number Name** |
|---|---|---|---|---|
| quecca | Q | 2022 | $10^{30}$ | Nonillion |
| ronna | R | 2022 | $10^{27}$ | Octillion |
| yotta | Y | 1991 | $10^{24}$ | Septillion |
| zetta | Z | 1991 | $10^{21}$ | Sextillion |
| exa | E | 1975 | $10^{18}$ | Quintillion |
| peta | P | 1975 | $10^{15}$ | Quadrillion |
| tera | T | 1960 | $10^{12}$ | Trillion |
| giga | G | 1960 | $10^{9}$ | Billion |
| mega | M | 1960 | $10^{6}$ | Million |
| kilo | k | | $10^{3}$ | Thousand |
| hecto | h | | $10^{2}$ | Hundred |
| deka | da | | $10^{1}$ | Ten |
| | | | $10^{0}$ | One |
| deci | d | | $10^{-1}$ | Tenth |
| centi | c | | $10^{-2}$ | Hundredth |
| milli | m | | $10^{-3}$ | Thousandth |
| micro | μ | 1960 | $10^{-6}$ | Millionth |
| nano | n | 1960 | $10^{-9}$ | Billionth |
| pico | p | 1960 | $10^{-12}$ | Trillionth |
| femto | f | 1964 | $10^{-15}$ | Quadrillionth |
| atto | a | 1964 | $10^{-18}$ | Quintillionth |
| zepto | z | 1991 | $10^{-21}$ | Sextillionth |
| yocto | y | 1991 | $10^{-24}$ | Septillionth |
| ronto | r | 2022 | $10^{-27}$ | Octillionth |
| quecto | q | 2022 | $10^{-30}$ | Nonillionth |

**NOTE:** In the above table, the abbreviation for the unit "micro" is the Greek letter μ, pronounced "mu." You may see "μg" written as "mcg" (micrograms) on food labels, over-the-counter drugs, nutritional supplements, and in magazines and newspapers. The μm (micrometer) is sometimes called a "micron."

The table at the left lists all the SI prefixes approved by the CGPM. (See p. xix for details about the CGPM.) The reason prefixes are added is the growth of science and technology. The "ronna" and "quecca" are needed because global data storage is doubling every year, and by the 2030s may exceed 1 yottabyte. Physicists have measured times as small as 247 zeptoseconds (the time it takes for light to cross a hydrogen atom), and forces as small as 42 yoctonewtons (a newton, abbreviated as "N," is the unit of force in the SI system). It seems probable that in 10 or 20 years, numbers will be in the quecca or quecto ranges. It is convenient to have prefixes in multiples of 1000, so the new prefixes will be appreciated when they are needed.

*The dates are the year that the CGPM approved the corresponding prefix (https://bit.ly/3UiGJsY). Prefixes without dates were in use before the CGPM was formed in 1875.

**Names commonly used in the United States (called the "short scale").

# HOW A RECHARGEABLE BATTERY WORKS

(To understand this material, reading Chapters 17 and 18 would be helpful.)

In Chapter 18, we discuss redox reactions and give some examples of the redox equations involved in discharging batteries to provide electricity to a light bulb, a motor, a smartphone, a laptop computer, or a hearing aid.

Here we discuss a very simple battery, the **Daniell cell**, shown in Figures 1 and 2 on the next two pages (https://bit.ly/2ZeC9TG). The Daniell cell is an example of a rechargeable battery. It is now used only for classroom demonstration purposes. The metal electrodes are made of zinc and copper. They are in a water solution of either $ZnSO_4$ or $CuSO_4$. Each beaker is called a **half-cell**, because it contains half of the reactants.

**AN ASIDE:** The term "cell" is very old. At the time when the Daniell cell was invented in 1836 by the English chemist and physicist **John Daniell** (1790-1845), they used a large number of these "cells" connected together, because one "cell" didn't supply enough electricity for their experiments. These assemblies were called batteries. Now, we usually call just one "cell" a battery. One "cell" batteries you might know are the AAA, AA, C and D batteries used in flashlights, remote controls for TVs, and such. In a few cases, the word "cell" is used even today. A flashlight that contains three D batteries is still referred to as a "3-cell D flashlight" or a "3-D cell flashlight." The common 9 V battery has 6 small batteries (called "cells") inside it that are connected.

The lead-acid or lead-storage battery in your car has 6 batteries inside it that are connected, and they are also called "cells." If your car has battery trouble, an auto mechanic might tell you, "One of the cells in your battery is bad." The large lithium-ion batteries used in electric vehicles consists of many smaller lithium-ion batteries that are connected together. These smaller individual batteries are also called cells.

**Discharging a Daniell cell**: In Figure 1, the Daniell cell is acting as a battery. Electrons are flowing out of the zinc electrode (the anode), through a resistance (the sawtooth inside the circle is the universal sign of resistance, such as a light bulb, electronic device, or motor), and into the copper electrode. The electrons come from the Zn metal, each atom giving up 2 electrons and putting $Zn^{2+}$ into solution. So when this battery discharges, the zinc electrode is dissolving. As the electrons flow into the copper electrode (the cathode) they combine with $Cu^{2+}$ ions, turning them into Cu metal, which deposits on the copper electrode. Depositing a metal on a surface using electricity is called **electroplating**. As the negative electrons flow out of the anode, the anode becomes positively charged. At the same time, the cathode,

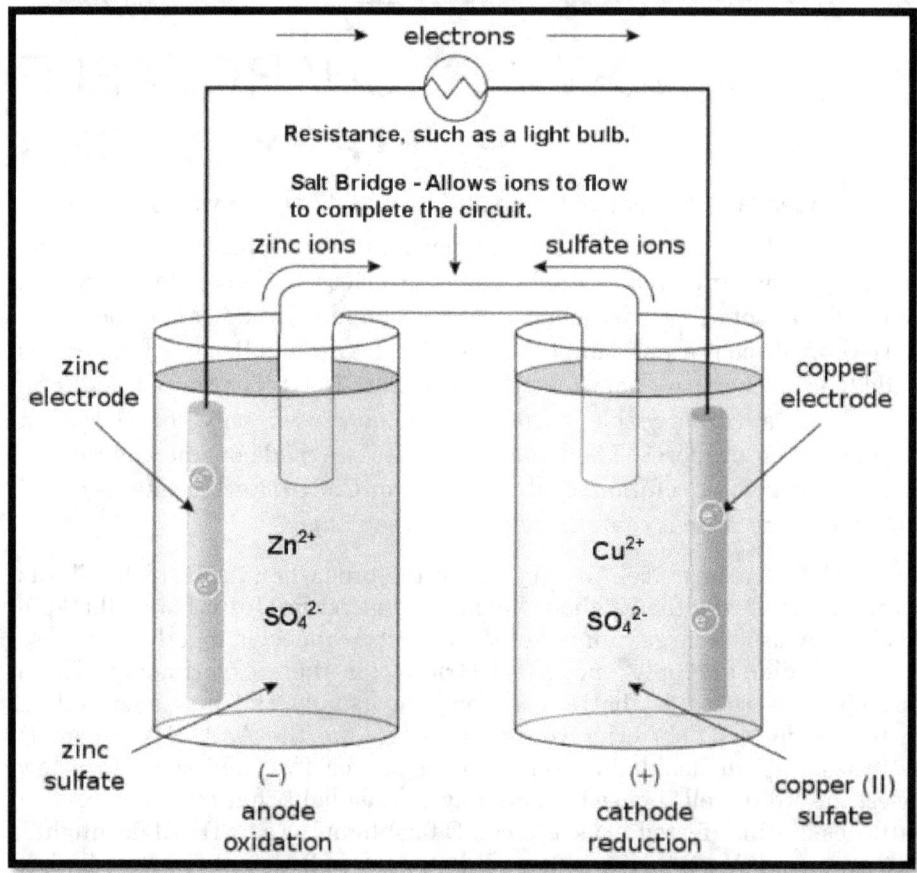

**Figure 1 Daniell Cell as a Battery.** *Attribution: By Rehua - Own work, CC BY 3.0, https://commons.wikimedia.org/w/index.php?curid=17902682*

which takes the electrons, becomes negatively charged. Almost instantly, the flow of electrons would stop, as they would be attracted to the positive anode and repelled from the negative cathode (opposite charges attract each other and like charges repel each other).

To allow a continuous flow of electrons, a **salt bridge** is used. It allows ions to flow between the anode and the cathode to keep them from building up an electrical charge, thus allowing the battery to keep working. As shown in Figure 1, $Zn^{2+}$ ions moving in the salt bridge remove positive charge from the anode, and the moving sulfate ions ($SO_4^{2-}$) in the salt bridge remove negative charge from the cathode.

The salt bridge in the Daniell cell is made from glass tubing that is filled with a water-based jelly-like substance such as agar-agar that has potassium chloride or potassium nitrate dissolved in it to make it conductive, allowing ions to move in the salt bridge. Both KCl and $KNO_3$ completely ionize in water to $K^+$ and $Cl^-$ ions, or $K^+$ and $NO_3^-$ ions.

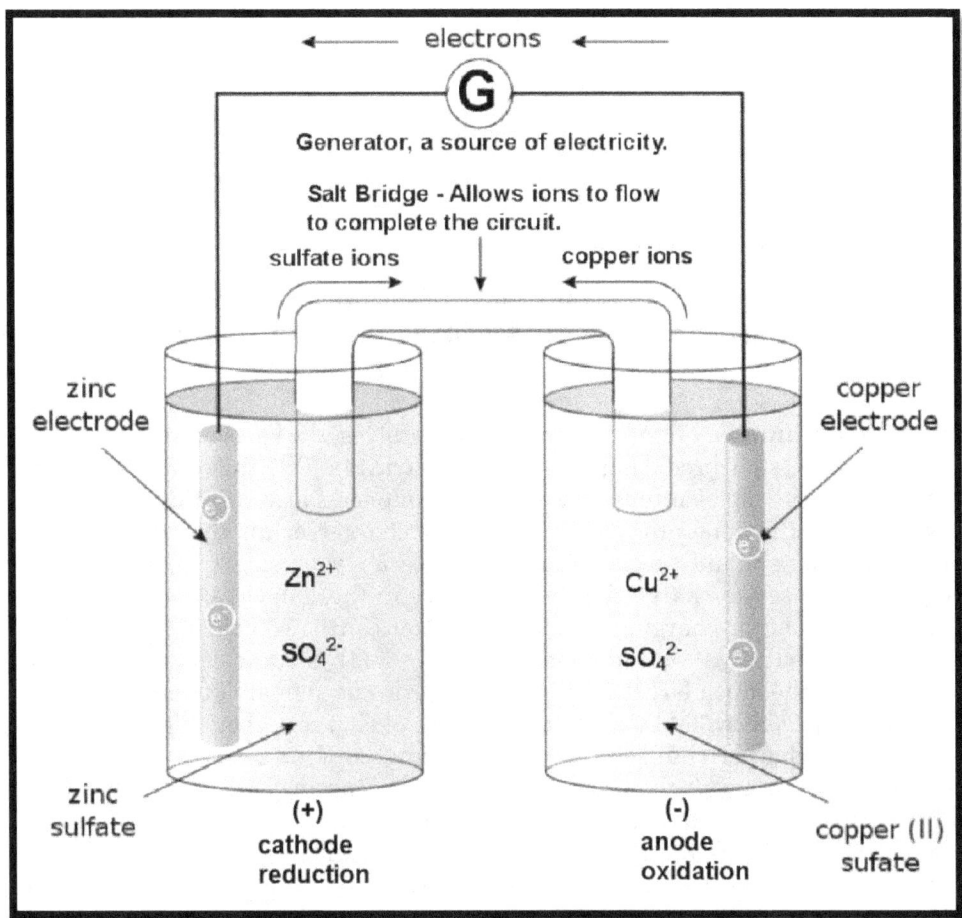

**Figure 2  Daniell Cell Being Charged.** *Attribution: By Rehua - Own work, CC BY 3.0, https://commons.wikimedia.org/w/index.php?curid=17902682*

The reason that these particular compounds are used is that their positive and negative ions move at about the same rate in the salt bridge, with the positive and negative ions moving in opposite directions. Thus they don't disturb the charge balance of the anode and cathode. They also don't react with any of the other chemicals in the battery. Small amounts of these ions may also flow into the half cells, but this doesn't affect the operation of the battery.

In batteries that are actually useful and not just used as a classroom demonstration, the salt bridge is replaced with a thin membrane called a **separator**, because it separates the anode and cathode. Only ions can cross the membrane; the solvent they are in can't cross the membrane. Separators are made from a suitably porous material, such as non-woven fabrics, plastic films, or ceramics. See https://bit.ly/3bWroc5.

The redox equations for the discharge of a Daniell cell, when it is acting as a battery, are (the spectator ion, $SO_4^{2-}$, is not shown):

$Zn_{(s)} \rightarrow Zn^{2+}_{(aq)} + 2e^-$  Oxidation at the anode.*

$Cu^{2+}_{(aq)} + 2e^- \rightarrow Cu_{(s)}$  Reduction at the cathode.

$Zn_{(s)} + Cu^{2+}_{(aq)} + 2e^- \rightarrow Zn^{2+}_{(aq)} + Cu_{(s)} + 2e^-$  Sum of half reactions.

$Zn_{(s)} + Cu^{2+}_{(aq)} \rightarrow Zn^{2+}_{(aq)} + Cu_{(s)}$  Sum without electrons.

$Zn_{(s)} + CuSO_{4(aq)} \rightarrow ZnSO_{4(aq)} + Cu_{(s)}$  Molecular equation.#

\* **NOTE:** (aq) means dissolved in water (aqueous), (s) is a solid.

# **NOTE:** $CuSO_{4(aq)}$ and $ZnSO_{4(aq)}$ are completely ionized in solution. The sulfate ion, $SO_4^{2-}{}_{(aq)}$, is a spectator ion. (See Example 4 on page 329 for a discussion of spectator ions.)

**Charging a Daniell Cell:** As the zinc anode dissolves, the battery runs down and will eventually stop producing electrons (electricity). But this battery can be recharged by putting electrons into it from an outside source. In Figure 2 on the previous page the generator can be a battery charger or another battery. The small wall charger and special charging cables for your smartphone or laptop computer are used to charge the Li-ion battery in these devices. Larger battery chargers are sold for charging regular car batteries (the so-called "lead-acid" or "lead-storage" battery — see problem 22 on page 344). Electric cars have a built-in charger for the much larger lithium-ion batteries used in all late model electric cars. See https://bit.ly/3M4XWTW for the types of chargers available.

A battery charger changes the voltage from 120 V or 240 V to the voltage needed for the battery. It then converts the AC (alternating current) to DC (direct current). For electric cars, there are some commercial charging stations that supply DC. See https://bit.ly/3p28zLa for more details. All batteries must be charged with DC, so that the electrons go in the proper direction to charge the battery.

When charging a Daniell cell, the zinc ions are getting electrons from the generator and electroplating on to the zinc electrode. The copper metal on the copper electrode is giving up electrons to the generator thus becoming copper ions. The copper electrode is dissolving. Note that in the salt bridge, the sulfate ions are coming out of the anode and the copper ions are coming out of the cathode. When the concentration of the zinc and copper ions is restored to their original values, the battery is fully charged and ready to be used again. The redox reactions for charging the Daniell cell are:

$Zn^{2+}_{(aq)} + 2e^- \rightarrow Zn_{(s)}$  Reduction at the cathode.

$Cu_{(s)} \rightarrow Cu^{2+}_{(aq)} + 2e^-$  Oxidation at the anode.

$Zn^{2+}_{(aq)} + Cu_{(s)} + 2e^- \rightarrow Zn_{(s)} + Cu^{2+}_{(aq)} + 2e^-$  Sum of half reactions.

$Zn^{2+}_{(aq)} + Cu_{(s)} \rightarrow Zn_{(s)} + Cu^{2+}_{(aq)}$  Sum without electrons.

$ZnSO_{4(aq)} + Cu_{(s)} \rightarrow Zn_{(s)} + CuSO_{4(aq)}$  Molecular equation.

**Important Things to Keep in Mind about Anodes and Cathodes:** Compare Figures 1 and 2. In Figure 1, the $Zn_{(s)}/Zn^{2+}_{(aq)}$ reaction takes place at the anode where oxidation is happening. In Figure 2, the $Zn_{(s)}/Zn^{2+}_{(aq)}$ reaction takes place at the cathode where reduction is happening. The $Zn_{(s)}/Zn^{2+}_{(aq)}$ reaction reverses direction as it goes from being an anode to a cathode. Also, in Figure 1, the $Cu_{(s)}/Cu^{2+}_{(aq)}$ reaction takes place at the cathode where reduction is taking place. In Figure 2, the $Cu_{(s)}/Cu^{2+}_{(aq)}$ reaction takes place at the anode where oxidation is taking place. The $Cu_{(s)}/Cu^{2+}_{(aq)}$ reaction reverses direction as it goes from being a cathode to an anode.

**Charging a Battery is Similar to Electrolysis:** On page 345, we discuss the electrolysis of aluminum oxide to give aluminum metal. If you look at the redox half reactions below, you will see that they are similar to those for the charging of a battery.

$Al^{3+} + 3e^- \rightarrow Al$   Reduction (at the cathode).   $Zn^{2+}_{(aq)} + 2e^- \rightarrow Zn_{(s)}$

$2O^{2-} \rightarrow O_2 + 4e^-$   Oxidation (at the anode).   $Cu_{(s)} \rightarrow Cu^{2+}_{(aq)} + 2e^-$

**NOTE:** Zinc metal is recovered for commercial purposes from its most common ore sphalerite (Zn,FeS – there are different proportions of Zn and Fe in different samples, hence the coma in the formula) by a rather complex process. See https://bit.ly/2xrBZfZ for details. Copper metal is recovered for commercial purposes from two types of ore, a sulfide, such as chalcopyrite ($CuFeS_2$) and an oxide, such as cuprite ($Cu_2O$) or tenorite (CuO). These are processed differently. See https://bit.ly/3ppWyR9 for details. Interestingly, much of the biblical King Solomon's wealth seems to have come from a copper mine. See https://bit.ly/3G7Yjce for details.

**You Might Wonder:** When the Daniell cell is discharging, thus acting like a battery, why does the zinc electrode dissolve? Why doesn't the copper electrode dissolve? The detailed explanation for this is difficult and beyond the scope of this book. It might be understandable after you have completed a general chemistry course. But a simple explanation is this: The tendency of a substance to take electrons in a battery has been measured and tabulated for many metals. It is called the **standard reduction potential (the unit is "volt"\*)**, which you will learn about in the electrochemistry lectures of a general chemistry course. A short table is at the right. Notice that copper is higher on the list than zinc, (0.34 V is larger than –0.76 V) which means that **copper ions want to take electrons (be reduced) in a battery more than zinc ions**.

| Standard Reduction Potentials | |
|---|---|
| Reduction Half Reaction | volts |
| $Ag^+_{(aq)} + e^- \rightarrow Ag_{(s)}$ | 0.80 |
| $Cu^{2+}_{(aq)} + 2e^- \rightarrow Cu_{(s)}$ | 0.34 |
| $2H^+_{(aq)} + 2e^- \rightarrow H_{2(g)}$ | 0.00 |
| $Fe^{2+}_{(aq)} + 2e^- \rightarrow Fe_{(s)}$ | –0.45 |
| $Zn^{2+}_{(aq)} + 2e^- \rightarrow Zn_{(s)}$ | –0.76 |
| $Al^{3+}_{(aq)} + 3e^- \rightarrow Al_{(s)}$ | –1.66 |

\*The unit "volt" is named after the Italian physicist **Alessandro Volta** (1745–1827). The abbreviation for the unit "volt" is "V." The term "potential" is essentially the same as voltage.

Three ways to say this when the **Daniell cell is acting as a battery** are:

(1) **Copper ions want to be reduced (take electrons) more than zinc ions.** The *Cu ions are reduced and the Zn metal is oxidized (in other words the zinc half reaction goes in the opposite direction to that shown in the above table of reduction potentials).*

(2) **Zn metal wants to be oxidized (give up electrons) more than Cu metal.** The *Zn metal is oxidized and the copper ions are reduced.*

(3) The *Zn metal is giving electrons to the Cu ions.* Or, the *Cu ions are taking electrons from the Zn metal.*

**NOTE:** Standard reduction potentials are all compared to the standard reduction potential of hydrogen, which is assigned a value of 0.00 V. The term "standard" refers to the temperature, concentration, and pressure (if gases are present) of the reactants. The standard temperature is 25.00 °C, the standard concentration is 1 M for each species dissolved in water, and the standard partial pressure is 1 atm for each gaseous species. These factors affect the voltage of a battery because the reduction potentials (voltages) of each of the half reactions of a battery is affected by their value. If the voltages of the half reactions change, the voltage of the battery will change, as you will learn in general chemistry.

Standard reduction potentials are so important in the study of electrochemistry that many have been measured over the years. For example, there are about 800 values of standard reduction potentials in the 91st edition (2010-2011) of the *CRC Handbook of Chemistry and Physics*. A short table of standard reduction potentials is at https://bit.ly/3My3DJq.

**Electronegativity (EN)**, a measure of the ability of an atom to attract electrons toward itself in a covalent bond, is discussed in Section 17-2. In Figure 17-1, we see that the EN value for copper is 1.9, and the EN value for zinc is 1.6. **Thus copper wants to take electrons to itself more than zinc does in covalent bonds (remember that EN also works for ionic compounds).** This is similar to what happens in the Daniell cell acting as a battery. **The electrons flow from the zinc electrode to the copper electrode.** There is a rough correlation relating how metals act in batteries and electronegativity. In fact, there is a rough correlation between most elements relating their EN and their standard reduction potential (https://bit.ly/3PdTdQc).

It is so important to understand what is happening in a Daniell cell that another summary follows:

**If the two metals, copper and zinc, are connected to each other in just the right way, as is done in a Daniell cell, there is an electrical force between the two that has the effect of the** *copper ions in solution pulling electrons away from the zinc metal electrode*, **turning the neutral zinc atoms in the zinc electrode into zinc ions, which then dissolve in the electrolyte. At the same time, when the** *copper ions in solution take those electrons, they turn into neutral copper atoms*, **which then electroplate onto the copper metal electrode.**

A deeper understanding of batteries must wait for a general chemistry course. A detailed, but advanced, explanation that you might want to look at if you are curious is at https://bit.ly/2VSUH9U. This is a link to an open access paper in the August 23, 2018 issue of the *Journal of Chemical Education* entitled "How Batteries Store and Release Energy: Explaining Basic Electrochemistry" by Klaus Schmidt-Rohr. The article includes a discussion of the Daniell cell and some other batteries.

### Resistance, Conductors, Semiconductors, and Insulators

The **resistance** of an electrical or electronic device to electricity (electron flow) causes the energy of the electrons to be reduced. This energy can be used in devices to produce heat or do work (there is always some unwanted heat produced). Examples are heating a toaster, lighting a bulb, turning a motor, recharging a battery, causing electrolysis, or operating electronic devices from watches to smartphones, televisions to computers. Except for superconductors (page 158), everything that electricity passes through has a resistance to it.

The resistance of a substance is generally placed in one of three groups: Conductors, semiconductors, and insulators. **Conductors** have a low resistance to electricity, **insulators** have a high resistance to electricity, and **semiconductors** have a resistance between that of conductors and insulators.

In general, metals are good **conductors** of electricity. Silver is the best conductor, followed by copper. Nonmetals in general are insulators.

An exception is the nonmetal graphite, which is made up of layered sheets of carbon atoms, https://bit.ly/2zUoAxS. Graphite is usually considered a conductor, but copper is a much better conductor by about 200 times. Diamonds (see page 56), also made of carbon, are very good insulators.* See https://bit.ly/2LJpDU4.

Semiconductors, such as silicon and gallium arsenide, are used in electronic components such as transistors and integrated circuits to control the flow of electrons. See page 258 for a discussion of large-scale integrated circuits, which would not be possible without semiconductors, thus allowing us to make modern computers, smartphones and many other electronic devices.

A table listing the electrical resistivity and conductivity of many substances is at https://bit.ly/2LECR4w. Notice that resistivity and conductivity are opposites of each other. The larger the resistivity, the smaller the conductivity. And the larger the conductivity, the smaller the resistivity.

---

*Metals are both good electrical conductors and good heat conductors. Diamonds are unusual in that they are electrical insulators, but very good heat conductors. They are about five times better at conducting heat than silver, the best metal heat conductor. See https://bit.ly/3fRTTu7 for more details about diamonds.

# THE DOUBLE SLIT EXPERIMENT WITH ELECTRONS

On page 257, we discussed how matter can be considered as both particles, and also as waves if it is moving. On page 285 we discussed how waves can undergo diffraction and interference. This essay will make use of these concepts and it would be advisable to review those pages.

In the figure on page 285, examples of a wave passing through one slit and two slits are shown. Waves passing through one slit undergo diffraction. Waves passing through two slits undergo both diffraction and interference. In this essay, we will discuss what happens when electrons pass through two slits. (Photons behave the same way, but since electrons have mass, it is more interesting to discuss them.)

An experiment to see what would happen when electrons passed through two slits was performed in 1989 by **Akira Tonomura** (1942-2012) and his co-workers at Hitachi, Ltd. in Japan.[1] Hitachi makes electron microscopes and Tonomura and his colleagues were experts at designing and working with them. They didn't use actual physical slits cut out of, say, metal or ceramic. Their specially designed electron microscope fired a beam of electrons at the equivalent of two slits which were made from an extremely fine wire filament (1 μm in diameter) placed between two conductive plates a centimeter apart. The wire had a positive charge relative to the plates. This arrangement is known as **biprism**. A diagram of the biprism is shown in Figure 1. Remember that the electrons are negatively charged and will be attracted by the positively charged wire and curve around it — 50% of the electron beam will go around one side of the wire and 50% will go around the other side — just as if they were going through two slits.

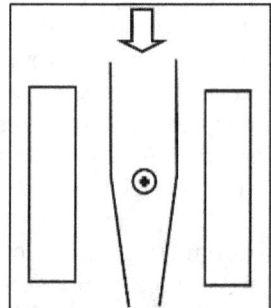

**Figure 1** The Biprism. The two rectangles are the conductive plates. The circle with the + sign is the positively charged wire filament. The two vertical lines which curve around the filament represent the beam of electrons. The arrow shows the direction the electrons are travelling.

After going around the wire, the separated beams of electrons are reunited into one beam. This beam of electrons is then focused and continues on its way to the recording screen, similar to a TV screen, where the electrons show up as white dots.

A diagram of Tonomura's setup is in his paper[1] and can also be found at https://bit.ly/2NXYWvU along with a touching memorial to him.

[1]"Demonstration of single-electron buildup of an interference pattern," **A. Tonomura, J. Endo, T. Matsuda, T. Kawasaki, and H. Ezawa,** *American Journal of Physics* 57, 117 (1989); doi: 10.1119/1.16104. The full paper can be found at https://bit.ly/2ZDVoVe.

The fascinating thing is that the experiment was designed so that the electrons arrived at the slits **one at a time**. Even so, as the electrons hit the recording screen, over time (about 20 min) they formed an interference pattern.[2] This is shown in the Figure 2.

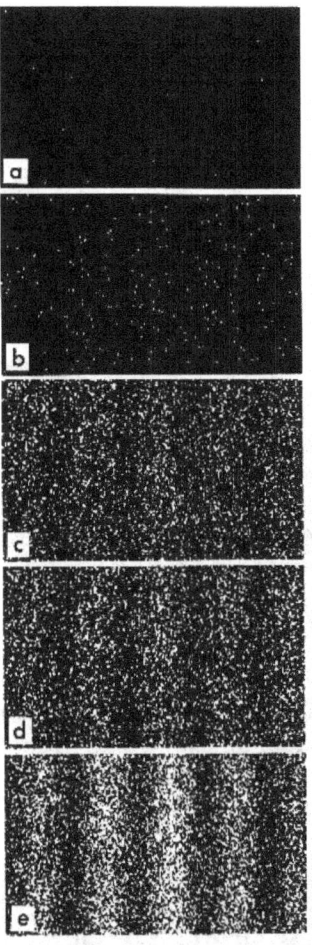

**Figure 2** The number of electrons (as white dots) in each box are: (a) 11, (b) 200, (c) 6000, (d) 40,000, (e) 140,000. *Figure is CC BY-SA 3.0. user:Belsazar - Provided with kind permission of Dr. Tonomura. https://bit.ly/2CXWQdp*

Since electrons behave as if they are both particles and waves, you might expect an interference pattern if many electrons passed through the slits at the same time and their waves interfered with each other. **But in this experiment, only one electron went through one slit at a time.** In the first box, (a), only 11 electrons struck the recording screen and they seem to hit at random. But as more and more electrons hit the recording screen, a clear interference pattern gradually emerges.

So you might ask: What was the wave of *one electron* interfering with if there were no other electrons nearby? And how did succeeding electrons know where to go to form the interference pattern?[3] **It seems like magic!**

Another experiment to study electrons going through two slits was done in 2012 by **Herman Batelaan** and his coworkers at the University of Nebraska. They used actual physical slits made from a gold-plated silicon nitride membrane. Their paper can be found at https://arxiv.org/abs/1210.6243. From this site, you can download the full PDF file of the paper. I suggest that you also open: "Ancillaryfiles (details): Supplementary_Movie_3.mov". Here you can watch a movie showing each electron as it hits the recording screen. At first it seems they are hitting at random, but after a while a beautiful interference pattern becomes visible. These authors call this a diffraction pattern, as do many scientists, but it is actually caused by a combination of diffraction and interference. Tonomura calls it an interference pattern.

Do physicists have any interpretation for what this magic could be? Maybe. Prof. **Brian Greene** of Columbia University (at https://bit.ly/3gvMSPk) says that the traveling electron can be described as a wave (called a probability wave, a strange wave indeed that cannot be discussed at this level). This wave, even from

---

[2]Other researchers have found interference patterns with protons, neutrons, atoms and molecules containing 114 atoms (https://bit.ly/2VxRfRk) and even going as high as 2000 atoms whose mass is over 25,000 Da, https://bit.ly/2XqhNaZ.

[3]Some physicists would not ask these questions in this way because of technical matters that are beyond the scope of this book.

a single electron, can, he proposes, go through both slits and interfere with itself.[4]

After many electrons, we eventually see the interference pattern that Tonomura and Batelaan found. It is not obvious what happens to the particle part of the electron (it has mass), as it is hard to believe that it can go through both slits at the same time. So the physical significance of what is happening still seems a mystery. (See Feynman's discussion below.)

In a separate set of double slit experiments (https://bit.ly/3VQYaT0), an electron detector was used to determine which slit each electron went though. **When scientists put in an electron detector (the "D" in Figure 4) to see which slit the electrons go through, the interference pattern disappeared, and what they got looks just like particles going through two slits.**

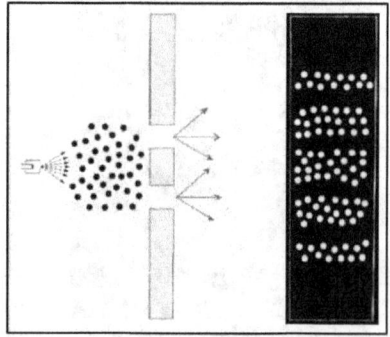

Now let's discuss the two experiments mentioned above with the help of two figures. Figure 3 **has no detector of electrons.** Figure 4 **has a detector of electrons.** The little device at the left of the figures is an electron gun. It shoots out a spray of electrons, shown as black dots.[5]

**Figure 3** Electrons (black dots) going through a double slit. At the far-right the recording screen shows the interference pattern that results. Turn the page 90° to see a vertical pattern as in Fig. 2.

**NOTE:** Don't confuse the detector with the recording screen.

Figure 3 shows the pattern that happens in the case where *the electrons are not detected until they hit the recording screen*. After going through the slits, the electrons form an interference pattern. Each part of the pattern is spread out because the electrons undergo diffraction (see page 285) as they go through the slits, just as waves do (see page 285), and come out in different directions, as the arrows show. And on the way to the recording screen, the electron's waves from the two slits interfere with each other. The interference pattern we get is similar to the ones that Tonomura and Batelaan got.

So in this experiment, *as seen from the interference pattern*, the electrons are acting as waves until they hit the recording screen,

---

[4]However, Prof. Greene says in his book, *The Elegant Universe*, page 108, "Nevertheless, the debate about what quantum mechanics really means continues. Everyone agrees on how to use the equations of quantum theory to make accurate predictions. But there is no consensus on what it really means to have probability waves … . But what appears certain is that no matter how you interpret quantum mechanics, it undeniable shows that **the universe is founded on principles that from the standpoint of our day-to-day experience, are bizarre.**" [Bold added.]

[5]In Figures 3 and 4, the black dots are only to show you that electrons are spraying out of the electron gun. (But remember that in the experiments described in this essay, only one electron went through one slit at a time.) It would be incorrect to think that we could actually see the electrons traveling from the gun to the double slit. We only see where the electrons went when they make their presence known as white dots on the recording screen.

where they make their presence known as white dots. The dots show that the electrons are now acting as particles. In this experiment, the electrons are acting as *both* waves and particles.

Figure 4 shows the pattern that would result in the case where *the electrons are detected*. After going through one slit or the other, the pattern shown by the electrons shows no interference. They presumably undergo diffraction as they go through the slits, just as waves do, and come out in different directions, as the arrows show. But something happens because of the detector (shown as a D). There is no longer interference of the waves coming from the two slits. **The electrons coming from the slits are no longer interfering with each other, and they form a pattern on the recording screen that is the same as if they acted only like particles.** (NOTE: One would get a similar pattern with a machine gun spraying bullets. To spread out, bullets would bounce off the walls of the slits, not diffract off them. See Section 13-1 for a discussion of machine gun bullets and seeing things.) The part of the detecting screen in the center has a high concentration of electrons because electrons from both slits can end up in the center of the screen. At the top and bottom of the screen, almost all the electrons that are seen go through the top slit or the bottom slit. For a wonderful essay (with great diagrams) on machine guns, electrons, and the double slit experiment, see https://bit.ly/3yLJVFq.

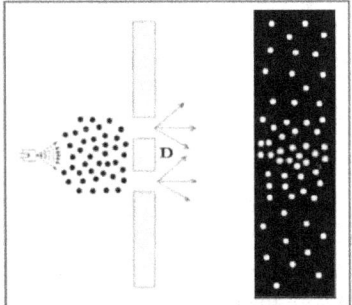

**Figure 4** Electrons (black dots) going through a double slit. In this experiment, an electron detector is shown as a D. At the far-right recording screen, there is no interference pattern.

Prof. **Richard Feynman** (1918-1988) of Caltech, who received the Nobel Prize in Physics in 1965, said this about the double slit experiment and the uncertainty principle (Sections 13-1 to 13-4):

1. It was suggested by **Heisenberg** that the then [in the 1920s] new laws of nature could only be consistent if there were some basic limitation on our experimental capabilities not previously recognized. He proposed, as a general principle, his *uncertainty principle,* which we can state in terms of our experiment as follows: "It is impossible to design an apparatus to determine which hole [he probably meant slit] the electron passes through, that will not at the same time disturb the electrons enough to destroy the interference pattern." If an apparatus is capable of determining which hole the electron goes through, it *cannot* be so delicate that it does not disturb the pattern in an essential way. **No one has ever found (or even thought of) a way around the uncertainty principle. So we must assume that it describes a basic characteristic of nature.**

2. **One might still like to ask: How does it work? What is the machinery behind the law? No one has found any machinery**

behind the law. No one can "explain" any more than we have just "explained." No one will give you any deeper representation of the situation. We have no ideas about a more basic mechanism from which these results can be deduced.** [Above quotes from Feynman R, Leighton R, Sands M (1965) *The Feynman Lectures on Physics, Volume III, Quantum Mechanics.* Quotes are taken from Sections 1-6 and 1-7 of the first link below.]

Here is a link to the section in these famous physics lectures about the double slit experiment: https://bit.ly/2DbOTkU. The full three volume set is available for free online at https://bit.ly/3gvYHVl. These books are written for undergraduate physics majors at Caltech. They are very famous and fascinating to look at.

It has been difficult for physicists to "explain" (in the Feynman sense) the double slit experiment. Prof. Greene's interpretation above is accepted by many physicists, but there are different interpretations.
See https://go.nature.com/37xUChJ and https://bit.ly/3oZ91JE.

Feynman's questions above quotes must be taken seriously. He was a "magician" as a physicist as described in the following quote:

> There are two kinds of geniuses, the "ordinary" and the "magicians." An ordinary genius is a fellow that you and I would be just as good as, if we were only many times better. There is no mystery as to how his mind works. Once we understand what they have done, we feel certain that we, too, could have done it. It is different with the magicians. They are, to use mathematical jargon, in the orthogonal complement [don't ask, but something like right angles] of where we are and the working of their minds is for all intents and purposes incomprehensible. Even after we understand what they have done, the process by which they have done it is completely dark. They seldom, if ever, have students because they [the magicians] cannot be emulated and it must be terribly frustrating for a brilliant young mind to cope with the mysterious ways in which the magician's mind works. Richard Feynman is a magician of the highest caliber. —**Mark Kac**, quoted by James Gleick in *Genius: The Life and Science of Richard Feynman* (Kac and Feynman were colleagues at Cornell and worked together on a research project.).[6]

[6]In addition to Gleick's wonderful biography of Feynman, you might also enjoy the graphic novel *Feynman* by Jim Ottaviani and Leland Myrick. Red-colored links to the videos of Feynman giving his Cornell Messenger Lectures in 1964 are at https://bit.ly/3TBc4Y1. The 6th lecture, "Probability and Uncertainty: the quantum mechanical view of nature" discusses the double slit experiment. In it he says, "I think that I can safely say that nobody understands quantum mechanics." (Your author attended these lectures. Feynman's Brooklyn accent is a put-on. He grew up in the Far Rockaway section of Queens, NY and in personal conversation spoke unaccented Standard American English.) You may get some idea about what Feynman means in the article by **Sean M. Carroll** (no relation) in *The New York Times*, Sept. 7, 2019 entitled "Even Physicists Don't Understand Quantum Mechanics." It is at https://nyti.ms/3e4W2jv. You might also enjoy the video lecture by Prof. Carroll at https://bit.ly/39W21a4 on the history of quantum mechanics. **NOTE:** The videos of Feynman's Messenger Lectures use the *Microsoft Azure Media Player* which allows concurrent high-lighted full text on the screen — a terrific type of sub-titles. It also allows many adjustments for optimum viewing (https://bit.ly/2OKoQTh).

# INDEX

Page numbers in **bold** indicate photographs or other graphics.

**A**aron, Hank, 159
Absolute zero, 198
  motion does not stop at, 198
Accuracy, of measurement, 99
Accuracy and precision, diagram illustrating, 100
  watches demonstrating different precision, **92**
Acetaldehyde and aldehyde dehydrogenase, 321
Acetate ion, 292
Acetic acid, 321
  oxidation number of atoms in, 316-318
  reaction with NaOH, 186
  reaction with sodium bicarbonate, 155
Acetic anhydride, used in making aspirin, 155
Acetone, oxidation number of atoms in, 316, 318
Acetylene, synthesis, 155
Acid-base neutralization, 178, 187
Acids, naming, 299
Actinide series, 247
Actinoid series, periodic table and, 247
Adding numbers, in scientific notation, 91
Addition, 66
  significant figures in, 101
Air pollution, and nitric oxide, 45
Airbags in vehicles:
  guanidinium nitrate used in, 303
  sodium azide used in, 303
  Takata Corporation use of ammonium nitrate in, 303
Alcohol and alcohol dehydrogenase, 321
Algebraic logic, use in scientific calculators, 36, 81
Alkali metals:
  bonding in, 278-282
  periodic table and, 248
  table of ions, 288
Alkaline earth metals:
  bonding in, 278-282
  periodic table and, 248
  table of ions, 288
Alloy, 345
Aluminum:
  burning of, 48
  combustion of, 147
  Hall–Héroult process for making, 345, 358
  from sodium and aluminum chloride, 155
  ion combined with Mg ion as antacid, 354
  reaction with ferric oxide, 151
Aluminum bromide, formation from elements, 156
Aluminum hydroxide, use as antacid, 354
Aluminum oxide, 48, 53, 345, 358
  as mineral corundum, 53
  as emery, 53
American diet, energy from refined sugar, 121
Ammonia, 51
  electronegativity of atoms in, 308
  fertilizer and, 189
  Lewis formula of, 268
  production of, 145
  structure of, 269
  synthesis of, 155
Ammonium Perchlorate Composite Propellant:
  toxicity of perchlorate, **142**
  used in solid fuel rocket boosters for Space Shuttle, **142**
Analytical balances, 101
Andromeda galaxy, **62**
Angular momentum:
  Orbital angular momentum, 231
  Spin angular momentum, 231
Ångström, Anders Jonas, 322
Ångström unit (Å), 322
Anode, 326, 345, 405-410
Antilogarithms, 351
Antimatter, 8
Aqua Lung, 202
Aquanauts, 213-214
Archimedes principle, 188
Argon, some uses of, 189
Arnold, Frances H., 218
Arrhenius, Svante, 187
Arsenate ion, 292
Ascorbic acid, 139
  determining molecular formula of, 168
ASML, maker of microchip machines, 258
Aspirin, synthesis, 155
Aston, F. W., 7
Astronaut Stephanie Diana Wilson, **108**
Astronauts, weight on moon, 1-2
*ate*, as ending, 291

417

Atmosphere, unit of pressure, 190
Atmospheric pressure, 190-191
Atomic bomb, 34, 121
Atomic hydrogen torch, 43
Atomic mass unit, 19. See
    also *unified atomic mass unit*
Atomic number, 5, 9
Atomic weight:
    as official unit, 141
    in units of grams, 125
    number of atoms contained in, 125
    related to mole, 126
    related to numbers of atoms, 123
    related to relative atomic mass, 141
    significant figures and, 31
    table, 27-30
Atoms, 1
    calculating number, from grams, 129
    calculating number, from grams
      of a substance, 135
    calculating number, from moles, 129
    calculating number, in mole of
      molecules, 133
    calculating number, in molecule, 133
    models of, 16
    notation for writing, 8
    of elements heavier than hydrogen, 10
Automobile lead acid battery, 344
Avogadro, Amedeo, 125, **140**
Avogadro Project, to define Avogadro's
    number, **122**
    Rabb, Sevelas, **157**
    Vocke, Bob, **157**
Avogadro's law and the ideal gas law, 217
Avogadro's number, 122, 126
    and atomic weight, 125, 126
Avogadro's constant, 126
    amount of substance, 136
    and molecules, 130
    feeling for size of, 127

**B**aking soda, 292
    reaction with acetic acid, 155
Balanced chemical equation:
    mass relationships from, 145
    mole identities from, 143
    mole interpretation of, 137
    mole relationships from, 145
Balances, analytical, 101
Balancing chemical equations, 45-50
    complex groups and, 50
    fractional coefficients and, 46
    hints on where to start, 49
    oxidation-reduction type, 325-340

whole number coefficients and, 46
Balancing redox equations:
    in acidic solution, 331-333
    in basic solution, 333-335
    examples, 335-340
    rules for, 330
    using half-reactions, 327-340
Balloons, hot air, **188**
    Archimedes principle and, 188
Barometer, mercury, 192
Barometric pressure, 191
Bartlett, Neil, 250
Base units in the SI system,
    table of, 115, 403
Baseball, wavelength at 90 mi/h, 257
Batelaan, Herman, 413
Batteries, various, 344:
    anode and cathode of, 326, 405-411
    electron flow in, 326, 405-411
    oxidation and reduction in, 326,
      405-411
    reactions of mercury, 325-326
    rechargeable, how it works, 405-411
    salt bridge, 406
    separator, 407
Batting average, calculating, 159
Bauxite, 53, 358
    purification by Bayer, 358
Beckman, Arnold O., 355
Bednorz, Georg, 158
Bertozzi, Carolyn, 218
Beryllium:
    isotope, 12
    Lewis formula of, 263
Beryllium hydride:
    Lewis formula of, 265
    structure of, 266
Bicarbonate ion, 292
Bicarbonate of soda, 292
Big Bang, 401
Binding energy, 32
BIPM, xix,
Birth control pills, 322
Bit.ly, shortened URLs, xv, xx
Blast furnace, 53
Bohr, Niels, **vi**, 16, 223, 245
Bohr's theory, failure of, 223
Bonding, in solutions, 59. See also
    *chemical bond*
Borane:
    Lewis formula of, 266
    structure of, 266-267
Borate ion, 292
Boron nitride, oxidation number of
    atoms in, 31
Boyle, Robert, 216
Boyle's law, 216, 217

Bragg, W. Lawrence, 281
Breathing, pressure-volume relationships, 201
Briggs, Henry, 348
Bromide ion, 289
Budil, Kimberly S., 34
Burbidge, Margaret, 401
Buret, **170**
   measuring volumes with, 102
   precision of, 102-104, 177
   used for performing titration, 177
Butane, determining molecular formula of, 164
Butyric acid, determining molecular formula of, 167

Cadmium, common positive ion of, 295
Calcium carbide, used to make acetylene, 155
Calcium carbonate, reaction with HCl, 148, 155
Calcium hydroxide, made from calcium oxide, 155
Calcium nitride, percent composition of, 168
Calcium oxide (quick lime), reaction to slaked lime, 155
Calculator, see also *Scientific calculator*:
   instructions for using scientific, 36
   logarithms and, 350
   scientific notation and, 81
Calibrating instruments, 99
Cannizzaro, Stanislao, 140, 217
Cancer, and mRNA vaccine, 323
Carbide cannon, 155
Carbide ion, 289
Carbon, range of oxidation numbers in, 319
Carbon-12, standard of atomic mass, 21
Carbon-14, formed by cosmic rays, 15
Carbonate ion, 292
Carbon dioxide, bonding in, 272, 275
   formation during combustion, 144
   greenhouse gas, 187, 305
   Lewis formula of, 272
   oxidation number of atoms in, 317
   photosynthesis and, 305
   structural formula of, 48
Carbon monoxide, from combustion, 144
Carborundum, 53
Carroll, Sean M., 416
Carson, Rachel, xv
Cassiopeia A supernova remnant, **front cover of book**, 400-401
   source of some elements in universe, 400-401
Cathode, 326, 345, 405-410

Cavendish, Henry, 7
Celsius, Anders, 193
Celsius degree, 193
   converting to degrees Fahrenheit, 193-195
   converting to kelvins, 19
Centigrade degree, 193
Centrifugal force, 2-3
Centimeter of water, as pressure unit, 192
Cerium(IV) sulfate, reaction with iron(II) sulfate, 185
CGPM, xix, 114, 404
Chandra X-ray Observatory, 400
Charles, Jacques, 216
Charles's law, 216, 217
Chemical Abstract Service (CAS), 247
Chemical bond, 259-282
   between hydrogen atoms, 41, 259-262
   covalent, 259-278
   ionic, 278-282
   use of $d$ orbitals, 278
Chemical equations:
   balancing of, 42-51
   balancing redox in acidic solution, 331-333
   balancing redox in basic solution, 333-335
   examples of balancing redox, 335-340
   mass relationships from, 145
   mole identities from, 143
   mole interpretation of, 137
   mole relationships from, 145
   rules for balancing redox, 330
   symbols separating reactants and products, 42
Chemical families:
   of periodic table, 245
   position in periodic table, 248
Chemical identity, 7
Chemical engineers, salary in 2022, 39
Chemistry professors (US):
   salaries in 1986, 24
   salaries in 2018, 39
Chemists (US) salaries in 1986, 37
Chemists (US) salaries in 2022, 39
Chlorate ion, 292
Chloride ion, 289
   in seawater, 186
Chlorine pentafluoride:
   Lewis formula of, 277-278
   structure of, 278
Chlorite ion, 292
Chloroacetic acid, oxidation number of atoms in, 318-319
Cholesterol, serum, 121
Chromate ion, 292

Chromium, common positive ions, 295
CIPM, xix
Clorox, preparation, 186
Cobalt, common positive ions, 295
Coefficients, in chemical equations, 45-46
Combined gas law, derivation, 209-210
Common names of compounds, 297
Complex fraction, simplifying, 176
Compound, 56
    empirical formula and weight of, 163
Compressed air, as used by scuba
    divers, 217
Conférence Générale des Poids et
    Mesures (CGPM), xix, 114
Conservation of atoms in chemical
    reactions, 45
Compton, Arthur Holly, 257
Conductors, electrical, 411
    metals as, 411
Converting from one unit to another, 113
Cooling of gas, effect on volume, 196-197
Copper, common positive ions, 295
    in Daniell cell, 405
    ores of (chalcopyrite, cuprite, tenorite), 409
Copper(II) sulfate pentahydrate, 185
    percent composition of, 168
Cornell Lectures by R. Feynman, 416
Cornell Lectures by L. Pauling, 265
Coronary heart disease, and serum
    cholesterol, 121
Coronavirus (SARS-CoV-2), **222**
    as cause of Covid-19 pandemic, 323
    Jason McLellan and Barney Graham, 323
    Katalin (Kati) Karikó and Drew
        Weissman, 323
    modified mRNA for vaccine, 323
    omicron variant, 322
    structure of RNA polymerase by
        cryo-EM, 323
    structure of "spike" protein and
        vaccine with 2P mutation by
        cryo-EM, 323
Cosmic rays, 7
    and isotope production, 12, 400
Cousteau, Jacques-Yves, 202, 213, 214
    *rapture of the deep*, 214
Covalent bonding:
    in 1st and 2nd periods elements, 264-274
    in 3rd period elements, 275-278
Covalent bonds, 261
    determining number in 2nd period
        elements, 270, 273
    of hydrogen molecule, 259-261

Covalent compounds, naming, 297-299
    order of naming, 298, 303
    use of prefixes when naming, 297
Crick, Francis, 323
Cryogenics, 322
Cryo-electron microscopy (cryo-EM) for
    determining molecular
    structure, 285, 322
Cryolite, 358
Crystallography, 285, 322, 323
    by electrons, 285, 322, 323
    by x-rays, 323
Cube root of number, 84
Cubing a number, 84
Curie, Marie and Pierre, **20**, 218
Cyanate ion, 292
    oxidation number of atoms in, 338
Cyanide ion, 292
    oxidation number of atoms in, 344

***d*** orbitals:
    periodic table and, 245, 248
    sketch, 228-229
    use in bonding, 278
Dalton, another name for the unified
    atomic mass unit, 19
Dalton, John, 19, 212
Dalton's law of partial pressures,
    19, 211-212
Daniell cell, 405-411
    charging, 408
    discharging (acting as a battery), 405-408
    redox equations for charging, 408
    redox equations for discharging, 407-408
Davisson, Clinton, 257
de Broglie, Louis, 257
Debye, Peter, 16
Decay, radioactive, 8
Decimals, and negative exponents, 76
Decompression after deep diving, 214
Deep sea diving, 213, 217
    toxicity of nitrogen, oxygen and
        helium, 214, 217
Democritus and Leucippus, *atomos*
    concept of matter, 1
Deuterium, 6-7
    as fusion fuel, 34
    isotopic symbol of, 9
Deuterium oxide, 43
Diamond:
    bonding in, 55, 56
    as electrical insulator, 411
Diatomic molecules of hydrogen and
    oxygen, 43

INDEX  421

Dichromate ion, 292
Diet, energy from refined sugar, 121
Diffraction of waves, 285
Dihydrogen, name for $H_2$, 43
Dihydrogen phosphate ion, 292
Dilution, of solution, 182
Dimensions, of physical quantities, 109
Dioxygen, name for $O_2$, 43
Dismutation, example of, 326
Disproportionation, example of, 326
Distributive law of algebra, 34
Division:
  of fractions, 68
  number of digits to be reported, 95
  of numbers in scientific notation, 73
DNA, elements for, in outer space, 303
Double bond, 272
  in oxygen molecules, 46, 272-273
Double slit experiment:
  with electrons, 412-415
  with water waves and light waves, 285
Doubly substituted water, 43
Druyan, Ann, 371
Dry cell (battery), 344

**E**instein, Albert, **20**
  mass-energy equation, 33
  photons of light, 257
Electrical attraction:
  between nucleus and electrons, 2-3, 252
  reason for forming ionic bonds, 281
Electrical balance, in redox equations, 330
Electrical charges, attraction and repulsion, 2
Electricity, 5
Electrode, 326
Electrolyte, 358
Electrolysis, 326, 358, 409
Electron configuration, 238
  for beryllium, 240
  for boron, 241
  for carbon, 241
  for fluorine, 242
  for helium, 239
  for hydrogen, 239
  for lithium, 239
  for neon, 242
  for nitrogen, 241
  for oxygen, 242
  for silicon, 243
Electron crystallography, 285
  using the cryo-EM method, 285, 322
Electron-electron interactions, effect on electron energy, 254

Electronegativity, 306-308, 410
  related to oxidation numbers, 308
  table of values, 307
Electrons, 1, 2
  charge of, 4
  energy of related to nucleus, 252
  in orbitals, rules for determining, 224-225
  mass of, 4
  maximum number in energy level, 226
  moles and, 136
  problem of seeing, 219-220
  rule for number in orbital, 227
  wavelength at 200 kV, 257
Electron spin, uncertainty principle and, 226
Electroplating, 326, 405
Element, 7, Section 4-1
  formed by filling of orbitals, diagram, 237
Element update, atomic numbers 104-118, 81
  information about on web, 81
Empirical formula, of compound, 163
Empirical weight, of compound, 163
End point of titration, 177, 179
Energy level:
  maximum number of electrons in, 226
  maximum number of orbitals in, 226
  number of orbitals in, 225-226
  orbitals in various, 233
Energy level diagram of orbitals, 235
Energy relationship of orbitals in one electron atom, 251
Enterprise, starship:
  photon torpedoes, 8
  warp drive, 8
Equivalence point of titration, 179
Ethane, burning of, 51
Ethanol, 53, 321
Ethyl alcohol, from fermentation, 156
Ethylene, empirical formula of, 163
Exact numbers, 105
  coefficients in chemical equations as, 107
  subscripts in chemical equations as, 106
Exponent of number, 64
Exponential notation, 64

*f* orbitals:
  periodic table and, 245-246, 248
  sketch of, 229
Fahrenheit degree, 193
  converting to Celsius, 193-195
Fahrenheit, Daniel Gabriel, 193

Faraday, Michael, vi
Fermi, Enrico, 7, 34
Ferric oxide, reaction with aluminum, 151
Fertilizer, from ammonia, 189
Feynman, Richard P, vi, xv, 415-416
   as a magician (of physics), 416
   biography of, 416
   graphic novel about, 416
   on the double slit experiment
      and the weirdness of quantum
      mechanics, 415-416
   quotes on honesty in science, xv
Fifth, alcoholic beverages and, 119, 121
Fission, nuclear, 34
Fixed notation, 402
Fixer in photography, 155
Flashlight battery, 344
Flerov, Georgy, 61
Fluoride ion, 289
Fluorine, as oxidizing agent, 342
   Lewis formula of molecule, 272
Forest fire, **40**
Formaldehyde, oxidation number of
      atoms in, 322
Formic acid, 321
Formula mass, as substitute for
      formula weight, 141
Formula weight, definition, 141, 166
   relationship to molecular weight, 168
   relationship to formula mass, 141
Fossil fuels, 305
Fractional abundance of isotopes, 30
Fractional coefficients, example of
      usefulness, 49
Fractions, multiplication and division, 68
Franklin, Benjamin, 1
Franklin, Rosalind, 323
Freezing point lowering:
   of bauxite, cryolite, and aluminum
      fluoride solution, 358
   of ice and salt solution, 358
   of water and alcohol solution, 358
   of water and ethylene glycol solution, 358
Freud, Sigmund, vi
Frisch, Otto, 34
Frontier Supercomputer, 323
Fuel cell, 326, 344
Fusion, nuclear, 34

**G**agnan, Émile, 202
Galaxy, Andromeda, 62
Gamma rays, wavelength and, 221
*Garbage in, garbage out* rule, 87
Gas, 184-214
   effect:
      of amount on pressure, 203
      of amount on volume, 204
      of temperature on pressure, 200-201
      of temperature on volume, 196-197, 200
   units of pressure, 190-193
Gasohol, 156
Gasoline, burning of, 48
Gay-Lussac, Joseph Louis, 217
Gay-Lussac's law, Section 12-9, 217
General Conference on Weights and
      Measures (CGPM), xix, 114
Germer, Lester, 257
Giant array, 56
   formula weight and, 166
*GIGO* rule, 87
Gleick, James, 416
Glucose:
   determining molecular formula of, 165
   fermentation of, 53, 156
   oxidation of, 148
   percent composition of, 162
Goeppert-Mayer, Maria, 218
Gold, common positive ions of, 295
Gonen, Tamir, 285, 322-323
Gosling, Raymond, 323
Grades on exams, calculating, 159-160
Graduated cylinder, **101**
   measuring volumes with, 102
   precision of, 102-104
Graham, Barney, 323
Grams:
   calculating from moles, 128
   calculating from number of atoms, 130
   unit of atomic weight, 124
Graphene, **opposite title page**
Graphite, 411
Gravitational field, and energy, 252
Gravity force, 2
Greene, Brian, 413-414
Greenhouse effect of $CO_2$, 187
Grouping of numbers in
      multiplication, 66
Groups, of periodic table, 245

**H**aber process to make ammonia, 155
Hahn, Otto, 34
Half-life of radioactive isotopes, 12
Half-reactions, 328
Halite crystals or Rock Salt, **57**
Hall, Charles Martin, 345
Hall (or Hall–Héroult) process for
      producing aluminum, 155, 345
Halogens, periodic table and, 249
Hamilton, Margaret, 258
Harmony, utility hub for International
      Space Station, 108

Head, Marti, 323
Heating a gas, effect on volume, 196
Heavy hydrogen, 7
Heisenberg, Werner, 220, 415
Heisenberg uncertainty principle, 220, 415
Heliox gas mixture for diving, 213
Helium, isotopes of, 11
  Lewis formula of, 262
  some uses, 189
  used in underwater breathing, 213-214
Héroult, Paul, 345
Hildebrand, Joel H., xv
History of structure of atom, 16
Hodgkin, Dorothy Crowfoot, 218
Hoffman, Darleane, 18
  synthesis of seaborgium-263, 18
Honesty in science, vi, viii, xv, 106, 371
Horizontal orbital diagram, 244
  for beryllium, 243
  for boron, 243
  for carbon, 243
  for first ten elements, 243-244
  for fluorine, 244
  for nitrogen, 244
  for oxygen, 244
  for phosphorus, 244
Huygens, Christiaan, 257
Hydrazine, as fuel for rocket propellant, 153
Hydride ion, 289
Hydrochloric acid:
  as stomach acid, 148
  oxidation number of atoms in, 311
  reaction with calcium carbonate, 148, 155
  reaction with potassium permanganate, 185
  reaction with zinc, 186
Hydrogen and discovery of, 7
  atom, 5
  covalent bonding in molecule, 259-261
  dihydrogen, name for $H_2$, 43
  isotopes of, 7
  Lewis formula of atom, 261
  Lewis formula of molecule, 261
  metallic above 500 GPa (in theory), 248
  orbital diagram for, 251
  production from sunlight, 327
  representations of molecule, 41-42
Hydrogen bomb, 34
Hydrogen carbonate ion, 292
Hydrogen chloride,
  Lewis formula of, 276
Hydrogen deuteride, 43
Hydrogen fluoride:
  determining molecular formula of, 166
  Lewis formula of, 269
  structure of, 270
Hydrogen-oxygen fuel cell, 344
Hydrogen peroxide:
  as oxidizer for rocket propellant, 53
  empirical formula of, 163
  naming and structure of, 292-293, 315
  oxidation number of atoms in, 315
Hydrogen phosphate ion, 292
Hydrogen sulfate ion, 292
Hydrogen sulfide:
  Lewis formula of, 276
  oxidation number of atoms in, 314
Hydrogen sulfite ion, 292
Hydrogenation, 155
Hydroxide ion, 292
*Hypo*:
  as prefix, 292
  in photography, 155
Hypochlorite ion, 292

Ice crystal, snowflake, **54**
*ide*, as ending, 291
Ideal gas law, 204-206
  combined, 210
  units used in, 206
  use in determining molecular weights, 209
Ideal gas law constant, 206
  units of, 206
  numerical value of, 206-207
Identities (unit conversions or conversion factors):
  from balanced chemical equation, 143
  relating physical quantities of, 112
  table of useful, 117
Identity element, in multiplication, 110
Imaginary numbers, 85
Indicator, used in titration, 178
Infrared, wavelength and, 221
Inorganic compounds, naming of, 287-300
Instruments, measurements and significant figures when using, 99
  watches demonstrating, **92**
Insulator, electrical, 411
Integrated circuit, very large scale, **258**
Interference of waves, 285
Interhalogen compounds, percent composition of, 167
Internal combustion engine, CO and $CO_2$ in exhaust, 144
International Space Station, 108
International System of Units, 114, 403-404

International Union of Pure and
  Applied Chemistry, xvi, xix, xx, 141,
  247, 287, 287, 299
    Commission on Isotopic Abundances
      and Atomic Weights, 141
Interstellar space, molecules in, 303
Inverse:
  of logarithm, 351
  of number, 84
  of operation, 84
Iodide ion, 289
Iodine heptafluoride, bonding in, 278
Ionic bonding, 278-282
Ionic compounds:
  reason for formation, 279, 281
  rules for naming, 288
Ionic equation, writing, 329
Ions, 279
    table:
      of Groups IA thru IIIA and
        ammonium, 288
      of Groups IVA thru VIIA, 289
      of polyatomic, 292
Iron:
  common positive ions of, 295
  production of iron in a blast furnace, 53
  rusting of, 51
Iron ore, burned in blast furnace, 53
Iron(II) sulfate, reaction with
  cerium(IV) sulfate, 185
Isotonic solution, 185
Isotope effect, 6
Isotopes, 6, 7
    artificial, 12
    discovery of lead isotopes, 169
    discovery of multiple stable isotopes
      of many elements, 8
    existence on earth and
      isotopic mass and binding energy, 33
    naming molecules with, 43 (footnote)
    naturally occurring, 12, 13-15
    radioactive, 7, 18, 169
    radioactivity, 8, 12, 18, 169
    substitution in atoms and molecule, 43
    table of mass and abundance for
      selected, 26
Isotopic mass, and binding energy, 33
Isotopic symbol, 8
*ite*, as ending, 291
IUPAC, xvi, xix, xx, 141, 247, 287, 287, 299
    Commission on Isotopic Abundances
      and Atomic Weights, 141

Joliot-Curie, Irène (daughter of the Marie
  and Pierre Curie), 218
Joyce, Joan, 159

Kac, Mark, 416
Karikó, Katalin (Kati), 323
Kelvin, Lord (William Thomson), 198
Kelvin temperature scale, 198
  converting to degrees Celsius, 198
Kerosine, burning of, 51
Kritcher, Andrea, 34

Lanthanide series, periodic table and,
  247
Lanthanoid series, periodic table and, 247
Large numbers, fun with, 91
Lasers, 53, **218**
LASIK eye surgery, **218**
Lavoisier, Antoine, **vi**, 7
Lawrence, Ernest Orlando, 18
Lawrencium, long-lived isotopes of, 18
Le Système International d'Unités, xviii,
  114
Lead-206, from decay of uranium, 12, 169
Lead-208, from decay of thorium, 169
Lead, common positive ions of, 295
Lead storage battery, 344, 408
Lewis formulas, 261
  hints for drawing, 262, 270
  of beryllium, 263
  of boron, 263
  of carbon, 263
  of carbon dioxide, 272
  of chlorine pentafluoride, 277-278
  of cyanate ion, 338
  of cyanide ion, 344
  of fluorine atom, 263
  of fluorine molecule, 272
  of helium, 262
  of hydrogen, 261
  of hydrogen chloride, 276
  of hydrogen sulfide, 276
  of lithium, 262-263
  of neon, 263
  of nitrogen atom, 263
  of nitrogen molecule, 272
  of organic molecules, 316
  of oxygen atom, 263
  of oxygen molecule, not possible,
    272-273
  of phosphine, 276
  of phosphorus, 264
  of phosphorus pentachloride, 276
  of second period elements, 263, 271
  of silane, 276
  of sodium, 264

of sulfur hexafluoride, 277
  using lines instead of dots, 274-275
Lehrer, Tom, periodic table song, 247
Lewis, Gilbert Newton, 261-262
Light, 220
  colors of visible, 221
Light wave, 220-222
  diffraction of, 285
  interference of, 285
Light year, distance, 400
Lightning, **opposite page 1**, 5
Limiting reagent (also called *limiting reactant*), 149
  determining, 151
  hint on determining, 153
Liquid protein diet, 140
Lithium:
  battery, 344
  isotopes, 11
  Lewis formula of, 262
  used to produce tritium, 8
Lithium fluoride, formation of ionic bond in, 282
Lithium hydride, Lewis formula, 265
Lithium hydroxide, removal of $CO_2$ from spacecraft and, 53
Lithium iodide battery, 324, 344
Lithium ion battery:
  advantages and disadvantages, **324**
  in smartphone and other devices, **324**, 408
Logarithm, natural, 353
Logarithms, 347-356
  calculator instructions for taking, 350
  e as base of natural logarithms, 353
  graph of, 349
  invention of, 348
  natural logarithms, 353

**M**agnesium citrate and oxide, 354
Magnesium hydroxide:
  reaction with HCl, 154
  use as an antacid or laxative, 354
Magnesium fluoride, formation of ionic bonds in, 282
Magnetism, 354
Maiman, Theodore H., 53
Mandela, Nelson, viii
Mantissa in logarithms, 365
  in scientific notation, 96
Manganese, common positive ions of, 295
Marathon race, 121
Margulis, Lynn, viii
Mass, 1, 231
  of electron, 4
  of neutron, 4, 33
  of proton, 4, 33
  unit of grams used for atomic weights, 124
Mass balance, in redox equations, 330
Mass number, 9
Mass spectrometer, 157
McLellan, Jason, 323
Mead, Margaret, vi
Measurements:
  accuracy and precision, 101
  percent uncertainty and, 93
Median, as an average, 39
Meitner, Lise, 34
Mendeleev, Dimitri, 232, 245
Mendeleev's 1871 periodic table, 232
Mercury column, pressure and, 190
Mercury, common positive ions of, 295
Mercury battery, 305-306, 325-326, 344
  toxicity of, **305**
Metalloids, 249
Metals:
  arrangement of atoms in solid, 56
  bonding in solid, 55
  description of and relation to periodic table, 249-250
Meters, pH, digital, **355**
Methane:
  burning of, 47
  combustion of, 147, 151, 154
  determining molecular formula of, 164
  electronegativity of atoms in, 308
  Lewis formula of, 267
  oxidation number of atoms in, 317
  structural formula of, 45
  structure of, 267-268
Methanol, 111, 321
Methyl isocyanide, a stinky gas, 216
Methylene radical, 267
Metric system, xviii, 114
  base units in SI system, 115, 403
  chaos in France before use, xix, 114
  table of SI prefixes, 115, 404
  table of useful identities (unit conversions or conversion factors), 117
Microplastics, in oceans, 305
Microscope, magnification of using visible light, 222
Microwave, wavelength and, 221
Milk of magnesia, antacid and laxative, 354
  pH of a solution, 354
  reaction with HCl, 154
Millimeters of mercury, unit of pressure, 190
  for blood pressure measurement, 191
Miramontes, Luis E., 322
Mixture, 58
Mohr titration, for determining chloride ion, 186

Molar Mass, 141
Molarity, 172
  calculations involving, 173-176
  formulas relating to moles, liters, grams, 175
  procedure for making solution, 172
Molecular equation, writing of, 329
Molecular formula, from percent composition and molecular weight, 164
Molecular mass, as substitute for molecular weight, 141
Molecular orbital, 260-261
Molecular orbital theory, 273
Molecular weight:
  and moles, 130
  calculating from atomic weights, 130
  determination using ideal gas law, 209
  relationship to formula weight, 168
  relationship to molecular mass, 141
Molecules, 41
  calculating number of, from grams, 131
  in interstellar space, discovered by Lucy Ziurys, 303
  of gas in motion, 189-190
  structure of, by cryo-EM, 285
Mole Day, 156
Mole identities from balanced chemical equation, 143
Mole interpretation of balanced chemical equations, 143
Moles:
  atomic weight and, 126
  Avogadro's number (constant) and, 126
  calculating from atoms, 129
  calculating from grams, 129
  calculating number of in a sample, 128
  derivation of word, 126
  elements and, 126
  interpretation of chemical reactions, 137
  molecules and, 130
  of electrons, 136
  official definition, 136
  of photons, 136
  related to gases, 203-204
Moles of atoms, calculating from moles of molecules, 132
Moles of molecules, calculating from grams, 131
Momentum, 231
Monoatomic oxygen and monooxygenase, 43, 297
Monatomic substance, 44
Moseley, Henry, 232, 245, 256
MRI machines, 158

mRNA vaccines, 323
Multiplication, 66
  grouping of numbers in, 66
  number of digits to be reported, 95
  of fractions, 68
  of numbers in scientific notation, 71
Müller, K. Alex, 158

**N**aming, of simple inorganic compounds, 287-300
NantEnergy, 304
Napier, John, 348
NASA, xxii, 108
National Ignition Facility, 34
National Institute of Standards and Technology (NIST), xvi, xix, xx, 157
Natural logarithm, 353
Negative exponents, and decimals, 76
Negative numbers:
  square root of, 85
  subtraction of, 78
Neptunium-237, from uranium-238, 15
Neutralization between acids and bases, 148, 178, 187
Neutral solution, 356
Neutrino, 8
Neutron, 3, 7, 8
  as nuclear glue, 11
  charge of, 4
  mass of, 4, 21
  mass of and binding energy, 33
Neutron stars, 4, 401
Newton, Isaac, 257
Nicad battery, 344
Nickel:
  catalyst used in hydrogenation, 155
  common positive ions of, 295
Nickel-cadmium battery, 344
NIST, xvi, xix, xx, 157
Nitrate ion, 292
Nitric acid, concentrated, 185
Nitric oxide:
  and air pollution, 45
  reaction to form nitrogen dioxide, 51
Nitride ion, 289
Nitrite ion, 292
Nitrogen gas in air, 45
  Lewis formula of atom, 263
  Lewis formula of molecule, 272
  oxides of, 298
  range of oxidation numbers in, 320
  triple bond in molecule, 46, 272
Nitrogen dioxide, and air pollution, 51
Nitrogen narcosis, 214, 217

Noble gases:
   compounds discovered, 250
   periodic table and, 249
Noddack, Ida, 34
Noether, Emmy, 231
Nomenclature, of simple inorganic compounds, 287-300
   update by IUPAC, 303
Nonbonded pairs of electrons, role in determining structure, 268-269
Nonmetals:
   description of and relation to periodic table, 249-250
   electrical properties of, 411
Norethindrone, 322
Notation for atoms, 8
Nuclear binding energy, 33
Nuclear fission, 34
Nuclear fusion, 34
Nuclear glue, neutrons as, 11
Nuclear reactor, 34
Nucleon, 32
Nucleus, 1
Nucleosynthesis, 400, 401
   Big Bang, 401
   cosmic ray fusion, 401
   dying low mass stars, 401
   exploding massive stars, 401
   exploding white dwarfs, 401
   links to articles by Jennifer A. Johnson, 401
   merging neutron stars, 401
   Sagan, Carl and, 401
Nylon, making in lab, **back cover**

Octahedral molecule, 277
Octane, burning of, 48, 154
Octet rule, 270, 273
   molecules not following, 276-278
Oganessian, Yuri, 61
Oganesson, element 118, page 61
Onnes, Heike Kamerlingh, 158
Oppenheimer, J. Robert, vi
Orbital angular momentum, 231
   of electron, 231
Orbital diagram, 238
   for beryllium, 240
   for boron, 241
   for carbon, 241
   for fluorine, 242
   for helium, 239
   for hydrogen, 239
   for lithium, 239
   for neon, 242
   for nitrogen, 241
   for oxygen, 242
   for silicon, 243

Orbitals, 223, 260-261
   approximate energy level diagram, 235
   diagram showing order of filling of elements, 237
   energy relationship in one-electron atoms, 251
   four basic types, for elements, 227
   maximum number in an energy level, 225
   order of filling diagram, 235
   order of filling in energy levels, 229
   order of filling summary, 236
   overlap of, 253
   rules for determining electrons in, 224-225
   shapes of, 224
   shielding, 253
   spacing in multielectron atoms, 235, 251
Order of naming in covalent compounds, 298, 303
Ores, reduction of, to metal, 340, 345, 358
Organic compounds, 287
   determining oxidation numbers of atoms from structural formulas, 316-319
Orion, constellation, 303
Orthomolecular medicine, 306
Ostwald, Wilhelm:
   Mol unit, German for mole, 126
   process, for making nitric oxide, 140
Overlap of orbitals, and shielding, 253
Oxalate ion, 292
Oxalic acid, reaction with potassium permanganate, 181
Oxidation, 305, 326, 340
Oxidation numbers, 306
   calculating, 309-311
   electronegativity and, 308
   keeping track of, 312
   of atoms in carbon compounds, 316-319
Oxidation-reduction equations, balancing of, 325-340
Oxidation-reduction reactions, 326
Oxide ion, 289
Oxides of nitrogen, table, 298
Oxidizing agent, 340
Oxyacids:
   corresponding negative ions, 300
   naming of, 299-300
Oxygen:
   as oxidizing agent, 342
   as reducing agent, 342
   double bond in molecule, 46, 272-273
   isotopic symbol of, 9

Lewis formula of molecule not
    possible, 272-273
poisoning by at high pressure, 214, 217
used in underwater breathing, 213-214
Oxygen difluoride, oxidation number
    of atoms in, 314

*p* orbitals:
    position in periodic table, 245, 248
    sketch of, 228
Parentheses, using, 66
Partial pressure, 212
Pascal, Blaise, 192
Pascal, unit of pressure, 192
Pauling, Linus Carl, xv, 169, 264, 265, 306
Payne-Gaposchkin, Cecilia, viii
*per*, as prefix, 292
Percent, 159
Percent abundance, of isotopes, 30
Percent composition, of compounds, 160
Percent uncertainty, of measurement, 93
Perchlorate ion, 277, 292
    as part of rocket propellant, 142
    toxicity of, 142
Periodic table, 245-251
    Day, 250
    interactive, 250
    regions of, 248
    relation to *s, p, d, f,* orbitals, 248
    TED talk on history of, xx
Periods, of periodic table, 245
Permanganate ion, 292
    reaction with HCl, 185
    reaction with oxalic acid, 181
Peroxide ion, 292
pH, 354
    calculations involving, 354-356
    history of, 354, 355
    table of pH values of common
        substances, 356
pH meters, digital, **355**
Pharmacy, over-the-counter items
    from shelves of, **286**
Phenolphthalein, as indicator, 178
Phlogiston theory, xvii
Phosphate ion, 292
Phosphide ion, 289
Phosphine, 276, 297, on Venus, 284
Phosphoric acid:
    reaction with KOH, 180, 186
    reaction with potassium hydroxide,
        180, 186
Phosphorus pentachloride:
    Lewis formula of, 276
    structure of, 276
Photograph 51 of DNA, 323
Photons, 220
    moles and, 136
    relation of wavelength and energy,
        221
Photosynthesis, 305
Physical constants, fundamental, 403
Pi Day, 156
Piston and cylinder arrangement, for
    studying the properties of gases,
    196
Pitchblende, ore of uranium, 15
Planck, Max, xvii, 20
Planetary model of atom, 16
Plaster of Paris, percent composition of,
    168
Plutonium-239, as fission fuel, 34
Plutonium-244, from colliding neutron
    stars, 15, 401
Polyatomic ions, balancing equations
    containing, 50
Polyatomic negative ions, 291, 292
Polymers, 19, 142
Positron, 8
Potassium deficiency, 140
Potassium hydroxide, reaction with
    phosphoric acid, 180, 186
Potassium permanganate:
    reaction with HCl, 185
    reaction with oxalic acid, 181
Powers and roots of numbers, 82
    scientific notation and, 85
Precision:
    in reading buret, 177
    in reading volumetric pipet, 178
    of measurement, 98
    of titration, 177
    significant figures and, 98
    watches demonstrating different, **92**
Precision and accuracy, diagram
    illustrating, 100
Pressure, 190
    units of, 190-193
Pressure of a gas:
    effect of amount on, 203
    effect of temperature on, 200-201
    effect of volume on, 201-203
Products, of chemical reaction, 42
Progesterone, 140
    structure of, by cryo-EM, 322, 323
Propane-oxygen fuel cell, 344
Proportional, directly and inversely,
    200
Proportionality, 199-200
    constant of, 200

INDEX  429

Propyne, hydrogenation of, 155
Protium, 5-7
   isotopic symbol of, 8
Proton, 3
   charge of, 4
   mass of, 4, 21
   mass of and binding energy, 33
   properties of, 4
Proxima Centauri, 400

**Q**uantum mechanics, 223
   double slit experiment and, 413-414
Quantum theory, 17, 223
   double slit experiment and, 413-414
Quicklime, 53
   reaction to form slaked lime, 155
Quotes, vi, viii, xv, 371

**R**abb, Savelas, 157
Radar, wavelength of, 221
Radical, as reactive species, 267
Radio waves, wavelength of, 221
Radioactive isotopes, 7-8, 18, 169
   short lived, none left on earth
      from stars, 401
Radionuclide(s), 7
Radium-226, 12, 17
Radon-222, 17
Rainbow, colors of, as related to
   wavelength, 221
Ramsay, William, viii
Rare earth series, periodic table and, 247
Ratio identities (unit conversions or
   conversion factors), of physical
   quantities, 111-114
Reactants, of chemical reaction, 42
Reaction equivalent, use in
   determining limiting reagent (also
   called *limiting reactant*), 150
Reciprocal of a number, 35
Red blood cells and pH, **346**
Redox equations, balancing of, 325-340
   balancing in acidic solution, 331-333
   balancing in basic solution, 333-335
   biological systems, in, 327
   examples of balancing, 335-340
   rules for balancing, 330
   used in analytical chemistry, 327
Redox reactions, 326. See also
   redox equations
Reducing agent, 340
Reduction, 326, 340
Rees, Martin, 371
Regular (or fixed) notation, 70
   converting to scientific notation
      for numbers less than one, 79-80
   converting numbers to
      scientific notation from, 69
   converting to scientific notation
      on calculator, 82
   rules for converting to and from
      scientific notation, 71
Relative atomic mass, as substitute for
   atomic weight, 141
Relative atomic weights:
   method of calculation, 25
   significant figures and, 31
   table of, 27-30
Resistance, 411
Resistivity, 411
Reverse Polish Notation (RPN),
   scientific calculators and, 36, 81
Richards, Theodore William, 169, 262
Riddles, 398
Rocket propellant, 53, 142
Roman numerals, **xxv**
Roots and powers of numbers, 82
   in scientific notation, 85
Rounding numbers, 104
Rubies, 48, 53
Rutherford, Ernest, 16

***s*** orbitals:
   position in periodic table, 248
   sketch of, 228
Sagan, Carl, viii, 8, 371, 401
Salicylic acid, used in making aspirin, 155
Saline solution, 185
Salt (NaCl) crystals, **57**
Salt bridge, 406
Sapphires, 48, 53
Science, honesty in, vi, xv, 106, 371
Scientific calculators:
   downloading for various devices, 402
   F ↔ E (fixed, or regular numbers, to
      exponential or scientific notation), 82
   downloading graphing calculators, 402
   instructions for using, 36-37, 80-87, 350-356
   logarithms and, 350-353
   powers and roots of numbers, 82-89
   purchase of, 402
   scientific notation and, 80-82
Scientific instruments, and significant
   figures, 98-101
Scientific notation, 65
   adding and subtracting numbers in, 91
   converting numbers to regular (or fixed)
      notation, 70
   converting numbers to regular (or fixed)
      notation on a calculator, 82

converting to regular notation for
    division of numbers in, 73
multiplication of numbers in, 71
negative exponents, 75
numbers less than one, 79-80
positive exponents, 69
rules for writing numbers in, 71
significant figures and, 96
SCUBA diver, **202**
SCUBA diving, 213
Seaborg, Glenn T., 18, 61, 245
Seaborgium, 18, 61
Sealab II, 213
    gas mixture used, 213
Seawater, determining chloride
    ion in, 186
Second period elements:
    covalent bonding in, 264-275
    Lewis formulas of, 263
    table of Lewis formulas of
        hydrogen compounds, 271
Seeing very small things,
    problem of, 219-223
Selenide ion, 289
Semiconductors, 258, 411
    silicon and gallium arsenide used
        in integrated circuits, 411
Semimetals, description of and
    relation to periodic table, 249-250
Serum cholesterol, related to risk of heart
    disease, 121
Shielding:
    of outer electrons by inner electrons, 253
    overlap of orbitals and, 253
Shock waves, 401
SI system of units, xviii-xix, 114, 403-404
Significant figures:
    accuracy and precision of, 98
    addition and subtraction of, 101
    atomic weights and, 31
    multiplication and division of, 95
    numbers with zeros and, 95-96
    percent uncertainty of, 93-95
    recognizing, 95
    scientific notation and, 96
Silane, Lewis formula of, 276
Silicate ion, 292
Silver, common positive ion of, 295
Silver bromide, in photography, 155
Silver nitrate, used in Mohr titration, 186
Silver tarnish, chemical formula of, 168
Simplest atom, 5

Single bond, definition, 272
Slaked lime, 53
    made from quicklime, 155
Slater, John C., 264
SmartPill™, 355
Soddy, Frederick, 7, 169
Sodium bicarbonate, 292
    reaction with acetic acid, 155
Sodium carbonate, 21
Sodium chloride:
    bonding in, 278-279, 281
    giant array, 57
    halite or rock salt, **57**
    structure of crystal, 57, 281
Sodium hydride, oxidation number of
    atoms in, 312
Sodium hydroxide:
    reaction with acetic acid (in vinegar), 186
    reaction with hydrochloric acid, 176-179
    reaction with sulfuric acid, 179
Sodium hypochlorite, preparation of, 186
Sodium perchlorate, oxidation
    number of atoms in, 314
Sodium thiosulfate, in photography, 155
Sodium vapor lamp, wavelength of
    light from, 121
Solute, 171
Solutions, 58, 171
    procedure for making known
        concentration of, 172
Solution stoichiometry, 176-182
Solvay Conference of 1927, **20**, 21
Solvay, Ernest, 21
Solvent, 171
Sørensen, Søren, 354, 355
Space Shuttle liftoff, **142**
Spectator ions, use in redox reactions, 329
Sphalerite, a zinc ore, 409
Spin, of electron, 226
    chemical bonds and, 260
Spin angular momentum, 231
    of electron, 226, 231
Spin state, of electron, 227
Square pyramidal molecule, 278
Standard reduction potential, 409, 410
Star Trek, 7
Sterno, 111
Strassmann, Fritz, 34
Strickland, Donna T., 218
Structure of the atom, history of, 16
Studying, tips on, xviii-xx
Subtraction:
    negative number and, 78
    numbers in scientific notation and, 91

significant figures and, **101**-104
Sulfate ion, 292
  oxidation number of atoms in, 314
Sulfide ion, 289
Sulfite ion, 292
Sulfur, range of oxidation numbers in, 322
Sulfur dioxide, 154
  oxidation number of atoms in, 313
  used to make sulfuric acid, 189
Sulfur hexafluoride:
  Lewis formula of, 277
  structure of, 277
Sulfuric acid, 189
  concentrated, method of diluting, 183
  reaction with NaOH, 179
Sulfur trioxide, 154
  oxidation number of atoms in, 313
Summit supercomputer, 323
Suntan, photon energy and, 221
Superconducting magnet, **158**
Superconductors:
  discovery of, 158
  high temperature, 158
Supernova, 41, 401
Supernova remnant, **cover photograph**
  identification of elements in, 256, 400-401
  origin of, 400
Superoxide ion, 293
  oxidation number of atoms in, 322
System of Units, International, xviii, xix, 403-404

Technetium-99 and $^{99m}$Tc, 14
Telluride ion, 289
Temperature:
  effect on pressure of gas, 201-203
  effect on volume of gas, 196-197
  units of, 193-195, 198
Tetrahedral angle, 268
  comparison with angle in ammonia, 269
  in methane, 268
Tetrahedral molecule, 268
Tetramethylammonium hydroxide (TMAH), 157
Tetraphosphorus or white phosphorus, 43
Tetrathionate ion, 292
Thermite reaction, 151
Thiocyanate ion, 292
Thiosulfate ion, 292
  oxidation number of atoms in, 314
Third period elements, covalent bonding in, 275-278
Thomson, George P. (son of J. J.), 257
Thomson, J. J., 7, 16
Thomson, William (Lord Kelvin), 198
Thyroxine, 139

INDEX  431

Tin:
  common positive ions, 295
  isotopes of, 22
Titin, 19
Titration, **170**, 176-177
  calculating molarity from, 176-182
  errors in performing, 101-104, 177, 179
Todd, Margaret, 7
Tonomura, Akira, 412 -413
Torr, unit of pressure, 191
Torricelli, Evangelista, 191
Transition metals:
  naming compounds containing, 293-297
  periodic table and, 245
  table of common ions, 295
Triangular planar molecule, 267
Trigonal bipyramidal molecule, 276
Trigonal pyramidal molecule, 269
Trimix gas mixture for diving, 213
Trioxygen, also known as ozone, 43
Triple bond, in nitrogen molecules, 46, 272
Tritium, 6, 7
  as fusion fuel, 34
  formed by cosmic rays, 15
  isotopic symbol, 9
  made in nuclear reactors from lithium, 7
TUMS® antacid tablets:
  reaction with HCl, 148, 155

Ultraviolet light:
  photon energy of, 221
  suntans and, 221
  wavelength and, 221
Uncertainty principle, 220
  ability to see electrons and, 222-223
  double slit experiment and, 415
  electron spin and, 226
Underwater living, 213-214
Unified atomic mass unit, 4, 19, 21-22
Units:
  conversions using more than one identity, 119
  converting from one to another, 113
  danger of miscommunication about units or reading them incorrectly, 109
  dividing, 111
  identifying with an "individualized" abbreviation, 135
  method of abbreviating, 109
  of physical quantities, 109, 403

table of some useful identities (unit conversions or conversion factors), 117, 404
Universal ideal gas constant, 206
Uranium, isotopes of, 15
    pitchblende, ore of uranium, 14, 15
Uranium-233, as fission fuel, 34
Uranium-235, as fission fuel, 34
Uranium-238, radioactive decay of, 15

**V**alence electrons:
    Lewis formula and, 261, 263
    periodic table and, 250
Valence shell:
    Lewis formula and, 263
    periodic table and, 250
Ventilator, for assisted breathing, 192
    unit of pressure for, 192
Venus, phosphine in atmosphere, 284
Very heavy hydrogen, 7
Vinegar:
    reaction with sodium hydroxide, 186
    reaction with sodium bicarbonate, 155
Visible light, colors and wavelengths, 221
Vitamin $B_{12}$, 120, 139
Vitamin C, 120, 139
    determining molecular formula of, 168
Vitamin E, 121
Vocke, Bob, 157
Voltaire, xx
Volume of gas:
    affect of amount, 204
    affect of pressure, 200-201, 202
    affect of temperature, 196-200
Volumetric flask, 172
Volumetric pipet, 178
von Meyer, Julius Lothar, 217

**W**atches of different precision, **92**
Water:
    as separate molecules in liquid, 57
    Lewis formula of, 269
    made from hydrogen and oxygen, 43
    of hydration, 168, 185
    oxidation number of atoms in, 312
    percent composition of, 162
    structural formula of, 44
    structure of, 269-270
Watson, James, 323
Wave, light, 220-222
    diffraction of, 285
    interference of, 285

Wavelength:
    of electron accelerated with 200 kV, 257
    of light, 220-222
    photon energy and, 221
    related to seeing an object clearly, 221-222
Weather maps, units used in, 191
Weather reports, aviation and TV, units used in, 191
Weight, 1
Weighted average, 22-25
Weinberg, Steven, 220
Weissman, Drew, 323
Welding, noble gases and, 189
Wilkins, Maurice, 323
Williams, Ted, 159
Wilson, Stephanie Diana, astronaut, **108**
    selected for Artemis Team, **108**
Wood alcohol (methanol), 111
Wunderlich, Carl, 193

**X**enon-124, longest measured half-life, 91
X-rays, wavelength and, 221
    crystallography and DNA, 323

**Y**east, fermentation, 53
Yonath, Ada, 218
Young, Thomas, 257
YouTube videos on chemistry, xx

**Z**ero kelvin, 198
Zero power, numbers to the, 76
Zinc:
    common positive ion of, 295
    in Daniell cell, 405
    reaction with HCl, 186
    sphalerite, ore of, 409
Zinc-air battery, rechargeable. **304**
Ziurys, Lucy, 303

www.ingramcontent.com/pod-product-compliance
Lightning Source LLC
Chambersburg PA
CBHW080449220526
45465CB00006B/2213